✓ B & T
28.50
4 Sep '79

D1306104

INDUSTRIAL APPLICATIONS
OF LASERS

Industrial Applications of Lasers

JOHN F. READY

HONEYWELL CORPORATE RESEARCH CENTER
BLOOMINGTON, MINNESOTA

ACADEMIC PRESS   New York   San Francisco   London   1978

A Subsidiary of Harcourt Brace Jovanovich, Publishers

Wingate College Library

TA
1677
R4

COPYRIGHT © 1978, BY ACADEMIC PRESS, INC.
ALL RIGHTS RESERVED.
NO PART OF THIS PUBLICATION MAY BE REPRODUCED OR
TRANSMITTED IN ANY FORM OR BY ANY MEANS, ELECTRONIC
OR MECHANICAL, INCLUDING PHOTOCOPY, RECORDING, OR ANY
INFORMATION STORAGE AND RETRIEVAL SYSTEM, WITHOUT
PERMISSION IN WRITING FROM THE PUBLISHER.

ACADEMIC PRESS, INC.
111 Fifth Avenue, New York, New York 10003

United Kingdom Edition published by
ACADEMIC PRESS, INC. (LONDON) LTD.
24/28 Oval Road, London NW1 7DX

Library of Congress Cataloging in Publication Data

Ready, John F., Date
     Industrial applications of lasers.

     Includes bibliographical references.
     1. Lasers--Industrial applications. I. Title.
TA1677.R4   621.389'6   77-6611
ISBN 0-12-583960-X

PRINTED IN THE UNITED STATES OF AMERICA

To my wife Claire.
Without her encouragement,
this book would not have been written.
Also, to my children,
John, Mike, Chris, Mary, Betsy, and Steve,
for their cooperation
while the book was being written.

077369

CONTENTS

PREFACE

This book is intended for people with specific problems to solve. They want to know if lasers can help them. Many previous books on lasers have provided excellent introductions to the subject of quantum electronics; relatively few have emphasized how lasers have been used for real applications in industry. This book attempts the latter task. It will certainly not solve the reader's problems; rather, it will show how lasers have been used in previous applications. It is hoped that this approach will stimulate the imagination of the reader, who can then evaluate the potential use of lasers to solve his or her own problem.

Thus, the book is not highly mathematical. Equations are given when they can be related to a specific result, but they are usually simply stated without long derivation. Chapter 17 on holography is an exception. I feel the mathematical development there is needed to understand how holograms "really work."

The book defines the jargon that accompanies lasers. It also includes much of the lore that is known and accepted among long-term laser specialists, but which is rarely collected and written down.

To make the book reasonably self-contained, chapters are included about several topics relevant to lasers themselves, but without regard to specific application. These include laser properties and types of practical lasers. Similarly, a discussion of accessories commonly needed in laser applications is included. The topic of laser safety is also very important for anyone contemplating a laser application.

The book for the most part emphasizes currently developed applications. Laser welding and cutting of metals, scribing of ceramics, and applications in alignment, surveying, and metrology are well established in industry.

These and other applications are described, with specific examples given to illustrate the capabilities of the techniques.

Future applications in an experimental status are also treated. These include items whose eventual importance is not fully established, but which could have big payoffs, such as laser-based communications systems and optical computer memories. Even farther off—but of enormous potential importance—are laser-assisted isotope separation and laser-assisted thermo-nuclear fusion.

A word about units is in order. I have used the units most commonly employed for each individual application. Thus, in the early chapters, the reader will find metric units exclusively. In chapters related to production applications (welding, drilling, etc.) the reader will find substantial use of English units. This has been done purposely, in the face of the growing movement toward metrification. Despite my personal preference for metric units, I have found it easier to work with production engineers using English units.

ACKNOWLEDGMENTS

I am indebted to the many people who aided in the production of this book. First, I thank all the individuals and organizations who gave permission to reproduce copyrighted material. I acknowledge the advice, encouragement, photographs, interest, ideas, and constructive criticism supplied by many colleagues at the Honeywell Corporate Research Center, particularly Rick Bernal, T. C. Lee, Fran Schmit, Obert Tufte, and Dave Zook.

I want to thank those who supplied photographs for use in the book: Fred Aronowitz of the Honeywell Systems and Research Center, Conrad Banas of United Technologies Research Laboratories, Edwina DeRousse of Coherent Radiation, Inc., Simon Engel of GTE Sylvania (now of HDE Systems, Inc.), Jo Ferrier of Electromask, Inc., Frank Gagliano of Western Electric, Phil Gustafson of Honeywell Avionics Division, Hans Mocker of the Honeywell Systems and Research Center, Don Neumann of GCO, Inc., Tom Schriempf of Naval Research Laboratories, Dick Stark of Spectra-Physics, Inc., Martin Stickely of ERDA, John Webster of Coherent Radiation, Inc., and Mike Yessik of Photon Sources, Inc.

Special thanks are due to those who labored diligently to produce the typescript and figures: Diane Ellringer capably typed the manuscript, and Martha Collier and Verna Squier produced the line drawings. I am very grateful and appreciative.

HISTORICAL PROLOGUE

The development of lasers has been an exciting chapter in the history of science and engineering. It has produced a new type of device with potential for application in an extremely wide variety of fields. A possible place to begin the history of lasers is with Albert Einstein, in 1917. Einstein developed the concept of stimulated emission on theoretical grounds. Stimulated emission is the phenomenon which is utilized in lasers. Stimulated emission produces amplification of light so that buildup of high-intensity light in the laser can occur. The fundamental nature of the stimulated emission process was described on a theoretical basis by Einstein.

This characterization of stimulated emission did not lead immediately to the laser. Much additional preliminary work was done in optical spectroscopy during the 1930s. Most of the atomic and molecular energy levels which are used in lasers were studied and investigated through that decade. In retrospect, it seems that by 1940 there was enough information on energy levels and enough development of optical materials to support the invention of the laser. All the necessary concepts had been developed.

During the Second World War, the attention of the technological community was diverted into the microwave region of the spectrum, and many technical advances concerned microwaves. This diversion probably caused the development of the laser to occur in a somewhat roundabout fashion. In 1954 a device called the "maser," standing for microwave amplification by stimulated emission of radiation, was invented by Charles Townes and his co-workers. The microwave maser operated using the principle of stimulated emission, and indeed this appears to be the first practical utilization of stimulated emission. There was much enthusiasm and development devoted to masers in the 1950s. However, this work has died out to some extent, and the main application of the maser today appears to be as a receiver in radioastronomy.

In 1958, Townes and Arthur Schawlow suggested that the stimulated emission phenomenon, which had been utilized in the maser, could also be used in the infrared and optical portions of the spectrum. This suggestion had the effect of returning scientific attention to optics, from which it had been diverted for almost two decades. The device was originally termed an optical maser, and much of the original literature on lasers refers to optical masers. However, that term was dropped in favor of the simpler "laser," standing for light amplification by stimulated emission of radiation. The word has endured.

After the suggestion of Townes and Schawlow, a number of laboratories began working on lasers. The first practical realization of the laser was the ruby laser, which was operated in mid-1960 by Theodore Maiman. The first ruby laser emitted a pulsed beam of collimated red light at a wavelength of 0.6943 μm. This laser, primitive though it was by today's standards, immediately captured the attention of the scientific and engineering community. It provided a source of light with characteristics different from those of conventional sources. It provided an extremely monochromatic source of light. The peak power of the pulsed ruby laser was high. For the first time, the laser provided a source of coherent electromagnetic radiation in the visible portion of the spectrum. Immediately many more laboratories joined in laser research.

The first gas laser, the helium–neon laser, was operated in 1961. This represented another significant advance. In its first manifestation, it operated in the infrared at a wavelength of 1.15 μm. This represented the first continuous laser. In 1962 the visible transition at 0.6238 μm of the helium–neon laser was developed. This represented the first continuous visible laser. The visible helium–neon laser has developed and proliferated until today it undoubtedly represents the most common type of laser.

Late in 1962 a still different type of laser was invented, the semiconductor laser, which operates using a small chip of gallium arsenide semiconductor. This laser was fundamentally different in construction from the earlier gas and ruby lasers.

During the interval from 1962 to 1968, much basic development in lasers occurred. Almost all of the basic types of lasers were invented in that era. In addition, almost all of the practical applications were suggested. The ability of the laser to melt and vaporize small amounts of metals was recognized very early. The laser's capability for welding, cutting, and drilling was identified. Uses such as communications, display, interferometry, holography, and all the multitude of practical applications that are in use today were identified. There was a great deal of enthusiasm for lasers in those early days.

However, the status of laser development itself was still rather rudimentary. The materials that were available were of poor quality. Lasers themselves were fragile devices, with poor reliability and durability. Even though all the applications and their potential benefits had been identified, lasers themselves had not yet developed enough to perform the applications well. In some quarters, disillusionment about the laser developed. In summary, the period up to approximately 1968 represents the beginning period of laser development, in which applications were identified but could not be carried out really well. In that era, all the basic types of lasers that are in use today were identified.

At some time in the late 1960s (perhaps around 1968, but extending over a period of a few years) this situation began to change. The engineering development of lasers improved. The existing laser types were improved so that by the early 1970s they had become much more durable. By the early 1970s industrial engineers could purchase reliable models of lasers and use them in practical applications. It had become possible to perform economically many of the applications that had been discussed in the earlier era. By 1974 the sales of lasers had grown to $500 million per year. Lasers had taken their place as a truly practical tool on production lines for cutting, welding, and drilling, in the construction industry for alignment, in the machine tool control business for distance measurement, and for many other types of measurements.

Still, some suggested applications remain in a development stage. For the future we can look for much expansion of current uses and for the practical utilization in industry of some of the applications that have been suggested, but still have not yet reached a truly industrial stage. In addition there are exciting possibilities for x-ray lasers, for laser-assisted thermonuclear fusion, and for laser-assisted isotope separation. Although all these items are still rather speculative, they represent exciting and important challenges for the future.

CHAPTER 1

FUNDAMENTALS OF LASERS

This chapter is meant to provide a fundamental background of information about lasers. It will cover the basic principles of how lasers operate and will define some of the terminology often encountered in laser work. The level of treatment is elementary; it is intended as an introduction for those who have not previously worked with lasers.

A. Electromagnetic Radiation

The light emitted by a laser is electromagnetic radiation.† The term electromagnetic radiation includes a continuous range of many different types of radiation. In this section we shall describe the spectrum of electromagnetic radiation, and show how laser light is related to other familiar types of radiation.

Electromagnetic radiation has a wave nature. The waves consists of vibrating electric and magnetic fields. The waves can be characterized by their frequency, i.e., by the number of cycles per second of oscillation of the electric or magnetic field. The distance between the peaks of the waves is the wavelength. The wave motion propagates with a characteristic velocity c ("the velocity of light") equal to 3×10^{10} cm/sec. The physical nature of the different types of electromagnetic radiation is the same. In all portions of the spectrum, it has the same velocity c, and the same electromagnetic nature. The only difference is in the frequency. The different regions of the spectrum are characterized by different ranges of the frequency of oscillation of the wave.

† The word "light" is used in an extended sense, including infrared and ultraviolet, as well as visible light. The word "radiation" should be interpreted as radiant energy and not as implying ionizing radiation.

There is a relation between the frequency f and wavelength λ valid for all types of electromagnetic radiation:

$$\lambda f = c \tag{1.1}$$

According to Eq. (1.1), the wavelength decreases as frequency increases.

Figure 1-1 shows the electromagnetic spectrum, with several familiar regions identified, along with both a wavelength and a frequency scale. At the low-frequency, long-wavelength end, are radio waves. As frequency increases and wavelength decreases, we pass through the microwave, the infrared, the visible, the ultraviolet, the x-ray, and the gamma-ray regions. The frequency varies by a factor of 10^{16} or more. The boundaries between the different regions are not sharp and distinct, but rather are defined by convention according to the ways in which the radiation interacts. We emphasize that all these types of radiation are essentially the same in nature, differing in the frequency of the wave motion.

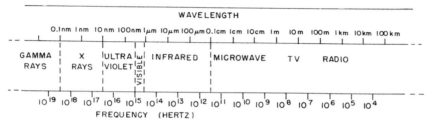

Fig. 1-1 The electromagnetic spectrum, with wavelength and frequency indicated for different spectral regions.

In addition to its wave characteristics, electromagnetic radiation also has a particlelike character. In some cases, light acts as if it consisted of discrete particlelike quanta of energy, called photons. Each photon carries a discrete amount of energy. The energy E of a photon is

$$E = hf = hc/\lambda \tag{1.2}$$

where h is Planck's constant, equal to 6.6×10^{-27} erg-sec. Equation (1.2) indicates that the photon energy associated with a light wave increases as the wavelength decreases.

In many of the interactions of light with matter, the quantum nature of light overshadows the wave nature. The light can interact only when the photon energy hf has a suitable value. In such interactions, the wave properties of the light appear to be secondary. It is difficult to reconcile conceptually the wave and particle pictures of light. For some experiments, for example, diffraction and interference, the wave nature will appear to predominate. For other experiments, such as absorption of light by atomic and molecular systems, the photon nature will appear to dominate. Thus, light can be absorbed between energy levels of a molecule only when the energy difference between the energy levels is equal to the photon energy.

The classic example of the photon nature of light is the photoelectric effect, in which electrons are emitted from a solid surface which absorbs the light. A surface will have a minimum amount of energy required for emission of an electron. This value is called the work function ϕ of the surface and is a function of the material. If the photon energy slightly exceeds the work function, that is if the wavelength is shorter than hc/ϕ, electrons can be readily observed. On the other hand, if the wavelength is slightly longer than this value, no photoelectric emission will be observed, even if the light is intense.

The fact that radiation has both a wave and a particulate character is referred to as the dual nature of radiation. This dual nature can be conceptually troublesome. Here we shall merely point out that in some interactions light will behave like a wave, in others like a particle.

For lasers, we are concerned with light in the near ultraviolet, visible, and infrared portions of the electromagnetic spectrum, i.e., with wavelengths in the approximate range 10^{-5}–10^{-2} cm† and frequencies of the order of 10^{13}–10^{15} Hz.

The extreme range of wavelengths covered by operating lasers runs from about 0.1 to 1000 μm, but at the far ends of this range, the existing devices are experimental laboratory systems. The region in which useful lasers for practical applications exist is about 0.3 to 10 μm. This region is shown in Fig. 1-2, which is an expanded portion of part of the electromagnetic spectrum. The wavelength of a number of popular lasers are noted. The wavelength of a given laser is sharply defined; the spread in wavelength for a laser of one material covers a miniscule fraction of the wavelength range shown in the figure.

Lasers can be constructed using a variety of different materials, each of which has its own distinctive wavelength. Figure 1-2 will serve to introduce some of these materials and to show where they occur in relation to each other and to the electromagnetic spectrum as a whole. These useful lasers, as listed in Fig. 1-2 include.

the CO_2 laser operating fairly far out in the infrared at 10.6 μm,

the neodymium: glass laser in the near infrared at 1.06 μm,

the gallium arsenide laser, in the near infrared at wavelengths around 0.85–0.90 μm,

---

† The wavelength of light is usually expressed in micrometers (μm) or nanometers (nm). The micrometer, equal to 10^{-4} cm, is sometimes referred to as a micron. The unit angstrom (Å), equal to 10^{-8} cm, is also sometimes encountered. The wavelength of green light may be given as 5.5×10^{-5} cm $= 0.55 \mu$m $= 550$ nm $= 5500$ Å.

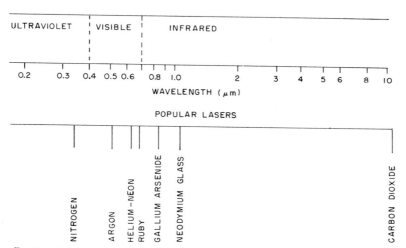

Fig. 1-2 Expanded portion of part of the electromagnetic spectrum, showing the wavelengths at which several popular types of lasers operate.

the helium–neon laser, emitting 0.6328 μm radiation, reddish-orange in color,

the argon laser, operating at several wavelengths in the blue and green portions of the spectrum, and

the nitrogen laser, an ultraviolet laser, operating at 0.3371 μm.

This enumeration does not exhaust the list of important lasers, but will serve as a reference point for some significant ones.

B. Elementary Optical Principles

In this section we shall briefly summarize several important optical phenomena that are commonly encountered in laser applications. This is not meant to be a comprehensive course in optics, but rather a brief review of some optical principles which shall be taken for granted later. These phenomena will already be familiar to students of optics. The three phenomena that we shall emphasize will be the polarization, diffraction, and interference of light beams. These phenomena are all based on the wave nature of light. Phenomena such as reflection, refraction, and index of refraction will be assumed to be familiar. The reader may refer to any elementary optics text.

An instantaneous snapshot of the wave amplitude of a linearly polarized beam of light is sketched in Fig. 1-3. The electromagnetic wave is characterized by an electric field and a magnetic field, perpendicular to each other and both perpendicular to the direction of propagation. The pattern

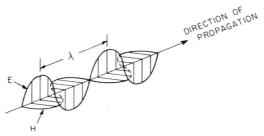

Fig. 1-3 Wave amplitude of a linearly polarized beam of light at one instant of time, showing the wavelength λ, and the components of electric field vector E and magnetic field vector H.

sketched in Fig. 1-3 will propagate along the direction of propagation at the velocity c. At a particular point in the beam of light, the electric field vector E will oscillate along a line in a plane perpendicular to the direction of propagation. This is the case of linear polarization, also called plane polarization. Unpolarized light, on the other hand, will contain many components of electric field vectors oriented in all directions perpendicular to the direction of propagation. In this case the instantaneous electric field vector seen by an observer at a fixed point would move randomly in the plane perpendicular to the direction of propagation.

If one combines two beams with linear polarizations at right angles, and with a constant phase difference, one has elliptically polarized light. The tip of the electric field vector may be considered to sweep around in an elliptical path. The ellipse reduces to a circle if the amplitudes of the two linearly polarized components are equal and if the phase difference is $\pi/2$ (or an odd integral multiple of $\pi/2$). The light is said to be circularly polarized. There may be two senses to the circular polarization: right-handed if the direction of rotation of the tip of the electric field vector is clockwise when one looks opposite to the direction in which the light travels, or left-handed if the rotation is counterclockwise.

Lasers beams are often linearly polarized. The polarization of a beam can be tested by inserting a polarizer in the beam path. Polarizers can take many forms, some of which will be discussed in Chapter 7. The most familiar is the common polaroid sheet. The polarizer has an oriented axis, called the transmission axis. It transmits light whose electric field vector is parallel to this axis. If the polarizer is inserted in a beam of linearly polarized light, maximum transmission will occur when the transmission axis is aligned with the direction of the electric field vector. When the polarizer is rotated, the transmitted light intensity will vary as the square of the cosine of the angle between the transmission axis and the direction of polarization. If the original beam is unpolarized, there will be no variation in the intensity of the transmitted light as the polarizer is rotated. However, the light transmitted by the polarizer will be linearly polarized and the intensity of the transmitted light will be one-half of the incident intensity.

We should remark on the possibility of using polarization effects to eliminate losses when light passes from a gas or vacuum into a transparent solid. Ordinarily a part of the light energy will be reflected and lost at the interface. This loss can be a serious problem in many optical designs. It can be eliminated by proper geometrical orientation of the interface and the polarization. In Fig. 1-4, the light is incident on the interface, assumed to be air–glass for specificity, at an angle θ to the normal to the surface. The direction of the light beam and the normal to the surface define a plane P. Two components of polarization may be considered, the so-called s-com-

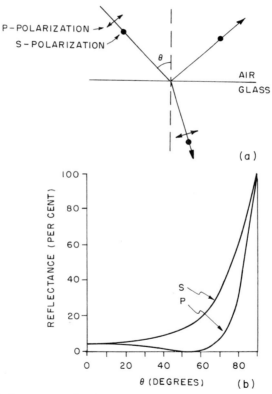

Fig. 1-4 Reflectance as a function of incidence angle θ for a light beam traveling from air into glass with index of refraction equal to 1.5. (a) defines the geometrical orientation and the p- and s-components of polarization. (b) defines the reflectance of each of these components as a function of the incidence angle θ, indicating a minimum at Brewster's angle for the p-component.

ponent, which is perpendicular to plane P, and the p-component, which lies in plane P.

Figure 1-4b shows the reflectance as a function of θ. At normal incidence, $\theta = 0°$, the loss is the same for each component, and is equal to 4%. The loss of the s-component increases with θ and approaches 100% at grazing angles near $90°$. The p-component however has a minimum of zero at an angle θ_B, called Brewster's angle. One has

$$\tan \theta_B = n \qquad (1.3)$$

where n is the index of refraction. For ordinary glass, $n \approx 1.5$, and $\theta_B \approx 56°$. Thus proper orientation of polarization and surface normal can eliminate the reflection loss when light enters a transparent solid.

Birefringence is another important phenomenon related to polarization. In certain transparent crystalline materials, the velocity of a light wave depends on its polarization. Such crystals are said to be birefringent, or doubly refracting. This is illustrated in Fig. 1-5. When the beam of unpolarized light enters the birefringent crystal, the two different components of

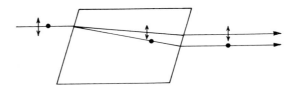

Fig. 1-5 Ray directions for two components of polarization in an unpolarized beam of light entering a birefringent crystal.

polarization propagate in slightly different directions as indicated. The difference arises because of the difference in velocity for the two different components of polarization. There will be two different beams of light emerging from the crystal, one polarized horizontally and one polarized vertically.

Birefringence can be demonstrated easily if one places a piece of birefringent crystal, e.g., calcite, on a sheet of printed paper. Two different images of the print will be seen through the crystal. Each image is produced by light of a different polarization. If one places a polarizer above the crystal and rotates it, the difference in polarization of the two images will be readily apparent.

In some materials birefringence may be induced by application of a voltage to the crystal. There are a number of such crystalline materials, for example potassium dihydrogen phosphate (commonly called KDP). This phenomenon is called the electrooptic effect.

The ability of birefringence to separate two perpendicular components of polarization in a light beam and to have them travel in different paths leads to many applications in devices commonly used with lasers. Since the birefringence may be controlled in some materials by application of voltage, many useful control functions are possible. Thus birefringence, either natural or electrically induced, is commonly used to construct polarizers, modulators, switches and light beam deflectors or scanners. These devices are very often used in connection with laser applications. Chapter 7 offers a more complete description of the application of birefringence in these devices.

A second phenomenon of importance is diffraction, which is the bending of light around the corners of an obstacle in the beam path, so that light enters the region which would be considered the geometrical shadow. As an example, we consider the shadow of a straight edge. According to geometrical optics, if an opaque edge is placed between a light source and a screen, the edge would produce a sharp shadow upon the screen. No light would reach the screen at points within the shadow, and outside the shadow the screen would be uniformly illuminated. However, careful observation of the shadow shows that the boundary region is characterized by alternate bright and dark bands, and that a small amount of light has been diffracted around the edge of the geometrical shadow. Diffraction effects have long been known, but they become important for a collimated laser beam. In many cases with conventional light sources, the relatively large beam spread of the light will tend to wash out the fringes formed by diffraction. The collimated beam of the laser makes diffraction much easier to observe. This has the twofold effect of providing practical applications based on diffraction, and the sometimes troublesome necessity of minimizing diffraction effects where they are not desired.

Perhaps the most important case of diffraction for lasers occurs when one considers a circular aperture. This is important because the exit aperture of the laser defines an obstacle that will diffract the beam. The practical implications of this will be considered in greater detail in Chapter 2. The angular spread of the light emerging from a uniformily illuminated circular aperture of radius a is proportional to $[J_1(ka \sin \theta)/ka \sin \theta]^2$, where θ is the angle between the normal to the aperture and the direction of observation, $k = 2\pi/\lambda$, and $J_1(x)$ denotes a Bessel function of order one. The behavior of $[J_1(x)/x]^2$ is similar to $[\sin x/x]^2$. This is an oscillating pattern of bright and dark rings. The light is most intense in the center of the pattern, $\theta = 0$, and falls to zero at the first dark ring, where $ka \sin \theta$ equals the first zero of $J_1(x)$. Succeeding smaller oscillations provide further bright and dark rings surrounding the center pattern. In order to determine the beam divergence, one may define the beam divergence as the angle corresponding to the first dark ring. The first zero of the Bessel function $J_1(x)$ occurs at $1.22\pi = 3.83$. This means that the angle of divergence θ_d will be given by

$$\theta_d \approx 1.22\lambda/2a \qquad (1.4)$$

As the aperture size is decreased, the angular spread of the diffracted light increases.

The third phenomenon of interest is that of interference. When two or more trains of waves cross one another, they interfere. The electric field at a given point will be a superposition of the waves from the different wave trains. The resultant electric field at any point and time is found by adding the instantaneous fields that would be produced by the individual wave trains. If the interference is such as to produce a low or zero value of the electric field, the wave trains are said to interfere destructively, and a dark region will occur. If the electric fields add to a relatively large value, they are said to interfere constructively and there will be a relatively bright region produced. Thus, interference effects can also produce bright and dark fringes.

For simplicity let us consider interference produced by light reflected from a thin wedged film, as is illustrated in Fig. 1-6. A ray of light is incident on the upper surface of the film. Part of the light is reflected at the first surface, and a part is transmitted, simultaneously undergoing refraction so that it is bent toward the normal to the surface. At the second surface, part is again reflected and part of this emerges from the film. This light will interfere with the light transmitted from the front surface. At points where the total optical path is equal to an odd integral number of half-wavelengths, the

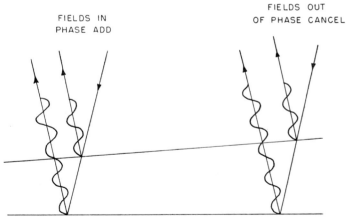

Fig. 1-6   Schematic diagram of interference produced in reflection of light from a thin wedged film, showsng constructive and destructive interference at different thicknesses of the film.

electric field vectors will be out of phase with each other and will add to zero. The condition for this is

$$2nd = (2N + 1)\lambda/2 \qquad (1.5)$$

where d is the physical thickness of the film, n is the index of refraction of the film, and $N = 0, 1, 2, 3. \ldots$ (We note that the optical thickness of the film is nd.)

At positions where the wedge is a quarter-wavelength thicker, the two reflected waves will be in phase with each other. At these positions,

$$2nd = N\lambda \qquad (1.6)$$

i.e., the optical path difference is an even integral number of half-wavelengths. Here N is again an integer. The electric fields will be in phase and will add to produce maximum light intensity. Thus, the pattern observed will be a system of bright and dark fringes, with the period of the fringes being the distance along the plate in which the thickness increases by a quarter-wavelength.

Interference fringes can be observed under many other situations. They have been known for many years, but once again the advent of the laser has made observation of interference fringes much easier. This leads to many practical applications.

C. Energy Levels

The concept of an energy level is an important idea for lasers. Laser operation takes place by transitions between different energy levels of an atomic or molecular system.

The simplest types of energy levels are those available to an isolated atom. The hydrogen atom will serve as a familiar example. The electron in the hydrogen atom can take up certain discrete orbits around the nucleus. From a simple classical point of view, one might expect a continuum of orbits to be available. However, quantum mechanics shows that only certain discrete, quantized orbits are possible. In the hydrogen atom these discrete orbits correspond to energy levels. The electron in a given orbit possesses a definite value of energy. The different energy values that the atom can have by virtue of the electron residing in the different quantized orbits are the energy levels of the hydrogen atom. Similarly, other isolated atoms all possess their own characteristic set of energy levels. A schematic diagram of a set of energy levels is shown in Fig. 1-7.†

† According to Eq. (1.2), energy is inversely proportional to wavelength. Hence, energy level diagrams can be expressed in units of reciprocal length (also called wave number).

Wingate College Library

When one considers a material such as a solid, a gas, or a liquid, the energy levels are no longer the energy levels of the individual atoms or molecules. An atom or molecule interacts with its neighbors and the energy levels of the individual atom or molecule are modified by the interaction. In a gas in which the density is relatively low, the interaction is not strong, and the energy levels are similar to the energy levels of the isolated atom or molecule.

In a condensed material, liquid or solid, the atoms are packed together and the interaction becomes strong. The energy levels of the individual atoms broaden and merge into a nearly continuous band, consisting of a large number of closely spaced energy levels. This is, for example, the origin of the familiar conduction and valence bands in a semiconductor. One important exception in solids is the case where the energy levels belong to an unfilled interior shell of electrons, surrounded by an outer shell of electrons. Examples are the transition metal elements, such as vanadium, chromium, manganese, and iron, and the rare earth elements, of which examples are neodymium, samarium, and europium. These electronic energy levels are shielded from the neighboring atoms by the exterior shells of electronic charge; the interaction does not broaden the energy levels so much as is usual in solids. The narrow energy levels in certain transition metals and rare earth elements are important in laser operation.

In addition to the electronic energy levels, molecules with more than one atom can possess additional vibrational or rotational energy levels. An example is the carbon dioxide molecule which consists of a linear arrangement of two oxygen atoms and a carbon atom with the carbon atom in the middle. The vibrational energy levels, for example, correspond to motion of the oxygen atoms relative to the carbon atom. Such energy levels are important in the operation of a number of gaseous lasers. Rotational energy levels correspond to rotational motion of an asymmetric molecule. Typically, the ordering of the energy is such that electronic energies are largest, vibrational energies are second, and rotational energies have the smallest magnitude.

D.   Interaction of Radiation and Matter

When radiation interacts with matter, it involves a change from one energy level to another. The energy difference between the two states must be balanced by absorption or emission of radiation. The following relation will hold:

$$hf = hc/\lambda = E_2 - E_1 \qquad (1.7)$$

where E_2 and E_1 are the energies of the two states involved (numbered so that $E_2 > E_1$), and hf is the photon energy.

We may distinguish three different ways in which radiation (or to be more specific, light) can interact with energy levels: fluorescence, absorption, and stimulated emission. These phenomena demonstrate the particulate nature of light. Whenever light is absorbed or emitted by an atomic or molecular system, it always involves a quantum of energy as expressed by Eq. (1.7).

1. Fluorescence

An atom in an upper energy level spontaneously decays to a lower level, emitting the energy difference as a photon with the appropriate frequency to satisfy Eq. (1.7). As an example, an electron in an orbit in a hydrogen atom may shift to a different orbit with a lower energy value. The difference in energies is emitted as a photon with frequency as specified by Eq. (1.7). The wavelengths corresponding to several transitions in the hydrogen atom are shown in Fig. 1-7.

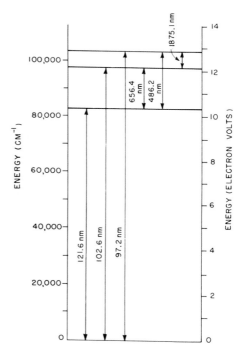

Fig. 1-7 Partial energy level diagram of the hydrogen atom, with wavelengths indicated for several transitions. Energy is given both in energy units (electron volts) and wave number units (reciprocal centimeters).

2. Absorption

Light of frequency f interacts with an atom or molecule in a low-lying energy level and raises it to a level of higher energy. The light is absorbed, its energy going into the atomic or molecular system. Unless the frequency satisfies Eq. (1.7) for some combination of energy levels, there will be no absorption. Since many materials have numerous, closely packed energy states, such a combination can exist for almost any wavelength and the material can be opaque. Absorption of light by materials is a familiar phenomenon.

3. *Stimulated Emission*

An atom in an upper energy level is stimulated by incoming radiation of frequency f satisfying Eq. (1.7) to give up its energy and fall to a lower energy level. This process is much less familiar than either fluorescence or absorption, but it is the process responsible for laser operation. It was first postulated by Albert Einstein, who showed that emission of light could be triggered by an incoming photon if an electron were present in a state with high energy. The photon must have the proper energy corresponding to the energy difference between the original state and a state of lower energy. The result of this process is that one photon interacts with the atomic system and two photons emerge. Both photons have the same frequency and travel in the same direction. Energy is extracted from the atomic system and appears as additional light of frequency f. The original light is still present, so we have amplified the light intensity in the stimulated emission process. Hence, the acronym LASER: Light Amplification by Stimulated Emission of Radiation.

These three processes are shown schematically in Fig. 1-8. The horizontal straight lines represent energy levels, the wavy arrows photons, and the vertical arrows the transition of the electron from one level to another.

We have now identified the basic stimulated emission process responsible for laser operation. In order to use this process, one requires the following things: some suitable material with appropriate energy levels, a

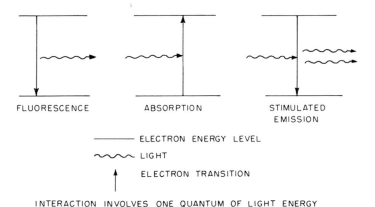

FLUORESCENCE   ABSORPTION   STIMULATED
                                             EMISSION

———————— ELECTRON ENERGY LEVEL

〜〜〜〜 LIGHT

↑ ELECTRON TRANSITION

INTERACTION INVOLVES ONE QUANTUM OF LIGHT ENERGY

Fig. 1-8   Schematic diagram of interaction of light with electronic energy levels in the processes of fluorescence, absorption, and stimulated emission.

method for producing a "population inversion" in the material, and a structure called a "resonant cavity." In the next sections we shall describe each of these requirements in turn.

E. Laser Materials

A laser needs some material with appropriate properties. The material is often called the "active medium." The active medium will have suitable energy levels for interaction with radiation. Laser action occurs at wavelengths at which the material has fluorescent emission, i.e., at frequencies satisfying Eq. (1.7) for some pair of energy levels.

Examples of materials with sets of energy levels that have proved useful include ruby (aluminum oxide crystal doped with a small amount of chromium), glass doped with a small amount of the rare earth element neodymium (Nd:glass), a type of garnet doped with neodymium (referred to as Nd:YAG), carbon dioxide gas, neon gas, ionized argon gas, and the semiconductor crystal gallium arsenide.

Potentially useful laser materials are studied spectroscopically to determine the energy levels. Some specific examples of energy levels are displayed in Fig. 1-9 for a solid material and in Fig. 1-10 for a gas. Figure 1-9 shows the important levels of ruby, with the laser transition at a wavelength of 0.6943 μm. Figure 1-10 shows the energy levels of neon, with three laser transitions identified. The levels are labeled with their spectroscopic designations (e.g., $^3s_2$), which need not concern us here, and with their energy† relative to the lowest energy level as zero.

Figures 1-9 and 1-10 also contain additional information regarding the production of the population inversion. We shall return to this later.

Spectroscopically, one desires materials with strong fluorescent efficiency, and narrow spectral widths for the fluorescent lines. It is not appropriate here to consider why the materials listed above are useful in lasers, while others with similar properties are less successful.

Some discussion of commonly encountered jargon follows. The fluorescent line which is used for laser operation has a finite spectral width. The distribution of frequencies in the fluorescent line defines the so-called fluorescent line shape. There are two commonly encountered line shapes— Lorentzian and Gaussian. The Lorentzian distribution function, which arises from the finite fluorescent lifetime of the emitting atoms, varies as

$$[(f - f_0)^2 + (\Delta f)^2]^{-1}$$

where f is the frequency, f_0 the center frequency of the line, and Δf the half

---

† Energy in electron volts (eV); 1 eV = 1.6×10^{-12} erg.

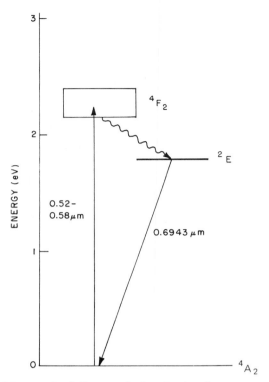

Fig. 1-9 Partial energy level diagram of ruby, showing the spectroscopic notation for the relevant energy levels for absorption of green light (0.52–0.58 μm) and spontaneous emission of red light (0.6943 μm).

width of the line. The Gaussian distribution function, which arises from Doppler shifts of moving molecules in a gas, varies as

$$\exp\{-(\ln 2)[(f - f_0)/\Delta f]^2\}$$

These distribution functions define the intensity of a fluorescent line as a function of frequency.

There is another distinction between line shapes: homogeneous or inhomogeneous. In a homogeneous line, any atom can emit over the whole width of the line; in an inhomogeneous line, any one atom can emit only in a small region of the line. Thus, for laser action at some frequency within the envelope of the fluorescent line, the entire population of atoms can contribute in a homogeneous line, but only a small fraction of the atoms can contribute in an inhomogenous line. When this fraction is used up, one says that "a hole is burned in the line" and the power output is limited by the fraction of

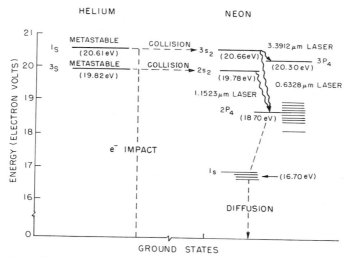

Fig. 1-10 Energy levels relevant to operation of helium–neon laser, with the spectroscopic designations indicated. The paths for excitation and laser emission are shown.

atoms available. Therefore, homogeneous lines are desirable. In general, Lorentzian lines are homogeneous; Gaussian lines are inhomogeneous. Specific examples of laser lines include

Gases: Inhomogeneous
Ruby at room temperature: Homogeneous
Ruby at low temperature: Inhomogeneous

F. Population Inversion

When light of intensity $I(z)$ and frequency f travels in the z direction through a medium with densities of atoms N_1 and N_2 in two energy levels† satisfying Eq. (1.7), one has absorption of the light according to

$$I(z) = I(0)e^{-\alpha z} \tag{1.8}$$

where α is the absorption coefficient and is given (for a Gaussian line shape) by

$$\alpha = (c^2/4\pi f^2 \tau)[(\ln 2)/\pi]^{1/2}(N_1/\Delta f) \tag{1.9}$$

where τ is the fluorescent lifetime of the material, and the other notation is as before. Note that c here is the velocity of light in the material, not the velocity of light in vacuum. A slightly different expression holds for a

---

† The levels are numbered so that level 2 lies higher than level 1.

Lorentzian line, but α is still proportional to $N_1/\Delta f$. Amplification by stimulated emission occurs as

$$I(z) = I(0)e^{+gz}$$

where g, the gain coefficient, is given by

$$g = (c^2/4\pi f^2 \tau)[(\ln 2)/\pi]^{1/2}(N_2/\Delta f) \qquad (1.10)$$

for a Gaussian line. We note the similarity of the forms for α and g. This means that absorption and stimulated emission are inverse processes. The net effect of passage through the material is

$$I(z) = I(0) \exp[(c^2/4\pi f^2 \tau)[(\ln 2)/\pi]^{1/2}(1/\Delta f)(N_2 - N_1)z] \qquad (1.11)$$

If $N_2 > N_1$, the light intensity will increase as it travels; if $N_1 > N_2$, the light intensity will decrease. To make the laser work, therefore, one needs $N_2 > N_1$. This situation is called a "population inversion." The population inversion is simply having more electrons in a high-lying energy level at which fluorescent emission starts than in the lower-lying energy levels at which the emission terminates. Without the population inversion, according to Eq. (1.11), there will be a net absorption at frequency f. With the population inversion, there will be light amplification by stimulated emission of radiation.

A population inversion is not the usual condition. When one has a material in thermal equilibrium at temperature T, one has

$$N_2/N_1 = \exp[-(E_2 - E_1)/kT] \qquad (1.12)$$

where k is Boltzmann's constant, equal to 1.38×10^{-16} erg/deg. This equation means that $N_2 < N_1$. To attain laser operation, one must upset the thermal equilibrium in some way to produce the unusual situation of a population inversion.

We now consider the so-called threshold condition, which relates the minimum population inversion required for a particular system. There are inevitable sources of loss associated with laser operation. One loss is the output through the two end mirrors, of reflectivities R_1 and R_2. (We assume here two mirrors at the end of the laser; we shall discuss this further in considering resonant structures.)

Another source of loss involves imperfections in the system, such as inhomogeneities in a ruby crystal. Such losses may be characterized by a loss coefficient β, such that the decrease in light intensity in a distance z is a factor $e^{-\beta z}$.

Using Eq. (1.11), the change in light intensity after a round trip through the active medium between two mirrors separated by distance L involves a factor

$$R_1 R_2 \exp(-2\beta L) \exp[2L(c^2/4\pi f^2\tau)[(\ln 2)/\pi]^{1/2}(N_2 - N_1)/\Delta f] \quad (1.13)$$

For a net increase in light intensity, this factor must be greater than unity. Taking logarithms, one gets

$$N_2 - N_1 > (4\pi f^2\tau \, \Delta f/Lc^2)(\pi/\ln 2)^{1/2}(\beta L - \ln R_1 R_2) \quad (1.14)$$

as the threshold condition, i.e., the minimum population inversion required to sustain laser action in a Gaussian line for a medium with the given properties. For a Lorentzian line the expression will differ in some inconsequential details. One sees that a small value of Δf i.e., a narrow spectral line width) is desirable to make it easy to get the required population inversion. Equation (1.14) is important in that it specifies the threshold at which the laser will begin to operate. One must take steps to supply the minimum population inversion $N_2 - N_1$ called for by this equation.

It may be shown that it is impossible to produce a population inversion by optical pumping in a system with only two levels. One therefore resorts to a system with either three or four levels. (The actual number of energy levels in any laser material is much more than four. However, only three or four will be directly involved in the laser operation.) Laser materials are specified as being "three-level systems" or "four-level systems." The distinction is this: in a three-level system the terminal level for the fluorescence is the ground level (i.e., the level with lowest energy). In a four-level system, the terminal level lies above the ground level. The situation is sketched in Fig. 1-11. Consider a typical three-level system. One achieves

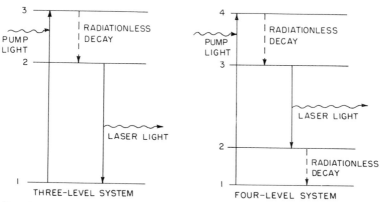

Fig. 1-11 Schematic diagram of excitation processes and laser emission for three-level laser systems, and four-level laser systems.

the population inversion by "pumping" with an auxiliary "pump light" source. One supplies light at a frequency $f_3 = (E_3 - E_1)/h$, which is absorbed to move atoms from level 1 to level 3. Then a (very fast) radiationless transition (accomplished by thermal vibrations of the lattice) will move the atoms to level 2. Stimulated emission then occurs between levels 2 and 1 at frequency $f_2 = (E_2 - E_1)/h$. If one pumps hard at frequency f_3 (i.e., makes the transition rate large for the transition from level 1 to level 3), one can have $N_2 > N_1$. Thus the population inversion is achieved. The three-level system has the drawback that state 1 is the ground state, so most of the atoms must be moved to state 2 and intense pumping is required. A four-level system removes this constraint. Radiationless transitions from level 4 to level 3 and from level 2 to level 1 are hopefully very fast. Level 2 then stays empty; to produce a population inversion it is only necessary to get a small number of atoms into level 3. Four-level systems are desirable in that they offer low thresholds. Most useful lasers are four-level systems. Ruby, a three-level system, is a notable exception.

With reference to Fig. 1-11, it is worthwhile to point out that the pumping to achieve the population inversion always occurs to the same level of energy, i.e., to the upper state for the fluorescent transition. When one pumps the system hard, one moves more and more atoms into this upper state, but does not move the atomic system to successively higher levels of energy.

How does one achieve the population inversion in practice? There are three main methods: "optical pumping," electron excitation, and resonant transfer of energy.

1. Optical Pumping

This method involves an auxiliary source of light that is absorbed to raise the electrons to high-lying energy levels. This has already been discussed with reference to Fig. 1-11. The ruby laser (see Fig. 1-9) employs optical pumping. One supplies light in the green portion of the spectrum (0.52–0.58 μm) which is absorbed to raise the chromium atom to the highest level shown in this three-level system. A radiationless decay to the upper fluorescent level for the 0.6943 μm transition follows. With intense pumping, the ground state will be depopulated, the population inversion will be produced, and laser action can occur.

Probably the most common such source is a flashlamp similar to those used in photography. The flashlamps commonly used in lasers are glass or quartz tubes filled with a gas at low pressure. Commonly xenon is used, although for some cases, higher peak brightness can be achieved by using lower atomic weight noble gases, such as krypton or helium. The lamps have

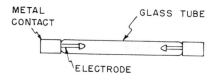

Fig. 1-12 Drawing of the structure of a linear flashlamp.

internal electrodes and can be prepared in a variety of shapes. Figure 1-12 shows a typical linear lamp. Helical and U-shaped lamps are also common. The range of performance is very broad. Electrical input energy and pulse duration and lamp size can be varied over a wide range.

When an electric current is discharged through the lamp, a high temperature plasma is formed which emits radiant energy over a wide range of wavelengths. The radiation generally has a complicated spectrum, and contains both line and continuum components. It is possible, within reasonable limits, to choose the gas and the operating parameters to match some reasonably broad portion of the spectrum to the absorption spectrum of the laser material being used.

The discharge in a flashlamp is initiated by causing a spark streamer to form between the electrodes. This is done by pulsing the lamp with a high voltage trigger pulse. Figure 1-13 shows a typical circuit including the high voltage trigger which is applied externally to the flashlamp. Once the flashlamp has been triggered, the main discharge capacitor C can discharge through the ionized gas. The series inductance is added to shape the current pulse through the lamp and to avoid ringing or reverse flow of current which is harmful to the lamp life. The lifetime of flashlamps will be considered in more detail in Chapter 5.

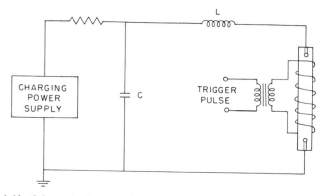

Fig. 1-13 Schematic diagram of a typical circuit for operation of a flashlamp.

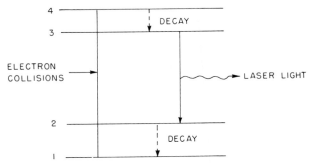

Fig. 1-14 Schematic diagram for excitation and laser emission for direct electronic excitation in a four-level laser system.

2. *Direct Electron Excitation*

This is shown schematically in Fig. 1-14. Energetic electrons are injected into the material and collide with ground state atoms raising them to energy level 4. A rapid radiationless decay to level 3 may follow. The population inversion is produced between levels 3 and 2. (Level 2 is originally empty.)

A specific example would be an electrical discharge through a gas. The situation would be very similar to a neon sign. The current flows between electrodes in the gas. Electrons are emitted by the cathode, and travel through the gas under the action of the applied electric field. The electrons collide with gas atoms or molecules. In such a collision, the kinetic energy of the electron is transferred to excitational energy of the gas atoms or molecules.

In order to produce a population inversion, the upper level 3 must be populated more than level 2 by collisions. This implies some suitable relation between the electron velocity spectrum, the cross sections for transferring energy to the different levels, and the rates at which the different levels decay. It may be difficult to find suitable combinations such that Eq. (1.14) will be satisfied. Because of this difficulty, the following variation, resonant energy transfer, is probably more often used.

3. *Resonant Transfer of Energy*

This method is used in gases. Two gases which have high-lying energy levels with approximately equal energies are required. Examples are He and Ne (see Fig. 1-10) or CO_2 and N_2. Electrons excite one of the gases to high-lying levels in collisional processes. These excited gas molecules make collisions with molecules of the second gas and transfer the excitation energy to the second gas. The molecules of the second gas end up in a high-lying level, with lower levels empty, so that the desired population inversion is established. Specifically, in the He–Ne laser, He atoms are excited by electrons. The excitation energy is transferred to Ne atoms and laser action occurs on the energy levels of Ne. In the CO_2–N_2 laser (usually called simply the CO_2 laser), the N_2 is excited and passes its excitation energy to CO_2, which produces the laser operation.

We note explicitly that energy is required to produce the population inversion. Energy is input to the active material in some form so as to raise atoms to high-lying energy levels. The energy output of the laser is always less than the energy supplied. Lasers are notoriously inefficient. Efficiencies of a few tenths of one percent are not uncommon.

G. Resonant Cavity

The final requirement for a laser is some sort of geometrical structure called a resonant cavity. We shall first assume that the resonant cavity consists of two parallel mirrors, and later consider some other possibilities.

To achieve laser action, one must provide a resonant cavity so that the light will make many passes through the active medium. Such an arrangement is needed so that the light can travel a long path in the medium. Figure 1-15 shows the structure schematically for a laser excited by a flashlamp. This figure assumes optical pumping and a rod of a solid crystalline material. In order to be specific, consider a ruby rod. For this case, the mirrors are reflecting coatings on the flat and parallel ends of the rod. Of course, many other configurations are possible.

Figure 1-15 shows all the elements for the laser—the active material,

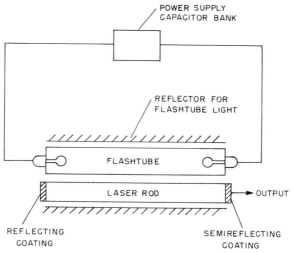

Fig. 1-15 Schematic diagram of an optically pumped solid laser.

the means for producing the population inversion (the flashlamp), and the resonant cavity (the mirrors on the end of the rod). For resonance, one must have an integral number of half-wavelengths between the mirrors:

$$q(\lambda/2) = d \qquad (1.15)$$

where d is the mirror separation and q is an integer much greater than unity. Typically there are many combinations of q and λ that satisfy this equation and for which λ lies within the fluorescent line. These are called the cavity modes, and laser action occurs only at these discrete wavelengths satisfying Eq. (1.15) for some integer q. Figure 1-16 shows the situation. The fluorescent line shape is shown, with the positions of the resonant cavity modes indicated. These modes are often called longitudinal modes. They are evenly spaced in wavelength with separation $\lambda^2/2d$, where d is the distance between the mirrors.

The situation is similar to the resonant modes of a microwave cavity, except the wavelength is of the same order of magnitude as the dimension of the geometrical structure in the microwave case. For microwaves, q in Eq. (1.15) would be of the order of unity. For visible and infrared light, the geometrical structure is much larger than the wavelength.

The spectral line shape of the emission from a laser can be complicated. Usually, there will be present in the laser output a number of different modes, each at a slightly different frequency, as shown in Fig. 1-16. Thus, the frequency spectrum consists of several discrete, separated frequency components, each due to the presence of a different longitudinal mode. We note that the frequency difference between the longitudinal modes is very small compared to the operating frequency of the laser, so that the laser is still almost monochromatic.

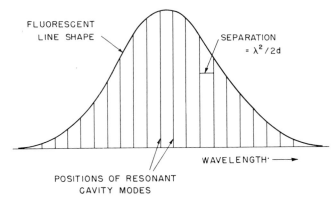

Fig. 1-16 Schematic diagram of the positions of resonant modes in a laser cavity, with λ the wavelength and d the cavity length.

A discussion of resonator modes should also include the spatial or transverse mode structure. This will be deferred until Chapter 2, where we will describe the spatial profile of the laser beam.

Now let us consider the growth of the radiation field between the mirrors. Light bounces back and forth between the mirrors, being amplified by stimulated emission on each pass through the active medium. The growth of light intensity is shown schematically in Fig. 1-17. Portions (a) and (b) show the original unexcited rod, and the excitation by the pumping light. Portion (c) shows fluorescence and the start of stimulated emission. Portions (d) and (e) show growth of the stimulated emission. The ends of the laser rod are flat and parallel and thus allow light to make many bounces between them along the axis of the rod. There will be some fluorescent light present in the rod because of fluorescent emission from the excited electrons. Some of this fluorescent light will move in a direction other than along the axis of the rod. It will reach the side of the rod, pass out, and be lost. Some of the fluorescent light will be in a direction along the rod axis. This light will be amplified by the stimulated emission process as it passes through the rod. At the semitransparent mirror at one end of the rod, some of the light will pass out as useful output of the laser. Some will be reflected back for another pass along the rod axis and a further buildup in intensity. This light will

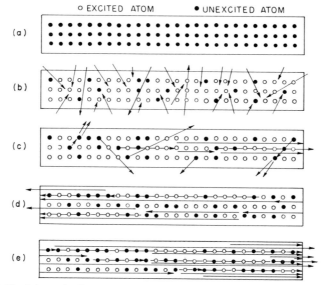

Fig. 1-17 Schematic diagram of excitation and stimulated emission in a solid laser rod: (a) Initial unexcited rod, (b) excitation by pump light, (c) fluorescence and the start of stimulated emission, (d) growth of stimulated emission preferentially along the rod axis, (e) continued amplification by stimulated emission and output from rod end.

pass back and forth through the rod many times. The result of this process is a regenerative cascade in which the energy stored in the upper energy levels of the electrons is rapidly converted into an intense, highly collimated beam of light. The growth of light intensity and the extraction of light through the semitransparent mirror is of course a continuous process, rather than an occurrence of separate bounces between the mirrors, as the simple description above might seem to indicate.

So far we have assumed that the two mirrors are plane parallel mirrors. Since plane parallel mirrors require exact angular alignment, the use of spherical mirrors, which are less sensitive to angular alignment, has become common. The study of resonant modes of curved mirror cavities has become quite detailed. If R_1 and R_2 represent the radii of curvature of the mirrors, and d their separation, one can have stable laser operation for many combinations of R_1, R_2, and d.

The mirror configurations may be judged on two criteria: stability and filling of the material by the light. Stability means that light rays bouncing back and forth between the mirrors will be reentrant. Good filling of the active material means that the spatial profile defined by the light rays fills all the volume of active laser material which is contained between the mirrors.

Plane parallel mirrors have good filling, because light rays may cover the entire volume between the mirrors. If perfectly aligned, the light rays are reentrant, but if the mirrors become even slightly misaligned, the light rays can walk off the edges of the mirrors after a number of reflections. Thus, plane parallel mirrors have marginal stability. Some combinations of spherical mirrors have good stability, in that a slight change from what was intended still leaves a reentrant path. However, not all spherical mirrors fill the volume effectively. An example is the confocal case, where $R_1 = R_2 = d$. The focal point for each mirror is at the midpoint of the cavity, so that the light rays will pass through a narrow waist halfway between the mirrors. They will not make use of the material which is off the center axis halfway between the mirrors. Thus a confocal configuration is said to have poor filling.

A number of possible mirror configurations are sketched in Fig. 1-18.

In practice, one often uses large radius mirrors ($R_1 \gg d$, $R_2 \gg d$) because they fall in a region that gives good stability and the beam spatial profile fills the active medium reasonably well. Plane parallel mirrors offer good use of the medium but this configuration has delicate stability. Hemispherical mirrors ($R_1 = d$, $R_2 = 0$) offer good stability, but the radiation does not fill the active medium well. Spherical mirrors ($R_1 = R_2 = d/2$) and confocal mirrors ($R_1 = R_2 = d$) have poor stability and do not fill the active medium. An example of an unstable configuration (one of many) is also shown in Fig. 1-18.

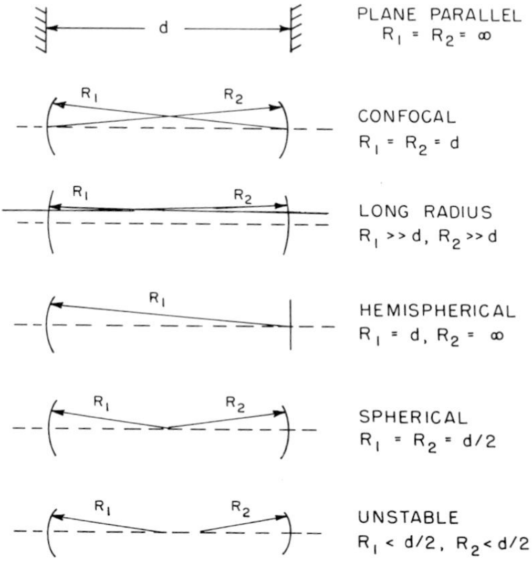

Fig. 1-18   A number of possible mirror configurations. R_1 and R_2 are the mirror radii and d is the mirror separation.

A commonly employed compromise between stability and filling is the long radius configuration, in which $R_1 \gg d$ and $R_2 \gg d$. The long radius geometry is probably the one most often used on modern commercial lasers.

In some conditions, unstable resonators are in fact used in lasers. Such resonators are most useful for lasers with a Fresnel number much larger than unity. The Fresnel number N_F is given by

$$N_F = a^2/d\lambda \tag{1.16}$$

where a is the radius of the laser medium, d the distance between the mirrors, and λ the wavelength. For long slender lasers, the Fresnel number may be of the order of magnitude of unity, and the stable, long radius configuration is useful. However, if the Fresnel number is large compared to unity, the spatial structure can become complicated and the beam divergence will be large. (See Chapter 2 for further discussion.) In this case, an unstable resonator may produce a more desirable spatial pattern.

The laser beam emerges around rather than through the output mirror. A typical unstable resonator configuration is shown in Fig. 1-19. In this situation both mirrors are high reflectivity mirrors. The loss out of the system tends to be rather high, so that for unstable resonators to be practical the gain must be reasonably high, typically more than 50 % per pass. Unstable resonators find most use in very high power gas lasers. Figure 1-19 also shows a typical method of coupling the beam out. Unstable resonators have

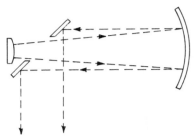

Fig. 1-19   Schematic diagram of the configuration of an unstable resonator.

probably been applied most often with CO_2 lasers. They are well suited for use with the high power CO_2 lasers, and have been used to produce well-collimated output beams of higher quality than would be available with stable resonators.

The reflecting surfaces commonly used in lasers are evaporated coatings consisting of multiple layers of dielectric materials. The technology of the mirrors used in lasers will be described in Chapter 7.

SELECTED REFERENCES

A.   Electromagnetic Radiation

C. L. Andrews, "Optics of the Electromagnetic Spectrum," Prentice-Hall, Englewood Cliffs, New Jersey, 1960, Chapter 1.

M. Born and E. Wolf, "Principles of Optics," 5th ed., Pergamon, Oxford, 1975, Chapter 1.

E. U. Condon, Electromagnetic Waves, *in* "Handbook of Physics" (E. U. Condon and H. Odishaw, eds.), McGraw-Hill, New York, 1967.

M. Garbuny, "Optical Physics," Academic Press, New York, 1965, Chapter 1.

H. L. Hackforth, "Infrared Radiation," McGraw-Hill, New York, 1960, Chapter 1.

F. W. Sears, "Optics," 3rd ed., Addison-Wesley, Reading, Massachusetts, 1949, Chapter 1.

B.   Elementary Optical Principles

M. Born and E. Wolf, "Principles of Optics," 5th ed., Pergamon, Oxford, 1975.

F. A. Jenkins and H. E. White, "Fundamentals of Optics," McGraw-Hill, New York, 1957.

M. V. Klein, "Optics," Wiley, New York, 1970.

F. W. Sears, "Optics," 3rd ed., Addison-Wesley, Reading, Massachusetts, 1949.

A. Sommerfeld, "Optics," Academic Press, New York, 1964.

C.   Energy Levels

W. S. C. Chang, "Principles of Quantum Electronics," Addison-Wesley, Reading, Massachusetts, 1969, Chapter 2.

H. M. Crosswhite (ed.), Atomic and Molecular Physics, *in* "American Institute of Physics Handbook," 3rd ed. (D. E. Gray, ed.), McGraw-Hill, New York, 1972.

R. M. Eisberg, "Fundamentals of Modern Physics," Wiley, New York, 1961, Chapters 10, 12, and 13.

M. Garbuny, "Optical Physics," Academic Press, New York, 1965, Chapter 3.
B. A. Lengyel, "Introduction to Laser Physics," Wiley, New York, 1966, Chapter 1.
H. W. Leverenz, "Luminescence of Solids," Wiley, New York, 1950, Chapter 1.
C. E. Moore, "Atomic Energy Levels," Circular of the Nat. Bur. of Std. 467, U.S. Government Printing Office, Washington, D.C., 1949.

D. Interaction of Radiation and Matter

M. Garbuny, "Optical Physics," Academic Press, New York, 1965, Chapter 3.
B. A. Lengyel, "Introduction to Laser Physics," Wiley, New York, 1966, Chapter 1.
A. Maitland and M. H. Dunn, "Laser Physics," North Holland Publ., Amsterdam, 1969, Chapters 1 and 3.
D. Park, "Introduction to the Quantum Theory," McGraw-Hill, New York, 1964, Chapter 10.
A. L. Schawlow and C. H. Townes, Infrared and Optical Masers, *Phys. Rev.* **112**, 1940 (1958).
J. R. Singer, "Masers," Wiley, New York, 1959, Chapter 2.
A. L. Schawlow, Optical Masers, *Sci. Am.*, p. 2 (June 1961).
A. Yariv and J. P. Gordon, The Laser, *Proc. IEEE* **51**, 4 (1963).
A. Yariv, "Quantum Electronics," 2nd ed., Wiley, New York, 1975, Chapter 8.

E. Laser Materials

W. R. Bennett, Gaseous Optical Masers, *Appl. Opt., Suppl. Opt. Masers*, p. 24 (1962).
G. Birnbaum, "Optical Masers," Academic Press, New York, 1964, Chapter 3.
A. L. Bloom, "Gas Lasers," Wiley, New York, 1968, Chapter 2.
W. S. C. Chang, "Principles of Quantum Electronics," Addison-Wesley, Reading, Massachusetts, 1969, Chapter 7.
A. Javan, W. R. Bennett, and D. R. Herriott, Population Inversion and Continuous Optical Maser Oscillation in a Gas Discharge Containing a He–Ne Mixture, *Phys. Rev. Lett.* **6**, 106 (1961).
Z. J. Kiss and R. J. Pressley, Crystalline Solid Lasers, *Appl. Opt.* **5**, 1474 (1966).
T. H. Maiman *et al.*, Stimulated Optical Emission in Fluorescent Solids. II. Spectroscopy and Stimulated Emission in Ruby, *Phys. Rev.* **123**, 1151 (1961).
A. Maitland and M. H. Dunn, "Laser Physics," North Holland Publ., Amsterdam, 1969, Chapter 1.
R. J. Pressley (ed.), "Handbook of Lasers," Chemical Rubber Co., Cleveland, Ohio, 1971.
A. D. White and J. D. Rigden, Continuous Gas Maser Operation in the Visible, *Proc. IRE* **50**, 1697 (1962).

F. Population Inversion

W. R. Bennett, Gaseous Optical Masers, *Appl. Opt., Suppl. Opt. Masers*, p. 24 (1962).
G. Birnbaum, "Optical Masers," Academic Press, New York, 1964, Chapter 2.
T. H. Maiman, Stimulated Optical Emission in Solids. I. Theoretical Considerations, *Phys. Rev.* **123**, 1145 (1961).
J. P. Markiewicz and J. L. Emmett, Design of Flashlamp Driving Circuits, *IEEE J. Quantum Electron.* **QE-2**, 707 (1966).
J. R. Oliver and F. S. Barnes, Rare Gas Flashlamps: The State of the Art and Unsolved Problems, *Proc. IEEE* **59**, 638 (1971).
A. L. Schawlow, Optical Masers, *Sci. Am.*, p. 2 (June 1961).
A. L. Schawlow, Advances in Optical Masers, *Sci. Am.*, p. 34 (July 1963).

W. V. Smith and P. P. Sorokin, "The Laser," McGraw-Hill, New York, 1966, Chapter 3.

A. Yariv, Energy and Power Considerations in Injection and Optically Pumped Lasers, *Proc. IEEE* **51**, 1723 (1963).

A. Yariv, "Quantum Electronics," 2nd ed., Wiley, New York, 1975, Chapter 9.

G. Resonant Cavity

W. R. Bennett, Hole Burning Effects in a HeNe Optical Maser, *Phys. Rev.* **126**, 580 (1962).

G. Birnbaum, "Optical Masers," Academic Press, New York, 1964, Chapter 6.

G. D. Boyd and J. P. Gordon, Confocal Multimode Resonator for Millimeter through Optical Wavelength Masers, *Bell Syst. Tech. J.* **40**, 489 (1961).

A. G. Fox and T. Li, Resonant Modes in an Optical Maser, *Proc. IRE* **48**, 1904 (1960).

A. G. Fox and T. Li, Resonant Modes in a Maser Interferometer, *Bell Syst. Tech. J.* **40**, 453 (1961).

H. Kogelnik and T. Li, Laser Beams and Resonators, *Appl. Opt.* **5**, 1550 (1966).

A. Maitland and M. H. Dunn, "Laser Physics," North Holland Publ., Amsterdam, 1969, Chapters 1, 4, 5, and 6.

B. J. McMurtry, Investigation of Ruby Optical Maser Characteristics Using Microwave Phototubes, *Appl. Opt.* **2**, 767 (1963).

A. E. Siegman, Unstable Optical Resonators for Laser Applications, *Proc. IEEE* **53**, 277 (1965).

A. E. Siegman, Unstable Optical Resonators, *Appl. Opt.* **13**, 353 (1974).

G. Toraldo di Francia, Optical Resonators, *in* "Quantum Optics" (S. M. Kay and A. Maitland, eds.), Academic Press, New York, 1970.

A. Yariv, "Quantum Electronics," 2nd ed., Wiley, New York, 1975, Chapter 7.

CHAPTER 2

PROPERTIES OF LASER LIGHT

Interest in lasers arises because laser light has unusual properties that are different from the properties of light from conventional light sources. The properties that we shall discuss in this chapter are monochromaticity (narrow spectral linewidth), directionality (good collimation of the beam), the spatial profiles of laser beams, the temporal characteristics of laser output, coherence properties, brightness (or radiance), the focusing characteristics, and the high power levels available. These properties are not all independent; for example, the focusing characteristics are due to the collimation and coherence of the beam. It is convenient to discuss these aspects separately. These properties, which can be very different from ordinary light sources, enable lasers to be used for the practical applications that will be described later.

A. Linewidth

Laser light is highly monochromatic, that is, it has a very narrow spectral width. The spectral width is not zero, but typically it is much less than that of conventional light sources. The narrow spectral linewidth is one of the most important features of lasers. Early calculations [1] indicated that the linewidth could be a small fraction of 1 Hz. Of course, most practical lasers have much greater linewidth.

In discussing linewidth, we must distinguish between the width of one mode of the resonant cavity and the width of the total spectrum covered by the laser. Many lasers operate in more than one longitudinal mode, so that the total linewidth will approach the fluorescent linewidth of the original material. Figure 2-1 illustrates the situation. The evenly spaced cavity modes are illustrated in (a). The frequency spacing is $c/2d$, with d the distance between the two mirrors. This is equivalent to a wavelength spacing of

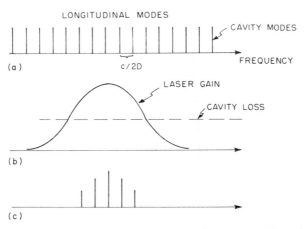

Fig. 2-1 Frequency spectrum of laser output. (a) shows the position of the resonant modes of the laser cavity, with c the velocity of light and D the mirror separation. (b) shows the gain curve for the fluorescent emission, with the cavity loss, relatively constant with frequency, also indicated. (c) shows the resulting frequency spectrum in which all modes for which gain is greater than loss are present.

$\lambda^2/2d$. Figure 2-1b shows the gain curve, and the loss curve superimposed. The gain curve has the same shape as the fluorescent line shape; the loss is constant over a small wavelength range. All modes for which gain is greater than loss can be present. Thus the spectral form of the output can be as shown in Figure 2-1c.

Some typical values are shown in Table 2-1 for some representative lasers. The fluorescent linewidth varies considerably for different laser materials. Also listed are typical lengths of the cavity for each laser and the number of longitudinal modes which could occur. The number of modes is calculated as the fluorescent linewidth Δf divided by the quantity $c/2d$, the spacing between longitudinal modes. In most cases, this will be an overestimate of the number of longitudinal modes that actually occur, because near the edges of the fluorescent line, the gain will not be high enough to sustain laser operation. For typical commercial helium–neon lasers, perhaps three or four modes will be in operation, rather than ten as the calculation indicates. CO_2 lasers present an interesting case, because at low pressure the fluorescent linewidth is narrow, of the order of 60 MHz. Thus, a low pressure CO_2 laser will usually operate in only one longitudinal mode. As the pressure increases to near atmospheric pressure, the fluorescent linewidth will broaden. Atmospheric pressure CO_2 lasers can operate in several modes. Neodymium:glass has an unusually wide fluorescent linewidth. In mode-locked operation the emission spectrum of a neodymium:glass laser can approach the fluorescent linewidth.

TABLE 2-1
Linewidth

Laser	Wavelength (μm)	Δf Fluorescent line width (MHz)	Typical cavity length (cm)	Number of modes under fluorescent line $\Delta f/(c/2d)$
He–Ne	0.6328	1700	100	~10
Argon	0.4880, 0.5145	3500	100	~20
CO_2 (low pressure)	10.6	60	100	~1
CO_2 (atmospheric pressure)	10.6	3000	100	~20
Ruby (room temperature)	0.6943	3×10^5	10	~200
Ruby (77° K)	0.6934	10^4	10	~6
Nd:glass	1.06	6×10^6	10	~4000

In summary, the total spectral width covered by many practical lasers will be almost as wide as the width of the fluorescent line—often around 10^9 Hz. We note that this is still much smaller than the operating frequency, $\sim 10^{15}$ Hz. Even a laser with many longitudinal modes simultaneously present in its output is still almost monochromatic. The individual modes have much narrower spectral width. In order to provide better monochromaticity, lasers are sometimes constructed so as to operate in only one longitudinal mode.

The most common method for providing single-mode operation involves construdtion of short laser cavities, so that the spacing between modes $(c/2d)$ becomes large and gain can be sustained in only one longitudinal mode. Other longitudinal modes will be outside the fluorescent line. This is illustrated in Fig. 2-2. This method is most commonly applied to helium–neon

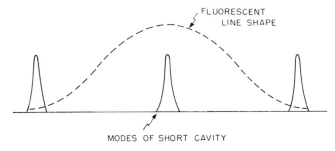

Fig. 2-2 Relation of line shape and resonant modes of a short cavity, which allows laser operation in a single longitudinal mode.

lasers. It has the disadvantage that the short cavity limits the power that can be extracted.

A second method for obtaining single-mode operation involves use of multiple end mirrors, as illustrated in Fig. 2-3(a). This defines cavities of two different lengths, L_1 and L_2. The length of the short cavity L_1 is chosen so that only one mode lies within the fluorescent linewidth. The laser must operate in a mode which is simultaneously resonant for both the cavities. This is illustrated in Fig. 2-3b. The lengths of the two cavities must be adjusted carefully, so that there is an overlap of two resonent modes. This places stringent requirements on mirror position and stability. However, this system allows operation of single-longitudinal-mode lasers at higher power. It is most commonly used with solid state lasers, such as ruby.

Even when operating in a single longitudinal mode, a laser still has a finite spectral linewidth. Since the width of a single mode is much less than the intermode spacing, it will be much smaller than for multimode operation.

Further frequency stabilization is obtained by vibrational isolation, by temperature stabilization (for example, by operating the laser in an oven),

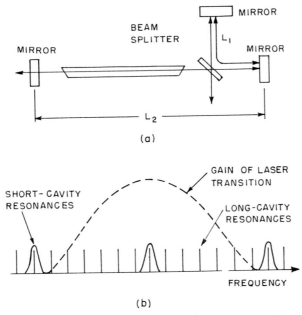

(a)

(b)

Fig. 2-3 Three-mirror method for obtaining operation in a single longitudinal mode. (a) The positioning of the mirror so as to define cavities of two different lengths L_1 and L_2. (b) The gain of the laser transition superimposed on the longitudinal mode spectra of the long cavity and of the short cavity. The laser can operate only in the one mode which is under the gain curve and which is simultaneously a mode of both cavities.

and by servocontrol of the mirror spacing according to the output power. In many practical cases, the width of a mode is dominated by variations in the length of the laser cavity caused by vibrations and temperature variations. The first steps for improving stability are use of materials with small thermal expansion coefficients, careful temperature control, isolation from mechanical and acoustic disturbances, and stabilization of gas-laser power supplies to maintain a constant index of refraction in the gas. With such care, single-mode lasers having a short-term spectral linewidth around 1 kHz are possible. Improved long-term stability is attained by servocontrol of the output to some stable reference device.

The output power of a single-longitudinal-mode laser has a local minimum when the length of the cavity is adjusted so that the laser frequency lies at the center of the spectral line. This local minimum is called the Lamb dip. The mirror spacing is varied by piezoelectric transducers, which are servocontrolled to center on the Lamp dip. More elaborate control is possible by using a saturated absorption in a molecular resonance line. This has so far been achieved only in experimental devices. This technique leads to fractional frequency instabilities of a few parts in 10^{13} over 1 sec periods.

One stable gas laser was constructed in a wine cellar in order to provide very good vibration isolation [2]. This work achieved a short-term frequency spread of the order of 20 Hz. This stability was achieved only by taking extremely great pains with the stability of the device. The frequency drift was as small as a few tens of hertz per second over periods of several minutes. This laser was an experimental model which has not been duplicated commercially.

Commercial single-mode helium–neon lasers are available which have the temperature stabilization and Lamb-dip servocontrol described previously. Direct measurements on the linewidth of such lasers are not available. In commercial models, there is a lower limit of 2 kHz set by the dithering servo on the Lamb dip. Thermal and vibrational instabilities may increase this to around 1 MHz. Thus a reasonable estimate for the frequency stability easily available in a laser is about 1 part in 10^9.

B. Beam Divergence Angle

One of the important characteristics of laser radiation is the highly directional, collimated nature of the beam. The collimation is important because it means that the energy carried by the laser beam can easily be collected and focused to a small area. For conventional light sources, for which the radiation spreads into a solid angle of 4π sr, efficient collection is almost impossible. The smallness of the beam divergence angle of a laser means that efficient collection is possible, even at fairly large distances from the laser.

The limitation on beam divergence angle is set by diffraction. This is a fundamental physical phenomenon, rather than an engineering limit which can be improved by better optical design. Diffraction provides a lower limit to the divergence of the beam. This lower limit is given approximately by the equation

$$\theta = K\lambda/d \tag{2.1}$$

where λ is the wavelength of the light, d the diameter of the aperture through which the light emerges, and K a numerical factor of the order of unity. For uniform beams, $K = 1.22$. For Gaussian beams (see Section 2C), $K = 2/\pi$. However, this value is valid only for the case of a Gaussian beam of infinite extent. In all practical cases the Gaussian beam will be truncated, i.e., the outer edges of the beam will be cut off. The effect of the truncating aperture will modify the diffraction limit somewhat.

Beams emerging from lasers which equal or approach the minimum value set by Eq. (2.1) are said to be diffraction-limited. This represents the best possible case.

The limiting aperture may be taken as the minimum beam waist inside the laser. The concept of the beam waist is illustrated in Fig. 2-4. Inside a laser cavity the field distribution of a transverse mode is concentrated near the center, not necessarily at the exact center of the cavity, but at the center of a so-called equivalent confocal cavity. The beam waist w_0 is given by

$$w_0 = (\lambda b/2\pi)^{1/2} \tag{2.2}$$

where

$$b = [2/(R_1 + R_2 - 2L)][L(R_2 - L)(R_1 - L)(R_1 + R_2 - L)]^{1/2}$$

with mirrors of radius of curvature R_1 and R_2 spaced by distance L. For many practical cases, the beam waists will be of the same order of magnitude as the diameter of the aperture of the laser. The diameter of the cavity is chosen so that the beam waist will nearly fill the active medium. This is especially true for gas lasers, where cavities are chosen so that the beam almost fills the plasma tube.

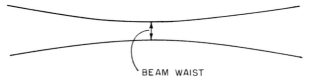

BEAM WAIST

Fig. 2-4   Definition of beam waist of a focused laser beam.

TABLE 2-2
Typical Beam Divergence Angle

Laser:	He–Ne	Ar	CO_2	Ruby	Nd:glass	Nd:YAG	GaAs
Beam divergence (mrad):	0.2–1	0.5–1	1–10	1–10	0.5–10	2–20	20×200

Table 2-2 lists typical beam divergence angles for some representative lasers. We note that in many cases there is a spread of approximately an order of magnitude. The spread is due to construction of lasers of different diameter. In the case of the higher-power lasers, particularly CO_2, ruby, and Nd:glass, the beam divergence angle will be larger than the diffraction limited value given by Eq. (2.1). The beam divergence angle tends to increase with increasing power output of the laser.

The values given in Table 2-2 are those representative of the raw beam emerging from the laser without further collimation. Collimation may be obtained by running the laser backwards through a telescope, i.e., by putting the light into the eyepiece of the telescope. A schematic diagram for this is shown in Fig. 2-5. The collimation of the laser beam will be the inverse of the magnification of the telescope used. Thus, one has

$$\theta_i d_i = \theta_f d_f \tag{2.3}$$

where the subscript i refers to the initial beam and the subscript f refers to the final beam emerging from the telescope, and where d and θ are, respectively, the beam diameter and beam divergence angle. This equation assumes that the laser beam is allowed to fill the telescope.

Let us consider a numerical example. If a helium–neon laser operating diffraction-limited in a TEM_{00} mode has a plasma tube with a diameter of 3 mm, the raw beam divergence angle will be approximately 2×10^{-4} rad according to Eq. (2.1). If this beam is collimated by a telescope with an output diameter of 10 cm, the beam divergence angle will be reduced to

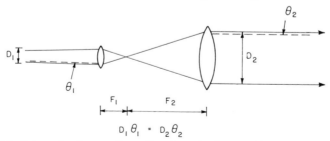

Fig. 2-5 Schematic diagram for collimation of a beam with initial beam divergence angle θ_1 and final beam divergence angle θ_2 by a telescope with two lenses of diameters D_1 and D_2, respectively, and focal lengths F_1 and F_2, respectively. The distance between the lenses is equal to the sum of the focal lengths.

approximately 6×10^{-6} rad. This means that at the distance of the moon, the beam would have expanded to a diameter approximately 1.5 miles.

The beam divergence angles as represented by Eq. (2.1) are those characteristic of lasers operating in the TEM_{00} mode. If multiple modes or more complicated higher-order modes are present, the beam divergence angle will be greater.

Let us consider propagation of a laser beam. The beam radius w as a function of propagation distance Z is given by

$$w = w_0[1 + (Z\lambda/\pi w_0{}^2)^2]^{1/2} \qquad (2.4)$$

where w_0 is the beam waist as previously described. From this equation we can see the distinction between the near field ($Z \ll \pi w_0{}^2/\lambda$) and the far field ($Z \gg \pi w_0{}^2/\lambda$). In the near field, the diameter does not spread much above the value at the beam waist. In the far field, the spreading angle θ will be given by Eq. (2.1).

C. Spatial Profiles of Laser Beams

There are certain distinctive spatial profiles that characterize the cross sections of laser beams. Gas lasers in particular usually have well-defined symmetry. The spatial patterns characteristic of gas lasers are termed transverse modes and are represented in the form TEM_{mn}, where m and n are small integers. The term TEM stands for "transverse electromagnetic." The transverse modes arise from considerations of resonance inside the laser cavity and represent configurations of the electromagnetic field determined by the boundary conditions in the cavity. In this respect, the transverse modes are similar to the familiar case of resonant modes in a microwave cavity. Because of the large difference in wavelength, many optical wavelengths are contained within the cavity, whereas typically only a few microwave wavelengths lie within the cavity.

The transverse modes can be represented in either rectangular symmetry or cylindrical symmetry. A full mathematical description of the modes requires the use of orthogonal polynomials. For present purposes we shall simply describe the configurations that commonly occur.

The notation TEM_{mn} can be interpreted in rectangular symmetry as meaning the numbers of nulls in the spatial pattern that occur in each of two orthogonal directions, transverse to the direction of beam propagation. Thus, some examples of the spatial distribution of the light intensity in some rectangular transverse-mode patterns are shown in Fig. 2-6. The TEM_{00} mode has no nulls in either the horizontal or vertical direction. The TEM_{10} mode has one null in the horizontal direction and none in the vertical direction. The TEM_{20} mode has two nulls in the horizontal direction and none in the vertical direction. The TEM_{11} mode has one null as one passes through the radiation pattern either horizontally or vertically. Such radiation patterns illustrated in Fig. 2-6 are commonly observed from many practical gas lasers. In many cases, a superposition of a number of modes can be present at the same time, so that the radiation pattern can become quite complicated.

It is desirable to obtain operation in the TEM_{00} mode. This transverse mode has been called the Gaussian mode and the intensity I as a function of radius r from the center of the beam is given by

$$I(r) = I_0 \exp(-2r^2/r_0^2) \qquad (2.5)$$

where I_0 is the intensity of the beam at the center and r_0 is the so-called Gaussian beam radius, i.e., the radius at which the intensity is reduced from its central value by a factor of e^2. The total power P is given by

$$P = \pi r_0^2 I_0/2 \qquad (2.6)$$

It is apparent that the simple spatial profile of the Gaussian TEM_{00} mode is desirable, because of its symmetry and also because the beam divergence angle is smaller than for the higher-order transverse modes.

Another fact of interest is that Gaussian beams propagate as Gaussian beams. The spatial profile of a Gaussian beam will retain its Gaussian form as the beam propagates and is transmitted through optical systems. Other higher-order modes will not retain their original near-field spatial distribution as the beam propagates or as the beam is transmitted through optical systems.

The beam divergence angle θ of a Gaussian beam is given by

$$\theta = (2/\pi)(\lambda/w_0) \qquad (2.7)$$

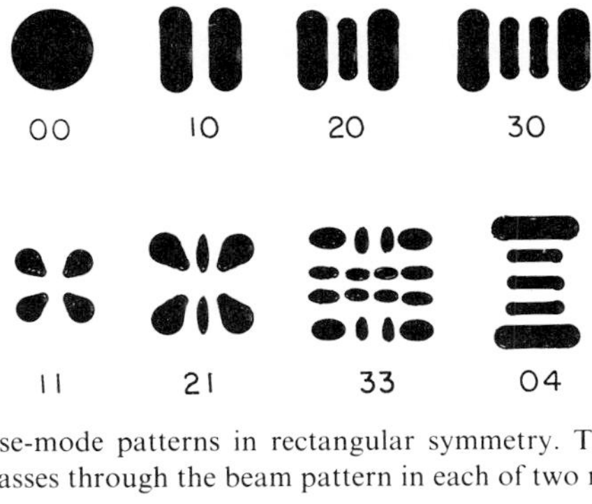

Fig. 2-6 Transverse-mode patterns in rectangular symmetry. The integers denote the number of nulls as one passes through the beam pattern in each of two mutually perpendicular directions.

If the observed beam spreading angle corresponds to this result (or perhaps exceeds it slightly) the beam is said to be "diffraction-limited." Only Gaussian beams are truly diffraction-limited. Other TEM_{mn} modes (m or $n > 0$) have larger beam divergence angles in the far field. One usually desires as small a beam divergence angle as possible.

In addition, there are solutions of the boundary conditions which allow cylindrical symmetry. In practice, gas lasers more often will operate in rectangular rather than cylindrical symmetry, even if the cross section of the laser tube is round. Some examples of cylindrical TEM modes are shown in Fig. 2-7. We note that the TEM_{00} mode is identical in rectangular and cylindrical symmetry.

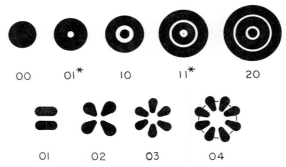

$$00 \qquad 01^* \qquad 10 \qquad 11^* \qquad 20$$

$$01 \qquad 02 \qquad 03 \qquad 04$$

Fig. 2-7 Transverse-mode patterns in cylindrical symmetry. The first integer defining the mode number indicates the number of nulls as one passes across the beam in a radial direction. The second integer specifies half the number of nulls in an azimuthal direction. The modes marked with an asterisk are linear superpositions of two modes rotated by 90° about the central axis.

The mode denoted by TEM_{01}^* represents a superposition of two similar modes rotated by 90° about the axis relative to each other. Thus, the TEM_{01}^* mode, which is often called the donut mode, is made up of a combination of TEM_{01} and TEM_{10}.

Sometimes gas lasers will operate in a higher-order transverse mode or perhaps in a number of such modes simultaneously. This is undesirable because of the beam divergence angle, which is larger than for the Gaussian mode. Slight imperfections of the laser, for example, misalignments of the mirrors, or dust particles, can cause a lack of symmetry and favor operation in high-order modes. Many laser manufacturers now design their gas lasers carefully, so that the plasma tube diameter is small enough to favor operation in the TEM_{00} mode. The diameter of the plasma tube is small enough that the beam waist of the TEM_{00} mode is entirely contained within the plasma, but the higher-order TEM_{mn} modes are not completely contained.

If a laser does in fact operate in a higher-order mode, it can often be made to operate in the Gaussian TEM_{00} mode by insertion of a circular aperture within the cavity. Apertures of progressively smaller size may be inserted until one reaches a diameter such that the laser will operate only in the TEM_{00} mode. This will generally be accomplished at some expense in power, but the accompanying decrease in beam divergence angle and the simplicity of the spatial pattern may be adequate compensation.

We note that the word "mode" has two meanings. The spatial patterns shown in Figs. 2-6 and 2-7 are the transverse modes of the laser. The cavity modes referred to in Fig. 2-1 are longitudinal (or axial) modes. It is possible for a laser to operate in several different longitudinal modes and several transverse modes at the same time. It is also possible to have the laser operate in a single transverse mode and several longitudinal modes, or in a single longitudinal and a single transverse mode.

In Chapter 1 we discussed the unstable resonators that are sometimes used with high-power gas lasers. For a long slim laser, with small Fresnel number, a single low-order transverse mode can extract all the power. However, if the Fresnel number N_F is much larger than unity, a larger number of higher-order modes can oscillate. The lowest-order mode will extract only approximately $1/N_F$ of the available power. This means that the laser will oscillate in a sizable number of high-order modes to extract all the energy from the laser medium and the beam divergence angle will be large. The unstable resonator is more easily adapted to output coupling in a low-order mode with a small beam divergence angle in the case of large Fresnel numbers. The lowest-order mode of the unstable resonator expands on repeated bounces to fill the entire cross section of at least one of the laser mirrors.

Figure 2-8 shows the integrated power as a function of the far field beam divergence angle from a CO_2 laser experiment [3] with an unstable resonator. This shows the far-field beam profile and indicates a well-collimated beam. The experimental points are in reasonably good accord with the theory that has been developed for unstable resonators.

The description so far has been relevant to the simple mode patterns observed commonly with gas lasers, such as the helium–neon laser and the CO_2 laser. Many solid lasers such as the ruby laser, the Nd : YAG laser, and the Nd : glass laser exhibit more complicated spatial patterns which are not describable in simple mathematical terms. The complicated structures are the result of inevitable imperfections within the laser rod, for example, inhomogeneities in the index of refraction and changes in the optical path length and in the birefringence of the laser material as it is heated by the pump lamp. An example of such a complicated spatial pattern observed in the operation of a ruby laser is shown in Fig. 2-9. It is an unfortunate fact

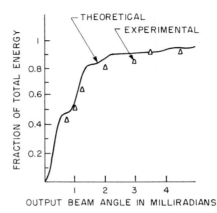

Fig. 2-8 Fraction of total beam energy contained within the indicated beam angle for a CO_2 laser operating with an unstable resonator. (From J. Davit and C. Charles, *Appl. Phys. Lett.* **22**, 248 (1973).)

of life that many high-power solid lasers provide complicated spatial structures such as this.

Single-transverse-mode operation has been obtained with solid lasers by limiting the power and by inserting apertures within the beam. It is also important to choose laser rods of high optical quality with as little variation in index of refraction as possible. A typical value for the diameter of the aperture required to achieve operation of a ruby laser in a TEM_{00} mode is 2 mm. As an example [4], a ruby laser of standard design was mode-controlled by means of a 2-mm-diam aperture. The position of the aperture was varied to use a good portion of the ruby and to maximize output. Peak powers of 2 MW were obtained in a 20 nsec-duration pulse. The wavefront was measured to be Gaussian. Single TEM_{00} mode operation of ruby lasers has also been attained by use of spherical mirrors, but proper adjustment of such a system is more difficult.

The radiation pattern from semiconductor lasers is of particular

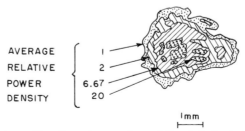

Fig. 2-9 Contour of irradiance in an unfocused Q-switched ruby laser pulse. Contours of equal average relative power density are indicated in arbitrary units.

interest. The radiation from semiconductor lasers typically emerges in a fan-shaped beam, with the broader dimension of the fan perpendicular to the narrow dimension of the junction. This pattern is the result of diffraction of the radiation through the narrow slit which is defined by the junction region in the semiconductor. This will be described in more detail in Chapter 3.

D. Temporal Behavior of Laser Output

There are a number of distinctive types of temporal behavior exhibited by lasers. An understanding of the different time sequences that can be exhibited by various lasers is important. In choosing a laser for a particular application, the availability of lasers with a particular pulse length and particular pulse repetition rate may be a significant factor.

Many types of lasers can be operated continuously. This includes most of the gas lasers, and in fact the most common method of operation for most gas lasers is continuous operation. In addition, the Nd:YAG laser, gallium arsenide semiconductor laser, and some dye lasers may be operated continuously. Ruby lasers have been operated continuously, but these were experimental models. Continuous operation of a ruby laser remains a rarity.

Even during continuous operation of typical lasers, there may be fluctuations in the power output. Figure 2-10 shows the output of a typical commercial helium–neon laser operating at 0.6328 μm, as viewed by a phototube. The lower trace is the ground level for the phototube and the upper trace shows the output of the phototube. The sweep speed is 20 μsec/ division. Fluctuations of the output of this laser may reach 10% of the average value. Components with frequency of the order of 1 MHz appear in the fluctuations. Power fluctuations in continuous lasers may arise from the effects of different modes going into and out of oscillation, or from thermal effects. The argon laser, in particular, is subject to fluctuations in its power output unless it is well stabilized thermally.

Gas lasers may be operated pulsed under some circumstances. Indeed, for the nitrogen laser, which operates in the ultraviolet at 0.3371 μm, the only method of operation available is pulsed operation. CO_2 lasers are often operated pulsed in order to extract high-peak power levels.

For solid lasers (e.g., ruby lasers), which are usually pumped by flash lamps, there are several distinctive regimes of pulse operation.

1. Normal Pulse Operation

No particular control is exerted on the pulse duration, other than by control of the duration of the flashtube pulse. With this type of operation,

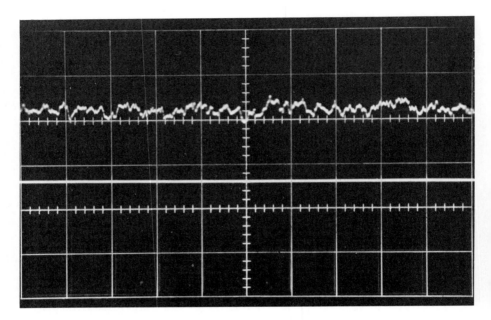

Fig. 2-10 Output of a typical commercial helium–neon laser operating at 0.6328 μm as viewed by a phototube. The lower trace indicates the ground level for the phototube signal and the upper trace shows the electrical output of the phototube at a sweep speed of 20 μsec/div. Amplitude fluctuations in the laser output are of the order of ±5%.

typical pulse widths fall in the range of 100 μsec to several milliseconds. The pulse duration may be controlled primarily by changing the inductance and capacitance of the power supply, so as to change the duration of the flashtube pulse. This type of operation is often called normal pulse operation, or free-running operation.

In many cases, normal-pulsed solid state lasers exhibit substructure within the pulse. The substructure consists of spikes with durations typically in the microsecond regime. These spikes have been called "relaxation oscillations." The spacing and amplitude of the relaxation oscillations are usually not uniform. Relaxation oscillations commonly occur in normal pulse ruby lasers. An example of the relaxation oscillation behavior for a normal pulse ruby laser appears in Fig. 2-11 [5]. The upper trace shows the pulse envelope, which lasts several hundred microseconds. The high frequency relaxation oscillations are superimposed on the general pulse envelope. The lower trace shows a portion of the pulse on an expanded time scale in order to demonstrate the behavior of the individual microsecond duration pulsations in the train of relaxation oscillations.

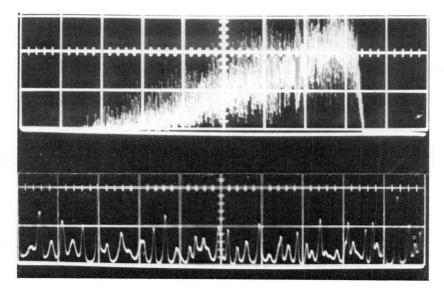

Fig. 2-11   Relaxation oscillation behavior for a pulsed ruby laser. The upper trace shows the pulse envelope at a sweep speed of 200 μsec/div. The lower trace shows an expanded version of the laser pulse at a sweep speed of 10 μsec/div. Time increases from right to left. (From J. F. Ready, "Effects of High Power Laser Radiation." Academic Press, New York, 1971.)

2. Q-Switched Operation

A factor called Q is often used to describe the properties of the cavity formed by the two mirrors. This factor Q characterizes the ability of the cavity to store radiant energy. A high value of Q means that energy is stored well within the cavity. A low value of Q means that energy present in the cavity will emerge rapidly. For example, if the output mirrors on the cavity are of high reflectivity, Q will be relatively high; if the mirrors are of lower reflectivity, whatever energy is present in the cavity will emerge rapidly through the mirrors, and Q will be low.

Let us consider a situation in which the Q of the cavity is high during the initial stages when the active laser material is excited, and then is suddenly reduced to a lower value. In this situation, a large amount of energy will be stored within the cavity initially, during the time when Q is high. Very little will emerge through the mirrors. Then after Q is switched to a lower value, the energy that has been stored will be extracted rapidly through the mirrors. Since a large amount of energy emerges in a short period of time, the power output of this laser will be high.

A variety of practical techniques have been devised to change the Q of the laser cavity. The technique is termed Q-switching. (Sometimes the synonymous terms Q-spoiled laser or giant pulse laser are used.) Q-switching has been employed with all types of lasers but is most common with the high power solid lasers such as ruby lasers, neodymium:glass lasers, and neodymium:YAG lasers. Q-switching has been employed in a number of instances with carbon dioxide lasers, but its use with other types of gas lasers or with semiconductor lasers is uncommon.

When a high-power solid laser, such as ruby, is pulsed in the usual fashion, without Q-switching (i.e., the light is allowed to emerge from the mirrors at its own natural pace, without modifying the Q of the cavity during the pulse), the laser is being operated in the free-running regime or as a normal pulse laser. Hereafter, we shall refer to pulsed ruby and similar lasers operating without Q-switching as "normal-pulsed lasers."

For a laser cavity consisting of a 100% reflecting mirror and an output mirror with reflectivity R, the factor Q may be expressed by the equation

$$Q = 2d\omega/(1 - R)c \tag{2.8}$$

where d is the distance between the mirrors, ω the angular frequency of the light, and c the velocity of light. According to this equation, when the reflectivity R of the output mirror is high, close to unity, the cavity Q will be high. When R is appreciably less than unity, the cavity Q will be low. Q-switching methods usually switch the effective value of the reflectivity of the output mirror from a high to a low value at a time when the laser active element has been pumped to a highly excited state.

A variety of practical means have been employed for Q-switching. These include rotating mirrors, electrooptic elements, bleachable dyes, and acoustooptic switches. We shall describe operation with a rotating mirror here in order to illustrate the basic concept of Q-switching. The other types of Q-switches will be described in Chapter 7.

Figure 2-12 shows schematically how one uses a mirror which consists of a rotating prism. The 100% mirror is a totally internally reflecting prism which is spinning. The axis of rotation lies in the plane of the paper. The type of prism commonly used is the Porro prism. A small amount of disorientation of the Porro prism about an axis perpendicular to the paper does not matter. Light will be reflected back in the same direction in which it came. Thus, this orientation will not be subject to disturbance by vibrations or other effects. If the prism is misaligned, the light which is built up in stimulated emission will not be able to reach the output mirror, and the effective output transmission is low. The pulse of light from the flashtube is synchronized with the rotation of the prism, so that while the laser rod is being pumped by the flashtube light, the prism is not parallel to the stationary mirror. There is no resonant cavity available. Laser operation is suppressed, and the population inversion in the rod can be increased to a high value, higher than that given by the normal threshold condition. Then in the course

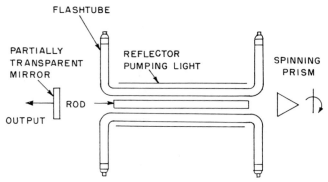

Fig. 2-12 Schematic diagram of a ruby laser Q-switched with a rotating prism.

of its rotation, the prism swings parallel to the stationary mirror, and the resonant cavity is established for a brief period of time. Then the laser rod, in a highly excited state, has a set of parallel mirrors, and laser operation can proceed in a very rapid pulse. The energy that has been stored will be emitted in a pulse of high power and shorter duration than if Q-switching had not been employed.

The Q-switched pulse yields a much higher peak power than the normal pulse, but involves extraction of a smaller amount of energy. Q-switching almost always involves some sacrifice in the energy of the pulse. However, the pulse length is shortened considerably, so that the peak power is increased. Table 2-3 compares the energy, power, and time duration available

TABLE 2-3
Typical Values for Ruby Lasers

Operation	Pulse energy (J)	Pulse duration (sec)	Peak power (W)
Normal pulse	10	500×10^{-6}	2×10^4
Q-switched pulse	1	50×10^{-9}	2×10^7

from typical ruby lasers. These are not the highest values recorded, but values that are easily attainable in laboratory models and are available in commercially produced ruby lasers. The numbers given in Table 2-3 are typical of the output of the same ruby laser with and without a Q-switching device.

3. *Mode-Locked Operation*

There may be further substructure to Q-switched laser pulses. This arises because of the phenomenon of mode-locking. If there are a number of longitudinal modes present in the laser output, these modes may interfere and lead to oscillation of the output. Thus, within the envelope of the Q-switched laser pulse, there may be a train of pulses. This situation is shown in Fig. 2-13, in which (a) shows a smooth envelope of a Q-switched laser pulse without mode-locking. A train of pulses arising from the mode-locking phenomenon is shown in (b). If only a single cavity mode is present in the laser output, the Q-switched pulse can have no substructure such as shown in (b). However, the appearance on an oscilloscope trace, as shown in (a), does not guarantee that the substructure is not in fact present. Many phototube–oscilloscope combinations are too slow to record the rapid fluctuations.

The minimum pulse duration that can be obtained by Q-switching techniques is of the order of 10 nsec. In order to obtain still shorter pulses, other techniques involving mode-locking are employed.

In most common lasers, a number of longitudinal modes, at slightly different frequencies, will be present simultaneously. Ordinarily, the phases of these modes will be independent of each other. Under some conditions, the relative phases of the modes can interact with each other and become locked together. This so-called mode-locking can be accomplished intentionally by introducing into the cavity a variable loss, for example, an electrooptic modulator (see Chapter 7) so that the gain of the cavity is modulated at the frequency $c/2d$, which is the difference in frequency between adjacent longitudinal modes. Under some conditions, mode-locking can occur

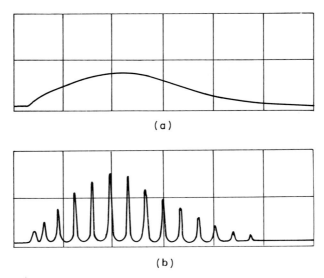

(a)

(b)

Fig. 2-13 Drawings of oscilloscope traces of Q-switched laser pulses. The traces are both at sweep speeds of 10 nsec/div. (a) shows a smooth envelope of a Q-switched pulse without mode-locking. (b) shows a train of pulses when mode-locking is present.

spontaneously. This often happens when a bleachable dye is used as the Q-switching element in a ruby or neodymium:glass laser.

In the time domain, the output of the mode-locked laser consists of a series of very short pulses. From a simple-minded point of view, one can consider that there is a very short pulse of light which is bouncing back and forth between the mirrors in the laser cavity. Each time this pulse of light reaches the output mirror corresponds to emission of a pulse.

Figure 2-13b shows the output when mode-locking does occur. There are a series of much shorter pulses separated in time by $2d/c$, the round trip transit time of the laser cavity. The peak power becomes much higher and the widths of these pulses are very short, usually too short to be distinguished by phototube–oscilloscope combinations. A typical value for the length of one of these single pulses is of the order of 10 psec.

In order to provide a single pulse with this duration, one may use an output coupler, for example, an electrooptic switch which will open when the train of mode-locked pulses has developed and remain open for a time long enough to allow one single pulse in a train of pulses to emerge. This single pulse is then run through an amplifier (which may consist of another laser rod, only without end mirrors) and amplified to a suitably high level. In this way, pulses with very high peak powers, of the order of trillions of watts, and short pulse durations, of the order of a few picoseconds, have been obtained. These pulses are currently of interest in thermonuclear fusion research, but have relatively little use in practical industrial applications.

Mode-locking will sometimes occur in Q-switched ruby and neodymium:glass lasers and sometimes not. Some lasers operate erratically, on one pulse showing mode-locking and on another pulse not exhibiting mode-locking. Conditions for reliably producing mode-locking have been identified, and include careful placement of the Q-switching elements within the cavity, so that reliable mode-locked operation can now be obtained.

We have so far discussed a variety of types of pulsed operation. Solid lasers, as for example, ruby and neodymium:glass, are most often operated pulsed. In many cases, the pulse repetition rate is low, perhaps one pulse per second or less. In the case of the very high peak power lasers, there may be long intervals between isolated pulses. The neodymium:YAG laser may be operated in a repetitively Q-switched mode at fairly high pulse repetition rates, up to kilohertz.

Gas lasers, e.g., helium–neon and argon, are usually operated continuously, but under some conditions could be pulsed. CO_2 lasers have been operated both pulsed and continuously. Gas lasers are conveniently pulsed simply by pulsing the power supply.

In order to summarize the situation, Table 2-4 shows commonly used methods of operation for a variety of types of laser. Typical pulse durations and typical pulse repetition rates are also given.

TABLE 2-4
Pulse Properties of Common Lasers

Type of laser	Commonly used method of operation	Typical pulse duration	Typical pulse repetition rate (pps)
Ruby	Normal pulse	1 msec	low, usually single shot
	Q-switched	50 nsec	low, usually single shot
Nd:Glass	Normal	1 msec	low, usually single shot
	Q-switched	50 nsec	low, usually single shot
	Picosecond	10 psec	low, usually single shot
Nd:YAG	Continuous	—	—
	Q-switched	500 nsec	up to 10,000
He–Ne	Continuous	—	—
Argon	Continuous	—	—
He–Cd	Continuous	—	—
N_2	Pulsed	10 nsec	100
CO_2	Continuous	—	—
	Pulsed (pulsed laser supply)	100 μsec	100
	Pulsed (TEA)	1 μsec	1–100
GaAs	Continuous	—	—
	Pulsed	200 nsec	10,000
Dye	Continuous	—	—
	Pulsed	10 nsec	100

E. Coherence Properties

The concept of coherence generally brings to mind the idea of an orderly train of waves. An analogy commonly used is that of ripples spreading out from a rock thrown into a pool of water. Actually the concept of coherence is more complicated than this. A considerable amount of mathematical development has gone into formulating the concept of coherence. It is not appropriate to review the details of the mathematics here. We shall simply state the result that the coherence can be specified in terms of a function called the mutual coherence function, which is denoted by $\gamma_{12}(\tau)$. The quantity $\gamma_{12}(\tau)$, which is a complex number, measures the correlation between the light wave at two points, P_1 and P_2, at different times t and $t + \tau$. Thus, the quantity $\gamma_{12}(\tau)$ has both spatial and temporal contributions.

The absolute value of $\gamma_{12}(\tau)$ lies between 0 and 1. The value zero corresponds to complete incoherence. If the absolute value is 1, the light is completely coherent. In practice, neither limiting value is ever achieved, but only approached. Thus, in all practical cases, one has partial coherence, i.e., $|\gamma_{12}(\tau)|$ is greater than 0 and less than 1. For single transverse modes in gas lasers, the absolute value of $|\gamma_{12}(\tau)|$ can be very close to 1, so that the light is close to being completely coherent.

The function $|\gamma_{12}(\tau)|$ can most easily be found by the production of fringe patterns, when light is allowed to interfere after traversing two different paths. The visibility V of fringe patterns is defined by

$$V = (I_{max} - I_{min})/(I_{max} + I_{min}) \qquad (2.9)$$

where I_{max} is the maximum intensity of the light at a bright fringe and I_{min} the minimum intensity of the light in a dark area of the interference pattern. If the two interfering waves have equal intensity, the visibility is equal to $|\gamma_{12}(\tau)|$. Thus, measurements of interference patterns provide a straightforward method of determining the degree of coherence of the light.

Coherence is important in any application in which the laser beam will be split into parts. Such applications are very common. They include distance measurements, in which the light is split into two beams which traverse paths of unequal distance, and holography, in which the light beams traverse different paths which may have approximately equal lengths, but which may have spatially different contributions.

If the coherence is less than perfect, the fringe patterns formed in the interferometer or the hologram will be degraded. Spatial coherence also determines how wide an area of the wavefront can be employed for making a hologram. Thus, coherence is of practical importance for many laser applications.

The function $\gamma_{12}(\tau)$ contains both spatial and temporal contributions. If one splits a beam of light into two parts, allows the two beams to traverse paths of different length, and then recombines the two beams, one is dealing with temporal coherence. The time delay τ equals $\delta L/c$, where δL is the difference in path lengths and c the velocity of light. Thus, one is in effect dealing with the function $\gamma_{11}(\tau)$, that is, the coherence in the light wave at one point in the wave field, but at times separated by τ. This gives the correlation between the light wave at one instant of time as it passes the chosen point and the light wave at a later time at the same point.

The concept of phase is crucial. The light wave at a point given by the position vector $\mathbf{r}$ may be represented by the equation

$$E(\mathbf{r}, t) = E_0(\mathbf{r}, t) \cos[\omega t + \phi(\mathbf{r}, t)] \qquad (2.10)$$

where E_0 is the amplitude, ϕ a phase function, and ω the angular frequency. The function E_0 is usually relatively constant. Most of the fluctuations in laser beams are phase fluctuations. If the phase function ϕ changes in a random manner, there will be imperfect correlation between the light wave at one instant and the same light wave at a later instant. Then $|\gamma_{11}(\tau)|$ would be reduced from unity. If there is no correlation at all in the value of ϕ, γ would approach zero.

The time period Δt in which the phase undergoes random changes is called the "coherence time," and it is related to the line width Δv of the laser by the equation

$$\Delta t \approx 1/\Delta v \qquad (2.11)$$

If measurements are made in a time short compared to the coherence time, the value of $|\gamma_{11}(\tau)|$ will be high. This is also true if the path difference is less than $c\,\Delta t$. For measurements that extend over longer times (or over longer path differences), $|\gamma_{11}(\tau)|$ will be reduced, approaching zero for times very long compared to the coherence time.

The presence of multiple longitudinal modes in the output of a laser broadens the spectral width and reduces the coherence length. This in turn reduces fringe contrast in an interferometric or holographic experiment. The influence of multimode oscillation is illustrated in Fig. 2-14, which shows the fringe visibility as a function of the difference in path lengths for a laser operating in one, two, three, four, and five longitudinal modes. This figure is obtained from the equation [6]

$$\text{visibility} = |\sin(N\pi\,\Delta L/2D)/N\sin(\pi\,\Delta L/2D)| \qquad (2.12)$$

In this equation N is the number of longitudinal modes, ΔL the difference in path lengths, and D the length of the laser cavity.

Figure 2-14 refers to a laser with a length of 1 m. This figure expresses the practical result that if a laser operating in multiple longitudinal modes is employed, the path difference that can be tolerated is limited. Thus, for applications such as holography, the permissible subject depth would be limited and for interferometry the distance over which a measurement could be made is limited. Single frequency operation of the laser would remove this restriction.

The concept of spatial coherence arises when one considers the function

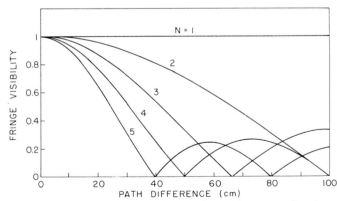

Fig. 2-14 Fringe visibility in an interferometric experiment as a function of the path difference for lasers operating in N longitudinal modes.

$\gamma_{12}(0)$. This is the result when light from two different points on the wave front are combined with no path difference. The properties of spatial coherence depend very much on the mode structure of the laser. The spatial coherence has a value very close to unity in a laser beam that consists of only a single transverse mode, as is commonly the case for well-stabilized gas lasers. This implies that the light beam has perfect spatial coherence over its entire cross section. However, if a laser beam has a number of transverse modes, the function $|\gamma_{12}(0)|$ can have any value between zero and unity, according to which points are examined in the cross section of the beam.

Many authors have reported experimental measurements of the coherence properties of lasers. For continuous gas lasers, the coherence time is approximately equal to the reciprocal of the spectral line width and the spatial coherence approaches unity for single-transverse-mode operation. He–Ne lasers in particular have been found to exhibit excellent coherence. A typical 1-m-long He–Ne laser operating in several longitudinal modes may exhibit a coherence time around 4×10^{-10} sec (coherence length $c \, \Delta t \approx 12$ cm).

The coherence length is relatively small because the presence of several modes broadens the spectral content of the laser output. If a He–Ne laser operates in a single longitudinal mode, the coherence length $c \, \Delta t$ can be much longer, many meters in some cases.

In one measurement involving a two-slit interference pattern [7], $|\gamma_{12}(0)|$ was in excess of 0.9985 for a continuous helium–neon laser operating in the TEM_{00} mode. When the laser was operated in a TEM_{01} mode, the value of $|\gamma_{12}(0)|$ was approximately 0.98. The decrease from unity was probably due to a weak admixture of higher-order modes. When a He–Ne laser is operated in multiple transferse modes, the measured value of $\gamma_{12}(\tau)$, as measured by a two pinhole interference experiment, depends upon the position of the pinholes in the cross section of the beam.

For pulsed lasers, the situation is more complex. In one experiment [8], spatial coherence of a pulsed ruby laser was demonstrated over a distance of approximately 0.005 cm. This laser operated in a manner such that only small regions ("filaments") in the ruby rod contributed to the laser operation. The experimental result corresponded to spatial coherence over one of the filaments.

In a study of spatial coherence of a Q-switched ruby laser [9], quantitative values of $|\gamma_{12}(0)|$ fell in the range 0.2–0.6. The exact value depended on the type of resonator employed. The higher values were obtained with resonators that encouraged operation in a small number of transverse modes.

The temporal coherence of pulsed lasers may be limited by the pulse duration or by shifts in the frequency of the laser during the period of emission. In an experiment [10] in which a pulsed ruby laser illuminated two slits with a variable path difference between the light reaching the two slits, the visibility of the fringes decreased for path differences greater than 2500 cm, and disappeared around 3000 cm. This corresponds to a coherence time around 0.1 μsec, whereas the duration of individual spikes was 0.56 μsec.

In summary, continuous lasers operating in a single longitudinal and single transverse mode can offer nearly ideal coherence properties. The coherence can be degraded by the presence of multiple modes or by variations introduced in pulsed operation. In Table 2-5, we summarize some of the results already described.

TABLE 2-5
Summary of Coherence Properties

Laser	Spatial coherence function $|\gamma_{12}(0)|$	Coherence time Δt	Coherence length $c\,\Delta t$
Continuous single-transverse mode	~1	1/linewidth	~10 cm for $N > 1$; many meters for $N = 1$ (N = number of modes)
Continuous laser with more than one transverse mode	Depends on points sampled	1/linewidth	~10 cm for $N > 1$; many meters for $N = 1$
Pulsed laser— multimode	~0.5	$\lesssim$ pulse length	$\lesssim c \times$ pulse length

Typical coherence lengths	
Laser	Coherence length (cm)
He–Ne (multimode)	20
He–Ne (single frequency)	100,000
Argon (multimode)	2
Nd:YAG	1
Nd:glass	0.02
GaAs	0.1

F. Brightness

The brightness of a source is the power emitted per unit area per unit solid angle. The relevant solid angle is that defined by the cone into which the beam spreads. Because lasers can produce high levels of power in very narrowly collimated beams, they represent sources of high brightness. (In radiometric work, this quantity is termed radiance. The term brightness is more widely used in laser work.)

A fundamental theorem of optics states that the brightness of a source may not be increased by any optical system. High brightness is essential for delivering high power per unit area to a target. The brightness of a laser can be limited by the presence of additional TEM_{mn} modes other than the TEM_{00} mode. Often, as laser power is increased, the number of modes increases and brightness remains fixed or increases only slightly. The quest to produce high power per unit area can thus involve improvement of mode characteristics and decreasing beam divergence angle just as much as increasing power.

The development of systems specifically designed for high brightness has led to diffraction-limited neodymium: glass lasers consisting of a single mode oscillator followed by amplifiers, such that the combination has yielded a brightness as high as 2×10^{17} W/cm^2-sr. More usual values are as follows: for a He–Ne laser, 10^6 W/cm^2-sr; for a Q-switched ruby laser 10^{11}–10^{12} W/cm^2-sr. For comparison, the sun has a brightness around 130 W/cm^2-sr.

G. Focusing Properties of Laser Radiation

In order to determine the available power density that can be produced by laser radiation, we must consider the spot size to which the beam can be focused. It is not possible to focus the beam to a mathematical point. There is always a minimum spot size which is determined ultimately by diffraction. Very often, of course, because of imperfections in the optical system, one will not be able to reach the limit set by diffraction, so that the spot size is larger than the following considerations indicate. However, in any optical system, there is an ultimate limit, which we will term the diffraction limit, which determines the minimum focal area and therefore the maximum irradiance which can be attained.

An important property of Gaussian beams is that they propagate as Gaussian beams, i.e., they have the same intensity distribution in both near and far fields. Gaussian beams are also uniphase, i.e., they have the same phase across the entire wavefront. A Gaussian beam can always be focused to the minimum spot size, of the order of the wavelength of the light. This property of coherent uniphase beams is very useful and is a property that distinguishes laser beams from incoherent light beams. Uniphase Gaussian beams can in principle be focused to smaller spots than incoherent beams.

If the original spreading of the beam was determined by diffraction effects originating at the aperture of the laser, and if the distance from the laser to the lens is small so that the beam has not spread much, we obtain the result

$$r_s = F\theta \tag{2.13}$$

for r_s, the size of the focal spot. This equation is commonly used as a convenient rule of thumb for estimating the minimum spot size obtainable with a laser beam with a known divergence angle θ and aberration-free lens of focal length F.

Since the diffraction-limited beam divergence angle is approximately $\theta = \lambda/D$, where D is the diameter of the limiting aperture, we have (assuming one fills the lens with the beam)

$$r_s = F\lambda/D = \lambda \times F\# \qquad (2.14)$$

where $F\#$ denotes the F-number of the lens. Since it is impractical to work with F-numbers much smaller than unity, the minimum value of r_s will be of the same order of magnitude as λ. This is the origin of the statement that a laser beam can be focused to a spot the size of the wavelength.

Now we give a specific numerical example which shows how the focusing properties of laser radiation are important in the production of laser effects. If a 10 mW gas laser with a Gaussian distribution has a spreading angle of the order of 10^{-4} rad, an F : 1 lens would produce a focused spot with an area of the order of 10^{-8} cm^2 according to the formulation above, and the power per unit area near the center would be of the order of 10^6 W/cm^2. The high power per unit area in the small focal area of the beam means that striking effects can be produced even when the total power in the beam is modest.

Let us consider an ordinary lamp with emission into 4π sr collected by a lens to produce 10^6 W/cm^2. One may show that the power per unit area that can be produced by a lens of focal length F is

$$4I_{\text{tot}}/\pi F^2$$

independent of lens diameter or distance from the source, where I_{tot} is the total power. Increasing the lens diameter would allow for collection of more of the light but at the same time would lead to an increase by the same ratio in the focal area because the effective beam spreading angle would be larger. The minimum focal length with which it is practical to work with reasonable size lenses is of the order of one centimeter. Lenses with shorter focal lengths have very low F-numbers and generally cannot be made free of spherical aberration. To get power per unit area of 10^6 W/cm^2 with a conventional light source, the source must emit total power of approximately 1 MW. Therefore, a modest laser can produce power per unit area comparable to what may be obtained with very powerful conventional light sources.

The preceding description of the diffraction-limited spot size represents an ultimate limit that is set by the principles of optics. This limit may be approached with good engineering, that is, if the lens is free of aberrations and if the spatial distribution in the beam is one of the ideal distributions that we have considered. If the spatial distribution is more complicated, as is often the case, or if the mode structure is complicated, or if the lens is less than perfect, one will not reach this ultimate limit. Typically, for ordinary ruby lasers, the minimum focal spot size produced with a simple lens is of the order of 300 μm. To reach a smaller spot size, the beam may be apertured, so that the effective divergence angle is reduced. Using this technique, focal spot sizes of the order of a few microns have in fact been produced. This involves a considerable loss of the total available power, and gives no increase in brightness.

In spite of their relatively poorer beam divergence angles as compared to gas lasers, high-power solid lasers, because of the very high peak powers, can easily produce a very high irradiance. A focal area of 10^{-3} cm^2 is typical for a ruby laser focused with a simple lens.

Gas lasers, which typically have more uniform spatial profiles, may be focused somewhat more easily. Low-power visible gas lasers operating in a TEM$_{00}$ mode may fairly easily be focused to focal diameters in the 1–2 μm range, near the diffraction limit. The depth of focus is proportionately small, of course. High-power gas lasers often operate in higher-order modes and cannot be focused to the diffraction limit. Typical values for minimum spot sizes for a high power CO$_2$ laser may lie in the 100 μm range.

In order to make effective use of a laser, good focusing of the beam is often needed. One generally desires a small focal area in order to minimize the heat-affected area in applications such as welding and hole drilling. In order to provide the smallest focal size, one must use a short focal length lens, according to Eq. (2.13). However, this may prove impractical on a production line because of limited depth of field. One must provide a lens with sufficient depth of field to allow for any vibration and lack of accuracy in positioning in the vertical direction. The depth of field Z is given approximately by

$$Z = \lambda F^2/a^2 \approx r_s^2/\lambda \tag{2.15}$$

where λ is the wavelength of the light, a the radius of the lens, and F the focal length. Thus, the depth of field increases with the square of the spot size. One must design an optical system that will provide a reasonable compromise between large depth of field and small focal area.

Lens aberrations must also be considered; aberrations will degrade the performance of the optical system. For a monochromatic, collimated laser beam incident along the axis of the lens, many of the aberrations considered in elementary optics are unimportant. The most important lens aberration is spherical aberration. In spherical aberration, the rays from a point source which enter the lens at different distances from the axis are not focused at a common point. The point image is spread to a blurred circle. Spherical aberration becomes large when the focal length of the lens becomes short. This fact sets a lower bound to the focal length.

Spherical aberration can be minimized by either of two techniques: use of aspherical lenses with specially ground surfaces or use of spherical lenses with specially chosen shapes. The aspherical lenses are expensive and offer no great advantage over best-form spherical lenses, so we shall consider only choice of shape of the spherical lenses.

As a reasonable practical choice, a planoconvex lens, with the convex side toward the beam, gives spherical aberration near the minimum. For CO_2 lasers, one often uses a germanium lens with index of refraction equal to 4. For this case the minimum spherical aberration occurs for a meniscus (convex-concave) lens, with the convex side toward the beam.

Figure 2-15 shows calculated results for focal spot diameter for a 1-in.-diam lens, of index of refraction 1.5, and a laser wavelength of 1 μm. The lower curve shows the diffraction limited spot size in the absence of spherical aberration. The upper curve shows the spot size set by the limitations of spherical aberration, when a best-form single-element lens is assumed. At small F-numbers, the effect of spherical aberration dominates, and the diffraction limit is not reached. In contrast to the predictions of Eq. (2.14), the spot size increases with decreasing F-numbers. The situation may of course be improved by the use of air-spaced multielement lenses of

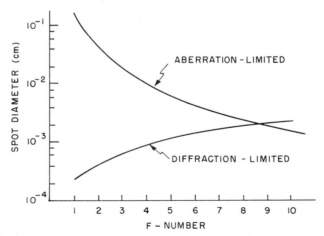

Fig. 2-15 Calculated results for focused spot diameter as a function of F-number of the focusing lens. These results are for a 1-in.-diam lens with index of refraction 1.5 and laser wavelength 1 μm. The lower curve shows the diffraction-limited spot size in the absence of spherical aberration. The upper curve shows the spot size set by the limitations of spherical aberration for a best-form single-element lens.

short focal length, and under favorable conditions, one can indeed focus a laser beam to a spot of single-wavelength dimension. In most practical industrial applications, one must usually retreat somewhat from this ultimate limit.

To summarize methods for focusing the laser beam to a small spot: One should choose a lens with as short a focal length as is consistent with the desired depth of field and the lens aberrations. For the given focal length, the shape of the lens must be chosen to give minimum spherical aberration.

Practical measurement of the focal spot size is also of importance. One obvious method would be to expose photographic film in the focal region and perform densitometric measurements on the optical density of the exposed film. This method has many problems with obtaining accurate measurements. One needs good knowledge of the sensitivity of the film and of its development characteristics. At high exposures, the film becomes saturated in the center of the spot so that one obtains a flattened profile. Overexposure can lead to overestimates of the focal spot diameter.

One method which has proved useful for determining focal spot size is to insert a sharp opaque edge, e.g., a razor edge, in the focal plane of the lens and move the sharp edge transversely across the focal area with a micrometer drive. A photodetector behind the focal area monitors the transmitted power in the beam as a function of the position of the edge. This technique yields a measurement of the beam profile from which the diameter of the focal region can be accurately deduced [11].

In particular, as applied to a Gaussian beam, one obtains

$$P_t = (P_0/2)[1 - \mathrm{erf}(2X/r_0)] \qquad (2.16)$$

where P_t is the transmitted power measured by the photodetector behind the opaque edge, P_0 the total incident power, r_0 the Gaussian beam radius, and X the position of the edge which is assumed to be driven in the x direction and which is measured on the micrometer. The notation erf denotes the error function, which is a tabulated function. Measurements of the total incident power and the transmitted power as a function of position of the opaque edge can easily be used to determine the Gaussian beam radius near the focal position.

The extension of this type of measurement to other beam profiles than the Gaussian is obvious. This technique has been effective for determining power distribution in a focused laser beam in the small region near the focus, and therefore for determining laser power density in industrial processing applications. The region of focal spot sizes of interest in many industrial applications is of the order of 0.001–0.01 cm, a region in which measurements made with a micrometer will provide adequate resolution.

H. Power

The aspect of lasers that has most caught the public imagination is their ability to produce very high levels of power. The highest powers come from pulsed lasers (such as ruby and Nd:glass) and are achievable only in short pulses.

We should distinguish between experimental devices and commercially available lasers. Experimental lasers can emit much higher levels of power than can be purchased from laser manufacturers. We shall describe the levels of power characteristic of different laser types in Chapter 4. Here we shall merely note some rough orders of magnitude. We note that the numbers given here are upper limits; most of the lasers in use operate at much lower levels.

For continuous output, the highest values are produced by CO_2 lasers. Large experimental CO_2 lasers have operated continuously near the 100 kW level, and commercial continuous CO_2 lasers around 10 kW are becoming available.

For pulsed lasers, solid state lasers produce extremely high levels of power. Commercial Q-switched Nd:glass lasers with peak power levels up to 10^{11} W are available (with nanosecond pulse durations). Experimental devices for laser-assisted thermonuclear fusion work have exceeded 10^{13} W peak power.

REFERENCES

[1]   A. L. Schawlow and C. H. Townes, *Phys. Rev.* **112**, 1940 (1958).
[2]   T. S. Jaseja, A. Javan, and C. H. Townes, *Phys. Rev. Lett.* **10**, 165 (1963).
[3]   J. Davit and C. Charles, *Appl. Phys. Lett.* **22**, 248 (1973).
[4]   J. E. Bjorkholm and R. H. Stolen, *J. Appl. Phys.* **39**, 4043 (1968).
[5]   J. F. Ready, "Effects of High Power Laser Radiation." Academic Press, New York, 1971.
[6]   R. J. Collier, C. B. Burckhardt, and L. H. Lin, "Optical Holography," Eq. 7.17. Academic Press, New York, 1971.
[7]   D. Chen and J. F. Ready, *Bull. Am. Phys. Soc.* **11**, 454 (1966).
[8]   D. F. Nelson and R. J. Collins, *J. Appl. Phys.* **32**, 739 (1961).
[9]   G. Magyar, *Opto-Electron.* **2**, 68 (1970).
[10]   D. A. Berkley and G. J. Wolga, *Phys. Rev. Lett.* **9**, 479 (1962).
[11]   *Laser Focus*, June 1973, p. 61.

SELECTED ADDITIONAL REFERENCES

A.   Linewidth

A. L. Bloom, "Gas Lasers," Wiley, New York, 1968, Chapters 3 and 4.
G. D. Boyd and H. Kogelnik, Generalized Confocal Resonator Theory, *Bell Syst. Tech. J.* **41**, 1347 (1962).

R. C. Duncan, Z. J. Kiss, and J. P. Wittke, Direct Observation of Longitudinal Modes in the Output of Optical Masers, *J. Appl. Phys.* **33**, 2568 (1962).

C. Freed, Design and Short-Term Stability of Single-Frequency CO_2 Lasers, *IEEE J. Quantum Electron.* **QE-4**, 404 (1968).

S. E. Harris, Stabilization and Modulation of Laser Oscillators by Internal Time-Varying Perturbation, *Appl. Opt.* **5**, 1639 (1966).

M. Hercher, Single-Mode Operation of a Q-Switched Ruby Laser, *Appl. Phys. Lett.* **7**, 39 (1965).

H. Kogelnik and C. K. N. Patel, Mode Suppression and Single Frequency Operation in Gaseous Optical Masers, *Proc. IRE* **50**, 2365 (1962).

B. J. McMurtry, Investigation of Ruby Optical Maser Characteristics Using Microwave Phototubes, *Appl. Opt.* **2**, 767 (1963).

G. E. Moss, High-Power Single-Mode HeNe Laser, *Appl. Opt.* **10**, 2565 (1971).

K. D. Mielenz *et al.*, Reproducibility of Helium–Neon Laser Wavelengths at 633 nm, *Appl. Opt.* **7**, 289 (1968).

T. G. Polanyi and I. Tobias, The Frequency Stabilization of Gas Lasers, *in* "Lasers, A Series of Advances," Vol. 2 (A. K. Levine, ed.), Dekker, New York, 1968.

P. W. Smith, Stabilized, Single-Frequency Output from a Long Laser Cavity, *IEEE J. Quantum Electron.* **QE-1**, 343 (1965).

B. Beam Divergence Angle

M. J. Adams and P. T. Landsberg, *in* "Gallium Arsenide Lasers" (C. H. Gooch, ed.), Wiley (Interscience), New York, 1969.

R. J. Collins *et al.*, Coherence, Narrowing, Directionality and Relaxation Oscillations in the Light Emission from Ruby, *Phys. Rev. Lett.* **5**, 303 (1960).

D. R. Herriott, Optical Properties of a Continuous Helium–Neon Optical Maser, *J. Opt. Soc. Am.* **52**, 31 (1962).

"Laser Focus 1976 Buyers' Guide," Advanced Technology Publ., Newton, Massachusetts, 1976.

D. C. Sinclair and W. E. Bell, "Gas Laser Technology," Holt, New York, 1969.

R. W. Waynant *et al.*, Beam Divergence Measurement for Q-switched Ruby Lasers, *Appl. Opt.* **4**, 1648 (1965).

C. Spatial Profiles of Laser Beams

J. A. Arnaud *et al.*, Technique for Fast Measurement of Gaussian Laser Beam Parameters, *Appl. Opt.* **10**, 2775 (1971).

A. L. Bloom, "Gas Lasers," Wiley, New York, 1968, Chapter 3.

A. L. Bloom, Far-Field Pattern of Gas Lasers Operating in the Lowest Order Gaussian Transverse Mode, *Appl. Opt.* **8**, 716 (1969).

G. D. Boyd and H. Kogelnik, Generalized Confocal Resonator Theory, *Bell Syst. Tech. J.* **41**, 1347 (1962).

J. F. Carr and S. S. Charschan, Obtaining Near Gaussian Intensity Distributions from Multi-mode Pulsed Ruby Lasers, *Appl. Opt.* **10**, 684 (1971).

E. S. Dayhoff and B. Kessler, High-Speed Sequence Photography of a Ruby Laser, *Appl. Opt.* **1**, 339 (1962).

V. Evtuhov and J. K. Neeland, Observations Relating to the Transverse and Longitudinal Modes of a Ruby Laser, *Appl. Opt.* **1**, 517 (1962).

R. J. Freiberg and A. S. Halsted, Properties of Low Order Transverse Modes in Argon Ion Lasers, *Appl. Opt.* **8**, 355 (1969).

T. V. George, L. S. Slama, and M. Yokoyama, Intensity Distribution of Ruby Laser Beams, *Appl. Opt.* **2**, 1198 (1963).

D. H. Holmes, P. V. Avizonis, and K. H. Wrolstad, On-Axis Irradiance of a Focused, Apertured Gaussian Beam, *Appl. Opt.* **9**, 2179 (1970).

H. Kogelnik and W. W. Rigrod, Visual Display of Isolated Optical Resonator Modes, *Proc. IRE* **50**, 220 (1962).

H. Kogelnik and T. Li, Laser Beams and Resonators, *Appl. Opt.* **5**, 220 (1966).

J. S. Kruger, Laser Modes—Some Basic Concepts, *Electro-Opt. Systems Design*, p. 11 (September 1972).

R. J. Pressley (ed.), Handbook of Lasers," Chemical Rubber Co., Cleveland, Ohio, 1971, Chapter 14.

A. E. Siegman, Unstable Optical Resonators, *Appl. Opt.* **13**, 353 (1974).

W. R. Watson and T. G. Polanyi, Radiation Patterns of Confocal HeNe Laser, *J. Appl. Phys.* **34**, 708 (1963).

R. W. Waynant *et al.*, Beam Divergence Measurement for Q-Switched Ruby Lasers, *Appl. Opt.* **4**, 1648 (1965).

H. Weichel and L. S. Pedrotti, A Summary of Useful Laser Equations—An LIA Report, *Electro-Opt. Systems Design*, p. 22 (July 1976).

A. Yariv, "Quantum Electronics," Wiley, New York, 1975, Chapter 7.

D. Temporal Behavior of Laser Output

R. R. Alfano and S. L. Shapiro, Ultrashort Phenomena, *Phys. Today*, p. 30 (July 1975).

R. J. Collins *et al.*, Coherence, Narrowing, Directionality and Relaxation Oscillations in the Light Emission from Ruby, *Phys. Rev. Lett.* **5**, 303 (1960).

R. C. Cunningham, A Guide to Mode Locking, *Electro-Opt. Systems Design*, p. 4A (April 1975).

A. J. DeMaria, D. A. Stetser, and H. Heynau, Self Mode-Locking of Lasers with Saturable Absorber, *Appl. Phys. Lett.* **8**, 174 (1966).

A. J. DeMaria *et al.*, Picosecond Laser Pulses, *Proc. IEEE* **57**, 2 (1969).

A. J. DeMaria, W. H. Glenn, and M. E. Mack, Ultrafast Laser Pulses, *Phys. Today*, p. 19 (July 1971).

A. J. DeMaria, Review of High-Power CO_2 Lasers, *Proc. IEEE* **61**, 731 (1973).

M. A. Duguay, S. L. Shapiro, and P. M. Rentzepis, Spontaneous Appearance of Picosecond Pulses in Ruby and Nd:Glass Lasers, *Phys. Rev. Lett.* **19**, 1014 (1967).

M. A. Duguay, J. W. Hansen, and S. L. Shapiro, Study of the Nd:Glass Laser Radiation, *IEEE J. Quantum Electron.* **QE-6**, 725 (1970).

M. A. Duguay and A. T. Mattick, Ultrahigh Speed Photography of Picosecond Light Pulses and Echoes, *Appl. Opt.* **10**, 2162 (1971).

M. A. Duguay, Light Photographed in Flight, *Am. Sci.* **59**, 550 (1971).

W. W. Duley, "CO_2 Lasers, Effects and Applications," Academic Press, New York (1976), Chapter 2.

V. Evtuhov and J. K. Neeland, Observations Relating to the Transverse and Longitudinal Modes of a Ruby Laser, *Appl. Opt.* **1**, 517 (1962).

A. Girard and A. J. Beaulieu, A TEA CO_2 Laser with Output Pulse Length Adjustable from 50 ns to Over 50 μs, *IEEE J. Quantum Electron.* **QE-10**, 521 (1974).

G. Kachen, L. Steinmetz, and J. Kysilka, Selection and Amplification of a Single Mode-Locked Optical Pulse, *Appl. Phys. Lett.* **13**, 229 (1968).

F. J. McClung and R. W. Hellwarth, Characteristics of Giant Optical Pulsations from Ruby, *Proc. IEEE* **51**, 46 (1963).

J. F. Ready, "Effects of High-Power Laser Radiation," Academic Press, New York (1971), Chapter 1.

P. W. Smith, Mode-Locking of Lasers, *Proc. IEEE* **58**, 1342 (1970).

E. B. Treacy, Measurement and Interpretation of Dynamic Spectrograms of Picosecond Light Pulses, *J. Appl. Phys.* **42**, 3848 (1971).

A. Yariv, "Quantum Electronics," Wiley, New York, 1975, Chapter 11.

E. Coherence Properties

N. P. Barnes and I. A. Crabbe, Coherence Length of a Q-Switched Nd:YAG Laser, *J. Appl. Phys.* **46**, 4093 (1975).

D. A. Berkley and G. J. Wolga, Coherence Studies of Emission from a Pulsed Ruby Laser, *Phys. Rev. Lett.* **9**, 479 (1962).

R. J. Collier, C. B. Burckhardt, and L. H. Lin, "Optical Holography," Academic Press, New York, 1971, Chapters 1 and 7.

M. S. Lipsett and L. Mandel, Coherence Time Measurements of Light from Ruby Optical Masers, *Nature (London)* **199**, 553 (1963).

G. Magyar, Some Measurements of the Spatial Coherence of a Giant Pulse Laser, *Opto-Electronics* **2**, 68 (1970).

D. C. W. Morley *et al.*, Spatial Coherence and Mode Structure in the HeNe Laser, *Brit. J. Appl. Phys.* **18**, 1419 (1967).

D. F. Nelson and R. J. Collins, Spatial Coherence in the Output of an Optical Maser, *J. Appl. Phys.* **32**, 739 (1961).

F. Brightness

W. F. Hagen, Diffraction-Limited High-Radiance Nd–Glass Laser System, *J. Appl. Phys.* **40**, 511 (1969).

D. C. Hanna, Increasing Laser Brightness by Transverse Mode Selection –1, *Opt. Laser Technol.*, p. 122 (August 1970).

G. Magyar, Simple Giant Pulse Ruby Laser of High Spectral Brightness, *Rev. Sci. Instrum.* **38**, 517 (1967).

G. Focusing Properties of Laser Radiation

A. L. Bloom, "Gas Lasers," Wiley, New York, 1968, Chapter 4.

A. L. Bloom and D. J. Innes, Design of Optical Systems for Use with Laser Beams, Spectra-Physics Laser Tech. Bull. No. 5, Spectra-Physics, Inc., Mountain View, California (1966).

L. R. Dickson and D. R. Cecchi, Determining Depth of Focus—Precisely, *Electro-Opt. Systems Design*, p. 20 (November 1970).

R. T. Pitlak, Laser Spot Size for Single Element Lens, *Electro-Opt. Systems Design*, p. 30 (September 1975).

J. F. Ready, "Effects of High-Power Laser Radiation," Academic Press, New York, 1971, Chapters 1 and 8.

H. Power

W. H. Christiansen and A. Hertzberg, Gasdynamic Lasers and Photon Machines, *Proc. IEEE* **61**, 1060 (1973).

A. J. DeMaria, Review of High Power CO_2 Lasers, *Proc. IEEE* **61**, 731 (1973).

W. W. Duley, "CO_2 Lasers, Effects and Applications," Academic Press, New York, 1976, Chapter 2.

W. F. Krupke and E. V. George, Lasers for Thermonuclear Fusion and Their Properties, *in*

"Industrial Applications of High Power Laser Technology" (J. F. Ready, ed.), Soc. of Photo-Opt. Instrumentation Eng., Palos Verdes Estates, California (1976).

"Laser Focus 1976 Buyers' Guide," Advanced Technology Publ., Newton, Massachusetts, 1976.

R. J. Pressley (ed), "Handbook of Lasers," Chemical Rubber Co., Cleveland, Ohio, 1971, Chapter 6, 7, and 11.

W. W. Simmons *et al.*, *IEEE J. Quantum Electron.* **QE-11**, 31 D (1975).

CHAPTER 3

PRACTICAL LASERS

In this chapter we shall describe the practical status of well-developed laser types and shall discuss the lasers that have reached commercial status. By commercial lasers, we mean those that are available for sale from a number of different suppliers. The properties that we shall describe in this chapter do not represent optimum values that have been obtained in laboratory experiments; rather we shall emphasize what one can buy relatively easily from commercial suppliers.

We shall discuss a number of different laser materials. This is a relatively small number of materials taken from the very large class of hundreds of materials in which laser action has been demonstrated in the laboratory. A question may arise about why are these materials better than other materials. The answer lies necessarily in a detailed examination of the rate constants describing the rates at which electrons move between the various excited states under the action of the optical or electrical pumping. A detailed examination of the good laser materials has usually revealed that the rate constants are such as to favor population of the upper laser level and not to provide deexcitation of that level through competing processes. Thus, reasonably convincing explanations for the high quality of the operation of these lasers have been developed. These explanations generally have been developed after the fact, i.e., after the laser has been demonstrated to operate well. Attempts at predicting good laser materials before the fact have been less successful. Many materials have been suggested on the basis of examination of excited electronic structure and the transitions between excited electronic levels. In some cases, although the predictions looked favorable, the materials have not become successful laser media. Therefore, for our purposes, we shall simply regard the materials discussed in this chapter as being the ones that happen to work well, or at least the ones which are most fully developed, without emphasizing too much why these are the successful materials, whereas other materials have been less success-ful.

It is possible that future developments in lasers may well change the list. New and emerging developments in lasers offer promise for operation of new types of lasers at different wavelengths. In the future, these may well supplant some of the lasers which are now commonly used. We shall, in the next chapter, discuss some of the prospects for future laser develop-ment.

A. Gas Lasers

Lasers which utilize a gaseous material as the active laser medium have become extremely common. There are many compelling reasons for the use of a gas. The volume of material can be large, in contrast to crystals and semiconductors. There is no possibility for damage to the material, such as occurs in high-power ruby and glass lasers. The material is homogeneous, so that problems involving optical inhomogeneity that are common in solid lasers are avoided. Heat can be removed readily by transporting the heated gas out of the region of the interaction, and in suitable types of gas lasers, high power can also be aided by flowing of the gas. There are distinct sub-divisions of gas lasers which we shall discuss here. These include the neutral gas laser (typified by the helium–neon laser), the ionized gas laser (typified by argon), and the molecular laser (typified by carbon dioxide).

1. Neutral Gas Lasers

A laser medium involving neutral gas species occurs in the positive column of a glow discharge in a medium such as a mixture of helium and neon. The excitation mechanism for a helium–neon laser has already been discussed in Chapter 1. Briefly, the passage of the electrical current, with a current density around 100 mA/cm^2, leads to excitation of the helium atoms through collisions with electrons in the electrical discharge. This excitation energy in turn is transferred through atomic collisions to the upper laser levels of the neon atoms, thus producing the population inversion.

The first neutral gas laser was demonstrated in 1961 with the operation of the 1.15 μm line in the helium–neon laser. The helium–neon laser remains the most common of all neutral gas lasers, although a number of other neutral gases have demonstrated laser action. The most notable ones are the noble gases, neon, argon, krypton, and xenon, with a variety of wavelengths available through the visible and near infrared portion of the spectrum.

Helium–Neon Lasers   Helium–neon lasers represent the most common of all types of lasers. It has been estimated that over 100,000 helium–neon lasers have been manufactured and sold commercially. It is probable that the number of helium–neon lasers in use exceeds the number of all other types of lasers combined. They represent a compact, portable and easily usable source of visible continuous laser light and as such have many uses for educational and demonstration purposes.

The energy level diagram for the helium–neon laser has already been shown in Chapter 1. The helium is excited by the passage of an electrical current through the gas mixture. The upper laser levels of the neon are excited by collisions with the excited helium atoms. There are three transitions that are common. These are at 0.6328, 1.15, and 3.39 μm. The 1.15 μm operation was the first observed in helium–neon lasers. However, it was rapidly followed by operation at 0.6328 μm. Because of the desirability of visible operation, most helium–neon lasers today are constructed to operate at this wavelength. We see from the energy level diagram (Fig. 1-10) for neon that there is competition between the 0.6328 μm line and the other lines. Thus, if the 1.15 μm line is operating, it tends to fill the terminal state for the 0.6328 μm line. Similarly, if the 3.39 μm line is operating, it uses up atoms that would otherwise contribute to the 0.6328 μm line. Thus, in many practical cases, one tries to suppress oscillation of the infrared lines. The 1.15 μm line may be suppressed by the use of mirrors which have relatively low reflectivity at 1.15 μm. However, the 3.39 μm line has extremely high gain. Even when low reflectivity mirrors are used, it is difficult to suppress this line. For this reason many helium–neon lasers incorporate magnets which provide an inhomogeneous magnetic field which varies widely over different parts of the laser tube. The inhomogeneous magnetic field broadens the profile of the 3.39 μm transition and thus reduces the gain available for that transition. The percentage reduction of the gain for the 0.6328 μm line is much less.

In addition, operation at some other visible wavelengths may be obtained by insertion of a prism in the helium–neon laser. The operation proceeds from the same upper level as the 0.6328 μm line but to a different lower sublevel. The dispersion of the prism causes the different wavelengths passing through it to travel in slightly different directions. The prism is rotated so that only one of these wavelengths will strike the mirror normally and be reflected directly back. In this way, any one of the several transitions can be favored over the others. One of the more important transitions that can be obtained in this way is a yellow line at 0.6118 μm.

The gain of the helium–neon laser at 0.6328 μm is relatively low, so that only very small losses can be tolerated in the laser cavity. Thus, the mirrors must be of high quality with low scattering losses. The output mirror typically has transmission around 1 or 2%. If the mirrors are external to the gas-filled laser tube, Brewster angle windows are necessary on the tube.

It is instructive to consider the output of the helium–neon laser as a function of mirror transmission. This is shown in Fig. 3-1 [1]. At zero transmission, the output is of course zero. With increasing transmission, the output increases, up to a maximum value. Above the maximum value, the total losses in the laser cavity, including the loss due to the output, become too great and the power decreases. At values of the transmission above 2.5%, the total losses exceed the gain, which in a helium–neon laser is small, so that the laser ceases operation. Because of the small gain in helium–neon, the optimum output coupling is low, less than 1% in this illustration.

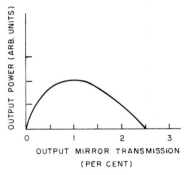

Fig. 3-1   Output of helium–neon laser as a function of mirror transmission. (From F. Petrů and Z. Veselá, *Opto-Electron.* **4**, 21 (1972).)

In the helium–neon laser, the gain is an inverse function of the diameter of the gas tube. This probably occurs because the lower level of the laser transition is depopulated by the collisions of the neon atoms with the walls of the tube. Thus, to maintain the population inversion and to keep the laser operating, the neon atoms must be able to collide readily with the walls of the tube. In order that the collision rate of the neon atoms with the walls be high, the diameter of the tube must be small. Typically diameters of a few millimeters are employed. This has several practical consequences. It limits the power output from helium–neon lasers because the volume of gas cannot be easily increased. In addition it affects the beam divergence angle, because the walls of the tube provide a limiting aperture. Thus, if one uses Eq. (2.1), one finds that a typical beam divergence angle for a helium–neon laser will be around 2×10^{-4} rad.

Most commercial helium–neon lasers are now constructed so as to operate in the TEM_{00} mode. This is done at least partially by ensuring that the ratio of the length of the tube to its diameter is such that the TEM_{00} mode is favored. Other higher-order modes have larger diameters and are cut off by the narrow aperture defined by the plasma tube.

Considerable engineering has gone into the design of cathodes for helium–neon lasers. The most popular cathode now seems to be a cold aluminum cathode which provides optimum lifetime. The lifetime of commercial helium–neon lasers exceeds 10,000 hr, according to manufacturers' statements.

The range of power outputs in commercially available helium–neon lasers extends from a few tenths of a milliwatt up to 50 mW, in nonfrequency stabilized devices. Such lasers typically have amplitude stability of the order of a few percent, that is, the power output varies by a few percent over short periods of time. We have shown a typical oscilloscope trace of amplitude noise in a commercial helium–neon laser in Fig. 2-10.

The pressure in the helium–neon laser tube depends on the diameter of the tube. Generally it is in the region of a few torr. The gas mixture is typically 90% helium and 10% neon. Early work with helium–neon lasers involved many studies of the gain of the laser as a function of gas mixture, gas pressure, tube diameter, and tube current. The commercial lasers are constructed to provide optimum gain for a given set of conditions.

Figure 3-2 shows the internal structure of typical small commercial helium–neon lasers. (a) is a photograph of the structure of a laser capable of emitting 0.5 mW in a TEM_{00} mode. The tube, visible in the front, is 11 in. long and has integrally sealed mirrors at each end. The larger structure visible behind the tube contains a gas reservoir. (b) shows a simple geometry suitable for mechanized production and rapid packaging. Stamped metal end plates are used both for electrical feedthroughs and for mirror seats. Earlier designs had usually involved glassblowing to produce a complex shape with pin seals and many protrusions.

For maximum power output, the total gas fill pressure P of the laser tube is given by the following equation

$$PD \approx 3.2 \tag{3.1}$$

In this equation, P is the pressure in torr and D the tube diameter in millimeters. This equation is an empirical equation, which represents a rule of thumb which is commonly used to determine the fill pressure in order to provide optimum output.

The voltage–current characteristics of the helium–neon laser tube are important. There is a region in which the voltage can be high but very little current flows. In order to provide current through the tube, one must first provide a sufficiently high voltage to break down the tube. In a typical small helium–neon laser, the breakdown voltage may be around 3400 V. After the breakdown voltage is applied, a plasma is created in the tube. This results in a large increase of current. Thus, in order to start the helium–neon laser operation, one requires an initial high-voltage pulse which momentarily

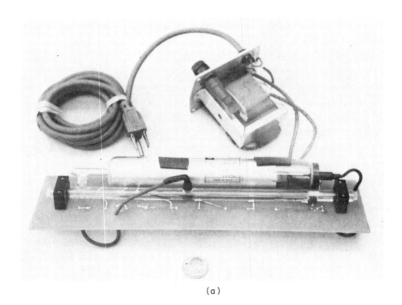

(a)

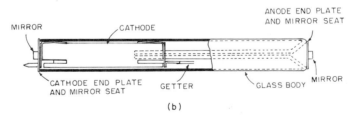

(b)

Fig. 3-2   (a) Photograph of the internal structure of a 0.5 mW commercial He–Ne laser. (b) Diagram of structure of He–Ne laser designed for volume production. (Courtesy of Spectra-Physics, Inc.)

increases the tube voltage above the breakdown voltage. After breakdown, the voltage required to maintain the discharge with currents of the order of several milliamperes is considerably reduced. A typical operating voltage might be around 1350 V. In this region, the plasma tube exhibits a negative resistance. This means that as the current through the tube increases, the voltage across the tube decreases. In order to restrict the current flow to a safe value, a ballast resistor is placed in series with the laser tube.

The output power as a function of tube current is shown in Fig. 3-3 for two different gas compositions [2]. For low tube currents the gas discharge is unstable. The plasma flickers on and off and there is no output. For currents greater than a few milliamperes, a stable plasma is produced and the laser output increases with increasing current. At still higher tube currents, the output power saturates and begins to decrease. This is probably the result of heating of the gas and of a decrease in the efficiency of excitation at increased current. In addition, operation at excessive values of current may decrease the tube life. Thus, one adds the ballast resistor in order to keep the tube current within the desired limit.

The output from the helium–neon laser at 0.6328 μm is a reddish-orange beam with a small divergence angle. One of the most striking characteristics about this beam is its appearance when the light strikes a diffusely reflecting surface, for example a painted wall or a piece of white paper. The beam has a granular, speckled appearance. This results from the coherence properties of the light beam. A random interference pattern is formed from the contributions of light reflected from various portions of the reflecting surface. On such a diffusely reflecting surface, there will be small hills and valleys, with dimensions of the order of one or a few wavelengths of light. The light scattered from such a surface may lead to constructive interference of the spatially coherent light. An observer will see a bright spot in the speckle pattern. At a different point, there will be destructive interference between all the components and an observer will see a dark spot in the speckle pattern.

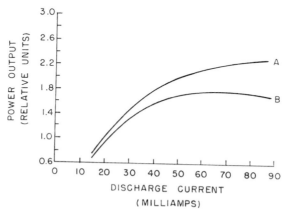

Fig. 3-3 Output of helium–neon laser as a function of discharge current. Curve A is for a gas mixture of 0.5 Torr of ³He and 0.1 Torr of Ne. Curve B is for 0.5 Torr of He and 0.1 Torr of Ne. (From A. D. White, *Proc. IEEE* **51**, 1669 (1963).)

The observation of this speckle pattern came as a surprise when continuous visible lasers were operated for the first time. It has practical implications in that the irregular illumination makes visible laser light a poor choice for applications involving illumination of surfaces. The speckle pattern is also a problem in holographic applications.

The effect of the speckle pattern is characteristic of the phenomena of diffraction and interference that are associated with all electromagnetic radiation. Speckle patterns can in principle be produced with conventional optical sources. However, it is only the laser, with its high degree of spatial and temporal coherence, that makes speckle patterns familiar.

For a typical nonfrequency-stabilized helium–neon laser, there may be four or five longitudinal modes present in the output. Figure 3-4 shows a drawing of the frequency spectrum of a typical helium–neon laser, as obtained with a high-resolution interferometer. In this case, four longitudinal modes were present. For a 100-cm-long tube, the spacing between the longitudinal modes, $c/2D$, will be 1.5×10^8 Hz. The coherence length of a laser with four modes present will be only 66 cm, according to Eq. (2.11). We note that the total fluorescent linewidth at the half-maximum points of the 0.6328 μm helium–neon laser transition is approximately 1.7×10^9 Hz.

Commercial frequency-stabilized helium–neon lasers are available. These lasers employ an oven and feedback control to keep the laser transition at the center of the Lamb dip, according to the discussion in Chapter 2. The stability of such lasers is approximately two parts in 10^7. Measurements carried out at the National Bureau of Standards on such commercial helium–neon frequency-stabilized lasers indicated a vacuum wavelength of 0.632991418 $\pm$ 0.00000003 μm. This will correspond to a coherence length in excess of 1000 cm.

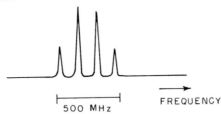

500 MHz   FREQUENCY

Fig. 3-4 Frequency spectrum of a typical helium–neon laser as obtained with a high-resolution interferometer.

2. *Ionized Gas Lasers*

An ion laser generally consists of a plasma with a glow discharge at high current density passing through it. The most common type of ion laser in practical use uses ions of noble gases, most often argon. A partial energy level diagram relevant to the argon ion laser is shown in Fig. 3-5. A number of laser transitions are indicated. The complete energy level diagram of the argon ion is complicated, and includes many more levels and more laser transitions than are shown in the figure. The figure does include most of the important laser transitions. The strongest transitions are at 0.4880 and 0.5145 μm. These energy levels are the levels of the argon ion, so that in order to operate an argon laser, the atoms must first be singly ionized. The ground state in this energy level diagram is the ground state of the argon ion, which lies almost 16 eV above the ground state of the neutral argon atom. In addition, the upper laser levels lie approximately 20 eV above the ionic ground state. Thus, a considerable amount of energy must be supplied to the neutral argon atom to raise it to the upper laser level of the argon ion.

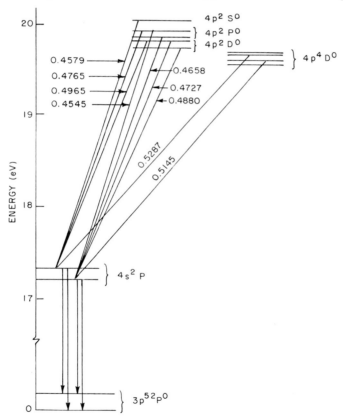

Fig. 3-5   Energy level diagram of the singly-ionized argon ion relevant to the operation of the argon ion laser. The spectroscopic notation for the energy levels is indicated. Various laser wavelengths are indicated in micrometers.

Unlike the case with helium–neon, the argon laser has higher gain, and greater output powers can be produced. The output scales nonlinearly with the current density, so that it is desirable to operate the argon laser with a narrow bore and a high current. Current densities above 100 A/cm² may be employed in argon lasers. The high current density produces heating and has significant effects on the construction of argon lasers.

The maximum output power reported for an argon laser is 150 W, but commercial models more often fall in the range from 2 to 10 W total output power. The total output power of an argon laser is usually specified as the sum of the powers for all the different wavelengths. Single-line operation on one of the wavelengths may be obtained through use of a prism in the laser cavity. Because of the dispersion of the prism, only one wavelength will be incident normally on the mirror, so that the laser can operate at only one of the possible wavelengths. This of course is accompanied by .a reduction in the output power. Table 3-1 shows typical values for the output powers available at a number of the different wavelengths in a typical commercial argon laser. Such a laser would be specified as a 4-W laser because it could emit 4 W when operated without the prism so that all the wavelengths would be present simultaneously. Selection of one particular line is accomplished by rotating the prism.

A magnetic field is commonly used in the construction of continuous argon lasers. The field is longitudinal; that is, the magnetic field is parallel to the axis of the laser. The main effect of the magnetic field is to increase the electron density in the plasma by constraining the electrons to move in a helical motion around the magnetic field lines. This prevents the loss of electrons to the walls.

Physical damage and erosion of the inside of the walls of the argon laser limit the maximum current. A common type of construction involves the use

TABLE 3-1
Representative Outputs

Nominal 4 W argon laser		Nominal 750 mW krypton laser	
Wavelength (μm)	Power (mW)	Wavelength (μm)	Power (mW)
0.5145	1400	0.6471	300
0.5017	200	0.5682	130
0.4965	300	0.5309	100
0.4880	1300	0.5208	100
0.4765	500	0.4825	50
0.4727	100	0.4762	70
0.4658	50	0.3507, 0.3564	200 (with special mirrors)
0.4579	150		
0.3511, 0.3638	100 (with special mirrors)		

of electrically isolated radiation-cooled segments supported in a quartz vacuum envelope to confine the discharge. The bore segments are constructed of a material to minimize erosion. Because of the high current densities, the argon laser plasma tube requires high temperature materials for the bore. Many different designs of argon laser construction have been described. Materials that are commonly used include graphite, quartz, or beryllium oxide. Beryllium oxide seems particularly good for withstanding the eroding effects of the electrical discharge. Its most important characteristic is its high thermal conductivity.

Argon lasers offer outputs in the ultraviolet at wavelengths of 0.3511 and 0.3638 μm. This makes the argon laser one of the few commercially available sources of ultraviolet laser radiation. Some commercial gas lasers have gas refill systems in order to compensate small changes in tube pressure that occur with time. Gas refilling can be accomplished through activation of a switch which opens a valve to the argon reservoir. The tube may be refilled to the desired optimum pressure. This option has extended the life of commercial argon lasers.

Krypton lasers are very similar in construction and characteristics to argon lasers. Table 3-1 gives a tabulation of the wavelengths of available krypton lasers. In single line operation, somewhat lower values of power are obtained than from an argon laser. The technology and lifetimes of krypton lasers seem to be somewhat inferior to those of argon lasers. The main reason for choice of a krypton laser would undoubtedly be to obtain a different variety of wavelengths, particularly wavelengths in the red and yellow portion of the visible spectrum at levels about 100 mW. Mixed gas lasers containing both argon and krypton are available and provide a composite choice of wavelengths.

The helium–cadmium laser is also an ion laser because it operates using ionized states of gaseous cadmium. In many of its characteristics however, the helium–cadmium laser appears much more similar to the neutral helium–neon laser.

The helium–cadmium laser is a representative of a type of metal vapor laser of which many examples have been demonstrated in the laboratory. The helium–cadmium laser is the most highly developed of such metal vapor lasers and it has reached a status of commercial availability. In a helium–cadmium laser, the laser action occurs between energy levels of the cadmium. The cadmium is in vapor form which is usually produced by an auxiliary heater. The excitation of the upper energy levels of the gaseous cadmium occurs in a manner very similar to the excitation of the neon levels in the helium–neon laser. In the electric discharge in the laser, helium atoms are excited by collisions with electrons. These excited helium atoms then collide with ground state cadmium atoms. The collisions produce excited levels of the cadmium ion. In a typical discharge in a mixture of helium and cadmium, the conditions are generally favorable for producing population inversions for two different laser lines. These lines have wavelengths of 0.4416 and 0.3250 μm. The 0.3250 μm line is particularly significant because it is the shortest wavelength ultraviolet laser line which is readily available.

The practical design of helium–cadmium lasers encounters a number of significant difficulties. The problem of maintaining a uniform distribution of cadmium metal vapor in the electrical discharge is perhaps the most important difficulty. The operation of the helium–cadmium laser requires production of a sufficiently high vapor pressure of cadmium metal, good control of the vapor in the discharge, the prevention of chemical reactions between the metal vapor and the laser tube itself, and finally the avoidance of deposition of a coating of cadmium metal on the windows and mirrors. There is a cataphoresis effect in this type of discharge. The cadmium metal vapor undergoes a flow toward the cathode in the electrical discharge. This makes it difficult to maintain a uniform pressure of cadmium. In order to counteract the cataphoresis, some designs of helium–cadmium lasers allow return paths for the cadmium to the region of the anode. In order to protect the windows and mirrors, there is generally a region of the tube with cold walls, separating the discharge region from the window. This allows the cadmium to condense on the cold walls before it reaches the window. This in turn, however, will eventually deplete the amount of cadmium present in the laser.

Considerable ingenuity in design and construction has been exercised to produce simple, reliable and compact helium–cadmium lasers with lifetimes quoted as exceeding 1000 hr. Continuous powers of up to 50 mW are available at the wavelength of 0.4416 μm. In order to maximize ultraviolet output, the mirrors must be replaced with mirrors which have optimum reflectivity for the ultraviolet. Continuous outputs up to 15 mW at 0.3250 μm are available.

As compared to the helium–neon laser, the output properties of helium–cadmium are fairly similar, including the available levels of power. It is likely that the helium–neon laser would be employed for most applications, except where the blue or ultraviolet wavelengths were specifically desired.

One further ion laser which should be mentioned is the xenon ion laser, which operates in the blue and green portions of the spectrum. Usually operated pulsed at high current density, this laser emits several lines in the region from 0.4955 to 0.5395 μm. Although pulse energy tends to be low, the peak power may be high enough to allow vaporization of thin films.

3. Molecular Lasers

So far all the lasers we have discussed operate on energy levels that are characteristic of electronic transitions. These energy levels correspond to transitions of electrons from one electronic energy level to another. In addition, there is another important possibility for energy levels which involves no change of electronic state. These are the vibrational and rotational energy levels of molecules. A polyatomic molecule may have a number of internal degrees of freedom which involve vibration and rotation of the molecule. According to the laws of quantum mechanics, the energies associated with the vibrations and rotations of polyatomic molecules are quantized, so that there is a spectrum of energy levels corresponding to these vibrations and rotations. If a population inversion is produced between two such vibrational or rotational levels, laser action may of course take place between those two levels. Typically the energies associated with electronic transitions are larger than the energies associated with vibrational transitions which in turn are larger than the energies associated with rotational transitions. Thus, molecular lasers tend to operate in the infrared portion of the spectrum. We emphasize that the transitions occur between vibrational sublevels of the ground electronic state.

Carbon Dioxide Lasers   The most important molecular laser uses CO_2 gas as its active medium. The carbon dioxide molecule is a linear molecule. The three atoms lie in a straight line with the carbon atom in the middle. There are three different types of vibrations which can occur in the carbon dioxide molecule. These vibrations are illustrated schematically in Fig. 3-6a. In the first mode of vibration, the carbon atom remains stationary, and the oxygen atoms oscillate about their equilibrium positions as illustrated by the arrows. At any given time, the oxygen atoms are moving in opposite directions along the line of symmetry. In the second mode of oscillation, which is a bending motion, all the atoms move in a plane perpendicular to the line of symmetry. The carbon atom moves in one direction while the two oxygen atoms are going in the opposite direction. The third mode is an asymmetric mode in which all the atoms move along the line of symmetry. At any given time, the carbon atom is moving in a direction opposite to the two oxygen atoms. The arrows denote the instantaneous displacements of the atoms from their equilibrium positions, which are denoted by the circles. The motion is a simple harmonic motion with all the atoms moving in phase and oscillating about their equilibrium positions.

There is, of course, energy associated with these vibrational modes and according to the rules of quantum mechanics, the energy is quantized. In the ground vibrational state, there are no vibrational motions. The excited states correspond to excitation of one or more quanta of vibrational energy. The molecule can vibrate in more than one mode at the same time and it can have more than one quantum of vibrational energy in each mode. The notation adopted to denote the states is in the form (ij^lk), where i, j, and k are

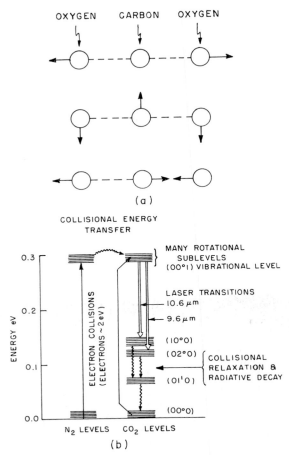

Fig. 3-6   (a) Schematic diagram of the possible vibrations of the carbon dioxide molecule.
(b) Energy levels relative to the operation of a CO_2–N_2 laser. The numbers for the vibrational
states are defined in the text.

integers denoting the quanta of vibrational energy in each of the three modes.
The integer i corresponds to the symmetric vibration, j to the bending
vibration, and k to the asymmetric vibration. The superscript l for the quantum number of the bending vibration is an additional quantum number
which arises because the bending vibration can occur in either of two perpendicular senses.

Thus, the state (10^00) represents a state with one quantum of symmetric
vibrational energy and no quanta in either of the other two states. In addition, each of the vibrational states is split into a number of rotational states.

The rotational structure is small compared to the spacing of the vibrational structure. A simplified energy level showing the vibrational energy levels which are of greatest importance for CO_2 laser operation is shown in Fig. 3-6b. We note explicitly that these are all sublevels of the ground electronic state of the CO_2 molecule. In addition, the rotation of the CO_2 molecule splits each of these levels into a number of sublevels.

The $(00^0 1)$ level is only 18 cm^{-1} above the first excited vibrational level of the nitrogen molecule, which can be excited by collisions with electrons having approximately 2 V of energy in the electrical discharge. This energy is transferred by collision with the CO_2 molecules to the $(00^0 1)$ state. This produces a population inversion between the $(00^0 1)$ state and the $(10^0 0)$ and $(02^0 0)$ states. These correspond to laser operation at 10.6 and 9.6 μm, respectively. The gain is higher for the 10.6 μm transition, so this will be the preferred transition. Because the two laser transitions share a common upper state, operation at 10.6 μm usually precludes operation at 9.6 μm. Therefore, the laser will generally operate at 10.6 μm unless some wavelength selecting mechanism, such as a grating, is employed. The lower laser levels relax by collision and by radiative emission. This eventually returns the CO_2 molecule to the ground vibrational level, the $(00^0 0)$ state.

The coupling of the vibrational and rotational motions of the CO_2 molecule leads to considerable additional substructure. Transitions may occur between any components for which the change ΔJ in rotational quantum number J is $\Delta J = \pm 1$. Thus, the output of the laser will have a considerable number of subtransitions.

The lines are denoted by the nomenclature P or R, corresponding to ΔJ equal to $+1$ or -1, respectively, followed by the rotational quantum number of the lower rotational level. Thus, the notation $P(22)$ refers to a line for which the upper state has $J = 21$ and the lower state has $J = 22$. Figure 3-7 shows some details of the rotational structure.

Figure 3-8 shows the gain distribution for the lines in the 10.6 μm transition, as a function of the rotational quantum number J of the lower state [3]. The separation of the P and R branches and the multiline nature of the emission are clearly illustrated.

Usually the output of a CO_2 laser will contain several of these lines operating simultaneously. Tuning to a single line is possible if one uses a rotatable diffraction grating in place of the total reflecting mirror. This allows any chosen line to operate individually, although at reduced power from multiline operation.

As the lower portion of Fig. 3-6 shows, the population inversion in the CO_2 laser is produced through collisions with nitrogen molecules. Early experiments showed laser output at a laser wavelength of 10.6 μm in pure CO_2; however, the power output was low. High power was obtained by

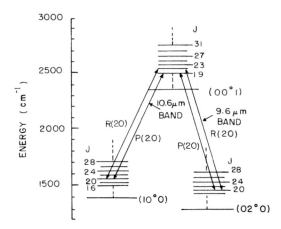

Fig. 3-7 Details of rotational substructure of CO_2 vibrational levels. The notation is defined in the text.

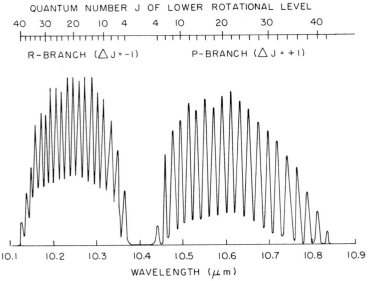

Fig. 3-8 Gain distribution in CO_2 as a function of wavelength and the quantum number J of the lower rotational level. (From U. P. Oppenheim and A. D. Devir, *J. Opt. Soc. Am.* **58**, 585 (1968).)

adding nitrogen to the mixture. The close coincidence in energy between the levels of nitrogen and CO_2 was shown in Fig. 3-6. Nitrogen molecules are excited in the gas discharge by collisions with electrons. They transfer their energy in collisions to CO_2 molecules. This produces a population inversion between the (00^01) state and the (10^00) state, which is initially empty. Laser operation then proceeds at the wavelength of 10.6 μm. The addition of helium to the gas mixture further increases the output power. The helium tends to deplete the population of the lower laser level by collisions and also tends to keep the gas mixture cool because of the high mobility of the helium atoms. Practical CO_2 lasers actually operate using a mixture of CO_2, N_2, and He although they are generally referred to simply as CO_2 lasers. Cooling of the gas mixture is desirable. If the temperature of the walls and the gas mixture is allowed to increase, the power output will decrease.

We note that CO_2 lasers generally are flowing gas systems. The gas flows through the laser and is exhausted. In a sealed-off CO_2 laser, the interaction of the electrical discharge with the gases causes decomposition and buildup of harmful products, probably CO. This is avoided most simply by continuously exhausting the gas. In applications where this is not feasible, for example, use of CO_2 lasers for communications systems on satellites, sealed-off CO_2 lasers have been designed which operate many thousands of hours. Such lasers involve the use of electrode materials which will getter the harmful decomposition products. For most industrial applications, the weight of the gas supply and the cost of the gas is not an important factor, so that flowing systems are used. Fast flow of the gas provides convective cooling and leads to the possibility of high output powers from relatively small volumes of gas.

Early studies on CO_2 lasers defined the operational parameters of CO_2 lasers pumped by electrical discharges along the length of a fairly long gas tube. The optimum pressures of the gases in a CO_2 laser were around 0.5 Torr for CO_2, 6 Torr for helium, and 1.5 Torr for nitrogen. The scaling indicated that the power output was directly proportional to the length of the laser tube, and only weakly dependent on the diameter. The diameter of the gas tube is chosen to be near the minimum permitted by diffraction of the TEM_{00} mode at 10.6 μm, in order that the laser should operate in the TEM_{00} mode. Thus, the diameter of the tube should scale as the square root of its length.

The CO_2 laser is also notable for its relatively high efficiency. Most lasers have low efficiency of conversion of electrical energy into optical energy. Typical efficiencies are perhaps a few tenths of one percent up to a few percent. The CO_2 laser is capable of operation in the range of 20–30% efficiency. Only the GaAs laser has higher efficiency (up to 40%).

With optimum coupling for a flowing gas CO_2 laser, energy extraction was found to be about 80 W/m of discharge length. Thus, continuous power outputs of a few hundred watts are reasonable to obtain. However, for conventional flowing gas laser systems, operating at pressures of a few torr, it is necessary to construct very long lasers in order to increase the power output much above a few hundred watts. This approach was pursued with construction of lasers which reached total lengths of 600 ft and yielded continuous output up to 8800 W.

In recent years other types of operation of CO_2 lasers have become common, yielding high powers in units that are smaller. New developments also yield other unusual characteristics. These developments include fast gas transport lasers, high-voltage pulsed lasers at high gas pressures (TEA lasers), electron-beam controlled lasers, gas dynamic lasers, and waveguide lasers.

Fast Gas Transport CO_2 *Lasers*   In order to consider fast gas transport lasers, we refer to the energy level diagram for the CO_2 laser. This indicates that the bottleneck for laser operation is the lower level, which must be depopulated by collisions. Early CO_2 lasers used flowing gas to carry away dissociation products produced by the electrical discharge. The dissociation products have a harmful effect on the laser operation; however, if the flow is very fast, the molecules reaching the lower laser level may be swept out and replaced with molecules in the upper laser level which have previously been excited. This will remove the bottleneck and will allow higher powers. This is the basis for the so-called gas transport laser in which the flow is very rapid.

In early models of the CO_2 laser, the gas flow was along the length of the gas tube. In order to provide a more rapid sweeping out of molecules in the lower state, a flow perpendicular to the optical cavity has often been adopted. Many different configurations have been developed to provide fast transverse flow of the gas mixture. In one example [4], a high-capacity blower blows the excited gas through the discharge region and through the optical cavity. The gas is cooled in a heat exchanger and recirculated back to the blower. The properties of this gas transport laser are shown in Fig. 3-9, which gives the output power as a function of the gas flow rates and of the electrical discharge power. The power increases very rapidly at high gas flow rates, until it saturates near 4000 ft^3/min. The output power can reach 1 kW at an input power near 14 kW. This entire unit can have a size around that of a desk, i.e., much smaller than the earlier high-power CO_2 lasers.

Fast gas transport lasers such as this have yielded continuous powers up to 27 kW with electrical efficiencies around 17% [5].

TEA CO_2 *Lasers*   Another important development involved high-voltage pulsing of the CO_2 laser at megavolt pulse voltages [6]. One method for excitation of such a laser involves a Marx bank of capacitors. This involves charging of capacitors in parallel at a voltage that does not exceed the breakdown voltage of the CO_2 laser tube. Then a spark gap is triggered, to connect the capacitors in series across the laser tube, and the voltage rises

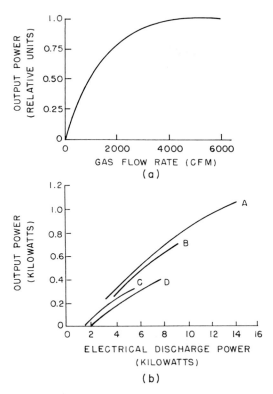

Fig. 3-9   Output of fast gas transport CO_2 laser. (a) shows the output as a function of gas flow rate. (b) shows the output as a function of electrical discharge power. The gas mixtures $(He/N_2/CO_2)$ in torr are for curve A: 5/11/2; curve B: 5/7/3; curve C: 5/5/2; curve D: 5/7/2. (From W. B. Tiffany, R. Targ, and J. D. Foster, *Appl. Phys. Lett.* **15**, 91 (1969).)

to a very high value. This method of pulsing the CO_2 laser can yield pulses of several joules, in pulses with durations of the order of some micro-seconds. Some models have become commercially available.

It was established that CO_2 lasers operated best with an electric field to pressure ratio of 10–50 V/cm Torr. For a 1-m-long discharge at 800 Torr, voltages in the neighborhood of 8×10^5–4×10^6 V are then required for proper operation. Since steady-state discharges at such high pressures are difficult to operate without having the discharge degenerate into an arc. CO_2 lasers operating at such high pressures have been operated pulsed. In addition, voltages around the megavolt region are easier to obtain under pulsed conditions with pulsed transformers such as a Marx bank.

To lessen the requirement for such extremely high applied voltage, transverse excitation has been used. In transversely excited lasers, the discharge is transverse rather than parallel to the optic axis. This arrangement reduces the electrode spacing from meters, as is typical in axially excited lasers, to centimeter dimensions. This simple and inexpensive technique has made it possible to assemble multimegawatt pulsed gas lasers fairly easily. These Transversely Excited Atmospheric pressure lasers have come to be known as TEA lasers.

The scaling of the output of the pulsed CO_2 laser with pressure indicates an increase as the square of the pressure. This is because both the rate of depopulation of the lower laser level and also the number of molecules available for laser action increases with pressure. The result is a square law dependence of output on pressure. However, as the pressure is increased, a plasma glow-to-arc transition becomes more probable. Such a transition is marked by a sudden increase in plasma current and a bright white arc in the discharge. When an arc discharge occurs, it will terminate laser action.

The main feature of the development of the atmospheric pressure CO_2 laser has been in the development of stable nonarcing discharges at high pressures. Early models of such lasers employed many resistors, typically of

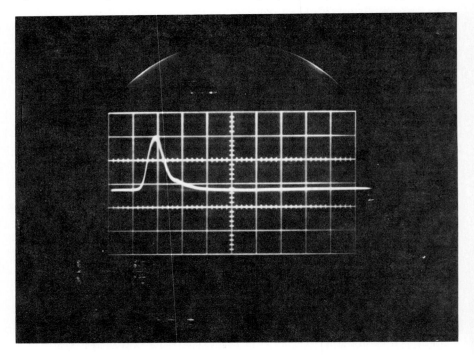

Fig. 3-10   Oscilloscope trace of CO_2 TEA laser pulse shape at a sweep of 200 nsec per division.

the order of 100/m. The resistors provided limitation for the current. More recent techniques have included preionization, special electrode shapes, and fast pulsing techniques so that the laser pulse ends before the arc develops. These techniques have made the TEA laser a useful tool, which can be operated reliably.

Currently available TEA lasers provide pulse durations in the submicrosecond regime, with total pulse energies ranging from 1 J to hundreds of joules. For the lower values of pulse energy, the pulse repetition rates range up to hundreds of pulses per second. Figure 3-10 shows an oscilloscope trace of a typical TEA laser pulse, with peak power around 10 MW and duration around 100 nsec. Figure 3-11 shows a photograph of a TEA laser, with various parts identified.

Fig. 3-11 Photograph of CO_2 TEA laser. The anode is the solid horizontal structure. The cathode is the horizontal structure with holes in it. The holes allow preionization by pins located behind the cathode. The optical path is horizontal, between the anode and cathode. Fans to circulate the gas mixture are visible in the foreground. Pulse forming circuitry is visible in the background.

Electron-Beam Controlled CO_2 *Lasers*   One difficulty with an ordinary discharge laser occurs because the free electrons in the plasma have two different functions. They must move through the discharge under the influence of the applied electric field, producing new electron–ion pairs by collision, to offset losses by recombination and diffusion to the walls. If the electric field is too low, the electron concentration decreases and the discharge goes out. The same electrons must excite gas molecules by collision to populate the upper laser level and produce the inversion required for laser action. The electron temperatures required for these two jobs are usually not the same. The laser works best at one value of electron temperature and the discharge may be stable at a different value. The new approach separates these functions. An externally produced high-energy electron beam fired into the laser region maintains the discharge.

This concept was first applied to the production of uniform discharges for gas lasers in 1971. It has now been used in several systems employing a variety of physical configurations.

Such a discharge uses a high-energy (100–200 keV) electron beam to ionize the gas. A field across the gas accelerates the resulting electrons and provides electrical excitation of the laser molecules. The discharge is non-self-sustaining without the electron beam. The provision of an external ionizing agent decouples charge production in the discharge from the electric-field or transport processes. In electrically excited CO_2 laser systems, the electric field which provides the electron velocity distribution that is optimum for pumping CO_2 to its upper laser level does not provide sufficient ionization to sustain the discharge. In the electron-beam controlled laser, separation of the source of ionization from the excitation process avoids the problems of self-sustained breakdown and arc formation. Adjustment of the electric field allows tailoring of the electron velocity distribution and molecular pumping efficiency.

Such a laser is sometimes called an "ionizer–sustainer" laser. One such system uses injection of a 130 keV electron beam into 40 liters of a $1:2:3$ mixture of $CO_2:N_2:He$ gas at 1 atm pressure. The laser produces fundamental mode output pulses up to 1200 J/pulse (a peak power of 50 MW in a pulse of 50 μsec duration) and multimode outputs up to 2000 J/pulse.

Gas Dynamic Lasers Another important development is that of the gas dynamic laser, in which the population inversion is produced not by an electrical discharge, but rather by expansion through a nozzle. Figure 3-12 shows a typical system.

The gas is heated to high temperature and pressure and allowed to expand through a supersonic nozzle. Heating may be obtained by several methods, such as by use of a shock tube or a plasma arc. The optical cavity is perpendicular to the flow. The gas flows past the optical cavity and is exhausted by a pump. As the gas passes through the region of the optical cavity, the pressure and temperature have cooled considerably because of expansion. The design of the nozzle is important for producing the desired conditions. A representative aspect ratio for a nozzle is indicated in Fig. 3-12.

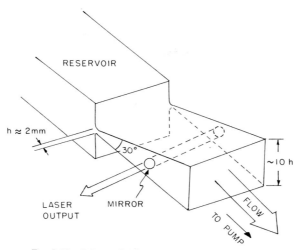

Fig. 3-12 Schematic diagram of gas dynamic laser.

The mechanism for production of the population inversion is shown in Fig. 3-13 [7]. (a) shows the fraction of the total energy contained in the various degrees of freedom, as a function of distance from the nozzle. Upstream of the nozzle, most of the energy is contained in translation and rotation of the molecules, with some fraction contained in the vibrational energy levels. The energy levels of interest in laser operation will be populated to some extent. The lower level is more populated than the upper level at thermal equilibrium. As the gas expands through the nozzle, the energy is transformed mainly into kinetic energy of the flow of the gas. If the gas cooled in thermodynamic equilibrium, the vibrational energy levels would become almost completely depopulated. However, the time constants of the processes are such that the lower level is depopulated more rapidly than the upper level. This is illustrated in Fig. 3-13b.

Some energy will remain frozen in the upper laser level because the relaxation rate of the upper level is too slow. The gas expands to a state where there are relatively few collisions which can remove molecules from the upper level. Thus, the upper level remains partially populated and a population inversion is produced for some distance downstream from the nozzle. It is important to remember that this population inversion is produced not by an electrical discharge but rather in the mechanics of the expansion process.

Such CO_2 lasers have been capable of continuous operation at levels up to 100 kW. These power levels far exceed those of other continuous gas lasers. One pulsed gas dynamic laser has been described [8] which uses a reservoir of pressure 1000 lb/in.^{2} at a temperature around 2000°K. The

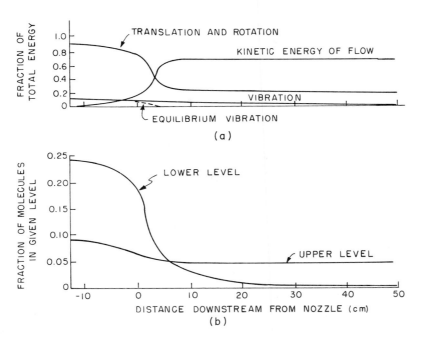

Fig. 3-13 Mechanism for production of population inversion in gas dynamic laser. (a) shows the fraction of the total energy contained in various degrees of freedom as a function of distance from the nozzle. (b) shows the fraction of the CO_2 molecules in the lower and upper laser levels as a function of distance downstream from the nozzle. (From E. T. Gerry, *Laser Focus*, p. 27 (December 1970).)

gas expands through two stages of nozzles, with the second stage containing 100 nozzles, 10 cm high, each with a throat of 1 mm. Multimode powers of 450 kW have been obtained for periods up to 4 msec.

Waveguide CO_2 Lasers Hollow dielectric waveguides have been employed both to confine the gas discharge and to guide the laser radiation in CO_2 lasers. Light propagation in a hollow structure with dielectric walls has well-defined low-loss modes of propagation which have been investigated in detail. The scaling laws for operation of CO_2 lasers indicate that the gain coefficient increases as the diameter of the gas discharge decreases. This makes it possible to build high gain devices with short length. The waveguiding effect makes it possible to use narrow gas bores without having large diffraction losses. Compact, continuous, single-mode carbon dioxide lasers have been constructed in waveguides with diameters of the order of 1 mm. Beryllium oxide is commonly used as the material for the dielectric wall. Output powers around 0.2 W/cm of length are possible with such lasers.

Thus, in a device only about 10 cm long, output powers in excess of 2 W can be obtained. Such small, rugged, waveguide CO_2 lasers offer an attractive source for many applications. Frequency tuning of such lasers has been demonstrated, with tuning over a range as large as 1.2 GHz. Such wide-band tunable lasers can find applications in communications, optical radar, pollution detection, and spectroscopy. The entire package can be small and portable, suitable for field use. The waveguide CO_2 laser represents the only available laser source capable of emitting continuous power levels as high as several watts in a small portable device.

Some details of construction of small waveguide CO_2 lasers are shown in Fig. 3-14. These devices, much smaller than conventional CO_2 lasers, which are generally at least 1 m long, are capable of emission of several hundred milliwatts continuously.

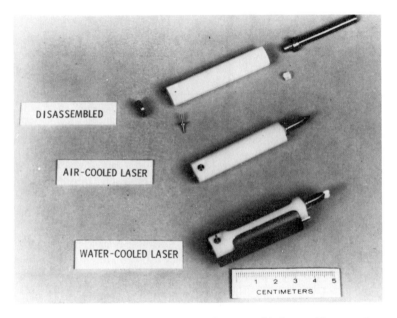

Fig. 3-14   Photograph of ultraminiature CO_2 waveguide lasers. (Photograph courtesy of H. W. Mocker.)

Summary of Advanced CO_2 *Lasers*   In the preceding subsections we have described a number of significant developments in CO_2 lasers. The description has been very brief and has only superficially covered the many possible configurations that have been developed for CO_2 lasers. The lesson is that CO_2 lasers are capable of operation in a number of different forms which yield higher outputs than the conventional configuration for a gas laser. These newer types of CO_2 lasers still have somewhat limited commercial availability, although models can be obtained from some manufacturers. Moreover, they are still undergoing development, so that we may expect to see more types of CO_2 laser becoming commercially available.

Other Molecular Lasers There are many other types of molecular lasers other than CO_2 lasers. Table 3-2 lists some of the important lines that are available to some commercial extent. We note that most of these lie fairly far in the infrared. The lasers illustrated in Table 3-2 are not nearly as common as the CO_2 laser and have only limited availability. However, they can be important sources for specialized purposes. We note that the wavelengths extend to extremely long values. At 337 μm the wavelength is over one-third of a millimeter. This is remarkable in that it represents electromagnetic radiation with an almost macroscopic wavelength. This wavelength is produced by techniques that can span the entire range through the infrared, visible and ultraviolet portions of the spectrum.

TABLE 3-2
Miscellaneous Molecular Lasers

Material:	N_2	CO	N_2O	H_2O	HCN
Wavelength (μm):	0.3371	5–6	10.5–11	28, 78, 118	311, 337

The nitrogen laser operating at 0.337 μm is worthy of special note. This laser operates only in a pulsed mode. It is of special importance because it represents the shortest high-power laser wavelength which can be obtained in a commercial system. The nitrogen laser emits pulses with durations of the order of 10 nsec and with peak powers up to 100 kW.

The carbon monoxide laser operates simultaneously on a large number of wavelengths between 5 and 6 μm. Carbon monoxide lasers appear to be developing rapidly. They have operated in most of the configurations which have been described for CO_2 lasers, including continuous electric discharge devices, ionizer–sustainer pulsed devices, TEA lasers, and gas dynamic lasers. The carbon monoxide lasers with highest continuous outputs have been cryogenic devices, but recent developments have demonstrated improved continuous operation at room temperature. Carbon monoxide lasers with fast gas flow can emit high power. Such lasers show considerable promise for future applications. They share many of the desirable characteristics of carbon dioxide lasers, and offer the advantage of a shorter wavelength. This fact would make them easier to use in applications such as cutting and welding of metals.

B. Solid State Lasers

The solid state laser is characterized by active media involving ions of an impurity in some solid host material. The laser material is in the form of a cylindrical rod with the ends polished flat and parallel. The pumping is by optical excitation. The prototype of the solid state laser is the ruby laser, the first laser which was operated. The term "solid state" is used in conjunction with lasers in a sense somewhat different from its common use. In connection with lasers, it has no connotation of semiconductors, but rather applies to lasers involving impurity ions in insulating host materials.

The ions that are commonly employed are either ions of the transition metals, such as chromium, manganese, cobalt and nickel, or of rare earth elements, i.e., the group of elements with atomic numbers from 58 to 71. These elements all share the common feature of an interior unfilled shell of electrons, which leads ultimately to a narrow fluorescent linewidth. According to Eq. (1.9), narrow fluorescent linewidth is favorable for laser operation, because it leads to high gain and to reduced requirements on the minimum population inversion necessary for laser operation.

The host materials in which these impurity elements are embedded tend to be hard, gemlike crystalline materials, or alternatively glasses. Such materials can readily be prepared and fabricated into cylindrical rods with polished ends. Many such materials have been demonstrated as lasers in the laboratory. Table 3-3 shows a very small sample of some such materials. This table gives a representative feeling for the ions that have been employed and the host materials in which they have been embedded. Of this variety of materials, the only ones that are widely used on a commercial basis are ruby, neodymium in glass, and neodymium in yttrium aluminum garnet (YAG). We shall describe each of these materials and its properties in greater detail.

In Table 3-3, the first column gives the ion, which is present as an impurity in a host material, the second column gives the host material, the third column gives a typical value for the concentration of the impurity, the fourth column gives the wavelength, the fifth column gives the temperature of operation, and the sixth column gives the threshold for laser operation which has been observed in terms of electrical energy input to the pulsed flash lamp. Since this threshold is dependent on a number of factors, such as crystal quality, ion concentration, the configuration of the optical resonator, and the shape and spectral output of the flash lamp, the values given for threshold may not necessarily be comparable on any consistent basis. The wide variety of host materials that are possible is represented particularly well by the number of host materials in which the neodymium ion has been used. The effect of the host material on changing the laser wavelength is

TABLE 3-3
A Selection of Solid State Laser Systems

Ion	Host	Concentration (%)	Laser wavelength (μm)	Temperature (°K)	Threshold (J)
U^{3+}	CaF_2	—	2.613	300	120
Cr^{3+}	Al_2O_3	0.05	0.6943	300	400
V^{2+}	MgF_2	—	1.1213	77	1070
Ni^{2+}	MgF_2	1.5	1.623	77	150
Sm^{2+}	CaF_2	0.01	0.7085	20	0.1
Dy^{2+}	CaF_2	0.03	2.36	77	1
Tm^{2+}	CaF_2	0.05	1.116	4.2	50
Nd^{3+}	CaF_2	—	$\begin{cases}1.0457 \\ 1.0461\end{cases}$	77 / 300	60 / 300 }
Nd^{3+}	$CaWO_4$	3.0	1.0584	300	0.5
Nd^{3+}	CeF_3	4–5	1.0639	300	16
Nd^{3+}	LaF_3	1.0	1.0407	300	17
Nd^{3+}	$PbMoO_4$	—	1.0586	295	60
Nd^{3+}	$YAlO_3$	3	1.0795	300	1.5
Nd^{3+}	$Y_3Al_5O_{12}$	3	$\begin{cases}1.0612 \\ 1.0642\end{cases}$	77 / 300	0.2 / 1 }
Nd^{3+}	YVO_4	2	1.0641	300	2
Nd^{3+}	$Ca_5(PO_4)_3F$	1.1	1.0629	300	4.4
Nd^{3+}	Silicate glass	5.0	1.060	300	1
Ho^{3+}	CaF_2	—	2.092	77	260
Er^{3+}	CaF_2	0.1	1.617	77	1000
Er^{3+}	$Y_3Al_5O_{12}$	—	1.6602	77	80
Er^{3+}	$Y_3Al_5O_{12} + 5\%\,Yb^{3+}$	1	1.6459	295	75

clear. The effect of increasing temperature for slightly changing the laser wavelength and for increasing the threshold energy is also clear. Many of the materials listed operate only at cryogenic temperatures. Addition of some additional rare earth elements which can absorb the pump lamp energy and transfer it to the active ion can sometimes improve the operation. This is illustrated in the case of erbium in an yttrium aluminum garnet lattice with addition of ytterbium.

Of the many lasers in Table 3-3, most are laboratory devices. We shall describe the few that are important for practical applications.

1. Ruby Lasers

Ruby represents the material in which laser operation was first demonstrated in 1960. Ruby is sapphire (crystalline aluminum oxide) in which a small percentage of the aluminum has been replaced by chromium. The chromium resides in the crystalline lattice of the sapphire as a trivalent ion.

The absorption spectrum of ruby is shown in Fig. 3-15 [9], for two different directions of the incident light relative to the c axis of the ruby crystal. Ruby is an anisotropic material, and has different optical properties along different crystalline directions. Ruby absorbs light in the green and blue portions of the visible spectrum and emits red fluorescence at a wavelength of 0.6943 μm. The emission has a spectrum approximately 0.0004 μm wide at room temperature. The energy level diagram of ruby has already been shown in Fig. 1-9. To summarize briefly, the absorption of green light raises an electron in the chromium ion to the broad upper level (denoted $^4F_2$ in spectroscopic notation) from which it relaxes very quickly to the upper level of the fluorescent transition (^2E). This state is relatively long-lived, with a lifetime of 2.7 msec at room temperature. If the exciting light is sufficiently intense and is delivered in a time short compared to 3 msec, most of the electrons find their way into the ^2E state. This produces a population inversion between the ^2E and the ground $^4A_2$ state, and laser action can proceed. Since the lower laser level is the ground state, ruby is a three-level laser system and more than 50% of the electrons must be excited in order to achieve a population inversion. This means that ruby requires fairly intense pumping, in contrast to four-level materials, in which the lower level for laser action is not the ground state.

The use of materials such as ruby for laser action depends on the electronic structure. The trivalent chromium ion has structure in which electrons are present in an outer shell while there is still an unfilled inner shell of electrons. The laser transitions in the chromium ion are transitions of

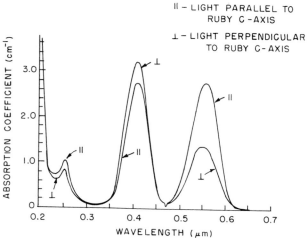

Fig. 3-15 Absorption spectrum of ruby. (From T. H. Maiman *et al., Phys. Rev.* **123**, 1151 (1961).)

electrons in the unfilled inner shell. The outer shell of electrons provides partial shielding from the electric field of the neighboring ions in the crystalline lattice. Thus, the spectroscopic linewidths of the fluorescent lines can be narrow compared to other materials in which perturbation by the electric field of neighboring ions broadens the spectral lines. Therefore, materials which have ions with unfilled inner shells in crystalline lattices make potentially good laser materials because of the possibility of having narrow lines. These include the transition metal elements and the rare earth elements. In both of these types of ions, there are unfilled inner shells of electrons surrounded by outer electrons. The resulting narrow linewidth is important, because it reduces the requirements on population inversion, according to Eq. (1.14).

In ruby which is used for laser purposes, the chromium concentration is relatively low, around 0.05%. This low concentration ensures that each chromium ion is relatively isolated from other chromium ions. As chromium concentration is increased in ruby, interactions between chromium ions on adjacent lattice sites tend to broaden the spectral lines.

The original observation of laser action in ruby was characterized by a millisecond duration pulse. This pulse showed characteristic substructure. The substructure consists of microsecond duration pulses, called relaxation oscillations, which have been illustrated in Fig. 2-11. The upper trace of that figure shows the envelope of the entire laser pulse, which contains much substructure. The lower trace shows the microsecond duration relaxation oscillations in greater detail. The relaxation oscillations arise because of the continuous pumping of the laser during the period of emission. A population inversion is established; stimulated emission begins and sweeps out the energy stored in the population inversion. This gives rise to one short pulse or spike of energy. The repumping of the population inversion by the pump light leads to another spike. Under some conditions, the relaxation oscillation spiking can be quite regular.

Q-switched operation is also common in ruby lasers. Q-switching using rotating mirrors was demonstrated very early. Because of the relative slowness of this Q-switching, most Q-switched ruby lasers now use either bleachable dyes or electrooptic cells.

Operation in a single transverse mode may be obtained at an expense of energy by aperturing the laser to a small diameter. The spatial structure of a ruby laser often is quite irregular. An example has been shown in Fig. 2-9, which gives the spatial structure of a Q-switched ruby laser. It is apparent that there are a number of hot spots where the power density is high. In order to achieve better spatial structure, it is necessary to choose a ruby of high optical perfection, to use an electrooptic Q-switch (see Chapter 7), and to insert a small aperture so that only a small volume of the ruby is used. The optical inhomogeneity in ruby can be relatively great, so that careful selection of a high-quality ruby rod is important. The bleachable dye Q-switches (see Chapter 7) tend to open irregularly over different spatial areas, because of the nonlinear nature of the bleaching process. On the other hand, electro-optic Q-switches open evenly over their entire aperture. The addition of a small aperture reduces the power output available from the ruby, but improves the mode structure greatly, so that operation in a single TEM_{00} mode is possible. Such lasers are often used for high-speed holographic applications.

Operation in a single longitudinal mode is also possible by insertion of an etalon in the laser. Such a single longitudinal mode laser has been described in the literature [10].

The original ruby lasers used helical flash lamps as illustrated in Fig. 3-16. More recent lasers tend to use linear flash lamps with close coupling between the linear flash lamps and the ruby. This design is more

Fig. 3-16 View of the interior of a ruby laser with a helical flash tube. The ruby is mounted along the axis of the helix. (Reprinted from *Electronics*, October 27, 1961; Copyright © McGraw-Hill, Inc., New York, 1961.)

efficient in the use of the pumping energy and leads to a lower threshold in terms of electrical energy necessary for laser operation.

Ruby lasers were widely utilized in the first few years of laser history, and were probably the most common type of laser in the early 1960s. With the rapid development of gas lasers in the late 1960s, particularly of the CO_2 laser, ruby lasers have become less common for many practical industrial applications.

The polarization of ruby laser light depends on the orientation of the crystalline axis in the laser rod. There is an axis of symmetry, called the c axis in ruby. If the ruby rod is cut so that the c axis is perpendicular to the rod axis, the ruby laser output will be polarized in a direction perpendicular to the direction of the c axis. If the ruby is cut in this direction, it may be readily observed that the apparent darkness of the ruby changes as the ruby is rotated. The ruby appears darkest when one is looking along the direction of the c axis. If the c axis is oriented along the axis of the laser rod, there will be no observable variation in darkness of the ruby as it is rotated. The laser output will be unpolarized for a ruby with this orientation.

The output characteristics of a typical ruby laser are shown in Fig. 3-17a. There is no output until the pump power has reached a minimum value, called the threshold. The threshold value for this laser, which employed a helical flashlamp, was near 1000 J. Typical threshold values for ruby lasers which use linear flashlamps, close-coupled to the ruby, are lower, perhaps a few hundred joules.

Above threshold, the output is an increasing function of input. Often a slope efficiency is defined. The slope efficiency is the percentage increase in output divided by the percentage increase in input, above threshold. For the example in Fig. 3-17, the slope efficiency is around 0.5%, and the total efficiency at the highest output is around 0.18%. More typical values for efficiency are perhaps a few tenths of 1% in modern ruby lasers. The limiting factor on output appears to be how much energy can be discharged into the flashlamps. There is a limited amount of energy that can be discharged through the flash tubes without damage. See Chapter 5 for discussion of maximum flashlamp inputs. Cooling a ruby laser below room temperature can increase the output energy. For typical operating conditions, cooling to 0°C can increase the output energy by perhaps 50% as compared to operation at 20°C. Operation at −100°C can increase the output by perhaps a factor of five.

The effect of reflectivity of the output mirror is shown in Fig. 3-17b. For a 6-in.-long ruby, the optimum output occurred near 65% reflectivity in room temperature operation. This value is considerably lower than the value for optimum output coupling for a helium–neon laser, because the gain per unit length in ruby can be much larger. In a shorter ruby, the

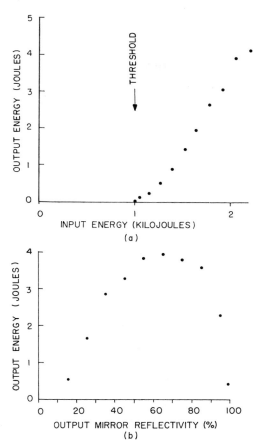

Fig. 3-17 (a) Output as a function of input energy for a typical ruby laser. (b) Output of ruby laser as a function of output mirror reflectivity.

optimum coupling could occur near 80% reflectivity. It has even been possible to operate ruby lasers with no output mirror. The 8% reflectivity at the ruby–air interface at the end of the rod can be sufficient to allow the laser to operate.

The dimensions of ruby rods are governed by crystal growing considerations. It is difficult to obtain ruby rods of good optical quality with length longer than 6 in. The diameter is limited by the necessity for having the pump lamp irradiate the volume of the ruby uniformly. If the ruby is too large in diameter, the pump light will be absorbed in the outer portions of the ruby, and the center will not be pumped. This sets a practical limit for diameters of ruby laser rods around $\frac{1}{2}$–1 in.

Output energies of typical ruby lasers are most often in the range from 1 to 10 J in a normal laser pulse. The highest value reported in the technical literature is 150 J [11] in a 17-cm-long ruby pumped by four linear flash-tubes. For Q-switched ruby lasers, powers in excess of 10^9 W have been reported. However, at power levels above a few hundred megawatts, the ruby tends to be damaged easily by the high optical fluxes. See Chapter 5 for additional details on this phenomenon.

2. Neodymium: YAG Lasers

Of the various laser hosts for the neodymium ion, yttrium aluminum garnet has come into widespread use. The choice of a host material depends on a number of factors. These include the ability to prepare the material in reasonably large samples of good optical quality, the hardness of the material, its ability to be polished easily, the reproducibility of preparation of the material, the absorption bands through which the pumping occurs and their relative efficiency, the thermal conductivity of the material, the fluorescent lifetime, the relative fluorescent efficiency and the possibility of operation at temperatures above cryogenic. All these things affect the ability of the particular system of laser host and neodymium impurity to emit reasonable amounts of energy in a laser pulse. The successful materials are those from which large amounts of energy can be extracted.

Yttrium aluminum garnet, almost universally called YAG, has the chemical composition $Y_3Al_5O_{12}$. YAG is a hard isotropic crystal. It can be grown to yield material of high optical quality, and can be polished to take a good optical finish. Although a number of other crystalline host materials are reasonably competitive, e.g., calcium tungstate and calcium fluorophosphate, YAG is probably the best commercially available crys-talline laser host. It offers low values of threshold and high values of gain. Its hardness (8.5 on the Moh scale), somewhat less than that of sapphire, allows for good optical polishing characteristics. It has high thermal con-ductivity, over ten times that of glass. This is one of the most important features. The absorption spectrum of neodymium in YAG contains many narrow lines.

The energy level diagram of the neodymium ion in YAG is shown in Fig. 3-18. This energy level diagram is mainly characteristic of the neody-mium ion. There are some small shifts with host material. The spectroscopic designations of the various levels are shown. The pumping is into a large number of levels that lie above 12,000 wavenumbers. There levels all relax nonradiatively to the $^4F_{3/2}$ level. The laser transition is between the $^4F_{3/2}$ and the $^4I_{11/2}$ level, which lies approximately 2000 wavenumbers above the ground state. Thus, this is a four-level laser system. Since the $^4I_{11/2}$ level will normally be unpopulated at reasonable temperatures, it is easy to obtain the population inversion with modest amounts of pumping energy. It is not, however, the low threshold which makes this host of particular

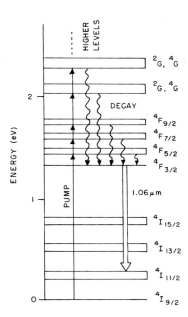

Fig. 3-18 Energy levels of the Nd ion. The spectroscopic notation is indicated for the levels. The excitation path and the laser emission are shown.

interest. There are other host materials that provide comparably low thresholds. It is rather the efficiency of energy extraction.

The output characteristics of Nd:YAG lasers may be discussed under three categories, namely, continuous, repetitively pulsed, and continuously pumped, repetitively Q-switched. In continuous operation, the laser is excited with either xenon lamps for power levels of the order of 10 W, or krypton lamps for power levels of the order of 100 W or more. The use of krypton lamps considerably increases the operating expense of the laser.

Some characteristics of Nd:YAG lasers for repetitive pulsing are given in Table 3-4 [12]. This gives some idea of the capability for tradeoffs between the average power, the pulse energy, the pulse duration, and the pulse repetition rate for such lasers.

A third type of operation is the continuously pumped, repetitively Q-switched method. The pump lamp remains on continuously, and a shutter between the laser rod and the mirror is opened for brief intervals, to allow Q-switched pulses at a high pulse repetition rate. Such operation usually involves an acoustic Q-switch (see Chapter 7). Some tradeoffs between average relative power, pulse duration, peak power and pulse repetition rate are shown in Fig. 3-19. Such lasers may be operated at pulse repetition rates up to 100 kHz, but at that level the average power is decreasing and the pulse duration is increasing. At low pulse repetition rates, the peak power is independent of pulse rate, so the average power increases with pulse rate. It can reach levels above 90% of what would be available in continuous operation with the same laser. Above 2 kHz, the peak power declines with

TABLE 3-4
Characteristics of YAG Lasers[a]

	Continuous	Pulse-pumped, helical lamp	Pulse-pumped, straight lamp(s)
Average power (W)	400	100	200
Pulse energy (J)	—	60	20
Pulsewidth (msec)	—	1–10	0.5–10
Peak power (W) (for 5 msec pulse)	400	12,000	4000
Pulse repetition rate (pulses/sec)	—	2	100
Lamp life	600 hr	3×10^5 pulses	10^6 pulses

[a] From M. J. Weiner, Society of Manufacturing Engineers Technical Paper MR 74-973 (1974).

increasing pulse rate, so that the average power goes through a maximum and then decreases.

Another variation of Nd:YAG lasers which has become commercially available involves frequency doubling. The output of the frequency-doubled Nd:YAG laser is in the green portion of the visible spectrum, at 0.53 μm. The technique of frequency doubling is described in Chapter 7.

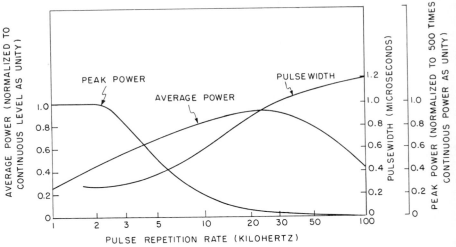

Fig. 3-19 Tradeoff of average power, pulse duration, and peak power as a function of pulse repetition rate for a repetitive Q-switched Nd:YAG laser. (Data from Quantronix, Inc.)

Peak pulsed powers around 100 MW are available in a short pulse. Such lasers may be used in applications where very small focal spot size is desired in material working applications, such as in micromachining.

3. Neodymium: Glass Lasers

Glass has a number of desirable characteristics as a laser host material. It can be made in large pieces of high optical quality and can be fabricated into a variety of sizes and shapes, ranging from fibers with diameters of a few micrometers to rods approximately 2 m long. The thermal conductivity of glass is lower than that of most crystalline hosts. Therefore, for high average power or high pulse repetition rate, the glass laser must have at least one small dimension for rapid heat removal. A second important difference between glass and crystalline laser hosts is that the emission lines of ions in glass are broader. This raises the threshold of the glass for laser action, because a larger population inversion is required to achieve the same gain, according to Eq. (1.13). On the other hand, the broad absorption lines in glass imply greater absorption of the pump light and allow greater energy storage. The broader emission line also allows mode-locking over a broader spectral range than is possible for other materials, so that glass lasers offer the best characteristics for the production of picosecond pulses. For systems requiring high energy outputs, or high radiance, or high peak power, glass systems generally outperform crystalline hosts. The highest values of efficiency in solid state lasers are obtained with glass.

Other differences between the neodymium: glass and neodymium: YAG lasers arise from differences in the thermal conductivity, in the fluorescent lifetime, and in the ability to fabricate large pieces of the materials. The longer fluorescent lifetime of neodymium: glass and also the ability to dope to high concentrations allows for greater energy storage per unit volume. In addition, glass may be fabricated into very large pieces, larger than is possible with YAG. Therefore, glass appears to be the material of choice for large laser systems where it is desired to extract the largest amount of energy. On the other hand, the higher conductivity of the YAG allows greater capabilities for average power. This includes both continuous operation and operation at high pulse repetition rates. Therefore, these two materials are both commonly employed in practical laser systems.

Neodymium has been used in a large variety of glasses. However, a glass with an alkali–alkaline earth base appears to offer the best combination of relatively long fluorescent lifetime, high fluorescent efficiency, and good durability. For example, a rubidium barium silicate glass has a fluorescent lifetime of 900 μsec.

Three typical regimes of operation may be identified for neodymium: glass lasers. These are the normal pulse mode, with pulse duration around 1 to 10 msec, Q-switched operation with pulse duration of some tens of nanoseconds, and picosecond pulse duration. The duration of the normal laser pulse may be varied by varying the duration of the pumping light pulse. The fluorescent lifetime of typical glasses is a function of the doping level of neodymium. Typically, the fluorescent lifetime begins to decrease as the neodymium concentration increases above approximately 2%. Therefore, optimum doping is around 2%. For pulse durations longer than 1 msec, repumping by the pumping source is required. The normal pulse neodymium : glass laser exhibits relaxation oscillation behavior such as has already been discussed relative to ruby lasers. Such lasers are available with multimode outputs up to 300 J.

Q-switching is accomplished either by electrooptic elements or by bleachable dyes (see Chapter 7). Since the half-wave voltage of electrooptic elements increases with increasing wavelength and since the dyes available for passive Q-switching appear not to be as effective as those for ruby lasers, operation in a simple Q-switched mode is not so often employed as with a ruby laser. Pulse durations of the order of 15 nsec in pulses containing a few joules of energy are available.

For very short pulses, one may select one of a train of mode-locked pulses. The glass laser is passively Q-switched and the series of mode-locked pulses is allowed to build up inside the cavity. When the power level has reached a desired value, one of the pulses may be switched out of the cavity by an electrooptic switch. Because of the broad fluorescent linewidth of the Nd : glass, the spectral width across which mode-locking may occur is of the order of 10^{12} Hz. The fully developed mode-locked pulse may have a duration of the order of 10^{-12} sec. The short pulse may then be amplified by passage through further glass laser rods. Power levels above 10^{12} W peak have been obtained in pulses of picosecond duration. Such systems are laboratory devices and are not readily available commercially.

Because glass can be fabricated in large pieces of arbitrary shape, glass lasers are sometimes used in a master-oscillator power-amplifier configuration. The laser pulse is produced in an original small laser (called the oscillator), which has external mirrors. The pulse is then allowed to pass through progressively larger stages of amplifiers. The amplifiers are rods of neodymium : glass which are optically pumped so as to produce a population inversion. They have no end mirrors and hence do not produce a laser pulse themselves. When the pulse from the oscillator travels through the amplifier, the energy stored in the population inversion is extracted in single-pass amplification. Thus, the pulse energy can be increased considerably without increasing the pulse duration. By increasing the size of the amplifier rod between stages, the power density is kept low enough so that damage to the glass does not occur (see Chapter 5). Such systems are mainly laboratory models or are being used in programs for nuclear fusion experiments. Although some such systems are being marketed, they have rather limited commercial availability.

C. Semiconductor Lasers

Semiconductor lasers employ structures that are much different from those of other lasers. A semiconductor laser uses a small ship of semiconducting material. In size and appearance, it is similar to a transistor. Thus, it is much different from gas and solid state lasers. Semiconductor lasers were first operated in 1962 and thus are almost as old as the other types of lasers. They do represent an entirely different approach to laser construction. They provide lasers with quite different properties than we have described heretofore.

The most commonly employed semiconductor material for lasers is gallium arsenide. We shall later describe some other semiconductors that have been used, but we emphasize that the most common type of semiconductor laser, and the only one that is widely available commercially, is gallium arsenide. The semiconductors silicon and germanium are perhaps the most widely known semiconductors. However, silicon and germanium have not yet been utilized as laser materials.

We begin by briefly reviewing some basic properties of semiconductors. Electrical conductivity in semiconductors occurs because of a relatively small number of free charge carriers which are capable of moving through the crystal under the action of an applied electric field. If the excess carriers are electrons, the semiconductor is called an n-type semiconductor. If the conductivity occurs because of motion of holes (a hole is a missing electron in one of the pairs of bonding electrons between atoms), the material is called p-type. At the interface between p-type and n-type material, a p–n junction is present. Such a junction forms a diode, which has been used in many solid state electronic applications. A p–n junction is also the region in which the laser action occurs in the semiconductor laser.

Semiconductor light emitting diodes (LEDs) are commonly used for display applications. Such LEDs are nonlaser devices. They use the same material as the laser, a p–n junction in gallium arsenide. However, they have lower radiance and are a device distinct from laser diodes.

A solid state laser, such as ruby, involves an isolated impurity ion embedded in a crystalline host lattice. The radiative transitions that produce laser action occur between discrete atomic energy levels of the impurity. The host lattice serves mainly to maintain the ions at fixed positions. In contrast, in the junction semiconductor laser, the energy levels involved in the laser action are characteristic of the entire crystalline lattice. These states are not discrete energy states, but are merged into energy bands, i.e., groups of energy states which lie so close together as to be continuous. The two energy bands of interest for a semiconductor laser are the valence band and the conduction band. The valence band is the highest band filled with electrons. The conduction band lies higher and is separated by a region of energy, the so-called energy gap, in which there are no allowed electronic states. Electrons can be excited by absorption of energy from the valence band to the conduction band. This leaves a hole in the valence band. Similarly, an electron can fall from the conduction band into a hole in the valence band. In this process of recombination, it emits the energy difference as radiation.

An energy band diagram of a *p–n* junction is shown schematically in Fig. 3-20. This diagram is characteristic of a semiconductor with a relatively large number of charge carriers. Such charge carriers can be produced by doping with impurities. For example, if a pentavalent impurity, such as antimony, is added to gallium arsenide, there will be an extra electron which is not needed to form bonding pairs. This electron will then be free to move and will contribute to *n*-type conductivity. Similarly, a trivalent impurity

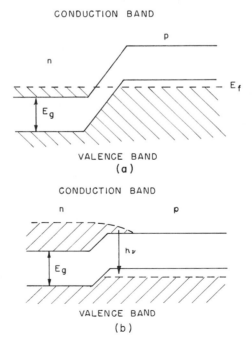

Fig. 3-20 Barrier voltage for a *p–n* junction in a semiconductor. (a) is without a forward bias and (b) is with a forward bias.

would lead to p-type conductivity. In the region of the junction, the bands undergo an energy shift as indicated. The electronic states are filled up to the Fermi level, denoted E_F. Thus, there will be electrons in the conduction band in the n-type material and there will be holes at the top of the valence band in the p-type material.

When a forward bias is applied to the junction, the barrier voltage is reduced, as shown in Fig. 3-20b. There will be a partial overlap between the regions where electrons are present in the conduction band, and where holes are present in the valence band. This narrow region near the junction thus has a population inversion. There are filled electronic states lying at energies above empty electronic states. In this region, radiative recombination can occur, with electrons falling across the gap and recombining with holes. The energy difference is emitted as radiation. Because of the population inversion, amplification by stimulated emission can occur. Thus a forward biased p–n junction can serve as the active material for a laser.

The thickness of the active region is small, around 1 μm. Thus, light propagating in the plane of the junction is amplified more than light traveling perpendicular to it. The laser can take the form of a rectangular parallelpiped, as shown in Fig. 3-21a. Two of the sides perpendicular to the junction are purposely made rough, so as to reduce the reflection. The other two sides are made optically flat and parallel, either by cleaving the crystal or by polishing it. The reflectivity of the air–semiconductor interface is usually high enough so that no additional mirrors are needed. The polished ends of the crystal themselves provide the required optical feedback. The laser emission thus occurs as indicated in Fig. 3-21a. The structure shown is the original structure used in the first semiconductor lasers. Later developments will be described below, in which this simple structure is somewhat modified. A photograph of a semiconductor laser diode is shown in Fig. 3-21b.

The most common operation of a semiconductor laser uses the forward-biased p–n junction. Electrons are injected from the n-type side of the junction and recombine at the junction region. Thus, there will be a net current flow. Such semiconductor lasers are often referred to as injection lasers.

In order for laser operation to occur, the population inversion must be great enough so that optical gain exceeds optical loss. Thus, the current density through the junction must exceed some minimum value. It must provide enough holes and electrons so that the radiation generated by their recombination exceeds the losses. The losses arise from several causes, including diffractive spreading of light out of the active region, transmission of light at the end of the junction, and absorption of light by free carriers in the junction. Thus, there is a threshold value of the current density for laser operation to occur.

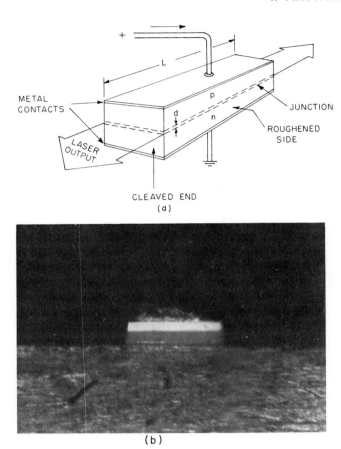

Fig. 3-21 (a) Schematic diagram of configuration of a semiconductor laser in a *p–n* junction of width *d*. (b) Photograph of a gallium aluminum arsenide semiconductor laser diode. The diode is the rectangular structure in the center. The horizontal line across the diode is the junction region. The diode is mounted on a copper block which covers the bottom half of the photograph. An electrical contact is visible at the top of the diode. The width of the diode is 0.47 mm.

Quantitatively, the threshold current density J_t (in amperes per square centimeter) can be expressed as

$$J_t = 6.3 \times 10^4 (n^2 d/\varepsilon) E^2 \, \Delta E(\alpha_D + \alpha_A + T/L) \tag{3.2}$$

where n is the index of refraction, d is the thickness of the junction, ε is the efficiency of conversion of hole–electron pairs in the junction region into photons, E is the photon energy in electron volts, ΔE is the spectral width of the fluorescent emission line in electron volts, α_D and α_A are the losses per unit length due to diffraction and absorption, respectively, T is the transmission of the end face of the diode, and L is the length of the diode. Since the loss due to absorption by free carriers increases rapidly with temperature, the threshold current scales approximately as the cube of operating temperature, between 77 and 300°K. Thus, GaAs injection lasers are often operated at liquid nitrogen temperature, although they may be operated at room temperature with increased current density. Above 100°K, the threshold current density for laser operation increases rapidly with increasing temperature. Because the electrical current through the junction dissipates heat which would tend to increase the threshold current, it is important to provide good cooling. Moreover, in order to minimize heating effects, it is common to operate GaAs lasers in short pulses at high pulse repetition rates. Although continuous operation can be employed, particularly for operation at cryogenic temperatures, pulsed operation is more common.

Pulse generators capable of providing narrow, high-current, pulsed outputs are available for use with injection lasers. Such units can yield drive currents up to 200 A, with nominal pulse widths around 100 nsec and with rise and fall times around 50 nsec. Short rise and fall times are required to minimize heating at current levels below the laser threshold. Variable internal clocking is available to provide variable pulse repetition rates. The overshoot of the current pulse must be small, since the peak inverse voltage that the laser diode can tolerate is small.

There is another possibility for excitation, namely, electron beam pumping. If a semiconductor is bombarded by a beam of high energy electrons, electrons can be produced in the conduction band and a population inversion produced. Such a laser does not need a p–n junction. This type of operation is less common than the injection method, but it does offer some attractive possibilities. One may scan the region in which laser emission occurs by scanning the electron beam. Such lasers have become commercially available in the form of an evacuated glass tube with an electron gun at one end, and a cadmium sulfide crystalline target at the other end. The target is a 1-in. square about 50 μm thick. Coherent emission occurs perpendicular to the plane of the cadmium sulfide target at a location opposite the point where the electron beam strikes the target. The electron beam is controlled by conventional deflection circuitry. Two-dimensional positioning of the laser source is thus possible on the crystal.

The remainder of the discussion on semiconductor lasers will emphasize the injection laser. The material for a semiconductor laser is often formed by cleaving the ends of the crystal to form the flat parallel faces needed for the optical cavity. Gallium arsenide cleaves easily along certain crystal planes. The sides are roughened to reduce stray reflection. The original material from which the laser is formed is an ingot containing a junction between *p*-type material and *n*-type material. Electrical contacts are applied to the top and bottom faces. A heat sink is needed to remove heat from the laser. Often the laser will be mounted on a copper block.

Typical dimensions of a GaAs laser are 1 or 2 mm. The light emitting region, the junction, from which the radiation originates, is a thin layer only a few micrometers thick.

Usually the mirrors for feedback and output coupling are the cleaved ends of the laser diode, without any further coating. The reflectivity at the interface between gallium arsenide and air is approximately 36%. If one desires to have output from only one end of the device, or if mirrors of higher reflectivity are needed to reduce the threshold for laser operation, the reflectivity may be increased by coating with metallic films.

It is important to consider the operating life of gallium arsenide laser diodes. We should consider the effects that degrade the laser output. Two different types of damage mechanisms have been identified. One involves catastrophic decrease in the laser output. The catastrophic damage may occur within a single pulse of the laser. It is associated with damage of the end surfaces of the laser. The damage is produced by the light output of the laser itself. In order to avoid this damage, the peak power output of the laser must be limited. For pulse durations around 100 nsec, the power output should be limited to no more than approximately 400 W/cm width of the junction. For longer pulse durations, the peak power must be reduced further.

There is also a gradual decrease in laser output with time. This is accompanied by an increasing threshold current. The damage is produced by high values of current flowing through the junction. This is a complex phenomenon which is complicated by random variations in the laser life. Early versions of semiconductor lasers had very limited lifetimes. In order to extend the life of the laser, the current density through the junction must be limited. Improved lifetime has been obtained in advances involving the use of heterostructures. Such structures can yield laser operation at reduced values of the current density. This extends the operating life of the laser. The problems of device reliability that plagued earlier gallium arsenide laser diodes have been largely overcome. The advances have involved reduction of crystallographic defects and strain, in addition to the use of heterojunction structures. Long-lived gallium arsenide lasers are marketed commercially. The operating lifetime of available semiconductor lasers now exceeds 1000 hr. Experimental devices have operated many thousands of hours without significant deterioration.

In order to understand the heterostructure, we first consider the simple *p–n* junction, the homojunction, which consists of a single material. Early semiconductor lasers were homojunctions which consisted of doped regions as shown in Fig. 3-22a. The p^+ region is a region of more highly doped *p*-type material. This region has a lower index of refraction than the *p*-type region. The variation in doping provides a variation in index of refraction as shown on the right side of the diagram. This provides confinement of the optical wave in the junction region, because the wave will tend to travel more slowly in the region of higher index of refraction. This is a wave-guiding phenomenon which tends to reduce diffractive losses, and hence to reduce the required current density. However, in the homojunction, the difference in index of refraction is not large and confinement is relatively poor. Figure 3-22b improves this situation by providing a gallium arsenide–aluminum gallium arsenide heterojunction. A fraction *x* of the gallium in

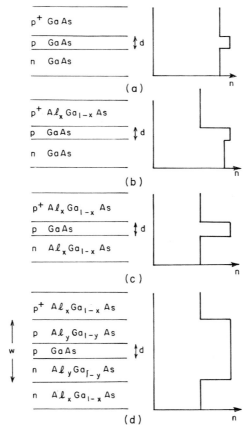

Fig. 3-22 Structure and index of refraction *n* for various types of junctions in gallium arsenide with a junction width *d*. (a) is for a homojunction. (b) is for a gallium arsenide–aluminum gallium arsenide single heterojunction. (c) is for a gallium arsenide–aluminum gallium arsenide double heterojunction with improved optical confinement. (d) is for a double heterojunction with a large optical cavity of width *w*.

the top layer has been replaced by aluminum. This structure provides a bigger change in index of refraction as compared to the homojunction. This larger change in index of refraction yields better characteristics as a dielectric waveguide. The confinement of the optical wave is better. The heterostructure reduces the light that spreads into the p^+ region because of the discontinuity in index of refraction at the heterojunction. This in turn leads to lower requirements on current density and therefore to improved life. This type of structure has been called a close confinement structure, because of the improved confinement of the optical wave in the narrow region of the junction. Aluminum gallium arsenide is used because of the good match in the lattice of gallium arsenide and aluminum arsenide. Also, the potential barriers at the junction region tend to confine the charge carriers to a small volume. This further reduces the required current density needed to obtain a population inversion.

Further advances have employed a second heterojunction, as is shown in Fig. 3-22c. This serves to confine the radiation even better by making an even larger change of index of refraction near the junction. The step in index of refraction has been increased on both sides of the junction. The double heterojunction structure, although it reduces the required current density, does have two disadvantages. The optical radiation is so well confined that the optical flux density is high. This leads to a higher probability of the catastrophic damage which occurs at high values of optical flux. In addition the region is so narrow that the beam divergence angle perpendicular to the junction becomes large. In order to overcome these two difficulties, a further modification, shown in Fig. 3-22d, has been used. This type of structure is called the large optical cavity (LOC). This uses regions of aluminum gallium arsenide of varying composition. The large optical cavity semiconductor structure was developed in order to compromise between structures that would provide high peak power with reduced threshold current. The basic feature of the LOC diode is that the region for carrier confinement is smaller than the low-loss waveguiding region for confinement of the optical radiation. The radiation is confined by the two outer hetero-junctions in the region of width w. Typically d may be in the range from 0.2–1 μm, and w may be several micrometers. Thus, the optical waveguide can be broader, so that the optical flux density does not become too great. This is achieved without sacrifice of the carrier confinement provided by the width of the narrow p-type gallium arsenide layer. The structures in Fig. 3-22 are all grown on the surface of a substrate of n-type GaAs. Liquid phase epitaxy has often been used to form the layered structure.

We list in Table 3-5 the required threshold current density for operation at 300°K. The marked improvement offered by the advanced structures is obvious.

TABLE 3-5

Typical Threshold Current Density for GaAs Laser Operation at 300°K

Homojunction	40,000 A/cm^2
Single heterojunction	10,000
Double heterojunction	1300
Double heterojunction, large optical cavity	600

Still another variant is the stripe geometry, illustrated in Fig. 3-23. There are a number of varieties of stripe geometry, but Fig. 3-23 will show the essential features. The stripe geometry is adopted so as to confine the radiation somewhat in the direction transverse to the junction. The SiO$_2$ insulator confines the electrical excitation to a narrow stripe down the length of the diode. This improves the spatial structure of the output beam, by reducing the tendency of the beam to break up into a number of filaments in a wide structure. It also makes the beam more round, because the diffraction limited beam divergence angle will be more nearly equal in the directions parallel and perpendicular to the junction. Typically the width of the stripe may be around 13 μm.

The properties of gallium arsenide lasers are somewhat different from those of most other lasers. The small dimensions of the junction in which the light is produced leads to a poorer collimation because of diffraction. The beam divergence angles are greater than the narrow beams of typical gas lasers or crystal lasers.

The spectral width of the radiation from the gallium arsenide laser is typically around 2–3 nm, larger than the spectral width of most other lasers. The GaAs laser has poorer coherence properties than other lasers. Therefore, a gallium arsenide laser can be regarded as a small, bright, area source of radiation. Although gallium arsenide lasers do not possess the properties of directionality and monochromaticity to the same degree as other lasers, they do have many important properties which make them attractive for some practical applications. They can be modulated easily at high frequencies by modulating the current through the junction. They are efficient, small, and rugged. They are much less expensive than other types of lasers.

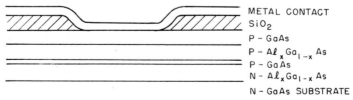

Fig. 3-23 Configuration of stripe geometry heterojunction semiconductor laser.

Gallium arsenide lasers emit radiation in the near infrared portion of the spectrum. The exact wavelength depends on the temperature at which the laser is operated, from about 0.845 μm at liquid nitrogen temperature to 0.905 μm at room temperature. Above room temperature, the threshold for laser operation rises, so that GaAs lasers would be difficult to operate. However, gallium arsenide lasers have been operated over the range of temperatures from liquid helium temperature to room temperature.

The wavelength at which the laser emits is mainly determined by the temperature of operation. However, other factors can have an influence. These factors include the exact nature and concentration of the doping elements used in the gallium arsenide, the driving current, and the presence of a magnetic field. Thus, for a particular laser, the wavelength of emission can vary slightly, and in fact can be tuned slightly.

As we mentioned earlier, operation of the gallium arsenide laser is characterized by a threshold current. Figure 3-24 shows the output spectrum of a typical GaAs laser as the current input is varied [13]. This figure represents continuous operation of a double heterojunction device at

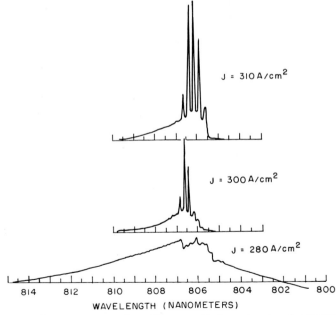

Fig. 3-24 Output spectrum of a gallium arsenide laser at various input current densities for continuous operation of a double heterojunction device at cryogenic temperatures. (From H. Kressel in "Laser Handbook," (F. T. Arecchi and E. O. Schulz-Dubois, eds.), Vol. 1. North-Holland Publ., Amsterdam (1972).)

cryogenic temperatures. At low current densities, there is a broad spectrum of spontaneous emission with a bandwidth around 100 nm. When the current through the junction is increased, stimulated emission begins when the optical gain exceeds the losses. The threshold current density will depend on the temperature, on the structure of the device, on the absorption losses in the material, on the reflectivity of the diode surface, and on the doping of the material. Threshold for this device was around 300 A/cm^2.

The threshold current at which laser operation begins is strongly temperature dependent. At the cryogenic temperature of 77°K, the threshold current in a gallium arsenide laser is about one-tenth the value at room temperature. Thus, cooling to cryogenic temperatures dramatically changes the operating characteristics of the laser.

When the threshold current is exceeded, the emission spectrum narrows and the intensity of the emission increases considerably. Figure 3-24 shows the spectrum both below threshold and above threshold. The letters on the curves refer to different values of current through the junction. Below threshold, only a broad spectrum characteristic of spontaneous emission is observed. When the current is increased slightly, strong narrow laser components are superimposed on the broad emission. As the current through the junction is increased, the intensity increases.

Gallium arsenide lasers are usually operated pulsed, with a relatively low duty cycle, of the order of 1%. This is because the current through the junction heats the diode, which in turn causes the operating characteristics to deteriorate. Therefore, pulsed operation is more common than continuous operation. However, gallium arsenide lasers have been operated continuously, both at liquid nitrogen temperature and at room temperature.

Table 3-6 gives a comparison of the performance of typical commercially available laser diodes at cryogenic temperature and room temperature. This table shows how the thresholds, driving currents, and pulse durations change with temperature. The peak power available from the laser does in

TABLE 3-6
Typical Properties of GaAs *Lasers*

	Cryogenic temperature	Room temperature
Wavelength	8450 Å	9050 A
Threshold current	0.7 A	7 A
Operating current	4 A	25 A
Pulse duration	1 μsec	200 nsec
Peak pulsed power	2.5 W	6 W
Duty factor	2%	0.1%
Power efficiency	40%	4%

fact increase as one goes to higher temperature. However, the duty cycle decreases and the average power that can be extracted from the laser also decreases. Tradeoffs between pulse repetition rate and pulse duration may be made, so long as one does not exceed the stated duty factor.

The power efficiency, which is a measure of the true electrical efficiency of the device (that is, the light output divided by the power dissipated in the device) can be high at cryogenic temperatures. It can approach 40%, which is a value higher than for any other type of laser. However, the efficiency also decreases at room temperature.

The width of the spectral band represented by the spontaneous emission is much wider than that of the stimulated emission. However, the stimulated emission produced by the laser is still much broader than that of conventional gas and crystalline lasers. It is of the order of 2–3 nm as compared to a typical spectral width around 0.001 nm for a He–Ne laser.

The emission spectrum is relatively complex and typically contains a number of longitudinal modes of the optical cavity. Figure 3-24 shows the presence of several modes at current densities above threshold. The spacing between longitudinal modes is relatively large, because of the short length of the optical cavity. However, the relatively large spectral width of the GaAs laser allows several modes to be present. The mode spacing Δv is

$$\Delta v = c/2L \tag{3.3}$$

where c is the velocity of light and L is the length of the cavity. The spacing $\Delta \lambda$ in wavelength is

$$\Delta \lambda = \lambda^2 \, \Delta v/c = \lambda^2/2L \tag{3.4}$$

where λ is the wavelength of operation. If $L = 0.2$ cm and $\lambda = 0.9 \times 10^{-4}$ cm, then

$$\Delta \lambda = 0.81 \times 10^{-8}/0.4 \approx 2 \times 10^{-8} \text{ cm} = 0.2 \text{ nm}$$

Thus, under the spectral width of 2–3 nm, there is room for 10–15 longitudinal modes to operate simultaneously.

One of the most important characteristics of gas lasers is the very small divergence of the emitted radiation. This characteristic is not shared by semiconductor lasers. The main reason is that light is emitted through the aperture defined by the small junction. Diffraction through the narrow dimensions of the junction spreads the beam into a broader angle than is observed with other types of lasers. Figure 3-25 illustrates this, and schematically gives a plot of the beam profile. Thus, the emission from a gallium arsenide laser tends to be an elliptical beam with a half angle of a few degrees in the direction parallel to the junction, and a larger value perpendicular to the junction.

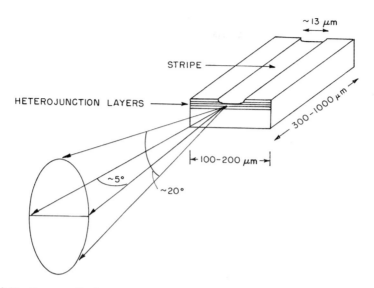

Fig. 3-25 Beam profile from a stripe geometry heterojunction gallium arsenide laser.

The use of the large optical cavity structure also helps here, because as the dimension of the region for beam confinement increases, the beam divergence angle due to diffraction decreases.

This relatively large beam spread means that the radiance (or brightness) of the GaAs laser is lower than that of other lasers. Radiance has previously been defined as the power emitted per unit area per unit solid angle. A typical value for the radiance of a GaAs laser is around 10^7 W/cm²-sr, much lower than the typical value of 10^{11} W/cm²-sr for a ruby laser. This value of radiance is still much higher than that of conventional light sources and of LEDs. This fact leads to the possibility of important applications.

The potential applications of gallium arsenide lasers are defined by its properties. Since gallium arsenide lasers do not share the narrow beam divergence angle and the narrow spectral line width that are commonly associated with lasers, the applications of gallium arsenide lasers are somewhat limited. They could not be used well for alignment because of the large beam divergence. They could not be used well for interferometric applications because of the large spectral width.

However, the other properties of small size, light weight, low power consumption, and high efficiency are important characteristics. They mean that gallium arsenide lasers can be considered for applications in the field, away from sources of electrical power. Two such applications that suggest themselves are range finding and communications. Both these applications take advantage of the fact that the gallium arsenide laser can be modulated at high frequency through the power supply.

Other Semiconductor Lasers In addition to gallium arsenide lasers, there has been a variety of other semiconductor lasers developed. Most of these lasers are in laboratory development and have not reached commercial status. Table 3-7 lists some semiconductor laser materials, and their wavelengths of operation. Most of these materials, with a few exceptions, have been operated only at cryogenic temperatures. The exceptions, which have been operated at room temperature, include GaAs, ZnO, and CdS. The shorter wavelength semiconductor lasers tend to operate only by electron beam pumping, whereas for the infrared semiconductor lasers, both injection and electron beam excitation are usually possible.

TABLE 3-7
Semiconductor Laser Materials

Material	Wavelength (μm)	Type of Excitation[a]
ZnS	0.33	E
ZnO	0.37	E
ZnSe	0.46	E
CdS	0.49	E
ZnTe	0.53	E
GaSe	0.59	E
CdSe	0.675	E
CdTe	0.785	E
GaAs	0.84–0.95	I, E
InP	0.91	I
GaSb	1.55	I, E
InAs	3.1	I, E
Te	3.72	E
PbS	4.3	I, E
InSb	5.2	I, E
PbTe	6.5	I, E
PbSe	8.5	I, E

[a] I = injection, E = electron beam.

In addition, mixtures of some of these materials have been formed, so as to allow operation at any wavelength within a range. For example, a mixture of indium arsenide and gallium arsenide can be made. The laser is denoted by $(In_xGa_{1-x})As$, where x denotes the fraction of indium arsenide. By varying x, the wavelength of operation may be varied continuously over the range from 0.84 to 3.1 μm. When $x = 0$, it becomes a gallium arsenide laser and operates at the usual wavelength of 840 nm. When $x = 1$, it becomes an indium arsenide laser and operates at 3.1 μm. Thus, the wavelength of operation may be varied over a wide range by varying the composition. Of course, once x is chosen and a particular wavelength is selected, that laser will operate only at that selected wavelength. To operate at a different wavelength, a different laser must be prepared with a different value of x. In this fashion, lasers have been prepared to cover the entire spectrum from 0.32 μm to past 30 μm. Figure 3-26 illustrates the ranges covered by some such mixed semiconductors. In this figure a shorter notation is employed, e.g., (InGa)As stands for mixed indium–gallium arsenide, which we denoted above as $(In_x Ga_{1-x})As$.

Such lasers can be tuned over small ranges by varying diode current, temperature, magnetic field, or hydrostatic pressure. For example, (PbSn)Te lasers have been tuned over the range 22–25 μm by a 10 kG superconducting magnet [14]. Lasers of composition $(PbSe)_{1-x}(SnTe)_x$ have been tuned over the range 9–12 μm by varying the temperature from 20 to 90°K, and over a total tuning range of 1 μm within this interval by varying the current in the diode [15]. Current tuning occurs by virtue of the small changes in operating temperature due to current variation. Current tuning (with

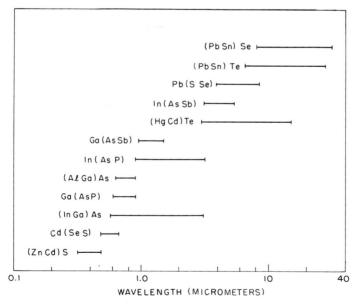

Fig. 3-26 Wavelength ranges covered by a number of semiconductor lasers of mixed composition. In this figure an abbreviated notation is employed. For example (InGaAs) stands for $(In_x Ga_{1-x}As)$. By varying x, a laser emitting at any wavelength within the indicated band can be prepared.

discontinuous jumps due to hopping between the longitudinal modes of the short cavity) was demonstrated over a range of about 0.07 μm for a (PbSn)Te laser operating near 8.8 μm [16].

D.   Organic Dye Lasers

Organic dye lasers represent another type of laser which is somewhat different in construction from the more familiar gas and solid state lasers. They employ liquid solutions of certain dye materials. The dye materials are relatively complex organic molecules, with molecular weights of several hundred. These materials are dissolved in organic solvents, commonly methyl alcohol. Thus, the active material for dye lasers is a liquid. Dye lasers are the only type of liquid laser which has reached a well-developed status.

Dye lasers were first demonstrated in the mid-1960s. Early devices encountered a number of problems. It was not until several engineering advances occurred in the early 1970s that dye lasers became really practical devices. Probably the most important feature that dye lasers offer is tunability. The monochromatic output of currently available dye lasers can be tuned over a broad range, from the near ultraviolet to the near infrared. Dye lasers are most likely to be chosen for applications where tunability is important. However, they probably will not often be used for applications in which one simply desires a visible laser output but the exact wavelength is not important. For such applications, (e.g., alignment), other types of lasers, such as He–Ne, are likely to offer better convenience and economy.

The dye materials that are used in lasers are similar to many familiar dyestuffs that are commonly employed as colorants in fabrics, plastics, soaps, cosmetics, etc. Common dye materials often contain a chain of carbon atoms with alternating single and double bonds. In order to introduce one specific material, let us briefly discuss the dye rhodamine 6G. Rhodamine 6G is one of the more important dye materials. It contains several benzene rings and has the chemical formula $C_{26}H_{27}N_2O_3Cl$ with a molecular weight around 450. It is soluble in methyl alcohol and in water. It can be used for coloring silk or paper pink. It represents a typical example of the dye materials that are used in lasers. The concentrations that are employed are relatively low. This indeed is a characteristic of dye materials, namely, that small amounts are effective in producing changes in the optical properties of material.

A typical energy level diagram of a dye material is shown in Fig. 3-27. Initially the entire population of the material is concentrated at the bottom of the ground level, S_0. When the dye material is irradiated with light

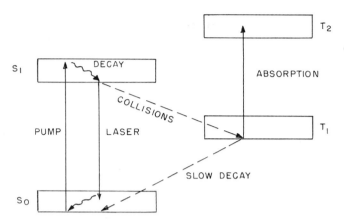

Fig. 3-27 Energy levels relevant to operation of dye lasers. S_0 is the ground singlet state, S_1 the first excited singlet state, and T_1 and T_2 the two lowest triplet states.

whose wavelength corresponds to the energy difference between S_0 and S_1, some of the ground state molecules are raised to level S_1. These levels are shown as broad bands which contain many vibrational and rotational sublevels, so that the many levels form a continuous band. When a dye molecule is raised to state S_1 by the pump light, it decays in a very short time to the lowest-lying sublevels of the state S_1. This decay occurs by nonradiative transitions which produce heat. The upper levels of state S_0 are initially empty. Thus, one obtains a population inversion between S_1 and the upper sublevels of S_0. These states are radiatively coupled, i.e., the transition from S_1 to S_0 occurs with emission of flourescent light. Because of the population inversion, gain by stimulated emission is possible, and laser action can occur. Because the laser transition occurs between the bottom of state S_1 and a level near the top of state S_0, the laser operation occurs at a wavelength shorter than that of the initial pumping light. We note that the fluorescent light can be emitted over a variety of wavelengths. Since many sublevels near the top of S_0 are empty, laser action could occur over a range of wavelengths corresponding to the energy difference between the levels. This leads to the possibility of tuning dye lasers.

There is a competing process which reduces the number of dye molecules available for laser operation. This is the presence of additional states, as indicated by states T_1 and T_2. These additional states are termed triplet states. The states S_0 and S_1 which provide laser operation are part of a different set of states, called singlet states.

When dye molecules are excited to the state S_1, some of them are capable of reaching state T_1 because of collisions. The molecules that reach T_1 tend to be trapped, because they remain in state T_1 for a relatively long time. The decay to the ground state S_0 from T_1 proceeds very slowly. As a result, state T_1 acquires a large population. This steals the molecules that could be operative in laser operation.

The situation is made even worse by the presence of state T_2. Absorptive transitions at the laser wavelength can take place between state T_1 and state T_2. It almost always happens in organic dyes that such triplet transitions occur at the available laser wavelength. The result is that as state T_1 becomes populated, the medium becomes more absorbing at the laser wavelength, because a fraction of the molecules accumulates in state T_1. This effectively shuts off the laser operation. The population of state T_1 occurs very rapidly, typically within less than a few microseconds. This means that dye lasers require the use of very fast pumping sources in order to provide a short pulse of laser operation, before T_1 becomes populated. Pulsed dye lasers are helped somewhat by use of quenching materials which speed up the return of molecules from T_1 to S_0. Oxygen is one such quenching molecule which helps to keep T_1 depopulated. However, even with quenching, no dyes have been discovered that allow the use of continuous pumping sources. Because of the troublesome triplet states, pumping of dye laser materials must occur in a rapid pulse.

Continuous dye lasers have been produced by rapidly circulating the dye molecules through a continuous pumping beam. The result is a dye laser which has continuous output. However, each molecule of the dye sees a brief pulse of pumping light as it passes through the pump beam. Then as molecules accumulate in state T_1 in a particular volume of fluid, that volume is circulated out of the laser cavity and is replaced by fresh fluid. The dye can be circulated by a pump. The circulation time is long enough so

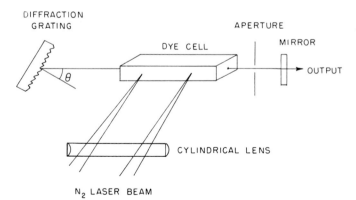

Fig. 3-28 Schematic diagram of pulsed dye laser excited by a nitrogen laser beam, with tuning by a diffraction grating.

that the state T_1 can decay before a particular dye molecule returns to the laser region.

In practice, lasers are used as the pumping sources for dye lasers. We shall discuss two configurations which have been employed. One geometry uses a pulsed nitrogen laser. This is shown in Fig. 3-28. The nitrogen laser operates only in very short pulses, with durations of the order of 100 nsec. Thus, the problem with population of state T_1 is eliminated. The laser pulse is inherently very short, of the order of a few nanoseconds, and there is not time for a triplet population to build up.

Dye lasers are inherently tunable because of the broad spectrum of empty sublevels in state S_0. Tunability is thus made possible by inserting a wavelength-selecting element in the cavity. In Fig. 3-28, the wavelength-selecting element is a diffraction grating which serves as one of the mirrors. The diffraction grating is a ruled grating with closely spaced rulings. The diffraction grating obeys the equation

$$N\lambda = 2d \sin \theta \qquad (3.5)$$

where N is a small integer, λ the wavelength, d the spacing between rulings, and θ the angle between the light and the normal to the grating. Only light reflected back into the direction from which it came can contribute to laser action. Thus, variation of θ can pick a particular value of wavelength. Therefore, rotation of the grating can serve to vary the wavelength of laser operation.

A second design is shown in Fig. 3-29. This represents a continuous dye laser pumped by an argon laser. The argon laser beam is focused to a small spot and the dye flows through the dye laser cavity. A typical velocity for the dye flow is around 10 m/sec. If the pump beam is focused to a diameter around 10 μm, the effective pumping pulse as the dye molecules pass through the pumping beam is around 1 μsec. The laser efficiency and output power

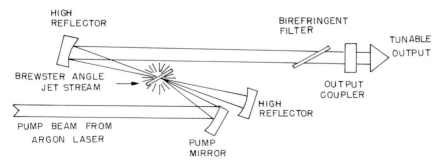

Fig. 3-29 Diagram of tunable, optically pumped dye laser using jet stream technology. (Courtesy of Coherent Radiation, Inc.)

are improved if the fluid flow is made as fast as possible and the diameter of the pump beam is made as small as possible. The dyes are forced to flow in a high speed jet, which is tilted at Brewster's angle to the incoming pump light.

This design uses a wavelength-selecting filter as the tuning element. As the filter is rotated, light of one particular wavelength will be able to reach the mirror and be reflected back through the laser cavity. In the design in Fig. 3-29, the curved pump mirror focuses the pump beam so that its beam waist coincides with the waist of the dye laser beam inside the dye stream.

If one employs either of the two configurations illustrated in Figs. 3-28 and 3-29, one may tune the laser by rotating the appropriate element, either the grating or the filter. A typical tuning range for one particular dye is several tens of nanometers. When the end of this tuning range is reached, one may physically change the dye material that is being used. This can be done by physically lifting out the reservoir of dye and replacing it with a reservoir containing a different dye. In this way, tuning across the entire

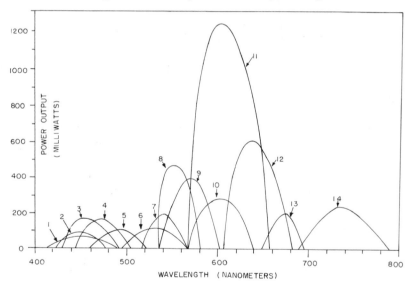

Fig. 3-30 Power levels available from continuous dye lasers as a function of wavelength. The identification of each curve gives first the name of the dye, second the pump power level from the argon laser pump, and finally the lines from the argon laser used for pumping. Curves: 1: carbostyril 165, 1.5 W, UV lines; 2: coumarin 120, 1.5 W, UV lines; 3: coumarin 2, 1.8 W, UV lines; 4: 7 diethylamino 4 methylcoumarin, 1.35 W, UV lines; 5: coumarin 102, 1.5 W, UV lines; 6: coumarin 7, 1.2 W, 476.5 nm line; 7: coumarin 6, 2.3 W, 488.0 nm line; 8: sodium fluorescein, 5 W, all lines; 9: rhodamine 110, 5 W, all lines; 10: rhodamine 6G, 2 W, all lines; 11: rhodamine 6G, 5 W, all lines; 12: rhodamine B, 5 W, all lines; 13: cresyl violet perchlorate, 5 W, all lines; 14: nile blue-A perchlorate, 2 W, 647.1 nm line of krypton ion laser. (Courtesy of Coherent Radiation, Inc.)

visible spectrum may be obtained. This is illustrated in Fig. 3-30, which shows available power levels which may be obtained in a continuous flowing dye laser pumped by an argon laser. At the short wavelength end, one uses the ultraviolet output of the argon laser. For other dyes, different wavelengths of the argon laser are used as indicated. The values of power that are required to give the particular level of operation are also indicated. Thus, one can tune between approximately 570 and 650 nm with rhodamine 6G as the dye material by simply rotating the filter. When the end of the tuning range is reached, the rhodamine 6G is replaced with rhodamine B and one may proceed further toward the red. When the end of the tuning range of the rhodamine B is reached, one uses cresyl violet perchlorate, and so on. In this way, one has the capability of tuning throughout the entire visible spectrum.

The power levels that are available in a continuous flowing dye laser approach 25% of the incident pump light. Throughout most of the visible spectrum approximately 100–200 mW can be obtained. In the yellow portion of the spectrum, using rhodamine 6G, over 1 W can be obtained.

The characteristics of a pulsed dye laser pumped with an ultraviolet nitrogen laser are shown in Table 3-8. The peak powers can reach tens of kilowatts. Tuning through the near ultraviolet and almost the entire visible spectrum is possible using various dye materials.

TABLE 3-8
Pulsed Dye Laser Capabilities

	Best values over 360–650 nm wavelength range
Peak power	50 kW
Pulse width	2 nsec
Pulse repetition rate	100 pps
Energy per pulse	10^{-4} J
Average power	10 mW
Spectral width	0.01 nm
Beam divergence angle	2 mrad ($\sim 2 \times$ diffraction limit)

The main advantage of dye lasers is their tunability in the visible portion of the spectrum. If one simply desired a visible laser, either continuous or pulsed, and did not care about the exact value of the wavelength, one would probably not choose a dye laser. There is, of course, the intermediate step of a laser pump. However, dye lasers do offer the very desirable feature of tunability. They allow one to choose a laser that operates at any desired wavelength in the visible portion of the spectrum. Thus, dye lasers offer great promise for many applications involving photochemistry, pollution detection, and spectroscopy.

REFERENCES

[1]   F. Petru and Z. Veselá, *Opto-Electron.* **4**, 21 (1972).
[2]   A. D. White, *Proc. IEEE* **51**, 1669 (1963).
[3]   U. P. Oppenheim and A. D. Devir, *J. Opt. Soc. Am.* **58**, 585 (1968).
[4]   W. B. Tiffany *et al., Appl. Phys. Lett.* **15**, 91 (1969).
[5]   C. O. Brown and J. W. Davis, *Appl. Phys. Lett.* **21**, 480 (1972).
[6]   A. E. Hill, *Appl. Phys. Lett.* **12**, 324 (1968).
[7]   E. T. Gerry, *Laser Focus*, p. 27 (December 1970).
[8]   W. H. Christiansen and A. Hertzberg, *Proc. IEEE* **61**, 1060 (1973).
[9]   T. H. Maiman *et al., Phys. Rev.* **123**, 1151 (1961).
[10] M. Hercher, *Appl. Phys. Lett.* **7**, 39 (1965).
[11] G. F. Hull, J. T. Smith, and A. F. Quesada, *Appl. Opt.* **4**, 1117 (1965).
[12] M. J. Weiner, Society of Manufacturing Engineers Technical Paper MR74-973 (1974).
[13] H. Kressel, *in* "Laser Handbook" (F. T. Arecchi and E. O. Schulz-Dubois, eds.), Vol. 1. North-Holland Publ., Amsterdam, 1972.
[14] I. Melngailis, *IEEE Trans. Geosci. Electron.* **GE-10**, 7 (1972).
[15] E. D. Hinkley, *Appl. Opt.* **15**, 1653 (1976).
[16] G. A. Antcliffe and J. S. Wrobel, *Appl. Opt.* **11**, 1548 (1972).

SELECTED ADDITIONAL REFERENCES

A. Gas Lasers

R. L. Abrams and W. B. Bridges, Characteristics of Sealed-Off Waveguide CO_2 Lasers, *IEEE J. Quant. Electron.* **QE-9**, 940 (1973).
J. D. Anderson, "Gasdynamic Lasers: An Introduction," Academic Press, New York, 1976.
A. J. Beaulieu, Transversely Excited Atmospheric Pressure CO_2 Lasers, *Appl. Phys. Lett.* **16**, 504 (1970).
A. J. Beaulieu, High Peak Power Gas Lasers, *Proc. IEEE* **59**, 667 (1971).
A. L. Bloom, Gas Lasers, *Appl. Opt.* **5**, 1500 (1966).
A. L. Bloom, "Gas Lasers," Wiley, New York, 1968.
M. L. Bhaumik, High Efficiency CO Laser at Room Temperature, *Appl. Phys. Lett.* **17**, 188 (1970).
D. J. Brangaccio, Construction of a Gaseous Optical Maser Using Brewster Angle Windows, *Rev. Sci. Instrum.* **33**, 921 (1962).
W. B. Bridges, Laser Oscillation in Singly Ionized Argon in the Visible Spectrum, *Appl. Phys. Lett.* **4**, 128 (1964).
W. B. Bridges *et al.*, Ion Laser Plasma, *Proc. IEEE* **59**, 724 (1971).
C. O. Brown, High Power CO_2 Electric Discharge Mixing Laser, *Appl. Phys. Lett.* **17**, 388 (1970).
P. F. Browne and P. M. Webber, A Compact 50-J CO_2 TEA Laser with VUV Preionization and the Discharge Mechanism, *Appl. Phys. Lett.* **28**, 662 (1976).
R. E. Center, High-Pressure Electrical CO Laser, *IEEE J. Quantum Electron.* **QE-10**, 208 (1974).
P. K. Cheo, CO_2 Lasers, *in* "Lasers, A Series of Advances," Vol. 3 (A. K. Levine and A. J. DeMaria, eds.), Dekker, New York, 1971.
W. H. Christiansen and A. Hertzberg, Gasdynamic Lasers and Photon Machines, *Proc. IEEE* **61**, 1060 (1973).
A. J. DeMaria, Review of High-Power CO_2 Lasers, *Proc. IEEE* **61**, 731 (1973).
W. W. Duley, "CO_2 Lasers, Effects and Applications," Academic Press, New York, 1976.

R. Fortin, Preliminary Measurements of a Transversely Excited Atmospheric Pressure CO_2 Laser, *Can. J. Phys.* **49**, 257 (1971).

C. G. B. Garrett, "Gas Lasers," McGraw-Hill, New York, 1967.

E. T. Gerry, The Gas Dynamic Laser, *Laser Focus*, p. 27 (December 1970).

J. P. Goldsborough, Stable Long Life CW Excitation of Helium–Cadmium Lasers by DC Cataphoresis, *Appl. Phys. Lett.* **15**, 159 (1969).

J. W. Goodman, Some Fundamental Properties of Speckle, *J. Opt. Soc. Am.* **66**, 1145 (1976).

F. E. Goodwin, T. A. Nussmeier, and F. C. Trimble, One-Year Operation of Sealed-Off CO_2 Laser, *IEEE J. Quantum Electron.* **QE-6**, 756 (1970).

E. I. Gordon and A. D. White, Similarity Laws for the Effects of Pressure and Discharge Diameter on Gain of He–Ne Lasers, *Appl. Phys. Lett.* **3**, 199 (1963).

O. S. Heavens, Some Recent Developments in Gas Lasers, *Contemp. Phys.* **17**, 529 (1976).

K. G. Herngvist, Low-Radiation-Noise He–Ne Laser, *RCA Rev.* **30**, 429 (1969).

A. E. Hill, Multijoule Pulses from CO_2 Lasers, *Appl. Phys. Lett.* **12**, 324 (1968).

A. E. Hill, Role of Thermal Effects and Fast Flow Power Scaling Techniques in CO_2–He–N_2 Lasers, *Appl. Phys. Lett.* **16**, 423 (1970).

A. E. Hill, Continuous Uniform Excitation of Medium Pressure CO_2 Laser Plasmas by Means of Controlled Avalanche Ionization, *Appl. Phys. Lett.* **22**, 670 (1973).

E. Hoag *et al.*, Performance Characteristics of a 10 kW Industrial CO_2 Laser System, *Appl. Opt.* **13**, 1959 (1974).

D. T. Hodges, Helium–Cadmium Laser Parameters, *Appl. Phys. Lett.* **17**, 11 (1970).

A. Javan, W. R. Bennett, and D. R. Herriott, Population Inversion and Continuous Optical Maser Oscillation in a Gas Discharge Containing a He–Ne Mixture, *Phys. Rev. Lett.* **6**, 106 (1961).

O. P. Judd, An Efficient Electrical CO_2 Laser Using Preionization by Ultraviolet Radiation, *Appl. Phys. Lett.* **22**, 86 (1973).

C. H. Knowles, Experimenters' Laser, *Popular Electronics*, p. 27 (December 1969).

E. F. Labuda, E. I. Gordon and R. C. Miller, Continuous Duty Argon Ion Lasers, *IEEE J. Quantum Electron.* **QE-1**, 273 (1965).

H. M. Lambertson and P. R. Pearson, Improved Excitation Techniques for Atmospheric Pressure CO_2 Lasers, *Electron. Lett.* **7**, 141 (1971).

R. V. Langmuir, Scattering of Laser Light, *Appl. Phys. Lett.* **2**, 29 (1963).

D. A. Leonard, Design and Use of an Ultraviolet Laser, *Laser Focus*, p. 26 (February 1967).

D. Malacara, L. R. Berriel, and I. Rizo, Construction of Helium–Neon Lasers Operating at 6328 Å, *Am. J. Phys.* **37**, 276 (1969).

R. L. McKenzie, Laser Power at 5 μm from the Supersonic Expansion of Carbon Monoxide, *Appl. Phys. Lett.* **17**, 462 (1970).

W. H. McMahan, High-Power, Visible-Output Gasdynamic Lasers, *Opt. Spectra*, p. 30 (December 1971).

K. D. Mielenz *et al.*, Reproducibility of Helium–Neon Laser Wavelengths at 633 nm, *Appl. Opt.* **7**, 289 (1968).

R. M. Osgood, W. C. Eppers, and E. R. Nichols, An Investigation of High Power CO Laser, *IEEE J. Quantum Electron.* **QE-6**, 145 (1970).

C. K. N. Patel, Continuous-Wave Laser Action on Vibrational–Rotational Transitions of CO_2, *Phys. Rev.* **136**, A1187 (1964).

C. K. N. Patel, P. K. Tien, and J. H. McFee, CW High Power CO_2–N_2–He Laser, *Appl. Phys. Lett.* **7**, 290 (1965).

C. K. N. Patel, Gas Lasers, *in* "Lasers, A Series of Advances," Vol. 2 (A. K. Levine, ed.), Dekker, New York, 1968.

C. K. N. Patel, High Power Carbon Dioxide Lasers, *Sci. Am.*, p. 23 (August 1968).

W. N. Peters and E. K. Stein, Helium Permeation Compensation Techniques for Long Life Gas Lasers, *J. Phys. E. (Sci. Instrum.)* **3**, 719 (1970).

R. J. Pressley (ed.), "Handbook of Lasers," The Chemical Rubber Co., Cleveland, Ohio, 1971, Chapters 6–8.

J. P. Reilly, Pulser/Sustainer Electric-Discharge Laser, *J. Appl. Phys.* **43**, 3411 (1972).

C. K. Rhodes and A. Szoke, Gaseous Lasers: Atomic, Molecular and Ionic, *in* "Laser Handbook," Vol. 1 (F. T. Arecchi and E. O. Schulz-Dubois, eds.), North-Holland Publ., Amsterdam, 1972.

J. D. Rigden and E. I. Gordon, The Granularity of Scattered Optical Maser Light, *Proc. IEEE* **50**, 2367 (1962).

J. D. Rigden, A Metallic Plasma Tube for Ion Lasers, *Proc. IEEE* **53**, 221 (1965).

W. H. Seelig and K. V. Banse, Argon Laser Emits 150 Watts CW, *Laser Focus*, p. 33 (August 1970).

W. T. Silfvast and L. H. Szeto, Simplified Low-Noise He–Cd Laser with Segmented Bore, *Appl. Phys. Lett.* **19**, 445 (1971).

D. C. Sinclair and W. E. Bell, "Gas Laser Technology," Holt, New York, 1969.

P. W. Smith, A Waveguide Gas Laser, *Appl. Phys. Lett.* **19**, 132 (1971).

C. L. Stong, How a Persevering Amateur Can Build a Gas Laser in the Home, *Sci. Am.*, p. 227 (September 1964).

C. L. Stong, More About the Homemade Laser, *Sci. Am.*, p. 106 (December 1965).

C. L. Stong, How to Construct an Argon Gas Laser with Outputs at Several Wavelengths, *Sci. Am.*, p. 118 (February 1969).

C. L. Stong, A Carbon Dioxide Laser is Constructed by a High School Student in California, *Sci. Am.*, p. 218 (1971).

C. L. Stong, An Unusual Kind of Gas Laser That Puts Out Pulses in the Ultraviolet, *Sci. Am.*, p. 122 (June 1974).

W. B. Tiffany, R. Targ, and J. D. Foster, Kilowatt CO_2 Gas-Transport Laser, *Appl. Phys. Lett.* **15**, 91 (1969).

C. J. Ultee and P. A. Bonczyk, Performance and Characteristics of a Chemical CO Laser, *IEEE J. Quantum Electron.* **QE-10**, 105 (1974).

C. B. Wheeler, Power Supplies for Continuous Gas Lasers and Similar Discharges, *J. Phys. E (Sci. Instrum.)* **4**, 159 (1971).

A. D. White and J. D. Rigden, Continuous Gas Maser Operation in the Visible, *Proc. IEEE* **50**, 1697 (1962).

A. D. White, Increased Power Output of the 6328 Å Gas Maser, *Proc. IEEE* **51**, 1669 (1963).

O. R. Wood, High-Pressure Pulsed Molecular Lasers, *Proc. IEEE* **62**, 355 (1974).

S. Yatsiv *et al.*, Pulsed CO_2 Gas-Dynamic Laser, *Appl. Phys. Lett.* **19**, 65 (1971).

B. Solid State Lasers

R. C. Benson and M. R. Mirarchi, The Spinning Reflector Technique for Ruby Laser Pulse Control, *IEEE Trans. Military Electron.*, p. 13 (January 1974).

R. B. Chesler, M. A. Karr, and J. E. Geusic, An Experimental and Theoretical Study of High Repetition Rate Q-switched Nd:YAG Lasers, *Proc. IEEE* **58**, 1899 (1970).

R. J. Collins *et al.*, Coherence, Narrowing, Directionality, and Relaxation Oscillations in the Light Emission from Ruby, *Phys. Rev. Lett.* **5**, 303 (1960).

M. A. Duguay, J. W. Hansen, and S. L. Shapiro, Study of the Nd:Glass Laser Radiation, *IEEE J. Quantum Electron.* **QE-6**, 725 (1970).

V. Evtuhov and J. K. Neeland, Pulsed Ruby Lasers, *in* "Lasers, A Series of Advances," Vol. 1 (A. K. Levine, ed.), Dekker, New York, 1966.

D. Findlay and D. W. Goodwin, The Neodymium in YAG Laser, *in* "Advances in Quantum Electronics," Vol. 1 (D. W. Goodwin, ed.), Academic Press, New York, 1970.

J. E. Geusic, H. M. Marcos, and L. G. VanUitert, Laser Oscillations in Nd-doped Yttrium Aluminum, Yttrium Gallium, and Gadolinium Garnets, *Appl. Phys. Lett.* **4**, 182 (1964).

T.-L. Hsu, Nd:YAG Laser Bibliography, *Appl. Opt.* **11**, 1287 (1972).

L. F. Johnson, Optical Maser Characteristics of Rare-Earth Ions in Crystals, *J. Appl. Phys.* **34**, 897 (1963).

L. F. Johnson, Optically Pumped Pulsed Crystal Lasers Other Than Ruby, *in* "Lasers, A Series of Advances," Vol. 1 (A. K. Levine, ed.), Dekker, New York, 1966.

W. Koechner, Multihundred Watt Nd:YAG Continuous Laser, *Rev. Sci. Instrum.* **41**, 1699 (1970).

W. Koechner, "Solid-State Laser Engineering," Springer-Verlag, Berlin and New York, 1976.

T. H. Maiman, Optical Maser Action in Ruby, *Brit. Commun. Electron.* **7**, 674 (1960).

T. H. Maiman *et al.*, Stimulated Optical Emission in Fluorescent Solids, II. Spectroscopy and Stimulated Emission in Ruby, *Phys. Rev.* **123**, 1151 (1961).

P. B. Mauer, Q-switch Dyes for Neodymium Lasers, *Opt. Spectra*, p. 61 (4th Quarter, 1967).

R. J. Pressley (ed.), "Handbook of Lasers," The Chemical Rubber Co., Cleveland, Ohio, 1971, Chapters 11 and 13.

E. Snitzer and C. G. Young, Glass Lasers, *in* "Lasers, A Series of Advances," Vol. 2 (A. K. Levine, ed.), Dekker, New York, 1968.

P. P. Sorokin *et al.*, Ruby Laser Q-switching Elements Using Phthalocyanine Molecules in Solution, *IBM J. Res. Develop.*, p. 182 (April, 1964).

M. L. Spaeth and W. R. Sooy, Fluorescence and Bleaching of Organic Dyes for a Passive Q-switch Laser, *J. Chem. Phys.* **48**, 2315 (1968).

R. J. Thornton *et al.*, Properties of Neodymium Laser Materials, *Appl. Opt.* **8**, 1 (1969).

C. G. Young, Glass Lasers, *Proc. IEEE* **57**, 1267 (1969).

C. G. Young, Report on Glass Lasers, *MicroWaves*, p. 69 (July 1968).

C. Semiconductor Lasers

H. C. Casey and M. B. Panish, Epitaxial Layer Cake, *Ind. Res.*, p. 57 (September 1975).

C. H. Gooch, "Gallium Arsenide Lasers," Wiley (Interscience), New York, 1969.

B. W. Hakki, Striped GaAs Lasers: Mode Size and Efficiency, *J. Appl. Phys.* **46**, 2723 (1975).

H. Kressel, Semiconductor Lasers, *in* "Lasers, A Series of Advances," Vol. 3 (A. K. Levine and A. J. DeMaria, eds.), Dekker, New York, 1971.

H. Kressel, Semiconductor Lasers: Devices, *in* "Laser Handbook," Vol. 1 (F. T. Arecchi and E. O. Schulz-Dubois, eds.), North-Holland Publ., Amsterdam, 1972.

H. Kressel *et al.*, Progress in Laser Diodes, *IEEE Spectrum*, p. 59 (May, 1973).

H. Kressel *et al.*, Light Sources, *Phys. Today*, p. 38 (May 1976).

H. F. Lockwood *et al.*, An Efficient Large Optical Cavity Injection Laser, *Appl. Phys. Lett.* **17**, 499 (1970).

M. B. Panish, Heterostructure Injection Lasers, *IEEE Trans. Microwave Theory Tech.* **MTT-23**, 20 (1975).

M. B. Panish, Heterostructure Injection Lasers, *Proc. IEEE* **64**, 1512 (1976).

R. J. Pressley (ed.), "Handbook of Lasers," The Chemical Rubber Co., Cleveland, Ohio, 1971, Chapter 12.

D. Organic Dye Lasers

M. Bass, P. F. Deutsch, and M. J. Weber, Dye Lasers, *in* "Lasers, A Series of Advances," Vol. 3 (A. K. Levine and A. J. DeMaria, eds.), Dekker, New York, 1971.

A. L. Bloom, CW Pumped Dye Lasers, *Opt. Eng.* **11**, 1 (1972).

G. Capelle and D. Phillips, Tuned Nitrogen Laser Pumped Dye Laser, *Appl. Opt.* **9**, 2742 (1970).

R. C. Cunningham, Dye Lasers Today and Tomorrow, *Electro-Opt. Systems Design*, p. 13 (December 1974).

O. G. Peterson, S. A. Tuccio, and B. B. Snavely, CW Operation of An Organic Dye Solution Laser, *Appl. Phys. Lett.* **17**, 245 (1970).

R. J. Pressley (ed.), "Handbook of Lasers," The Chemical Rubber Co., Cleveland, Ohio, 1971, Chapter 9.

B. B. Snavely, Organic Dye Lasers: Headed Toward Maturity, *Electro-Opt. Systems Design*, p. 30 (April 1973).

C. L. Stong, A Tunable Laser Using Organic Dye Is Made at Home for Less than $75, *Sci. Am.*, p. 116 (February 1970).

CHAPTER 4

STATUS OF LASER DEVELOPMENT:
PRESENT AND FUTURE

Chapter 3 discussed details of many of the practical types of lasers that are now available. This chapter will describe the status of commercial laser development, with emphasis on what can be purchased, rather than on properties that are available only in experimental devices. The chapter begins with a tabulation of presently available laser equipment, emphasizing the levels of power than can be obtained in commercial models. We then discuss prospects for future developments, emphasizing those types of lasers which seem likely to reach commercial status within a reasonable time.

A. Tabulation of Laser Properties

In this section we tabulate the values of laser output that are available from commercial systems. This tabulation is meant to give an idea of representative values of laser output that are available with reasonable ease. Continuous and pulsed lasers are tabulated separately in Tables 4-1 and 4-2. For continuous lasers, the various materials are listed, the wavelengths are given, and ranges for power are specified. The power levels that are specified are given both as a typical level and also as the highest range which is readily available commercially. Of course, all these values have been exceeded by laboratory models, but it is not the purpose of this tabulation to attempt to give the highest values ever obtained. Beam divergence angles are also specified as typical values. We note that the beam divergence angle often increases with increasing power output. The values for helium–neon, argon, krypton and helium–cadmium lasers are usually TEM_{00} mode power outputs. For the other lasers, the powers at the high

TABLE 4-1

Tabulation of Values for Commercial Lasers: Continuous Lasers

Material	Type of material	Wavelength (μm)	Power (W)		Typical beam divergence angle (mrad)
			Typical	High range	
He–Ne	Neutral gas	0.6328	0.005	0.05	0.8
Argon	Ionized gas	0.4880, 0.5145, and others	2	15	0.8
Krypton	Ionized gas	0.6471 and others	0.5	1	0.8
He–Cd	Ionized gas	0.4416 and 0.3250	0.010 (0.005 @ 0.3250 μm)	0.075 (0.015 @ 0.3250 μm)	0.8
CO_2	Molecular gas	10.6	300	6000	2
Nd:YAG	Solid state	1.06	10	1000	5
GaAs	Semiconductor	0.85–0.905	0.02	1	25 × 125
Dye	Liquid solution	Tunable 0.4100–0.7800	0.2	1	1

TABLE 4-2
Tabulation of Values for Commercial Lasers: Pulsed Lasers

Material	Type	Type of pulse	Wavelength (μm)	Pulse energy (J)		Pulse duration (sec)	Pulse rep. rate (pps)	Average power (W)	Typical beam divergence (mrad)
				Typical	High				
CO_2	Molecular gas	TEA	10.6	2.0	400	2×10^{-7}	to 300	50	5
CO_2	Molecular gas	High-voltage pulsed	10.6	0.75	0.75	5×10^{-4}	200	100	4
Nitrogen	Molecular gas	High-voltage pulsed	0.3371	0.002	0.01	10^{-8}	100	1	10
Ruby	Solid state	Normal	0.6943	5	120	5×10^{-4}	1/30	—	5
Ruby	Solid state	Q-switched	0.6943	1	15	1.5×10^{-8}	1/30	—	5
Nd:YAG	Solid state	Normal	1.06	1	100	10^{-3}	100	100	5
Nd:YAG	Solid state	Q-switched	1.06	0.1	1	1.5×10^{-8}	to 5×10^4	10	5
Nd:Glass	Solid state	Normal	1.06	10	300	10^{-3}	1/30	—	5
Nd:Glass	Solid state	Q-switched	1.06	2	50	2×10^{-8}	1/30	—	5
GaAs	Semiconductor	Current pulsed	0.85–0.905	10^{-5}	3×10^{-3}	10^{-7}	to 5000	6×10^{-3}	25×125
Dye	Liquid solution	Pumped by pulsed laser	0.3600–0.6500 (tunable)	10^{-4}	1	2×10^{-9}	100	0.01	2
Nd:YAG	Solid state	Frequency-doubled	0.53	0.15	0.15	1.5×10^{-8}	20	3	4
Xenon	Ionized gas	Current pulsed	Several lines, 0.5–0.54	10^{-4}	10^{-4}	5×10^{-7}	10	10^{-3}	5
Nd:YAG	Frequency-doubled	Q-switched	0.532	0.15	—	15×10^{-9}	20	3	4

end of the energy range are usually for multimode operation. The values of power available for TEM_{00} operation may be lower than those specified in the table.

For the pulsed lasers, the type of pulse is specified, the wavelength is given, and a range of pulse energies is given. A typical energy for which many models are available is specified and the highest level which appears to be readily commercially available is also specified. Typical values of pulse duration and pulse repetition rate are specified. The average power is listed separately. This value may be less than the product of energy and pulse repetition rate, because the energy per pulse may decrease at the higher values of pulse repetition rate. The average power is a valuable specification because it gives the tradeoff between pulse energy and pulse repetition rate that characterizes a maximum value for the particular type of laser. The average power is not specified for ruby or Nd:glass lasers, since it is not a significant parameter, because of the typically low pulse repetition rates.

Typical beam divergence angles are also specified. As with the continuous lasers, the beam divergence angle often increases with increasing power and energy. The higher levels of power are usually attained in multimode operation. Values for operation in the TEM_{00} mode may well be lower. The different types of pulsed operation for the CO_2 laser, the ruby laser, the Nd:YAG laser and the Nd:glass laser have been discussed in Chapter 3.

The table of pulsed lasers does not include the extremely high power ($\sim 10^{13}$ W) picosecond duration systems that are being constructed in a number of laboratories for studies involving laser-assisted thermonuclear fusion. Although such lasers can be important, and can be purchased, they are not easily commercially available.

For all the lasers in Tables 4-1 and 4-2, one should note that the cost usually increases with the power output, when other features are similar. The power–price tradeoff of course means that most of the units sold are near the lower end of the range in power. The price usually scales up with power at a rate less than linearly. For example, for helium–neon lasers in the range from a few milliwatts to 50 mW, the price increases about as the three-fourths power of the laser output.

B. Future Developments

In this section we discuss some of the prospects for future developments in laser technology. We begin by discussing conventional lasers, i.e., types which are already reasonably well developed and are now available. These include the helium–neon, ruby, CO_2, Nd:YAG, and gallium arsenide lasers. Such lasers have been available for a number of years, and have reached some level of maturity. For most of these lasers, the most important developments in recent years have involved reliability. Early in the history of lasers, a laser was a rather fragile, delicate device, which required considerable care and maintenance. That situation has changed. Lasers have become reliable, durable and reasonably economic to operate. One of the most outstanding examples of this has been the gallium arsenide laser, which in early models was very short-lived. The development of the heterostructures discussed in Chapter 3 has considerably improved this situation, so that gallium arsenide lasers with much longer useful operating lives are now available.

The power outputs available from several of the conventional lasers has not increased much in recent years, and for these lasers there do not appear to be great prospects for further increases. These include, in particular, the helium–neon, the argon, the ruby, and the Nd:glass lasers. For these relatively mature lasers the prospects for considerable advances in power output levels in the future do not seem very bright. These lasers appear to have reached some status of maturity, with the technology of their construction being reasonably well developed.

Two exceptions to this judgment are the CO_2 laser and the Nd:YAG laser. These lasers have shown advances in recent years, with new varieties of construction and new methods of operation, which have led to new types of devices and to higher levels of output. Some of these advances have been discussed in previous sections. Based on the rate of these increases, it appears likely that CO_2 lasers and Nd:YAG lasers will continue to develop in the future. In particular, it seems likely that higher levels of continuous power will become more easily available. As an example, it appears probable that multikilowatt CO_2 lasers will come into much more widespread use.

As examples of developing lasers which could find many applications, we describe tunable lasers with wider ranges of tunability than those now available, chemical lasers, and ultraviolet lasers. Many such lasers have been demonstrated as experimental devices. We may expect them to reach commercial status in time to influence many practical uses.

1. Tunable Lasers

We have already discussed some of the possibilities for tuning of lasers in the sections on dye lasers and semiconductor lasers in Chapter 3. These are the only commercial lasers which offer any reasonable tuning capabilities at present. However, it appears that further advances involving different tuning concepts are likely to make tunable lasers available over a broad spectral range. The current situation is that dye lasers, discussed earlier, allow continuous tuning through the entire visible spectrum. Semiconductor lasers use current tuning and magnetic field tuning. This allows some capabilities for tuning over relatively small, discontinuous segments in the infrared. A variety of devices under development offer the possibility for tuning all the way from the ultraviolet to the far infrared. Some of these devices are summarized in Table 4-3. The devices in Table 4-3 that have not been previously described are the so-called parametric oscillator, the spin–flip Raman laser, and the use of nonlinear optics to produce harmonics of continuous visible transitions. It is not appropriate in the limited space available here to give a complete description of all these phenomena, but we shall give a short description of each of these concepts.

TABLE 4-3
Tunable Lasers

Type	Total tuning range	Tuning range for one material	Comments
Dye lasers	350–800 nm	30 nm	Needs another laser
Semiconductor lasers	IR	0.1 μm (10 cm^{-1})	Discontinuous, cryogenic
Spinflip Raman laser	IR	~5 μm	Needs another laser and large magnet
Parametric oscillator	2–20 μm	Several μm	Needs pump laser
Nonlinear effects using variable wavelength driver	UV	~20 nm	Needs tunable laser pump

Nonlinear-optical-device concepts rely on the properties of a nonlinear crystal. In a noncentrosymmetric crystal, a significantly large dielectric polarization, proportional to the square of the strength of an applied laser field, can be induced. The nonlinear polarization can radiate, permitting harmonic generation or sum- and difference-frequency generation when fields at two frequencies are applied. For a substantial buildup of radiation at these new frequencies, one must satisfy a "phase matching" condition: The wave vectors of the applied fields and the generated field must have the same relationship as their frequencies. For sum-frequency generation, if the angular frequencies and wave vectors of the ith wave are ω_i and $\mathbf{k}_i$, and the subscripts 1, 2, and 3 refer to the first and second incident wave and the output wave, then one must have

$$\omega_1 + \omega_2 = \omega_3 \quad \text{and} \quad \mathbf{k}_1 + \mathbf{k}_2 = \mathbf{k}_3$$

The usual way to satisfy this condition is to select the propagation direction and crystal temperature so that crystal birefringence offsets the effects of dispersion.

If we have a tunable laser in one region of the spectrum, we can get tunable outputs at shorter or longer wavelengths with sum-frequency and difference-frequency generation. Dye lasers are often the original tunable source.

Sum-frequency generation has been accomplished by mixing ruby laser and dye laser outputs. Between 100 and 200 kW were generated in a 7 nsec pulse. This technique, with ammonium dihydrogen phosphate (ADP) as the nonlinear material, can provide an output tunable down to about 235 nm.

Second harmonic generation implies that

$$\omega_1 = \omega_2 = \omega_3/2$$

Second harmonic generation using a dye laser with an ADP crystal has given megawatt pulses with about 20 mJ of energy tunable from 280 to 290 nm.

The optical parametric oscillator closely resembles microwave parametric oscillators and amplifiers, and its operation is also closely related to difference-frequency generation. The basic device is a nonlinear crystal placed between two wavelength-selective mirrors to form an optical cavity. A laser field at a frequency ω_p (the pump frequency) is applied to the crystal, usually through one of the end mirrors. Initially the pump radiation mixes with photon noise in the crystal, leading to a buildup of radiation at two frequencies, ω_s (the signal frequency) and at a so-called idler frequency, $\omega_i = \omega_p - \omega_s$, which are mutually phase-matched for difference-frequency generation with the pump. If losses in the cavity are less than the gain of the buildup process, oscillation occurs. To change the phase-matched wavelengths, thereby tuning the oscillator, the indices of refraction of the crystal are varied with temperature, crystal rotation or electric field.

Parametric oscillators operate continuously, as well as pulsed, and thresholds for continuous operation can be as low at 2.8 mW. Continuous wave devices, however, are difficult to operate, need careful geometric design and tend to be unstable.

A lithium niobate parametric oscillator, pumped with a ruby laser, has yielded peak power up to 340 kW and has had up to 45% conversion efficiency of the pump radiation into signal radiation [1]. The average power can be as high as 350 mW near 2.1 μm, with average power-conversion efficiency of 70% when a lithium niobate crystal inside a repetitively Q-switched Nd:YAG laser cavity is used. No single optical parametric oscillator has been tuned over a very wide region because of difficulties with mirror coatings and materials.

The spin–flip Raman laser is a decive that used a fixed-frequency laser (at present, a CO or CO_2 gas laser) to pump a semiconductor crystal in a magnetic field. The pump-laser photons lose energy when they collide with an electron in the crystal and flip its spin from one orientation in the magnetic field to a different orientation. The downshifted Raman photon is separated in energy from the pump photon by the change in electron spin energy, which is proportional to the external magnetic field. Consequently, the Raman photon frequency depends on magnetic field. At sufficiently high pump power, stimulated emission of Raman photons can exceed losses, and exponential gain and oscillation occur.

In the most widely studied spin–flip Raman laser, n-type indium antimonide is the semiconductor crystal. This laser has been operated both pulsed (CO_2 or CO laser as pump) and continuously (CO pump).

Tuning over the range from 9.0 to 14.6 μm has been demonstrated in a device using InSb pumped by 10.6 μm radiation from a CO_2 laser, using magnetic fields up to 100 kG. Mercury cadmium telluride is another promising material for use in spin–flip Raman lasers. It offers the possibility of broad tuning ranges with smaller values of magnetic field.

The availability of lasers tunable over a broad spectral range would offer many advantages. It would allow the user to pick any wavelength desired over a very broad range. This could have great utility in such applications as photochemistry, or in spectroscopy. The source would be highly monochromatic.

2. Chemical Lasers

Chemical lasers represent another laser type that appears likely to reach practical status in the near future. Chemical lasers offer the desirable characteristic of possible operation without an electrical input. One allows chemical reactants to flow together and react. The population inversion is produced by the excitation energy produced in the chemical reaction. In principle, it should be possible to operate a chemical laser without any external source of electrical power. All the required energy would be produced in the chemical reaction. One of the leading examples of chemical lasers can be summarized by the set of reactions

$$F + H_2 \longrightarrow HF^* + H$$
$$H + F_2 \longrightarrow HF^* + F$$
$$HF^* \longrightarrow HF + h\nu$$

In the first reaction, a free fluorine atom is required to initiate the reaction. One of the persistent problems in chemical lasers has been the necessity of such free atoms, and methods for producing them efficiently. The excited HF molecule (denoted HF*) produced in this reaction is in an excited state which can be the upper level of the laser transition. The third reaction above indicates the transition to the lower laser state, which is not populated by the chemical reaction. This is accompanied by emission

TABLE 4-4
Chemical Lasers

Chemical reactants	Reactions	Laser molecule	Wavelengths (μm)	Reported output	Chemical efficiency (%)
H_2-F_2	$F + H_2 \longrightarrow HF^* + H$ $H + F_2 \longrightarrow HF^* + F$	HF	2.6–3.6	4500 W cw 2300 J/pulse	10
D_2-F_2	Similar to HF system	DF	3.6–5.0	—	—
H_2-Cl_2	Similar to HF system	HCl	3.5–4.1	—	—
CS_2-O_2	$O + CS_2 \longrightarrow CS + SO$ $SO + O_2 \longrightarrow SO_2 + O$ $O + CS \longrightarrow CO^* + S$ $S + O_2 \longrightarrow SO + O$	CO	4.9–5.7	25 W cw	2.5
$D_2-F_2-CO_2$	$F + D_2 \longrightarrow DF^* + D$ $D + F_2 \longrightarrow DF^* + F$ $DF^* + CO_2 \longrightarrow DF + CO_2^*$	CO_2	10.6	560 W cw	5
C_3F_7I	$C_3F_7I \xrightarrow{photolysis} C_3F_7 + I^*$	I	1.32	65 J/pulse 1.2 GW peak	—

of light energy, hv. Thus, the population inversion is produced automatically whenever the chemicals react and yield some excited state molecules as the end product. Some electrical energy may be required for initiation, that is, for production of the original free atoms, but one the reaction has begun, further free atoms are produced and these reactions may continue cyclically.

Table 4-4 shows some chemical laser systems that have been investigated. We note that the CO_2 laser may be operated as a chemical laser. The chemical efficiency is the percentage of the total available chemical reaction energy that emerges as laser light.

Table 4-4 also includes an example of an iodine laser which operates somewhat differently. The compound C_3F_7I is broken up by photolysis by ultraviolet light. This yields excited iodine atoms, which are in an excited state suitable for laser action. The iodine laser, a pulsed system, offers some attractive possibilities for high power outputs in the near infrared.

The most highly developed of the chemical lasers are the hydrogen fluoride laser, operating at a variety of wavelengths around 3 μm and the carbon monoxide laser, operating at a variety of wavelengths around 5 μm. Both these lasers have undergone substantial development. It appears likely that high power systems may reach commercial availability in the future. Such lasers would offer an alternative wavelength for many applications. Operation at 2.8 μm, for example, could enhance metalworking, as compared to use of the CO_2 laser, because the reflectivity of metallic surfaces is not so high there as at 10.6 μm. Continuous chemical lasers with power outputs in the kilowatt range have been demonstrated in the laboratory, and it appears likely that their power capabilities will increase in the future. Such lasers can, in principle, operate without any electrical input simply by flowing the chemicals in. As such, they could provide a high-power laser for remote installations without the necessity of providing electrical power.

3. Ultraviolet and x-Ray Lasers

One of the important thrusts of laser development has been toward the ultraviolet and x-ray region of the spectrum. The current availability of ultraviolet lasers is relatively limited, with only a small number of continuous lines available from argon, krypton, and helium–cadmium, and the pulsed line at 0.3371 μm available from nitrogen. Recent developments have indicated the possibility of high power ultraviolet lasers. Such lasers operate using the molecular excimers of gases, such as argon, krypton, and xenon. These noble gases exist stably only as monatomic molecules. However, some excited states or Ar_2, Kr_2, and Xe_2 may be bound states. These are called excimers. The definition of an excimer is a molecule which is bound in an excited state, but which is not bound in its ground state. High pressure rare gas excimers exhibit a molecular emission in the vacuum ultra-violet region of the spectrum, and offer high conversion efficiency from electron kinetic energy to light output. The excitation is produced by the interaction of fast electrons. A set of reactions suitable for excimer production for a Xe_2 laser is given schematically in Fig. 4-1. The upper excited state for laser operation, Xe_2^* is produced in a complicated chain of collisions between ions (Xe^+), molecule ions (Xe_2^+), atoms (Xe), molecular excimers (Xe_2^*), and free electrons (e). The final emission of laser radiation,

$$Xe_2^* \longrightarrow 2Xe + h\nu$$

produces free xenon atoms which are again available for further interaction. At each stage, collisions and absorption of radiation may send the reaction backwards along the chain, as indicated by the leftward arrows. Possible losses by diffusion are also indicated.

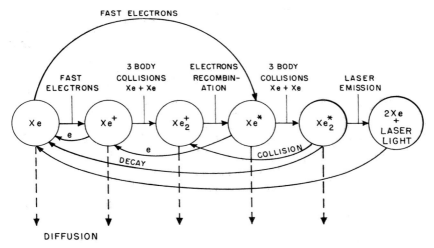

Fig. 4-1 Schematic diagram of processes occurring in the excitation of a xenon excimer laser.

In practice, an excimer laser consists of a high-pressure cell of the gas (several hundred psi pressure). Dielectric mirrors are mounted within the gas cell. The excitation is achieved by a pulsed beam of high-energy electrons, which are deposited in the gas. Some experiments have used a 70 kA pulse of 1 MeV electrons.

Table 4-5 lists several excimer species from which laser operation has been achieved, and the wavelength of the peak power output. Some excimers produce radiation in the visible, but perhaps the greatest interest involves

TABLE 4-5
Noble Gas Excimer Lasers

Excimer	Peak wavelength (μm)	Typical output characteristics	
Ar$_2$*	0.1261	Peak power	10^5 W
Kr$_2$*	0.1457	Pulse energy	10 mJ
Xe$_2$*	0.1722	Pulse duration	10 nsec
XeO*	0.5578	Efficiency	1%
XeF*	0.3540	Linewidth	0.8 nm
XeBr*	0.2818		
KrF*	0.2484		

vacuum ultraviolet operation. Typical values for output parameters of noble gas excimer lasers are also listed in Table 4-5. These lasers are all experimental devices so far, but they offer the possibility for high-power systems operating in the vacuum ultraviolet.

The desirability of x-ray lasers has been known for many years. Many applications would flow from the availability of a coherent source of x radiation. Many proposals for x-ray lasers have been put forward and many laboratories are engaged in attempts to produce x-ray lasers. To date, no one has been successful. However, the intensity of the pursuit makes it likely that x-ray lasers will be developed in the future. One persistent problem is, of course, the lack of mirrors to form a resonant cavity for x-ray lasers. One of the active media suggested for such lasers is the plasma produced in the interaction of high-power laser pulses with surfaces. Such plasmas exhibit intense x radiation. There is reason to expect that a population inversion may be available for some of the high-lying states of multiply ionized atoms in the plasma.

REFERENCES

[1]   E. O. Ammann et al., Appl. Phys. Lett. 16, 309 (1970).

SELECTED ADDITIONAL REFERENCES

A.   Tabulation of Laser Properties

F. T. Arecchi and E. O. Schulz-DuBois, "Laser Handbook," Vol. 1, North-Holland Publ., Amsterdam, 1972.
A. J. DeMaria, Review of CW High-Power CO_2 Lasers, Proc. IEEE 61, 731 (1973).
W. W. Duley, "CO_2 Lasers, Effects and Applications," Academic Press, New York, 1976.
T. L. Hsu, A Nd:YAG Laser Bibliography, Appl. Opt. 11, 1287 (1972).
"Laser Focus 1977 Buyers' Guide," Advanced Technology Publ., Newton, Massachusetts, 1977.

A. K. Levine and A. J. De Maria (eds.), "Lasers, A Series of Advances," Vol. 3, Dekker, New York, 1971.

M. B. Panish, Heterostructure Injection Lasers, *Proc. IEEE* **64**, 1512 (1976).

R. J. Pressley (ed.), "Handbook of Lasers," Chemical Rubber Co., Cleveland, Ohio, 1971.

I. J. Spalding, Lasers—Their Applications and Operational Requirements, *Optics and Laser Technology*, p. 263 (December 1974).

O. R. Wood, High-Pressure Pulsed Molecular Lasers, *Proc. IEEE* **62**, 355 (1974).

B. Future Developments

R. L. Aggarwal *et al.*, High-Intensity Tunable InSb Spin-Flip Raman Laser, *Appl. Phys. Lett.* **18**, 383 (1971).

E. R. Ault, R. S. Bradford, and M. L. Bhaumik, High-Power Xenon Fluoride Laser, *Appl. Phys. Lett.* **27**, 413 (1975).

M. L. Bhaumik, R. S. Bradford, and E. R. Ault, High Efficiency KrF Excimer Laser, *Appl. Phys. Lett.* **28**, 23 (1976).

M. L. Bhaumik, High Efficiency UV Lasers, *Laser Focus*, p. 54 (February 1976).

N. Bloembergen *et al.*, The Future of Lasers, *Phys. Today*, p. 23 (March 1972).

R. Burnham, F. X. Powell, and M. Djeu, Efficient Electric Discharge Lasers in XeF and KrF, *Appl. Phys. Lett.* **29**, 30 (1976).

R. L. Byer *et al.*, Second Harmonic Generation and Infrared Mixing in $AgGaSe_2$, *Appl. Phys. Lett.* **24**, 65 (1974).

A. N. Chester, Chemical Lasers, A Status Report, *Laser Focus*, p. 25 (November 1971).

A. N. Chester, Chemical Lasers: A Survey of Current Research, *Proc. IEEE* **61**, 414 (1973).

T. A. Cool, T. J. Falk, and R. R. Stephens, $DF-CO_2$ and $HF-CO_2$ Continuous-Wave Chemical Lasers, *Appl. Phys. Lett.* **15**, 318 (1969).

T. A. Cool and R. R. Stephens, Efficient Purely Chemical CW Laser Operation, *Appl. Phys. Lett.* **16**, 55 (1970).

T. A. Cool, J. A. Shirley, and R. R. Stephens, Operating Characteristics of a Transverse-Flow $DF-CO_2$ Purely Chemical Laser, *Appl. Phys. Lett.* **17**, 278 (1970).

R. C. Cunningham, The Near-Term Laser Outlook, *Electro-Opt. Systems Design*, p. 22 (December 1976).

R. A. Gerber *et al.*, Multikilojoule HF Laser Using Intense Electron Beam Initiation of H_2-F_2 Mixtures, *Appl. Phys. Lett.* **25**, 281 (1974).

W. F. Hagen, Ultraviolet Lasers, *Ind. Res.*, p. 48 (May, 1972).

R. L. Herbst, R. N. Fleming, and R. L. Byer, A 1.4–4 μm High-Energy Angle-Tuned $LiNbO_3$ Parametric Oscillator, *Appl. Phys. Lett.* **25**, 520 (1974).

J. J. Hinchen and C. M. Banas, CW HF Electric-Discharge Mixing Laser, *Appl. Phys. Lett.* **17**, 386 (1970).

T. Hirschfeld, Tunable Laser-Good and Bad, *Opt. Spectra* p. 22 (August 1974).

K. Hohla and K. L. Kompa, Gigawatt Photochemical Iodine Laser, *Appl. Phys. Lett.* **22**, 77 (1973).

S. F. Jacobs, M. Sargent, and M. O. Scully (eds.), "High Energy Lasers and Their Applications," Addison-Wesley, Reading, Massachusetts, 1974.

S. F. Jacobs *et al.* (eds.), "Laser Induced Fusion and X-Ray Laser Studies," Addison-Wesley, Reading, Massachusetts, 1976.

S. F. Jacobs *et al.* (eds.), "Laser Photochemistry, Tunable Lasers, and Other Topics," Addison-Wesley, Reading, Massachusetts, 1976.

P. W. Kruse, Observation of First Stokes, Second Stokes and Anti-Stokes Radiation from a Mercury Cadmium Telluride Spin-Flip Raman Laser, *Appl. Phys. Lett.* **28**, 90 (1976).

Y. S. Liu, W. B. Jones, and J. P. Chernoch, High-Efficiency High-Power Coherent UV Generation at 266 nm in 90 Phase Matched Deuterated KDP, *Appl. Phys. Lett.* **29**, 32 (1976).

N. Manyuk, G. W. Iseler, and A. Mooradian, High-Efficiency High-Average-Power Second-
Harmonic Generation with $CdGeAs_2$, *Appl. Phys. Lett.* **29**, 422 (1976).

H. Mirels and D. J. Spencer, Power and Efficiency of a Continuous HF Chemical Laser, *IEEE
J. Quantum Electron.* **QE-7**, 501 (1971).

C. K. N. Patel, T. Y. Chang, and V. T. Nguyen, Spin-Flip Raman Laser at Wavelengths up
to 16.8 μm, *Appl. Phys. Lett.* **28**, 603 (1976).

G. C. Pimentel, Chemical Lasers, *Sci. Am.*, p. 32 (April 1966).

S. K. Searles and G. A. Hart, Stimulated Emission at 281.8 nm from XeBr, *Appl. Phys. Lett.*
27, 243 (1975).

R. J. Seymour and F. Zernicke, Infrared Radiation Tunable from 5.5 to 18.3 μm Generated
by Mixing in $AgGaS_2$, *Appl. Phys. Lett.* **29**, 705 (1976).

D. R. Sokoloff et al., Extension of Laser Harmonic-Frequency Mixing into the 5-μ Regions,
Appl. Phys. Lett. **17**, 257 (1970).

D. J. Spencer et al., Preliminary Performance of a CW Chemical Laser, *Appl. Phys. Lett.* **16**,
235 (1970).

D. J. Spencer, H. Mirels, and T. A. Jacobs, Comparison of HF and DF Continuous Chemical
Lasers; I. Power, *Appl. Phys. Lett.* **16** 384 (1970).

J. I. Steinfeld (ed.), "Electronic Transition Lasers," MIT Press, Cambridge, Massachusetts,
1976.

R. W. Waynant and R. C. Elton, Review of Short Wavelength Laser Research, *Proc. IEEE*
64, 1059 (1976).

A. Yariv, "Quantum Electronics," Wiley, New York, 1975, Chapter 17.

CHAPTER 5

CARE AND MAINTENANCE OF LASERS

High-power lasers can be self-destructive devices. Crystals, mirrors, and other optical components exposed to laser light at high power density all show degradation, more or less rapidly. This fact is of practical economic interest to users of lasers. The laser rods, ruby and glass, used for high-power pulsed lasers are particularly expensive. Frequent replacement of laser mirrors can also be an economic hardship. The power within the optical cavity can be much higher than the power emitted from the laser, so that materials used within the laser are very much at risk.

We note that the power per unit area, rather than the absolute value of the power, is the determining factor. Thus, many pulsed lasers are now designed in a master-oscillator–power-amplifier configuration. The pulse is generated at relatively low power in the oscillator and is then amplified in amplifier stages with large cross section, so that the power per unit area never becomes too high.

Because of the economic importance of damage to lasers, we shall devote this chapter to describing the damage processes and to a discussion of some of the precautions and maintenance that should be employed in laser use.

A. Damage and Deterioration of Lasers

The generation of very high power pulses, often performed with ruby or Nd:glass lasers, can produce damage to the laser rods themselves and also to the mirrors and other optical components in the beam. The damage phenomena are particularly important for Q-switched lasers and mode-locked lasers, with short pulse durations. Lasers with longer pulse duration are less susceptible to damage. The common lower-power gas lasers, for example, helium–neon and argon, are not vulnerable to dramatic catastrophic damage of the type discussed in this section. Such lasers undergo gradual deterioration, which we shall discuss later.

1. Damage in Transparent Materials

High-power laser light can produce damage in materials that are nominally transparent to the light at low intensity. The initiation of optical damage or optical breakdown occurs at some threshold of laser power density. At values below the threshold, the light is transmitted, apparently without effect on the transparent material. When the laser power density is increased above the threshold value, breakdown occurs, and produces damage in the material. The damage can take the form of pitting, cracking and vaporization of material. It is often accompanied by a bright flash and a sharp sound. Thereafter, the optical component, whether window, laser rod, or mirror, has its performance much degraded for use as a laser component.

As an example of the catastrophic effects of optical breakdown at a nominally transparent surface, Fig. 5-1 shows a photograph of a potassium chloride surface, damaged by a high-power CO_2 laser pulse. The power per unit area was above 10^8 W/cm^2. At lower power densities, potassium chloride is transparent to CO_2 laser light and no damaging effects occur. As one gradually increases the laser power density, one reaches a threshold where there is a bright spark. Cracking and material removal such as illustrated in this figure occur.

The breakdown threshold depends on the nature of the material, impurities and imperfections present in the material, and the state of surface finish. The breakdown also shows some statistical fluctuations. The spread of the statistical fluctuations can be narrow enough that one can define a breakdown threshold.

There is a voluminous literature describing observations of laser-induced damage in optical materials, and also much material on investigation of the mechanisms of the damage. It is not the intent of this section to provide a complete review of the phenomena involved, but we shall summarize some of the pertinent conclusions.

Early work on identifying damage mechanisms was hindered by a number of problems. These included uncertainties in the structure of the laser pulse, the presence of small absorbing particles in the materials, and a phenomenon called self-focusing. In self-focusing, the electric field of the light wave causes a change in the index of refraction of the material, so that the light wave is constrained to move in a small narrow filament through the material. This phenomenon means that the laser power density can be much higher than the value expected without self-focusing. In reality, many of the early observations on laser damage threshold were probably measurements of the threshold for self-focusing.

In many other cases, the breakdown has been initiated by the presence of small absorbing inclusions. Preparation of material of high perfection, and use of finely focused laser beams, so that the probability of hitting an inclusion is small, have led to measurements of the intrinsic breakdown threshold of various materials. The intrinsic threshold is the threshold characteristic of the material itself, without self-focusing or the influence of absorbing inclusions. Moreover, it is now generally believed that the

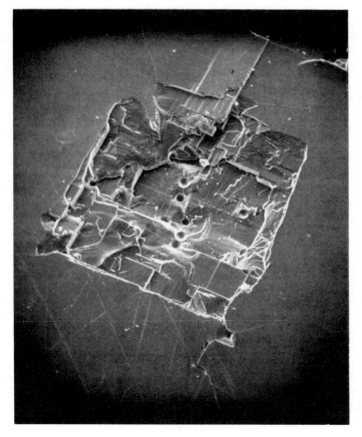

Fig. 5-1 Scanning electron microscope photograph of damage on the surface of KCl produced by a 1-μsec-duration CO_2 laser pulse delivering about 10^8 W/cm^2 to the surface. At lower power densities, the KCl is transparent to CO_2 laser radiation, and no damage occurs. The width of the area shown in this photograph is 1 mm.

optical breakdown (at least under conditions of intrinsic breakdown) is similar to dielectric breakdown caused by a dc electric field.

The mechanism for breakdown includes avalanche or cascade ionization, in which electrons are accelerated in the electric field of the light wave, gain energy, collide with other electrons, and ionize them. This process yields an absorbing plasma into which a significant fraction of the laser energy can be coupled, and eventually produces damage effects in the material.

The breakdown threshold expressed in terms of the electric field in the light wave may be related to the breakdown threshold for dc dielectric breakdown by the relation

$$E_T(\omega) = E_T(0)(1 + \omega^2\tau^2)^{1/2} \tag{5.1}$$

where $E_T(\omega)$ is the electric field needed to cause breakdown at frequency ω, and τ is an effective electron collision time, which may be approximately 10^{-15} sec. Studies have shown that the breakdown values of electric field for various alkali halides are essentially the same at dc, 10.6 μm, and 1.06μm [1].

Another observation has been that the threshold for damage often is lower at the surface of the material than in the bulk material. This has been related to the presence of small imperfections, such as scratches, pits and voids, near the surface [2]. The discontinuity in dielectric constant at a surface can enhance the electric field near the surface. The enhancement depends on the geometry of the defects which may be present in the surface. The enhancement of the electric field leads to a lower apparent breakdown threshold. Defects such as small scratches and pits can be produced by abrasive polishing. Suitably polished surfaces free of defects have shown breakdown thresholds comparable to those of the bulk material.

This discussion will serve as a brief introduction to the phenomena encountered in catastrophic breakdown of nominally transparent materials by high laser power. We now turn to questions of degradation of various components used in lasers.

2. *Laser Rods*

Ruby and Nd:glass lasers are of special interest for generation of short, subnanosecond-duration pulses. It is in this regime that damage problems become most acute. The laser rods themselves can be damaged by the high light intensity. The damage often takes the form of cracks or voids within the rod or of pits on the end of the rod. Once damaged, the output power from the laser decreases, or the laser fails to operate. There is a threshold for damage, below which effects are small. Figure 5-2 shows data on the damage threshold for a common type of glass often employed in high power Nd : glass lasers. This figure represents a glass with a relatively high threshold for damage. The data is expressed in terms of the laser light energy per unit area. The threshold for damage is the value of energy density at which damage begins. The threshold decreases as the pulse duration decreases. The points identified on the curves indicate the laser power per unit area. The peak power per unit area that can be extracted without damage increases as the pulse duration is shortened.

The data presented in Fig. 5-2 indicate that if a given energy density is required, one should at low peak powers and make the pulse length as long as possible. However, if certain peak power level is required, the pulse length should be shortened in order to reduce the total energy density.

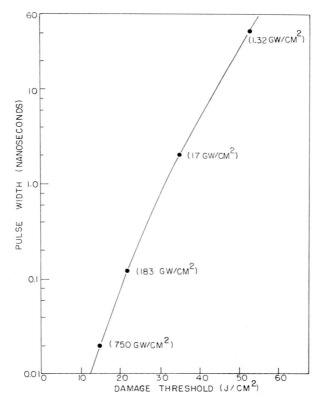

Fig. 5-2 Threshold values for optically induced damage in ED-2 glass as a function of laser pulse duration. The points noted on the curve give the thresholds in terms of power per unit area. This glass is a popular type of Nd-doped glass for high-power lasers. (Data from Owens-Illinois, Inc.)

Damage may occur both internally within the rod and at the end of the rod. Usually damage begins near the output end of the rod and progressively grows inward toward the opposite end with further pulsing of the laser. The damage threshold shows a great deal of scatter because of variation among individual specimens.

The damage commonly manifests itself as internal bubbles or voids which may be viewed by their scattering of light when a helium–neon laser beam is directed down the laser rod. The data shown in Fig. 5-2 is representative of the production of damage internal to the rod within a single shot. Once such damage has been produced, the output of that particular laser rod is considerably decreased.

It is also possible to produce damage from a number of shots, even when one remains below the threshold for damage in a single shot. Table 5-1 gives some data for the number of shots that can be expected before damage is observed in a ruby rod [3]. These numbers are appropriate to a pulse duration of 10 nsec. These data represent the number of times that laser action could be obtained from the ruby before there was a decrease of 30% of the output power, under constant excitation conditions. At levels of the order of 3×10^9 W/cm^2, damage can be produced within a single pulse. As the power density is decreased, the number of shots that can be extracted from the ruby increases considerably. Figure 5-3 shows a photograph of a ruby with internal voids, produced by about 20 shots at a power level of 200 MW [4].

TABLE 5-1
Number of Shots before Damage (30% Decrease in Output) Occurs in Ruby Rods, for 10 nsec Duration Laser Pulse[a]

Power density (W/cm^2)	3×10^9	10^9	3×10^8	10^8	3×10^7
Number of shots	1	25	500	5000	100,000

[a]From F. P. Burns, IEEE Spectrum, p. 115 (March 1967).

For Nd : glass rods, platinum inclusions within the glass are capable of initiating damage. Early neodymium-doped glass produced in platinum crucibles showed low damage thresholds. The use of platinum crucibles has largely been discontinued. Ceramic melters for laser glass are commonly used now and the threshold energy density for damage has been considerably increased, with typical values as represented in Fig. 5-2.

Catastrophic damage of the type described in Section 5.A.1 and in the present section is important for lasers with very high peak power and short pulse duration. An example is the program on laser-assisted thermonuclear fusion, where extensive efforts to reduce damaging effects are underway. These phenomena are probably less often encountered in commercial

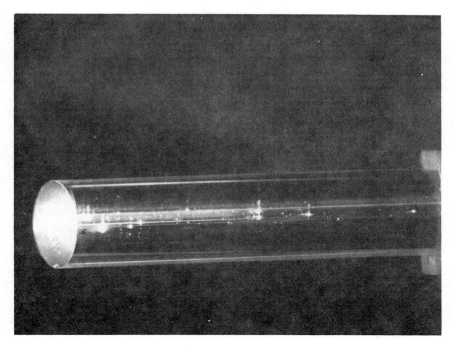

Fig. 5-3 Damage in a 5/8-in. diam ruby rod occurring after 20 shots at 200 MW. The damage is viewed by scattering of a He–Ne laser beam traversing the rod axis. (From J. F. Ready, "Effects of High Power Laser Radiation," Academic Press, New York, 1971.)

lasers, in which the manufacturers ensure that components are not exposed to laser powers above their breakdown thresholds.

3. Mirrors and Windows

Damage in mirrors is also important. For visible and near infrared lasers, thin film dielectric mirrors are used, as described in Chapter 7.

For the earliest lasers, the mirrors were usually metallic coatings, often vacuum-deposited or chemically deposited silver. These mirrors tended to degrade rapidly; for example, on a ruby laser a chemically deposited coating of silver on the ends of the ruby rod would begin to develop holes in areas where the laser power was high, within a very small number of shots. In addition, such mirrors usually had unacceptably high values of loss due to absorption and scattering.

Most modern laser mirrors are composed of multilayer dielectric coatings. Such coatings are made by successive evaporation of a number of thin layers of dielectric material. Each layer has a thickness of one quarter of a wavelength, for the particular wavelength at which the laser is to operate. Two different materials are used in alternate layers. The two materials are chosen to have alternately high and low index of refraction at the particular wavelength. A fuller description of the properties of multilayer dielectric coatings used as mirrors is given in Chapter 7.

Damage thresholds for some thin films used as quarter-wave anti-reflection coatings and for multilayer dielectric coatings with high reflectivity are tabulated in Table 5-2 [5]. The data represent films prepared by techniques used in good commercial practice. The data were obtained with a ruby laser emitting spikes about 80 to 100 nsec long with peak power in the range of 2 to 20 MW. The beam was focused on the surface with a 42 mm focal length lens.

TABLE 5-2
Damage Thresholds for Films [a,b]

Material	Damage Threshold (J/cm^2)	
MgF_2	300–360	
SiO_2	250	
ZrO_2	115–280	
TiO_2	115–280	
ZnS	23	
Al	$\lesssim 5$	For comparison
$ZrO_2 + MgF_2$	110–190	Prepared at 10^{-4} Torr
$ZrO_2 + MgF_2$	270–290	Prepared at 10^{-5} Torr
$(ZrO_2 + MgF_2)^3$	120–340	3 layers
$(ZrO_2 + MgF_2)^7$	150–270	7 layers
$(ZrO_2 + MgF_2)^{10}$	110–250	10 layers
$(TiO_2 + SiO_2)^7$	120–140	7 layers
$(ZnS + MgF_2)^{10}$	60	10 layers

[a] A. F. Turner, *in* "Damage in Laser Materials: 1971" (A. J. Glass and A. H. Guenther, eds.), NBS Special Publication 356, p. 119. U.S. Dept. of Commerce, 1971.

[b] Quarter-wave films on glass substrates at focus of 42 mm lens. Q-switched ruby laser with peak power 2–20 MW in series of spikes 80–100 nsec duration.

The data show considerable scatter between individual samples. The damage threshold measured on different samples of the same material can vary as much as a factor of two. However, there are definite trends that emerge. Some materials show higher damage threshold than others. Films made with magnesium fluoride show consistently higher damage threshold than films made with titanium dioxide. Multilayer films of two components show damage thresholds typically intermediate between the damage threshold of the constituent materials. Increasing the number of layers tends to reduce the damage threshold slightly. A value for an aluminum metallic coating is included for comparison.

In one application, metal mirrors are still used. This is for the high reflectivity mirror in a CO_2 laser. This use is feasible because the reflectivity of metals becomes very high at the CO_2 laser wavelength of 10.6 μm. Gold-coated copper or molybdenum mirrors are often used. They exhibit reasonably good damage characteristics inside high-power CO_2 lasers.

Data on the damage thresholds for CO_2 laser mirrors struck by a 600 nsec duration pulse are given in Table 5-3 [6]. These are thresholds for damage in a single shot. Metal mirrors such as Cu and Mo were found to be durable.

TABLE 5-3
Damage to CO_2 Laser Window and Mirror Materials[a]

	Material	Damage threshold (600 nsec pulse) (J/cm^2)
Mirrors	Coatings of wide bandgap materials (ThF_4)	30–65
	Coatings II–VI semiconductors (CdTe, ZnTe)	1–2
	Metals (Cu, Mo)	35
Windows	KCl	>75
	ZnSe	27–41
	CdTe	1.2–2.6

[a] A. I. Braunstein et al., in "Laser Induced Damage in Optical Materials: 1973" (A. J. Glass and A. H. Guenther, eds.), pp. 151, 157. U.S. Dept. of Commerce, 1973.

Damage to mirrors can also occur cumulatively, appearing as pits on the surface or as a damaged coating after many shots, even when no effect was observable after a single shot. Figure 5-4 shows a photograph of a coated germanium mirror for a CO_2 TEA laser, after several hundred thousand laser shots, at a power density around 10^6 W/cm^2. The laser output had decreased about 30% with the output mirror in this condition.

Some applications of high-power CO_2 lasers require high power density to be delivered to a distant target. These applications are limited by the lack of suitable transparent window materials at 10.6 μm. Distortion of the window may degrade system performance at powers well below those required to damage the window. A spatially inhomogeneous laser beam causes a temperature gradient which produces nonuniform changes in the thickness and index of refraction of the window, causing it to become a lens having aberrations, birefringence, and a finite focal length. The increased thickness at the center gives a positive focal length. The effect of the index change gives

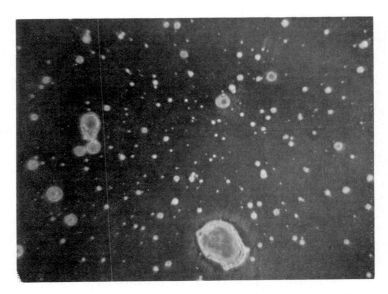

Fig. 5-4 Damage occurring on germanium mirror used as the output mirror in a CO_2 TEA laser. This damage developed after several hundred thousand shots at 10 MW output over a 1-in. area and reduced the laser power by about 1/3. The width of the area covered by this photograph is 3.8 mm.

generally a negative contribution to the focal length for ionic crystals and a positive contribution for covalent crystals. Thus, the net distortion with ionic window materials can be small because of partial cancellation of the effects due to thickness and refractive index.

Catastrophic damage can also occur in CO_2 laser window materials. Some threshold values are given in Table 5-3 for a 600 nsec duration pulse. Alkali halides, such as KCl, with very low absorption at 10.6 μm, seem to offer high resistance to damage, as well as to thermal distortion.

4. *Flash Tubes*

The flash tubes that are employed for pumping high-power lasers also have a finite lifetime. The lifetime is reduced as the input energy is increased. At some value of the energy, which depends on the diameter of the flash tube and the length of the arc, the flash tube will explode in a single shot. Figure 5-5 shows the explosion energies for flash tubes as a function of the diameter of the flash tube and the duration of the discharge of electrical energy through the flash tube [7]. This pulse length may be adjusted by varying the circuit parameters, including the capacitance and voltage of the power supply and the value of the circuit inductance. The explosion energies

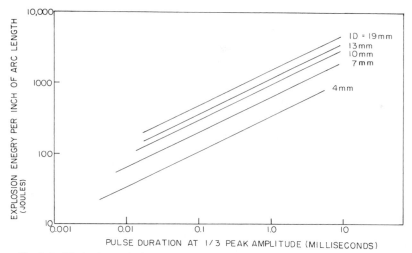

Fig. 5-5   Flashtube explosion energy as a function of pulse duration, with tube inside diameter as a parameter. The energies defined by the curves cause the flashtube to explode in a single shot. (From G. A. Hardway, *MicroWaves* **5**, 46 (1966).)

are expressed in terms of the energy input to the flash tube per unit length of the flash tube. This value will cause the flash tube to explode within a single pulse. In order to obtain reasonably long life from flash tubes, one should operate at a value below 50% of the single-shot explosion energy. When flash tubes are run at low values of input energy, perhaps 10 to 20% of the explosion energy, the lifetime may be expected to be millions of shots. However, the desire to obtain high output power from the laser usually requires larger inputs to the flash tube. Some tradeoff must generally be made between lifetime of the flash lamp and the output of the laser.

Figure 5-6 gives the results of computer calculations on the lifetime that may be expected for xenon flashlamps which pump a ruby laser rod [8]. In order to interpret this curve, one specifies the lamp life in terms of the number of shots desired. One also specifies the stored energy desired in the laser rod. This is given as the ordinate in joules of output per unit length of the rod. One should read across at the desired value of stored energy and obtain the intersection with the line giving the number of shots desired for the life of the flash tube. Then the input energy required for the flash tube and the pulse duration for the flash tube pulse can be read at the position of this intersection.

5.   *Summary*

At very high levels of power, lasers become essentially self-destructive. Some damage must be expected. There must be a suitable operating budget

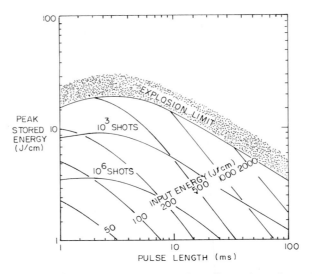

Fig. 5-6 Stored energy per unit length in a 1-cm-diam ruby rod containing 0.05% chromium by weight, pumped by a 10-mm-diam xenon lamp, as a function of pump pulse length. Lamp lifetime and input energy per unit length are given as parameters. (From J. Holzrichter, High Power Solid and Liquid Lasers, Presented at the University of Tennessee Space Institute, short course on High Power Lasers and Their Applications, Tullahoma, Tennessee, September 25–29, 1972.)

for providing replacement of damaged materials and for maintenance of the laser. Usually the most serious damage would be to a laser rod, because the costs of the other elements are generally lower than the cost of the rod. These components, for example, the mirrors, can be replaced at less expense, although if the replacement becomes frequent, this can have a serious economic effect also.

In applications requiring high peak power and short pulse duration, the lifetime of the components should be estimated beforehand in order to determine probable operating costs. The data presented here indicate some probable lifetimes for certain operating conditions. In general, the lifetime may be extended by reducing the power per unit area. Thus, the user must make his own tradeoff between operating cost and peak power density.

For lower-power lasers, lifetimes of thousands of hours are common, with eventual failure arising from factors like gas depletion or leaks, rather than from damage produced by the laser beam. Some data relative to lifetimes and failure mechanisms of a number of common laser types are summarized in Table 5-4 [9].

TABLE 5-4

Lifetimes and Failure Mechanisms for Various Laser Types[a]

Laser	Mean life (hr)	Range of life (hr)	Failure mechanisms
He–Ne	12,270	1224–44,000	Epoxy outgassing, leaks, helium loss
He–Cd	1975	800–2100	Cadmium depletion, helium entrapment
Argon	2056	1350–6000	Bore erosion, gas cleanup, outgassing, envelope integrity
CO_2 (sealed)	3041	200–11,000	Leaks, electrode degradation, helium loss, dirty mirrors
CO_2 (TEA)	>2000	—	Thyratron wearout, spark gap degradation, damage to optics
GaAs	4000	—	Distorted beam patterns, dead areas in beam

[a] Data from T. R. Gagnier, *Laser Focus*, p. 86 (October 1976).

B. Care and Maintenance

In many lasers of relatively low power, e.g., small He–Ne lasers, the tubes are sealed and there is little that the user can do to extend their life beyond what the manufacturer has already done.

The main items of interest in high-power solid lasers are the rods of ruby or glass, the mirrors, and the flash tubes. We have already discussed the mechanisms of damage in some of these elements and presented some quantitative data for damage thresholds. In this section, we discuss some practical methods for minimizing the damage and for keeping lasers operating at their optimum output.

1. Laser Rods

The main consideration is to keep the power per unit area below the threshold at which damage occurs. It must be recognized that the damage thresholds are generally specified in terms of an average power per unit area, integrated over the spatial cross section of the beam and over the temporal duration of the pulse. In many cases, instantaneous values for the power may be higher. This is particularly true in the case of mode-locked lasers. The peak value of the power may be much higher than the value averaged over the pulse. Thus, if it is not desired to have the very short pulses associated with mode-locking, one should suppress mode-locking in order to minimize damage. Damage in lasers which exhibit mode-locking does in fact occur much more easily.

Also, one should ensure that the spatial profile of the beam is relatively uniform, and does not contain hot spots where the power becomes very high. Bleachable dye Q-switches are particularly troublesome in this regard, because they tend to bleach unevenly and produce irregular beam profiles. The use of electrooptic or acoustooptic Q-switches produces beams with smoother spatial profiles.

For the very high power lasers which are of interest for thermonuclear fusion research, the general procedure has been to produce a pulse of relatively small power in a master oscillator with small cross section. The beam is then allowed to pass through several stages of amplification. Each amplifier has progressively larger cross sections, so that the power per unit area never becomes too high. A schematic diagram of one such a device is shown in Fig. 5-7, with values for the energy, power, clear aperture, and power per unit area at the exit of each stage [10]. The original pulse is produced by a mode-locked Nd:YAG laser, with a duration around 200 psec. It is then amplified by passage through rod amplifier stages, that consist of Nd:glass rods without end mirrors. Then it is further amplified by passage through three stages which contain Nd:glass discs. Each of these stages contains a number of discs which are oriented at Brewster's angle. This configuration allows efficient cooling. It also allows efficient excitation of the relatively thin discs by pump light incident on the faces of the discs. If one tried to excite a laser rod of large diameter, the pumping light would be absorbed near the edges and the excitation would not be spatially uniform across the rod. Between stages, the beam is expanded by lenses so as to fill the aperture of the next stage. In this fashion, the power per unit area in each

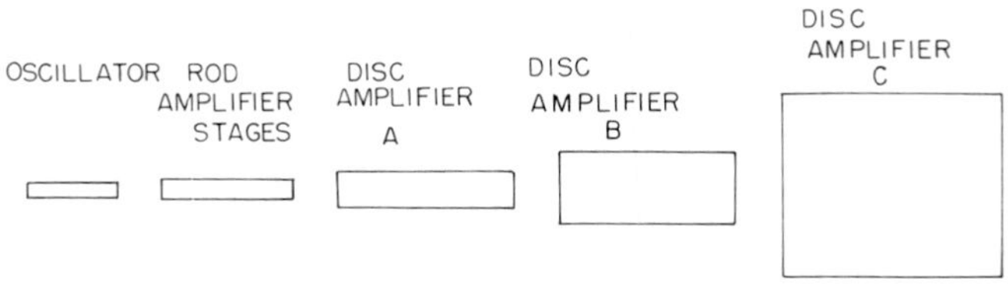

	OSCILLATOR	ROD AMPLIFIER STAGES	DISC AMPLIFIER A	DISC AMPLIFIER B	DISC AMPLIFIER C
ENERGY (J)	0.0015	1.09	5.48	50.0	274.0
POWER (W)	1.5×10^7	5.2×10^9	2.6×10^{10}	2.4×10^{11}	1.3×10^{12}
APERTURE (CM)		2.5	3.6	8.5	20.0
PEAK POWER DENSITY (W/CM^2)		3×10^9	4×10^9	6×10^9	10×10^9

Fig. 5-7 Schematic diagram for a master-oscillator power-amplifier configuration for a high-power laser.

stage is kept below the threshold for damage. The values of power per unit area given in Fig. 5-7 were calculated from the measured peak power and beam profile, and are somewhat higher than if the beam had a uniform spatial profile across the aperture. Further planned expansions of large glass lasers envision up to 20 such chains similar to that shown in Fig. 5-7. These devices would have final output energies of 10,000 J in 1000 psec.

In such devices, which are often fired at targets that are at least partially reflecting, any reflected light which returns into the laser will also be amplified as it goes backwards through the amplifier stages. Going in the reverse direction, the areas become progressively smaller, and one would quickly exceed the threshold power per unit area. In such applications, one must take great care to decouple the laser from any reflected light.

2. Mirrors

Mirrors are particularly vulnerable in high-power lasers. The multilayer dielectric mirrors which are commonly used will degrade in some hundreds or thousands of shots at high power levels. In order to obtain higher damage thresholds than are available with dielectric coated mirrors, etalons are sometimes employed. The etalon consists of a stack of plates of material with fairly high refractive index, e.g., sapphire. The plates are aligned very carefully so that the surfaces are all parallel. The reflection occurs through a cooperative interference effect. The reflectivity R that may be obtained is given by

$$R = (n^{2N} - 1)^2/(n^{2N} + 1)^2 \qquad (5.2)$$

where n is the index of refraction and N the number of plates. Table 5-5 gives the value of the reflectivity as a function of the number of plates for sapphire reflectors with an index of refraction equal to 1.78. Even a single plate gives 27% reflection, which may be suitable for operation as the output mirror in a high-power laser. As the number of plates is increased, the alignment and fabrication problems become difficult, but very high reflectivities may be obtained.

TABLE 5-5
Reflectivity of Sapphire Etalon

Number of plates	1	2	3	4	5
Reflectivity	0.27	0.67	0.89	0.96	0.99

All surfaces must be kept scrupulously clean and free of dust. Any dust particles present on any surfaces, including mirrors, ends of the rod, Q-switching elements, or plates within a resonant reflector will be vaporized and can produce a spot of damage on the surface. In some cases, immersion of the end of the rod in a liquid, e.g., water, has been used as a means of reducing damage at the ends of the laser rods. As an example of the effect of dust, a CO_2 laser mirror coating survived irradiation by 1000 W/cm^2 for 20 sec when it was clean. When it was purposely covered by dust, it damaged below 100 W/cm^2 [11].

Fingerprints, of course, are to be avoided on optical surfaces. Fingerprints degrade the optical performance and are difficult to remove.

The mirrors and other surfaces in lasers commonly have vacuum deposited coatings, either reflective coatings or antireflection coatings. These coatings must be kept clean. If absorbing particles or films are present on a coating, they will absorb energy and cause degradation of the coating. Careful cleaning of the coatings is necessary to ensure long life and good operation of the laser.

Before we discuss cleaning, it is important to point out some of the different types of coating materials that have been used. These include hard coatings, semihard coatings and soft coatings. Hard coatings include hard refractory materials, such as silicon dioxide and titanium dioxide. Such coatings are very durable, nonwater soluble, and resistive to scratching. Such coatings are commonly used as mirrors and antireflection coatings in modern lasers operating in the ultraviolet, visible, and near infrared portions of the spectrum. Semihard coatings are materials such as zinc sulfide or thorium fluoride, which are often used on CO_2 laser mirrors. Such materials are not as hard or durable as the materials listed earlier, but provide low absorption in the infrared. They may not be resistant to water and are less resistive to scratching. Soft coatings are not often used on present-day lasers, although some mirrors produced in the early history of lasers were of the soft type. Such materials can be water soluble and are not resistive to scratching.

Hard coatings may be cleaned by the following procedure: They may be washed in warm soapy water with a mild liquid detergent, if there are large amounts of contaminants on the surface. Such cleaning may not always be necessary. The final cleaning step (whether washing is necessary or not) involves the use of a solvent. Different solvents have been recommended by different workers. Satisfactory results have been obtained using acetone, ethyl alcohol, or methyl alcohol. The alcohols seem to leave smaller amounts of film, but they can dissolve oil from the fingers and deposit that oil on the surface. In addition, the use of pure ethyl alcohol involves governmental restrictions.

In order to clean the surface, one should hold a piece of lens tissue above the surface and place a few drops of solvent on the paper, using an eye dropper. Then the lens tissue should be lowered onto the surface and pulled across the surface in one smooth motion. If the surface is not clean, the procedure should be repeated, using a clean sheet of lens tissue. The lens tissue is dropped onto the surface and dragged across it so that the dry portion of the tissue removes any solvent residue.

For semihard coatings, one should not use soap and water. Instead, if there are large amounts of dirt on the surface, the surface can be flooded with acetone, which can then be blown off with a jet of dry nitrogen. Then, for the final cleaning step, a piece of lens tissue should be folded several times, and the thick folded edge squirted with a few drops of solvent. The folded edge can then be wiped across the surface with a single, continuous, smooth stroke. If the part is not clean, this step may be repeated, using a new piece of lens tissue for each wipe.

Bare metallic surfaces are sometimes used in CO_2 lasers also. Copper and gold surfaces in particular are very soft and can be easily damaged by contact. For such coatings, one should use noncontact flushing with acetone, followed by blowing the acetone off with dry nitrogen.

If older soft coatings are to be cleaned, they may be damaged by any solvent. Such mirrors have not been commonly used since the mid 1960s. However, if there is an older laser with a mirror with a soft coating which requires cleaning, perhaps the best method is simply to blow any contaminants off the surface with a jet of dry nitrogen.

The use of collodion as a cleaning material for optical surfaces has gained favor in some quarters. Collodion is deposited on the optical surfaces and the solvent is allowed to evaporate. This leaves a thin tough film on the surface which is stripped off after several minutes. It carries away all surface particles and leaves no surface film.

In one study [12], the effects of various films and surface treatments on the reflectance of CO_2 laser mirrors was investigated. A satisfactory cleaning could be obtained by flooding the surface with methanol and then blowing the liquid off with a Freon-12 duster.

Proper alignment of laser mirrors is also needed. This is essential after replacing mirrors, cleaning mirrors, or periodically during operation to keep the laser operating at maximum power.

Poor alignment can degrade the profile of the optical beam and cause reduced power output. Some solid lasers (e.g., gallium arsenide and some ruby lasers) have mirrors deposited directly on the ends of the laser medium. Such mirrors are permanently aligned. However, most lasers have mirrors separated from the active medium. This includes solid lasers (such as ruby) with separate external mirrors and all gas lasers. For these lasers, the mirrors must be aligned parallel to each other, and, in the case of the solid laser, parallel to the ends of the laser rod also. In addition, there may be other components, such as Q-switches, which are inserted between the mirrors and the laser medium. These also require alignment.

Helium–neon lasers are sometimes used to align the mirrors and other components inside a laser. Optical surfaces can be lined up parallel, two at a time, by adjusting their orientation until the multiple reflections of the helium–neon laser beam from the two surfaces coincide. Use of the helium–neon laser has the disadvantage that the beam diameter is narrow. External optics are needed to expand the beam to fill the aperture of the laser to be aligned. Also, alignment of the beam parallel to the axis of the cavity of the laser may be difficult. There are usually no built-in viewing optics in the helium–neon laser. Because of these drawbacks, a specialized device called an autocollimator is commonly used. The autocollimator is basically a telescope focused at infinity. A reticle in the telescope is illuminated by a bright incoherent source of light, such as a tungsten lamp. An image of the reticle is projected as a beam of parallel light. If there is a reflecting surface in the path of the beam, an image of the reticle will be reflected back, and can be observed in the focal plane of the telescope. The arrangement is shown in Fig. 5-8. The image will be displaced depending on the relative angle of orientation of the reflecting surface and the axis of the telescope. If the axis of the telescope is perpendicular to the reflecting surface, the image of the reticle will be observed in the center of the field of view.

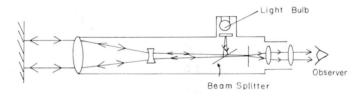

Fig. 5-8   Schematic diagram of autocollimator.

Autocollimators are commercially available. They are easy to use and are valuable tools in aligning optical components and mirrors in lasers. In order to use the autocollimator for alignment, one aligns the reflecting surfaces in the laser, two at a time. It is not strictly necessary that the auto-collimator be perpendicular to the surfaces. If the reflection from each surface is made to fall at the same position, the surfaces will be aligned parallel. Since one will usually be aligning through partially transmitting mirrors, one may well have multiple images. The user must distinguish which images are the primary reflections from the surfaces to be aligned. This may be done by blocking the surfaces one at a time. The situation is shown in Fig. 5-9. In this figure, which represents a solid laser with a ruby rod and two external mirrors, mirror 1 is lined up with one end of the ruby rod using the auto-collimator. When mirror 1 is being aligned, the undesired reflections

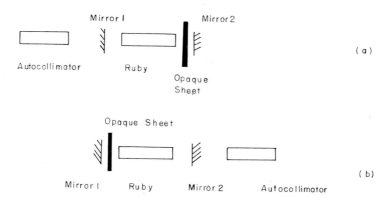

Fig. 5-9 Schematic diagram for alignment of ruby laser. (a) Left mirror being aligned with rod face. (b) Right mirror being aligned.

involving mirror 2 may be eliminated by placing a sheet of opaque material between the ruby and mirror 2. Then mirror 2 is lined up with the opposite end of the ruby rod, with mirror 1 blocked off. The ruby will have been polished with its ends accurately parallel. Thus, this procedure yields complete alignment of the system.

A complicating factor arises. There may be a number of reflected images from the various surfaces. These include reflections from both front and back surfaces of each mirror, both ends of the ruby, and also multiple reflections, where the reflection from one surface makes one or more bounces through the system before returning to the autocollimator. This situation is shown in Fig. 5-10, which shows reflections which lead to the various images. The images which arise from multiple reflections are usually fainter, and are shown by dotted lines. The example again uses a ruby laser for specificity but is applicable to other types of lasers.

An example relative to Fig. 5-10 will show how the different images may be distinguished. If one is lining up mirror 1 with the front end of the ruby rod, and the path from the ruby to mirror 2 is blocked, one may observe six images, as identified in Fig. 5-10.

(1) Reflection from the first surface of the mirror.

(2) Reflection from the second surface of the mirror.

(3) A double bounce between these two surfaces. The two surfaces of the mirror often are offset with a slight wedge angle.

(4) Reflection from the ruby rod.

(5) A double bounce between the first surface of the mirror and the ruby.

(6) A double bounce between the second surface of the mirror and the ruby.

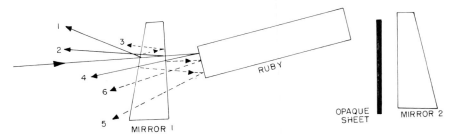

Fig. 5-10   Identification of images observed in autocollimator. The solid lines (1, 2, 4) represent images resulting from reflection at a single surface. The dashed lines (3, 5, 6) represent images resulting from reflections at more than one surface.

In Fig. 5-10, the wedge angles and the misalignment of the ruby are exaggerated in order to show the angular separation more clearly. Refraction at the mirror surfaces is neglected. In addition, there may be other images corresponding to further multiple bounces of the light between the various reflecting surfaces. However, these usually are faint and probably will not interfere. Reflections from the rear surface of the ruby will overlap those from the front surface because of the parallelism of the two ends of the ruby.

One desires to align the second surface of the mirror and the front of the ruby, i.e., reflections (2) and (4) listed above. How is one to distinguish between them? If one inserts an opaque surface between the front of the ruby and the mirror, only reflections (1), (2), (3) will be present. Since the mirror surface is chosen to be highly reflective for red ruby laser light, reflection (2) will be red in color and will probably be more intense than the other two. This allows identification of the image due to reflection (2). Then the blocking surface is removed and three additional images appear. One may distinguish them by their relative displacement as the mirror is rotated slightly. Since reflection (4) does not depend on the orientation of the mirror, it will remain stationary as the mirror is rotated. Reflections (5) and (6) depend on double bounces involving the mirror and will rotate. This allows one to distinguish the image corresponding to reflection (4). Then one simply aligns the mirror so that the images corresponding to reflections (2) and (4) overlap. Then the procedure is repeated with the other end of the rod.

The preceding procedure is given as an example. Similar procedures can be developed to align surfaces in any laser, two at a time, using an autocollimator. It requires some experience to distinguish the different images observed in the autocollimator and to determine which are the correct ones to bring into alignment. However, with some practice, this can be done accurately.

For lasers which have a number of optical components which require aligning or which operate at a low pulse repetition rate, a device such as an autocollimator is needed to provide alignment. However, if the laser operates continuously or at a high pulse repetition rate, it may be possible to eliminate these steps and obtain optimum alignment simply by optimizing the power. This may be done, for example, with continuous gas lasers with external mirrors. Many commercial lasers provide two-axis micrometer drives for each mirror in order to adjust the mirror alignment. If the laser is not operating, one runs the mirror adjustment through a number of positions until some position at which it operates is obtained. This may be done systematically by setting the x-adjustment at some position and sweeping the y-adjustment. Then the x-adjustment is set to a new value and the y-adjustment reswept. Eventually one will find a place where the laser will be in operation. Then the mirrors are adjusted one at a time to optimize the power. It is usually necessary to work back and forth between the mirror adjustments, first adjusting one axis for maximum power, then adjusting the second axis, and then returning to the first axis and readjusting. Shortly one can zero in on an optimum position at which the power output is maximized.

In some cases it is possible to obtain a local maximum in this way, even though the best position has not been obtained. This is possible if the laser medium is lined up at an angle to the axis of the mirrors. Once a local maximum power has been obtained, one should check to make sure that this off-axis condition does not hold. This can be checked by displacing one of the mirror alignments from the local maximum and then sweeping the second mirror alignment. The maximum power output obtained during this sweep should be lower than the optimum previously obtained. If a higher value is found during the sweep, one should continue displacing the first mirror and sweeping the second until an absolute maximum is located.

3. Gas Refill

Gas lasers of course do not have the gas damaged in the same way as damage occurs in high-power solid lasers. Moreover, in CO_2 lasers, the gas is flowed through the laser and exhausted. However, in He–Ne lasers and argon lasers, depletion of the gas supply usually causes the laser output to drop after some thousands of hours of operation. Many commercial argon lasers have refill reservoirs which may be activated as needed. Otherwise the remedy is to send the tube back to the manufacturer for refill. Some workers have been able to rejuvenate He–Ne lasers with low output by backdiffusing helium into the tube. Typically, keeping the laser in a helium atmosphere at 760 Torr for a few days has had some effect in restoring the output [13].

4. Semiconductor Lasers

Semiconductor laser degradation has been discussed in Chapter 3, with respect to configurations for improved lifetime. Manufacturers of semiconductor lasers have made great progress in improving their lifetime. The most important measure available to the user to insure optimum life is not to exceed the maximum current rating of the semiconductor laser.

5. Laser Power Supplies

The high-voltage power supplies often used with lasers are subject to degradation, particularly if there is a pulsed discharge. Electrode material is sputtered and the electrodes become pitted. The variety of possible power supply designs precludes all but the most general of comments. Electrode surfaces must be periodically inspected and refurbished in order to optimize operation of lasers which use high current flows. An example would be CO_2 TEA lasers, for which occasional cleaning and polishing of pitted electrode surfaces may be needed.

6. Operating Costs

All lasers have some operating costs. These include expendables, such as electricity, coolants, and gases, as well as repair and maintenance of parts which damage or degrade. The cost may be low, a few pennies per hour, for low-power He–Ne lasers. For lasers with modest material-processing capabilities, such as CO_2 or Nd:YAG lasers, operating costs may be a few dollars per hour. Very high power lasers, which can cause optical damage to their components, can be very expensive to operate.

Whatever the application, the user should estimate the costs for his systems. He should include the care and maintenance required, the possible periodic replacement of degradable items, and any expendables used. This procedure is essential in the economic analysis of any laser application in industry.

REFERENCES

[1] D. W. Fradin, E. Yablonovitch, and M. Bass, *Appl. Opt.* **12**, 700 (1973).
[2] N. Bloembergen, *Appl. Opt.* **12**, 661 (1973).
[3] F. P. Burns, *IEEE Spectrum*, p. 115 (March 1967).
[4] J. F. Ready, "Effects of High Power Laser Radiation." Academic Press, New York, 1971.
[5] A. F. Turner, *in* "Damage in Laser Materials: 1971" (A. J. Glass and A. H. Guenther, eds.), NBS Special Publication 356, p. 119. U.S. Dept. of Commerce, 1971.
[6] A. I. Braunstein *et al.*, *in* "Laser Induced Damage in Optical Materials: 1973" (A. J. Glass and A. H. Guenther, eds.), pp. 151, 157. U.S. Dept. of Commerce, 1973.

[7] G. A. Hardway, *MicroWaves* **5**, 46 (1966).

[8] J. Holzrichter, High Power Solid and Liquid Lasers, presented at University of Tennesse Space Institute, short course on High Power Lasers and Their Applications, Tullahoma, Tennessee, September 25–29, 1972.

[9] T. R. Gagnier, *Laser Focus*, p. 86 (October 1976).

[10] W. W. Simmons *et al.*, *IEEE J. Quantum Electron.* **QE-11**, 31D (1975).

[11] T. T. Saito, G. B. Charlton, and J. S. Loomis, *in* "Laser Induced Damage in Optical Materials: 1974" (A. J. Glass and A. H. Guenther, eds.), NBS Special Publication 414, p. 103. U.S. Dept. of Commerce, 1974.

[12] J. R. Buckmelter, *Laser Digest*, Air Force Weapons Laboratory, p. 106 (May 1974).

[13] K. W. Ehlers and I. G. Brown, *Rev. Sci. Instrum.* **41**, 1505 (1970).

SELECTED ADDITIONAL REFERENCES

A. Damage and Deterioration of Lasers

M. Bass and H. H. Barrett, Avalanche Breakdown and the Probabilistic Nature of Laser-Induced Damage, *IEEE J. Quantum Electron.* **QE-8**, 338 (1972).

L. G. DeShazer, B. E. Newman, and K. M. Leung, Role of Coating Defects in Laser-Induced Damage to Dielectric Thin Films, *Appl. Phys. Lett.* **23**, 607 (1973).

D. W. Fradin, Laser-Induced Damage in Solids, *Laser Focus*, p. 39 (February 1974).

D. W. Fradin and D. P. Bua, Laser-Induced·Damage in ZnSe, *Appl. Phys. Lett.* **24**, 555 (1974).

A. J. Glass and A. H. Guenther (eds.), Damage in Laser Materials: 1971, Nat. Bur. Std. Spec. Publ. 356, U.S. Dept. of Commerce (1971).

A. J. Glass and A. H. Guenther (eds.), Laser Induced Damage in Optical Materials: 1972, Nat. Bur. Std. Spec. Publ. 372, U.S. Dept. of Commerce (1972).

A. J. Glass and A. H. Guenther (eds.), Laser Induced Damage in Optical Materials: 1973, Nat. Bur. Std. Spec. Publ. 387, U.S. Dept. of Commerce (1973).

A. J. Glass and A. H. Guenther (eds.), Laser Induced Damage in Optical Materials: 1974, Nat. Bur. Std. Spec. Publ. 414, U.S. Dept. of Commerce (1974).

A. J. Glass and A. H. Guenther (eds.), Laser Induced Damage in Optical Materials: 1975, Nat. Bur. Std. Spec. Publ. 435, U.S. Dept. of Commerce (1976).

J. F. Ready, "Effects of High Power Laser Radiation," Academic Press, New York, 1971, Chapter 6.

E. Yablonovitch, Optical Dielectric Strength of Alkali–Halide Crystals Obtained by Laser-Induced Breakdown, *Appl. Phys. Lett.* **19**, 495 (1971).

B. Care and Maintenance

W. B. Alexander, Specifying Glass for Laser Applications, *Electro-Opt. Systems Design*, p. 12A (April 1975).

W. B. Alexander, Uses of Glass in Laser Applications, *Electro-Opt. Systems Design*, p. 23A (April 1975).

Cleaning and Care of Optical Components, from "The Optical Industry and Systems Directory," Vol. II, Optical Publ., Pittsfield, Massachusetts, 1976, p. E-178.

P. N. Everett, Technique for Aligning Laser Mirrors Using Gas Laser, *Rev. Sci. Instrum.* **37**, 375 (1966).

D. W. Gregg, Liquid Immersion for Reducing Damaging Effect of Laser Giant Pulses to Dielectric Mirror Coatings, *Appl. Phys. Lett.* **8**, 316 (1966).

G. A. Hardway, Why Lasers Fail-And What to Do about It, *MicroWaves*, p. 46 (April 1966).

How to Clean Optics, Booklet published by Coherent Radiation, Palo Alto, California.

Introduction to Flashlamps, Tech. Bull. 1, ILC Tech., Sunnyvale, California.

K. G. Leib and R. S. Eng, Reactivation of a Helium–Neon Laser by Diffusion, *J. Sci. Instrum.* **44**, 313 (1967).

S. Levy and R. H. Wright, Replication as a Window Cleaning Technique, *Rev. Sci. Instrum.* **42**, 1737 (1971).

C. F. Padula, Laser System Alignment; Recovering the Lost Joules, *Electro–Opt. Systems Design*, p. 24 (February 1973).

M. J. Soileau and V. Wang, Improved Damage Thresholds for Metal Mirrors, *Appl. Opt.* **13**, 1286 (1974).

W. J. Spawr and R. L. Pierce, Metal Mirror Selection Guide, *Opt. Laser Technol.* **8**, 25 (1976).

P. A. Staats and H. W. Morgan, Comment: Rejuvenation of He–Ne Lasers, *Rev. Sci. Instrum.* **42**, 1380 (1971).

CHAPTER 6

LASER SAFETY

There are some definite hazards that accompany the use of lasers. Most obvious is the possibility of damage to the eye. Skin burns can also be produced by high-power lasers. The subject of laser safety has evoked much emotional controversy over a period of years. Standards for laser safety are still in an evolutionary stage.

With careful evaluation of the hazards and a well-designed safety program, the potential dangers can be reduced to an acceptable level for scientific and industrial use. There are many other hazardous pieces of equipment which are commonly used; lasers are not unique in this respect. The use of lasers can yield desirable results at a level of risk which is consistent with other types of equipment.

In what follows, we shall first describe some of the physiological effects of laser radiation, with emphasis on the levels of irradiation at which harmful effects have been observed. We then discuss safety measures, emphasizing the approach of the American National Standards Institute. Finally, we describe the regulatory bodies that have adopted rules relative to laser safety.

We shall not describe medical applications of lasers. Although such applications, which include treatment of diabetic retinopathy, attachment of detached retinas, and removal of tattoos, are surely important, they do not fall within the realm of industrial applications.

In what follows, we shall emphasize the very real hazards of high-power lasers. We should not forget that laser light is light, and human beings have always been exposed to light. We are not dealing with a new agent whose effects are unknown and mysterious. Laser light, because of its unusual properties, can be focused to a small spot on the retina in the eye and thus could be a greater eye hazard than a nonlaser source of the same power, but

basically the hazards from lasers are not qualitatively different from those
of other high-intensity light sources.

A. Physiological Effects

The structures of the body most at risk for laser light are the retina (the
photosensitive surface at the back of the eyeball), the cornea (the front
transparent layer of the eye), and the skin. The retina can be damaged by
visible (0.4 < λ < 0.7 μm) lasers and near-infrared (0.7 < λ < 1.4 μm)
lasers. The light from ultraviolet (λ < 0.4 μm) and far-infrared (λ > 1.4 μm)
lasers does not reach the retina, but can damage the cornea. The skin can
be affected by lasers of any wavelength.

1. Effects on the Eye

Only light in the wavelength range from 0.4 to 1.4 μm can penetrate
through the anterior structures of the eye and reach the retina. Figure 6-1
[1] shows two curves: the transmission of light through the optical media in
the front of the eye and reaching the retina, and also a curve showing
the product of transmission through the ocular media and absorption in the

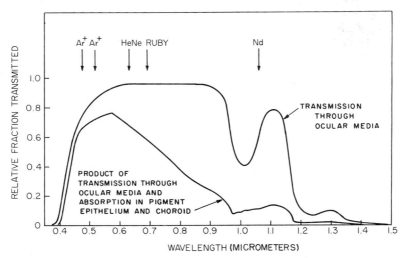

Fig. 6-1 Spectral characteristics of the human eye. The top curve gives the percentage
transmission through the anterior structures of the eye, i.e., the fraction of the incident light
which reaches the retina. The lower curve shows the product of the transmission and the absorption in various layers of the retina; it gives the fraction of the incident light absorbed in the
retina. It defines the relative risk for the retina. The wavelengths of some common lasers are
noted. (From W. J. Geeraets and E. R. Berry, *Am. J. Ophthal.* **66**, 15 (1968).)

various layers of the retina. This latter curve essentially defines the hazard to the retina as a function of wavelength. The wavelengths of some common lasers are also indicated for reference. Light of wavelengths shorter than about 0.4 μm or longer than 1.4 μm will not affect the retina. Of course, if it is intense enough, it can damage the structures in which it is absorbed, usually the cornea. Thus, a CO_2 laser is capable of damaging the cornea.

The retina is the organ which is the most at risk, because of the focusing action of the lens of the eye. A collimated beam of laser light entering the eye would be focused to a small area on the retina, so that the power density in the focal spot would be much higher than in the incident beam. Since the diameter of the beam on the retina could be as small as 20 μm, a very large increase in power density is possible. The appropriate factor for enhancement of power density is $(d_p/d_r)^2$, where d_p is the diameter of the pupil of the eye and d_r is the focal diameter of the spot on the retina. A representative case might be $d_p \approx 0.5$ cm, $d_r \approx 20 \times 10^{-4}$ cm. This would yield a peak power density on the retina about 6×10^4 times larger than the power density entering the eye. Thus, the retina could be damaged at levels where other structures of the body would not be endangered.

Figure 6-2 [2] shows some data on the laser power per unit area required to form minimal lesions (burns) on the retina, as a function of duration of exposure. These data express the experimentally determined threshold for damage in terms of power per unit area arriving at the retina. The values shown in Fig. 6-2 represent the power density incident *on the retina* required to form a minimal barely observable burn on the retina. The threshold is the value at which the observers could barely detect small burns when

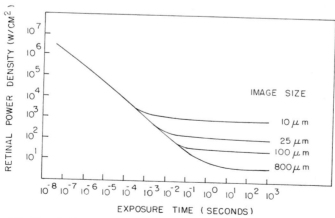

Fig. 6-2 Threshold power density needed to cause minimum opthalmoscopically observable lesions on the retina of rabbit eyes, for retinal image sizes as indicated. (From A. M. Clarke *et al., Arch. Environ. Health* **18**, 424 (1969).)

observing the retina through an ophthalmoscope. Of course, at higher values than the threshold, more serious damage occurs, including production of large burns and craters in the retina, ejection of retinal material into the vitreous medium of the eye, and hemorrhage within the eyeball.

The data in Fig. 6-2 were obtained by studies on rabbit eyes. The threshold power for an observable lesion is a strong function of both pulse duration and retinal image size.

The dependence of the threshold for retinal damage on the diameter of the spot on the retina is of interest. At very short exposure times, thermal conduction is not important. The threshold is independent of spot size. At longer exposure times, a higher power density is required to form a lesion for a small spot, because of thermal conduction. Thermal conduction can remove heat from a small area more easily than a large area.

The data presented in Fig. 6-2 do not completely define the hazard of laser radiation for the retina. Data in Fig. 6-2 were obtained mainly with light sources near the red end of the visible spectrum. As we shall see, sources near the blue end of the visible spectrum can have more damaging effects. There is very little data available on the effect of lasers with pulse duration in the picosecond regime. The extrapolation of data from rabbit eyes to human eyes is not certain. The effects of laser light on the ability of the eye to see well as part of a living organism have not been completely established. The effects of repeated subthreshold pulses are unknown. There may be effects that occur below the threshold for observable lesions. Another difficulty at longer exposures is the variation of damage threshold with image size, because it is difficult to predict the exact focal size on the retina in an accidental exposure. Nevertheless, there is enough data to allow establishment of guidelines. The results shown in Fig. 6-2 have been employed in the establishment of some of the laser safety standards that have been developed.

The threshold values given in Fig. 6-2 may easily be exceeded by commonly available lasers. Figure 6-3 shows the power per unit area that would be delivered to the retina by direct viewing of pulsed ruby lasers as a function of distance from the laser, for three different conditions of atmospheric visibility. One set of curves represents a pulse with 30 nsec duration and 1 J total energy. The second set of curves represents a pulse with 200 μsec duration and 2 J total energy. These values are representative of small, easily available ruby lasers. The threshold value for each pulse duration from Fig. 6-2 is shown for comparison. This indicates that the beam from such a laser can be dangerous even at long distances from the laser.

The data in Fig. 6-2 were accepted for a number of years as the definitive data for laser power density required to produce a minimum retinal burn, and served as the basis for a number of laser safety standards. The data in

Fig. 6-2 were obtained with light near the red end of the visible spectrum, and the burns appeared to be consistent with a thermal model, in which the observed damage is produced by heating.

Later work with lasers at wavelengths throughout the visible spectrum has indicated that photochemical activity is needed to account for results

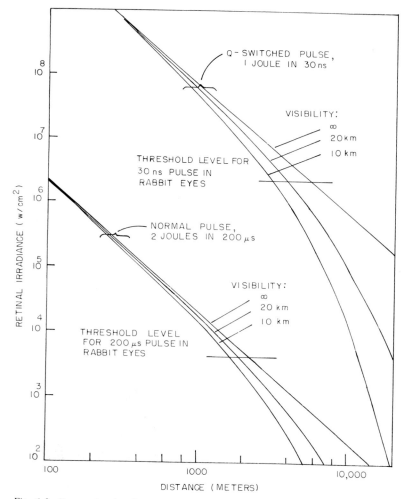

Fig. 6-3 Power density (irradiance) delivered to the retina as a function of distance from the laser. Results are shown for two different ruby laser pulses, with parameters as indicated. For each case, results are given for three different conditions of atmospheric visibility, as indicated. The threshold levels for minimum retinal lesions in rabbits are shown by the short horizontal lines, for each of the two pulse durations.

near the blue end of the visible spectrum [3]. Figure 6-4 shows results for the wavelength dependence. The figure plots the reciprocal of the threshold retinal irradiance required to produce a minimum retinal burn in an anesthetized rhesus monkey, for several exposure times. The threshold irradiance is much lower in the blue than in the red. The difference cannot be accounted

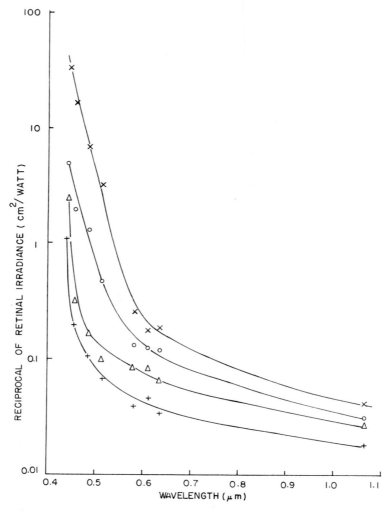

Fig. 6-4   Retinal sensitivity to threshold damage in the rhesus monkey, as a function of wavelength. The ordinate gives the reciprocal of the power density (irradiance) needed to cause a threshold lesion. Pulse durations: +, 1 sec; Δ, 16 sec; o, 100 sec; ×, 1000 sec. (From W. T. Ham, H. A. Mueller, and D. H. Sliney, *Nature* **260**, 153 (1976).)

for in a purely thermal model, and it appears that photochemical action influences the damage in the blue. In a practical sense, this means that for equal power, a blue laser will be more dangerous to the retina than a red laser. Laser safety codes are being revised to take these new findings into account.

Another area about which little is known is the possibility of cumulative effects from successive exposure. One study [4] used a GaAs laser to expose the eyes of rhesus monkeys. The total energy required to produce a threshold lesion (50% probability level) was 562 μJ when the laser was operated at 40 pps and 222 μJ at 1000 pps. This result demonstrated a cumulative effect as the pulse repetition rate was increased. There do not appear to be any data relevant to repeated exposures at low repetition rates, less than one per second.

Even if one neglects possible cumulative effects and considers only single exposures, problems of protection against the retinal hazard expressed by Figs. 6-2 and 6-4 are not simple, and there is not a complete consensus of opinion as yet. For lasers with long exposure durations (greater than about 0.1 sec), there is especially great disagreement. This is because of the possibility of at least partial protection from damage at levels near threshold through the natural aversion response to bright light. Different observers disagree on how effective this will be in protecting human eyes in transient exposures.

The laser that has provoked the most controversy has been the helium–neon laser operating in the range of a few milliwatts. This laser is commonly used in many applications, and its output is near the level suggested for maximum permissible exposure. Figure 6-5 shows data on the threshold

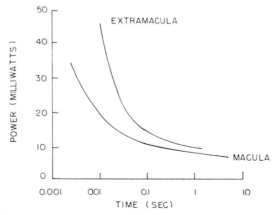

Fig. 6-5   Threshold values of He–Ne laser power needed to cause minimum retinal lesions in rhesus monkeys, as a function of exposure duration. The ordinate gives the total laser power entering the eye. (From P. W. Lappin, *Arch. Environ. Health* **20**, 177 (1970).)

level for lesions on the retina of monkeys due to exposure to continuous He–Ne lasers [5]. The macula is the small sensitive region in which vision is most acute. The minimum power which was found to cause a lesion was 7 mW. The monkeys' eyes were rigidly restrained, so it is unlikely that an accidental human exposure would produce as much effect.

For the CO_2 laser operating at 10.6 μm, radiation does not penetrate to the retina. It is absorbed at the cornea and can cause corneal burns. For minimal burns near threshold, the effect is to cause an opacity of the cornea, which heals within a few days. Some data on the threshold CO_2 laser power density required to cause corneal opacities are shown in Fig. 6-6. These data were obtained by workers using rhesus monkeys. We note in comparing Figs. 6-2 and 6-6, that the power density in Fig. 6-2 is defined at the retina, and could be higher than that at the cornea by a factor as large as 10^5. Thus, for equal total powers in the beam, a visible laser is likely to be more dangerous than an infrared laser because of the focusing action of the eye.

In this section we have defined the hazard to the eye, in terms of a number of parameters of the exposure, e.g., power density, wavelength, exposure duration. What has been defined is the threshold power density for a minimum observable burn, in a single exposure. There are some data indicating microscopic histological changes that can occur at lower levels than the burn threshold, so safety codes usually apply some safety factor. The effect of chronic or multiple exposures is almost completely unknown.

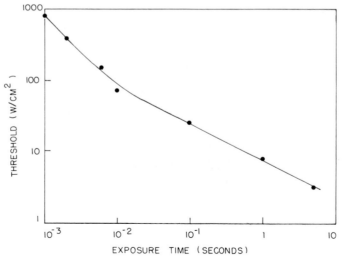

Fig. 6-6 Threshold CO_2 laser power density needed to cause minimum damage to the cornea of rhesus monkeys, as a function of exposure duration. (Data due to B. E. Stuck, Frankford Arsenal.)

The data are mostly based on animal experiments, but the small amount of data on human beings (accidental exposures and volunteers) has indicated that these threshold values are relevant for the establishment of laser safety codes.

In terms of the functioning of the eye as part of a living organism, the effect of a small burn on the retina will depend greatly on its location. A small burn on the macula, the small sensitive region in which vision is most acute, could severely reduce visual acuity. The same small burn on the periphery of the retina would have less noticeable effects. Small retinal burns show at least partial healing over a period of time, with some recovery of the accompanying loss of visual acuity.

2. Effects on the Skin

High-power lasers can also burn skin. The problem is less acute than for the eye, because one does not encounter the increase in power density associated with the focusing action of the eye, and because at the threshold level, a small burn on another exposed part of the body is likely to be of less concern than a similar burn on the retina or cornea. Still, laser skin burns are a subject of valid concern. High-power lasers now in use in industry are potentially capable of inflicting frightful burns.

The interaction of laser radiation with the skin depends on both wavelength and skin pigmentation. Figure 6-7 shows the reflectivity of the skin

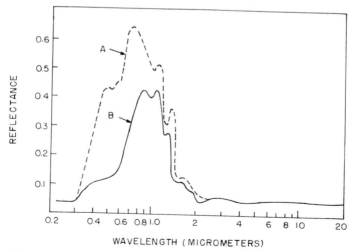

Fig. 6-7 Reflectance of human skin as a function of wavelength. Curve A is for lightly pigmented skin; curve B is for heavily pigmented skin. (From D. H. Sliney and B. C. Freasier, *Appl. Opt.* **12**, 1 (1973).)

as a function of wavelength [6]. In the visible, the skin is quite reflective, with the exact value depending on the pigmentation. Thus, much of the visible light incident on the skin will be reflected. In the far infrared, the skin becomes highly absorbing; the reflectivity is low, independent of pigmentation.

For the light which is not reflected, there is a considerable variation in transmission through human tissue. There is a maximum in transmission near 1.15 μm; nearly 20% of the incident light at this wavelength will be transmitted through the human cheek [7]. Transmission drops off as one moves away from this wavelength; at wavelengths greater than 1.4 μm and less than 0.5 μm, transmission through the cheek is virtually zero. In contrast, at 10 μm, absorption is very high, and occurs in a very thin surface layer. In water the extinction length, within which 90% of the incident light is absorbed, is about 0.03 mm at 10 μm wavelength [8].

Thus, laser skin burns are perhaps a consideration mostly for CO_2 lasers. The reason is threefold:

(1) CO_2 lasers are the most common of high-power lasers, capable of emitting at levels high enough to cause skin burns.

(2)   Reflectivity of the skin is low at 10 μm so almost all the energy is absorbed, whereas in the visible and near infrared a significant fraction can be reflected.

(3)   The absorption coefficient is very high, so the energy is deposited in a thin layer. In contrast, for visible and near infrared lasers, absorption will occur over a more extended depth, so that heating effects will be less severe.

This discussion does not imply that other lasers cannot be hazards for skin burns, but CO_2 lasers are probably the most significant current hazard. For large CO_2 lasers used in industry, even a transient passage through the beam could produce severe burns.

The threshold level at which a CO_2 laser can produce skin burns is defined in Fig. 6-8 [9]. These data are relevant to the shaved skin of anesthetized white pigs. The criterion for threshold as given in the figure was a mild reddening of the skin 24 hr after exposure.

When the data in Fig. 6-8 are compared to Fig. 6-6 for threshold levels for CO_2 laser damage to the cornea, we note that the threshold levels are similar. The cornea is not particularly more sensitive to damage by a CO_2 laser than are other portions of the skin. It is apparent that a small burn on the cornea could be more troublesome than a similar small burn elsewhere on the skin.

Near the threshold level for skin burns, thermal mechanisms appear to dominate and the damage appears to arise from thermal effects. There is

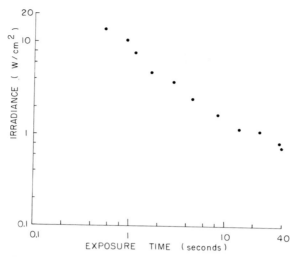

Fig. 6-8 Threshold levels for minimum burns on the skin of white pigs as a function of time, for CO_2 laser radiation. (Data from A. S. Brownell and B. E. Stuck, Frankford Arsenal, AD Report 785, 609 (1972).)

no evidence that carcinogenic effects or other unexpected effects occur. At high levels, well above threshold for minimal burns, shock effects can be excited. The rapid absorption of energy can produce a shock front which travels through the body and which can damage an organ removed from the impact point. Medical investigators using high-power ruby and Nd:glass lasers observed such effects early in the history of lasers. Adoption of a safety program which eliminates threshold skin burns will automatically eliminate damage from laser-induced shock waves.

3. *Other Hazards*

Most emphasis in laser safety has been put on the beam emitted by the laser. There are other possible hazards associated with lasers, e.g., electrical shock hazards, poisonous or corrosive substances used in the laser or in associated equipment like modulators or Q-switches, and possible noxious vapors produced in laser vaporization of some target materials. The most dangerous aspect of a laser is the electrical power supply associated with it. A number of people have indeed been electrocuted by contact with laser power supplies. Electrical hazards are, of course, not unique to the laser. Precautions similar to use of any high-voltage electrical equipment are required. A complete laser safety program must consider not only hazards arising from the beam, but also all the ancilliary hazards, some of which have just been mentioned briefly.

B.   Laser Safety Practices and Standards

The fundamental laser safety practice is not to allow laser radiation at a level higher than a maximum permissible exposure level to strike the body. Maximum permissible exposure levels have been defined in laser safety standards, as we shall describe later, for both eyes and skin, and for various wavelengths and exposure durations.

The safety measures may involve physical barriers or engineering controls. Physical barriers are materials opaque at the laser wavelength which are positioned so that the laser beam cannot strike the body. Engineering controls are defined steps which ensure that the laser will not be activated until the beam path is clear. Both types of control measures have application. Physical barriers may be most appropriate in a production environment, where the workers may be less skilled in optical practice. The entire work station can be enclosed so that there is no possibility of the beam striking a human being. Engineering controls may be more appropriate for technical measurements, where the workers tend to be more skilled and where a rigid enclosure would be incompatible with the measurement.

The most common safety measure for eye protection is to use protective eyewear of sufficient optical density to reduce the exposure to the eye to the level where there is no danger. Laser safety eyewear is specified by values of the optical density at various laser wavelengths for which it is to be used. Optical density (O.D.) is defined by

$$\text{O.D.} = -\log_{10}(I/I_0) \tag{6.1}$$

where I_0 is the incident intensity and I is the transmitted intensity. Thus, O.D. 5 means attenuation by a factor of 10^5.

Safety eyewear should be employed wherever the possibility exists for exposures at levels above the maximum permissible exposure. For industrial use, possibly the best rule is never to allow direct viewing of the beam, at least for pulsed lasers and for continuous lasers operating above some level.

For pulsed lasers and for high-power continuous lasers, the beam path should be designed so that the operator cannot get his eye in the beam, and so that there are no specular reflections of the beam outside the protected area. Protective eyewear would be required in cases where diffuse illumination could be a hazard.

There is no one material which provides effective protection for all laser wavelengths and still allows effective vision. There are laser protective glasses effective for a laser of any particular wavelength, and a number of commercial sources offer such glasses. Ordinary safety eyeglasses with side shields will attenuate CO_2 laser beams considerably, but there may be danger of the glass cracking in high-power beams.

The luminous transmission is very important. If luminous transmission is low, the user may not find the eyewear comfortable. Laser safety eyewear is available both as spectacles and as goggles. Goggles offer more complete protection, with no possibility of radiation entering the eye around the edge. Goggles do cut off side vision and are sometimes unacceptably hot or heavy. Spectacles (with side shields) are light and probably more acceptable to the wearer than are goggles. However, spectacles ground to a user's prescription are uncommon and are expensive. Goggles will fit over the wearer's ordinary prescription glasses.

It should be mentioned that exposure of laser safety eyewear to high levels of laser radiation can cause damage in the form of melting, bleaching, cracking, or shattering. Some laser safety eyewear has shown structural changes after several seconds of direct exposure to laser beams delivering power density around 6–12 W/cm^2 [10]. Changes involved both lens and frame materials. In some cases a portion of the incident beam was transmitted through holes produced in the materials. Users of protective eyewear should determine the failure point of their eyewear, either from the manufacturer or by direct test.

Although laser protective eyewear has been offered by a number of suppliers for many years, advances are still needed. Desirable developments include better luminous transmission while retaining high optical density at the laser wavelength, better compatibility with prescription eyewear, and improved comfort for user acceptability.

The preceding discussion has applied to direct exposure of the eye by laser radiation. The reflected laser beam can also be dangerous, especially because reflections sometimes occur in unexpected directions. We distinguish between specular reflection (e.g., from a mirror or shiny piece of metal) and diffuse reflection (e.g., from a piece of white paper).

The possibility of specular reflection of the beam must be considered carefully. Dangerous reflections can occur from many types of objects in the beam path–glass surfaces, wristwatches, bottles, etc. In metalworking, one often deliberately terminates the beam at a shiny target. A specular reflection is as dangerous as the direct beam and the same precautions hold as for the direct beam.

For diffuse reflection, one must consider the exposures at the retina, since for distances of the order of room dimensions, the power density at the retina is independent of distance from the spot. This occurs because the power density at the cornea decreases with the distance from the diffusely reflecting surface, but the focal area on the retina decreases with distance in the same ratio. This makes a maximum permissible exposure at the cornea much harder to specify for diffuse reflection than for direct illumination or for specular reflection. High-power lasers can easily reach levels where the reflection from a diffusely reflecting surface can be dangerous.

Skin burns are perhaps less of a problem. Most of the common lasers one encounters, e.g., helium–neon, are incapable of producing a skin burn. Large lasers used for metalworking can indeed produce fearsome burns, but such installations are relatively easily enclosed. The most appropriate measure for a laser which is a real hazard for skin burns could be to completely separate the beam from people by means of physical barriers.

Operation of high-power lasers in closed areas with restricted access is desirable. Warning lights activated when the laser is energized, suitable warning signs, audible signals when appropriate, and closed doors to the laser room with interlocks to the laser power supply are all desirable features.

A laser installation should have a safety officer responsible for evaluating and controlling potential hazards of the specific installation. Periodic eye examinations for laser personnel are to be recommended. Some institutions provide eye examinations for their laser personnel on a routine basis.

Finally, safety training for personnel who will operate lasers is much to be recommended. Comprehensive laser training programs for industrial personnel have been developed [11].

The ANSI Standard

Various guidelines and standards have been proposed for the maximum permissible levels of exposure to laser radiation. The guidelines are not meant to be fine lines between danger and safety, but are evaluations of the current status of experimental studies (mostly experiments on animals), and represent a level below which no harmful effects are known to occur. The guidelines are meant to be applied with thoughtful evaluation to specific circumstances of laser use.

The maximum permissible levels for eye exposure are specified at the cornea, rather than at the retina, because in practice it is much easier to measure the power or energy density at the cornea.

One set of standards that has particular significance is that developed by the Z-136 Committee of the American National Standards Institute (ANSI) [12]. This code is a voluntary code. It is perhaps the most influential and widely adopted of the laser safety standards. Moreover, the legally binding regulations that have been adopted or may be adopted by various agencies have often had similarities to the ANSI code and appear to have drawn from it. Thus, we shall describe the ANSI code, since it is especially significant.

There are several areas contained in the ANSI code. These concern:

(1) definition of maximum permissible exposure levels to laser radiation,

(2) classification of lasers into classes according to level of hazard, and

(3) definition of safety practices relative to each class.

Some specified levels of maximum permissible exposure (MPE) for the eye are shown in Fig. 6-9. These values for the MPE are expressed as average values over limiting apertures, which are specified for various wavelength regions and types of exposure. For ocular exposure by visible lasers, (0.4–0.7 μm), the limiting aperture is 7 mm; for other wavelengths

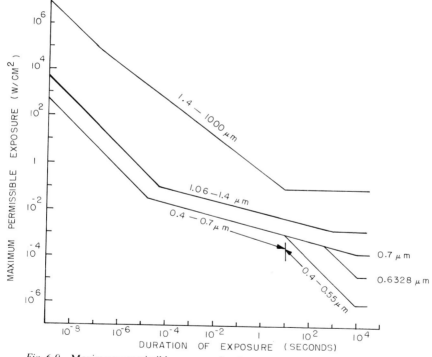

Fig. 6-9 Maximum permissible exposure for the eye as a function of exposure duration. The levels are relative to single exposures from a collimated laser beam, with wavelengths as indicated on the curves. The levels are expressed as power per unit area on the cornea, averaged over a limiting aperture. The appropriate aperture is 0.7 cm for wavelengths between 0.4 and 0.7 μm, and 0.1 cm for other wavelengths between 0.2 and 100 μm. (The figure is derived from information presented in Standard Z136.1–1976, "For the Safe Use of Lasers," American National Standards Institute, New York (1976).)

less than 100 μm, the limiting aperture is 1 mm. The curves in Fig. 6-9 are relevant to a single exposure for direct ocular exposure for a collimated beam. The curves in Fig. 6-9 cover only a fraction of the cases described in the ANSI document. Other curves would be relevant for extended sources, for example. In addition, there are several special qualifications, for which the original document should be consulted.

The curves in Fig. 6-9 do define the MPE for some of the most commonly encountered lasers. For wavelengths between 0.7 and 1.06 μm, a specified wavelength-dependent multiplicative factor can be applied to the MPE relevant to 0.7 μm. Other wavelength-dependent values of MPE are specified for ultraviolet wavelengths (<0.4 μm).

For the visible wavelength region (0.4–0.7 μm), the MPE is independent of wavelength for exposure duration less than 10 sec. For times longer than 10 sec, the MPE becomes wavelength dependent. The two limiting curves (a curve for 0.7 μm and one for 0.4–0.55 μm) are shown; curves for wavelengths between 0.55 and 0.7 μm will lie between the limiting cases. One important example, 0.6328 μm, is shown. These reduced values of MPE for shorter wavelengths are influenced by the results indicating possible photochemical mechanisms of retinal damage. These mechanisms, which are more pronounced near the blue end of the visible spectrum, were described earlier in this chapter.

The MPEs given in Fig. 6-9 strictly apply to single exposures. For repetitively pulsed lasers, the MPE is reduced by a factor which decreases as pulse rate increases. Values of the factor are 100% at 1 pps, 31.6% at 10 pps, 10% at 100 pps, and 6% at 1000 pps.

The ANSI recommendation for MPE of the skin to laser radiation is shown in Fig. 6-10, as a function of exposure duration for several wavelength regions. Values in the ultraviolet are strongly wavelength dependent in the region between 0.302 and 0.315 μm. Two curves are shown in this wavelength interval for two specific wavelengths. In addition, the total ultraviolet exposure of the cornea is restricted to less than 1 W/cm^2.

The values in Fig. 6-10 are to be interpreted as average values over a specified limiting aperture, which is 1 mm over the wavelength range from 0.2–1.00 μm. We note in comparing Figs. 6-9 and 6-10 that the MPE for the skin and for the cornea are the same for the long wavelength region (1.4–1000 μm).

A second important topic in the ANSI document is the division of lasers into four classes, according to degree of hazard.

The philosophy of the four classes appears to be as follows:

Class I   Exempt—emission of power or energy levels below the levels at which any harmful effects are known to occur.

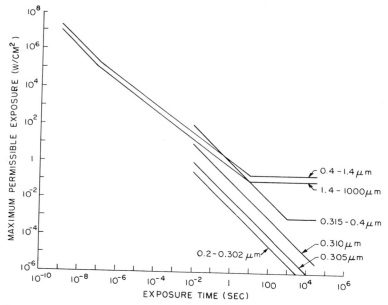

Fig. 6-10 Maximum permissible exposure for skin exposure to a laser beam, with wavelength as indicated, as a function of exposure duration. The exposures are expressed over a limiting aperture, which is 0.1 cm for wavelengths between 0.2 and 100 μm. (The figure is derived from information presented in Standard Z136.1–1976, "For the Safe Use of Lasers," American National Standards Institute, New York (1976).)

Class II   Low power visible—for *visible, continuous* lasers only.

Class III   Medium power—emission at levels where harmful effects from exposure to the direct beam could occur, but for which diffuse reflections should not be dangerous.†

Class IV   High power—emission at levels where harmful effects from diffuse reflections could occur.

The classification implies that above some level, even diffuse reflections could be hazardous. Figure 6-11 shows values of the maximum radiant exposure on a diffusely reflecting surface which will not produce hazardous reflections. Many commonly available lasers can exceed the levels shown in this figure.

The exact classification schedule is somewhat complicated, and depends on both wavelength and on the duration of the possible exposure. It is beyond

† A subclass, denoted IIIa, is also created, to include members of Class III whose output lies between 1 and 5 times the lower limit for the class, and whose power density does not exceed prescribed values. The practical significance appears to involve mostly visible continuous lasers in the 1–5 mW regime.

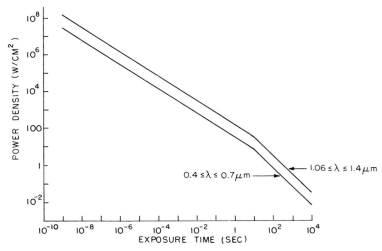

Fig. 6-11 Maximum levels of power density incident on a diffusely reflecting surface which will not produce hazardous reflections, as a function of exposure time. Results for wavelength λ between 0.7 and 1.06 μm may be inerpolated by a wavelength-dependent factor. (The figure is derived from information presented in Standard Z136.1–1976, "For the Safe Use of Lasers," American National Standards Institute, New York (1976).)

the scope of this chapter to present the full classification scheme in detail; the full ANSI document should be consulted for details of the classification system. For reference, the relevant classifications are given in Table 6-1 for a few common types of lasers.

TABLE 6-1
Examples of Classification of Common Lasers

Laser	Wavelength (μm)	Upper Limit			
		Class I	Class II[a]	Class III[b]	Class IV
He–Ne (continuous)	0.6328	6.8 μW	1 mW	0.5 W	>0.5 W
Argon (continuous)	0.5145	0.4 μW	1 mW	0.5 W	>0.5 W
CO_2 (continuous)	10.6	0.8 mW	—	0.5 W	>0.5 W
CO_2 (TEA—100 nsec pulse)	10.6	80 μJ	—	10 J/cm^2	>10 J/cm^2
Nd:YAG (continuous)	1.064	0.2 mW	—	0.5 W	>0.5 W
Nd:YAG (Q-switched— 20 nsec pulse)	1.064	2 μJ	—	0.16 J/cm^2	>0.16 J/cm^2

[a] Class II is relevant only to continuous, visible lasers.

[b] Subclass IIIa is for lasers in Class III with output between one and five times the lower limit of the Class III, provided also that the appropriate maximum permissible exposure over the limiting aperture is not exceeded.

The table gives the upper limit for Classes I, II, and III. Membership in Class I is reserved to lasers whose emission limit is below the maximum for that class. Membership in Classes II and III is interpreted as being for those lasers whose emission level is above the limit for the next lower class and below the upper limit for the given class. Class IV lasers are those which exceed the upper limit of Class III. For continuous lasers (emission duration >0.25 sec) the levels are given in units of power. For pulsed lasers, the levels are expressed in units either of energy or energy density per pulse. The limits in Table 6-1 for pulsed lasers are relative to single pulses. For repetitively pulsed lasers, the single-pulse energy and the average power must both fall below the upper limit for a given class if the laser is to be in that class.

Some specific recommendations for laser safety practices were also incorporated in the ANSI report. The report discussed four classifications of lasers, as defined above, according to the degree of hazard, and specified control measures for each class. Various control measures are described relevant to each type. The control measures are summarized briefly in Table 6-2.

This discussion is a brief summary meant to illustrate the classification of lasers and their control measures, and is not intended to be a complete description of a laser safety program. The ANSI recommendations on classification of lasers and on the control measures are quite detailed. In formulating a laser safety program, one should refer to the complete document.

TABLE 6-2
Summary of Control Measures in ANSI Standard[a]

Laser class	Name	Control measures
I	Exempt	None
II	Low power visible	Warning labels
III	Medium power	Eleven specified measures, including eye protection, use in controlled areas, beam enclosures, warning labels, and operator training
IIIa	1–5 mW cw subclass	The eleven measures for Class III are advisory, not mandatory
IV	High power	Seventeen specified measures, including the eleven for Class III, plus six more stringent measures on eye protection, area control, and warning systems, enclosed beam path, remote firing, and monitoring and key-switch interlocks

[a] Simplified version. See complete ANSI document for full explanation.

C. Regulatory Bodies

The ANSI laser safety standard is voluntary. There are a number of organizations which enforce nonvoluntary laser safety standards and practices. The individual employer comes immediately to mind.

Many employers enforce laser safety standards; in many cases the standard is the ANSI standard. In this section, we shall briefly describe the laser safety codes deriving from three sources:

(1) The Bureau of Radiological Health (BRH), which is part of the Food and Drug Administration (FDA), which in turn is part of the Department of Health, Education and Welfare.

(2) The Occupational Safety and Health Administration (OSHA), which is part of the Department of Labor.

(3) Various states.

1. *Bureau of Radiological Health (BRH)*

The BRH has published performance standards for laser products [13]. These standards took effect August 2, 1976. All laser products sold after that date must comply with the standards. There is no retroactive provision applying to laser products sold before that date.

In order to understand the BRH regulations, one should note that the burden of meeting the standards falls on the manufacturer† of laser products, not on the user. In the words of the preliminary remarks to the standards, "The FDA has no authority over the manner in which electronic products are used . . . the standard does not prohibit the purchase of other laser products, of any classification, for any purpose." The user will find, however, that the cost of laser products increased as of August 2, 1976 as a result of equipment, such as key interlocks and emission indicators mandated by the BRH. For large laser systems, the additional cost is a small fraction of the system cost. For some small helium–neon lasers, it amounted to a substantial fraction of the cost of the laser.

The BRH has classified lasers into four classes. The classification is similar in philosophy to that of ANSI. In addition, practical application of the classification schedule often will put a given laser in the same numbered class for both the ANSI and the BRH schedules. Nevertheless, the BRH classification schedule is quite detailed, with factors that depend on both the wavelength and the duration of emission. Classification of a given laser in practice requires one to have the detailed regulations at hand.

The regulations require the manufacturer to provide certification of his products, based on acceptable measurement procedures. This require-

---

† The regulations also apply to a buyer who builds the laser into a system for resale.

ment is providing great impetus for the development, calibration, and standardization of methods for accurate measurement of laser power.

Also, additional safety-related items are to be added to certain classes of lasers manufactured after August 2, 1976. These items include safety interlocks, remote control connectors, key-actuated master controls, emission indicators to show when the laser is operating, labels indicating laser type and class, and permanently attached beam attenuators.

In addition, several use categories were established: medical laser products, surveying, leveling and alignment laser products, and demonstration laser products. Restrictions were placed on lasers "manufactured, designed, intended or promoted" for any of these uses.

2. Occupational Safety and Health Administration (OSHA)

There have been indications in the laser community that OSHA is developing laser safety standards. Such standards would, of course, affect the user of lasers in a much more direct manner than the BRH regulations.

As of late 1977, OSHA had not promulgated standards of laser safety. There is an exception to this statement: Regulations relative to the use of lasers in construction were published in 1974 [14]. These regulations set maximum levels for exposure of employees to laser light intensity. Still, the great majority of laser operations in industry are not yet covered by explicit OSHA regulations.

It seems certain that such regulations will eventually be published. A classification scheme has been developed that may serve as a basis for eventual OSHA regulations. Lasers in use have been classified according to this scheme, which is similar to the classification schemes both of ANSI and BRH [15]. When publication of the OSHA standard does occur, there may be a period for interested parties to comment on the regulations and then perhaps some revision. It is possible that there will be a period of one or two years between initial publication and the time at which the regulations take effect.

Further, there has been speculation that OSHA standards on laser safety would be somewhat similar to the ANSI standards. This must remain in the realm of conjecture until OSHA standards are actually promulgated.

3. State Laws

There are also many individual states that have passed laws regarding the registration and control of lasers. The individual restrictions vary widely from one state to another. It is imperative that the user of a laser check the restrictions applied by his own state. A summary of the individual state laws as of April 1975 has been published [16].

REFERENCES

[1]   W. J. Geeraets and E. R. Berry, *Am. J. Ophthal.* **66**, 15 (1968).
[2]   A. M. Clarke *et al.*, *Arch. Environ. Health* **18**, 424 (1969).
[3]   W. T. Ham, H. A. Mueller, and D. H. Sliney, *Nature* **260**, 153 (1976).
[4]   R. W. Ebbers, *Am. Indust. Hyg. J.* **35**, 252 (1974).
[5]   P. W. Lappin, *Arch. Environ. Health* **20**, 177 (1970).
[6]   D. H. Sliney and B. C. Freasier, *Appl. Opt.* **12**, 1 (1973).
[7]   C. H. Cartwright, *J. Opt. Soc. Am.* **20**, 81 (1930).
[8]   J. G. Bayly, V. B. Kartha, and W. H. Stevens, *Infrared Phys.* **3**, 211 (1963).
[9]   A. S. Brownell and B. E. Stuck, Frankford Arsenal, AD Report 785,609 (1972).
[10] K. R. Envall *et al.*, "Preliminary Evaluation of Commerically Available Laser Protective Eyewear," HEW Publication (FDA) 75-8026, March 1975.
[11] J. F. Smith, J. J. Murphy, and W. J. Eberle, Society of Manufacturing Engineers Technical Paper Mr 75-575 (1975).
[12] Standard Z-136.1—1976, "For the Safe Use of Lasers," American National Standards Institute, 1430 Broadway, New York, New York 10018.
[13] Federal Register, Vol. 40, No. 148 (Thursday, July 31, 1975).
[14] Federal Register, Vol. 39, No. 122 (Monday, June 24, 1974).
[15] Laser Hazard Classification Guide, HEW Publication (NIOSH) 76-183, July 1976.
[16] R. J. Rockwell, *Electro-Opt. Systems Design*, p. 12 (April 1975).

SELECTED ADDITIONAL REFERENCES

A. Physiological Effects

D. O. Adams, E. S. Beatrice, and R. B. Bedell, Retina: Ultrastructural Alterations Produced by Extremely Low Levels of Coherent Radiation, *Science* **177**, 58 (1972).

E. S. Beatrice and G. D. Frisch, Retinal Laser Damage Thresholds as a Function of Image Diameter, *Arch. Environ. Health* **27**, 322 (1973).

A. S. Brownell, Skin and Carbon Dioxide Laser Radiation, *Arch. Environ. Health* **18**, 437 (1969).

A. M. Clarke *et al.*, Laser Effects on the Eye, *Arch. Environ. Health* **18**, 424 (1969).

A. M. Clarke, W. J. Geeraets, and W. T. Ham, An Equilibrium Thermal Model for Retinal Injury from Optical Sources, *Appl. Opt.* **8**, 1051 (1969).

A. M. Clarke, Ocular Hazards, *in* "Handbook of Lasers" (R. J. Pressley, ed.), Chemical Rubber Company, Cleveland, Ohio, 1971.

R. W. Ebbers and I. L. Dunsky, Retinal Damage Thresholds for Multiple Pulse Lasers, *Aerospace Medicine* **44**, 317 (1973).

R. W. Ebbers, Retinal Effects of a Multiple-Pulse Laser, *Am. Ind. Hyg. Assoc. J.* **35**, 252 (1974).

M. Eleccion, Laser Hazards, *IEEE Spectrum*, p. 32 (August 1973).

J. K. Franks and D. H. Sliney, Electrical Hazards of Lasers, *Electro-Opt. Systems Design*, p. 20 (December 1975).

G. L. M. Gibson, Retinal Damage from Repeated Subthreshold Exposures Using a Ruby Laser Photocoagulator, *Aerosp. Med.* **44**, 433 (1973).

L. Goldman *et al.*, Effect of the Laser Beam on the Skin, *J. Invest. Dermatol.* **40**, 121 (1963).

L. Goldman, "Biomedical Aspects of the Laser," Springer-Verlag, Berlin and New York, 1967.

L. Goldman and R. J. Rockwell, "Lasers in Medicine," Gordon and Breach, New York, 1971.

L. Goldman, "Applications of the Laser," Chemical Rubber Co., Cleveland, Ohio, 1973.

L. Goldman et al., Studies in Laser Safety of New High-Output Systems, 1. Picosecond Impacts, Opt. Laser Technol., p. 11 (February 1973).

L. Goldman et al., Studies in Laser Safety of New High Output Systems, 2. TEA CO_2 Laser Impacts, Opt. Lasers Technol., p. 58 (April 1973).

W. T. Ham et al., Ocular Effects of Laser Radiation, Part I, Acta Ophthal 43, 390 (1965).

W. T. Ham et al., Retinal Burn Thresholds for the Helium–Neon Laser in the Rhesus Monkey, Arch. Ophthal. 84, 797 (1970).

W. T. Ham et al., Ocular Hazard from Picosecond Pulses of Nd:YAG Laser Radiation, Science 185, 362 (1974).

W. T. Ham, H. A. Mueller, and D. H. Sliney, Retinal Sensitivity to Damage from Short Wavelength Light, Nature (London) 260, 153 (1976).

C. H. Knowles, Safe Helium–Neon Lasers Advance Understanding of Light, Phys. Teacher, p. 69 (February 1972).

P. W. Lappin, Ocular Damage Thresholds for the Helium–Neon Laser, Arch. Environ. Health 20, 177 (1970).

H. M. Leibowitz and G. R. Peacock, Corneal Injury Produced by Carbon Dioxide Laser Radiation, Arch. Ophthal. 81, 713 (1969).

D. MacKeen, S. Fine, and E. Klein, Safety Note: Toxic and Explosive Hazards Associated with Lasers, Laser Focus, p. 47 (October 1968).

M. A. Mainster, T. J. White, and J. H. Tips, Corneal Thermal Response to the CO_2 Laser, Appl. Opt. 9, 665 (1970).

W. L. Makous and J. D. Gould, Effects of Lasers on the Human Eye, IBM J. Res. Develop. 12, 257 (1968).

K. W. Marich et al., Health Hazards in the Use of the Laser Microprobe for Toxic and Infective Samples, Am. Ind. Hyg. Assoc. J. 33, 488 (1972).

N. A. Peppers et al., Corneal Damage Thresholds for CO_2 Laser Radiation, Appl. Opt. 8, 377 (1969).

J. F. Ready, "Effects of High-Power Laser Radiation," Academic Press, New York, 1971, Chapter 7.

D. H. Sliney and B. C. Freasier, Evaluation of Optical Radiation Hazards, Appl. Opt. 12, 1 (1973).

D. H. Sliney, D. C. Vorpahl, and D. C. Winburn, Environmental Health Hazards of High-Powered Infrared Laser Devices, Arch. Environ. Health 30, 174 (1975).

D. H. Sliney, Health Hazards from Laser Material Processing, Soc. of Manufacturing Eng. Tech. Paper MR75-581 (1975).

R. Tinker, The Safe Use of Lasers, Phys. Teacher, p. 455 (November 1973).

A. M. Vassiliadis et al., Thresholds of Laser Eye Hazards, Arch. Environ. Health 20, 161 (1970).

M. L. Wolbarsht (ed.), "Laser Applications in Medicine and Biology," Vol. 1, Plenum Press, New York, 1971.

M. L. Wolbarsht (ed.), "Laser Applications in Medicine and Biology," Vol. 2, Plenum Press, New York, 1974.

M. L. Wolbarsht and D. H. Sliney, Needed: More Data on Eye Damage, Laser Focus, p. 10 (December 1974).

H. C. Zweng, Accidental Q-Switched Laser Lesion of Human Macula, Arch. Ophthal. 78, 596 (1967).

B. Laser Safety Practices and Standards

Am. Conf. Govt. Ind. Hyg., Threshold Limit Values for Chemical Substances and Physical Agents in the Workroom Environment, ACGIH, Cincinnati, Ohio (1975).

American National Standards Institute, Standard Z136.1-1976 for the Safe Use of Lasers, ANSI, New York (1976).

S. S. Charschan (ed.), "Lasers in Industry," Van Nostrand–Reinhold, Princeton, New Jersey, 1972, Chapter 10.

I. L. Dunsky, W. A. Fife, and E. O. Richey, Determination of Revised Air Force Permissible Exposure Levels for Laser Radiation, *Am. Ind. Hyg. Assoc. J.* **34**, 235 (1973).

S. M. Michaelson, Standards for Protection of Personnel Against Nonionizing Radiation, *Am. Ind. Hyg. Assoc. J.* **35**, 766 (1974).

G. E. Myers, Operating Lasers : Don't be Half-Safe, *Electro-Opt. Systems Design*, p. 30 (July 1973).

C. H. Powell and L. Goldman, Recommendations of the Laser Safety Conference, *Arch. Environ. Health* **18**, 448 (1969).

D. H. Sliney and R. Yacovissi, Control of Health-Hazards from Airborne Lasers, *Aviation, Space Environ. Med.* **46**, 691 (1975).

J. F. Smith, J. J. Murphy, and W. J. Eberle, Industrial Laser Safety Program Management, Soc. of Manufacturing Eng. Tech. Paper MR75-575 (1975).

C. H. Swope and C. J. Koester, Eye Protection Against Lasers, *Appl. Opt.* **4**, 523 (1965).

C. W. Swope, Design Considerations for Laser Eye Protection, *Arch. Environ. Health* **20**, 184 (1970).

D. C. Winburn, Selecting Laser-Safety Eye Wear, *Electro-Opt. Systems Design*, p. 46 (October 1975).

D. C. Winburn, Safety Considerations in the Laser Research Program at the Los Alamos Scientific Laboratory, *Ann. N.Y. Acad. Sci.* **267**, 135 (1976).

C. Regulatory Bodies

R. W. Courtney, Illinois Laser System Registration Law, *Laser J.*, p. 19 (July/August 1970).

Department of Health, Education and Welfare, Performance Standard for Laser Products, *Fed. Register* **40**, No. 148 (July 31, 1975).

Department of Labor, Construction Safety and Health Regulations, *Fed. Register* **39**, No. 122 (June 24, 1974).

Laser and Laser Safety Regulations, Spectra-Phys. *Laser Rev.* (February 1976).

R. L. Mortensen, Perspective on Laser Safety, *Electro-Opt. Systems Design*, p. 66 (August 1976).

R. L. Mortensen, Lasers and Construction Safety, *National Safety News*, p. 68 (April 1976).

R. J. Rockwell, Current Status of State Laser Safety Legislation, *Electro-Opt. Systems Design*, p. 12 (April 1975).

Uniformity in Safety Codes, *Laser Focus*, p. 42 (October 1973).

CHAPTER 7

LASER COMPONENTS AND ACCESSORIES

In this chapter we shall describe several components and various types of accessory equipment. These devices are often used either in laser construction or together with lasers in a system designed for a particular application. These components and accessories represent separate fields of technology, many of which were well developed prior to the advent of the laser. In many cases, the demands of laser technology have spurred further rapid development in these associated areas. Since they are often closely connected with laser construction or with laser applications, a complete description of laser technology must include a description of these associated areas. Because these technologies are well developed in their own right, the description will necessarily be somewhat abbreviated.

Specific components used in lasers will include the technology of mirrors, polarizers, Q-switches, and harmonic generators, and the availability of infrared transmitting materials for lenses and windows for infrared lasers. Accessory devices often used with lasers in a system for an application include detectors, modulators, and beam deflectors. In order to describe these devices, we must include a discussion of the fields of electrooptics, acoustooptics, and nonlinear optics.

A. Mirrors

In Chapter 1 we described the resonant cavity, which provides the multiple reflections of light back and forth through the active medium. The resonant cavity consists of two mirrors. We now describe what these mirrors actually are.

The mirrors most often used in lasers are evaporated coatings consisting of multiple layers of dielectric materials, vacuum deposited on a transparent substrate. Early in the history of lasers, thin metallic coatings, such as silver or aluminum, were often used. They were prepared by either vacuum or chemical deposition, and in the case of solid state lasers, could be directly on the end of the laser rod. However, these coatings inevitably had high scattering losses and also were easily subject to damage. Almost all mirrors now used in practical laser systems are multilayer mirrors. They use the principal of interference to produce high reflectance values. The coatings consist of alternating layers of thickness equal to one-quarter of the wavelength of light for which high reflectivity is desired. The materials are alternately of high and low index of refraction, as illustrated in Fig. 7-1. These mirrors are produced by evaporating the materials in vacuum, so that the substrate is coated by the evaporated material. The high and low index materials are evaporated alternately, and the thickness of each layer in the stack has an optical thickness of one-quarter of a wavelength.

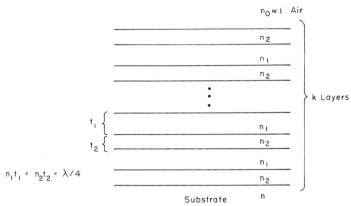

Fig. 7-1   Schematic diagram of a multilayer coating, in which n_1 and n_2 are the indices of refraction of the two materials and t_1 and t_2 their physical thicknesses. The optical thickness of each layer is one-quarter wavelength. The index of refraction of the substrate is n and the index of refraction n_0 of air is approximately equal to unity.

Usually an odd number of layers is used, with the high index material next to the substrate. The reflectivity R of such a set of k layers is given by the equation

$$R = (n_2^{k+1} - n_0 n n_1^{k-1})/(n_2^{k+1} - n_0 n n_1^{k+1}) \tag{7.1}$$

where n_2 and n_1 are, respectively, the indices of refraction of the high and low index materials, $n_0 \approx 1$ is the index of refraction of air, and n the index of refraction of the substrate material. Figure 7-2 shows a plot of the reflectivity as a function of the total number of layers in the stack for several different values of index of refraction. The low index of refraction is chosen as 1.46, representative of SiO_2. Various values for the high index are shown.

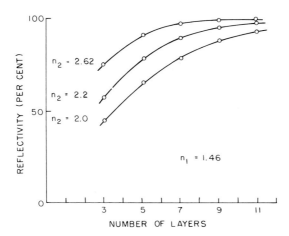

Fig. 7-2   Reflectivity as a function of the number of layers in a multilayer stack for different values of the index of refraction n_2 and with n_1 equal to 1.46, representative of SiO_2.

The value of 2.62 is representative of titanium dioxide. The reflectivity becomes very high ($\sim 100\%$) for a reasonably small number of layers.

The reflectivity can be adjusted to any desired value by choosing materials with appropriate values of index of refraction and by adjusting the number of layers. Scattering losses in carefully prepared mirrors can be very low, of the order of 0.1%. Thus, mirrors with reflectivity as high as 99.8 or 99.9% are available using multilayer dielectrics. Such mirrors are more durable than metallic mirrors, and are the common mirrors generally employed in presentday lasers.

A wide variety of dielectric materials have been employed for such coatings. Examples of some materials include titanium dioxide, silicon dioxide, silicon monoxide, cerium oxide, zirconium dioxide, zinc sulfide, cryolite (sodium aluminum fluoride), magnesium fluoride, and thorium fluoride.

The reflectivity as a function of wavelength for the mirrors for a typical helium–neon laser is shown in Fig. 7-3, for both the high reflectivity mirror and for the output mirror, which has somewhat lower reflectivity.

In addition, the rear surfaces of the mirrors are often antireflection coated, with a quarter-wavelength of a suitable material to reduce the reflectivity at the laser wavelength. The technique of antireflection coating is commonly employed in optical instruments, for example, in binoculars, in order to reduce undesired reflections from the air–glass interface.

There is one instance in which metallic mirrors are commonly used. This is for the high reflectivity mirror in far infrared lasers, especially CO_2 lasers. Because the mirrors often must withstand high levels of radiation,

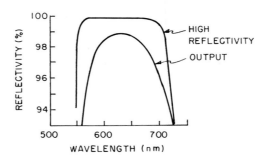

Fig. 7-3   Reflectivity as a function of wavelength for mirrors for a typical helium–neon laser.

they are relatively massive and are made of metals with high thermal conductivity. Internal channels for liquid cooling may be provided in some cases. For high-power lasers, such liquid cooling reduces thermal distortion and increases the damage threshold.

The metals most commonly used are high-purity copper and molybdenum. Copper mirrors can withstand higher flux densities without damage than molybdenum. Molybdenum suffers less thermal distortion than copper, but is more susceptible to corrosion than copper if water cooling is used. The surface of high-purity copper is subject to tarnishing, so the copper mirrors are often coated with another metal, such as gold or electroless nickel.

All types of mirrors described above, both the multilayer dielectric and the metallic mirrors, are available from a number of suppliers, with a choice of radii of curvature to accommodate the various configurations for the resonant cavity, as described in Chapter 1.

B.   Polarizers

A widely known type of polarizer is the so-called dichroic polarizer which consists of a polarizing material embedded in a plastic sheet. Dichroic materials exhibit high transmission for light polarized in one direction and high absorption for light polarized perpendicular to that direction. When a polarizing sheet is placed in a beam of unpolarized light, the transmitted light is plane-polarized and has an intensity one-half of the incident beam. Such dichroic polarizing sheets are inexpensive, and are commonly used in classroom demonstrations of polarized light. The dichroic sheets do have some disadvantages. Since they absorb half the incident power, they cannot be used with high-power laser beams. The quality of the polarized light is often not sufficiently good for exacting laser work. The extinction ratio, the ratio of the intensity of the transmitted beam to that of the absorbed beam, is not so high with the plastic sheet polarizers as it is with high-quality crystal polarizers which use birefringence.

The phenomenon of birefringence has been described in Chapter 1. Prism polarizers operate by physically separating the two different components of polarization. This is possible because of the difference in velocity of the two different components. Figure 7-4 shows one type of polarizing prism. The crystal is split into two triangular parts, separated by a small air gap. Both components of polarization enter the crystal and travel in the same direction, but have different velocities. When the light reaches the air gap, the index of refraction for one of the rays is large enough that the angle between the beam and the surface normal exceeds the critical angle for total internal reflection. That polarization is reflected, as indicated. The other component of polarization strikes the interface at an angle less than the critical angle, and is transmitted, as indicated. The device is constructed with two triangular pieces of material so that the transmitted ray will not

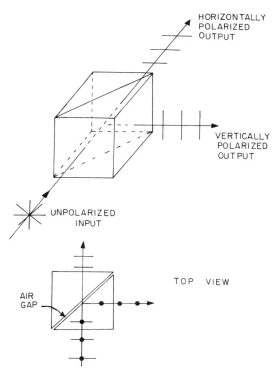

Fig. 7-4 A type of polarizing prism in which a birefringent crystal is split into two triangular parts separated by a small air gap.

be deviated. The crystalline material, with its two indices of refraction, and the angle of the interface are chosen so that one component is reflected at the interface and the other component is transmitted. There are many other designs that yield polarizing prisms, but this discussion illustrates the general principle.

Some polarizers have cement between the two pieces. Generally, cemented polarizers are unsuitable for work with lasers, because the high intensity beam will damage the cement. Thus air-spaced polarizers are usually used with lasers. Crystalline polarizing prisms are generally of high optical quality and offer better extinction ratios between the two components of polarization than do sheet polarizers.

C.   Infrared Materials

In many applications of lasers the beam must pass through a transparent window or be focused by a lens. If the laser wavelength is in the visible, near-ultraviolet, or near-infrared, familiar materials may be used for the window or lens. Glass or quartz transmit in these regions. One may use the easy availability and well-developed technology of glass and quartz. Farther in the infrared, glass and quartz are no longer transmitting and one is obliged to look for other materials. The long wavelength limit of transmission is around 2.7 μm for many optical glasses and around 4.5 μm for quartz. Thus glass or quartz may be used with Nd:glass or Nd:YAG lasers, but not with CO_2 lasers. There are a number of materials which offer good transmission at 10.6 μm. Because the technology of such materials is less known than that of glass and quartz, it is worth reviewing the availability of materials for lenses and windows at 10.6 μm. Such materials include alkali halides and a number of semiconductors. Some of the more commonly used materials are listed in Table 7-1. The alkali halides, notably sodium chloride and potassium chloride would appear to be almost ideal. They offer low absorption at 10.6 μm. The index of refraction is low, so that reflection losses are small. Their cost can be low. They offer the additional benefit of transmission in the visible, so that an infrared system employing these optics can be lined up with visible light. The one very significant disadvantage, which greatly restricts their use, is the fact that they adsorb water vapor from the atmosphere. This will cloud the surfaces, so that unless special care is employed to provide a low humidity environment, their life will be short.

Irtran II and Irtran IV are polycrystalline, hot-pressed materials which are available at moderate cost, have good stability and some visible transmission. The index of refraction of these materials is higher than that

TABLE 7-1

Windows and Lens Materials for CO_2 Lasers

Material	Long wavelength limit (μm)	Advantages	Disadvantages
NaCl & KCl	12, 15	Very low absorption, visible transmission, low cost, low index	Stability, water absorption
Irtran II	14	Moderate cost, visible transmission.	Damage, absorption loss.
Irtran IV	22	Moderate cost, visible transmission.	Damage, absorption loss.
Ge	23	Low absorption.	High index, thermal run-away, high cost
GaAs	18	Low absorption	High index, very high cost
CdTe	30	Very low absorption	High index, very high cost
ZnSe	20	Visible transmission, very low absorption	High index, very high cost

of the alkali halides, so that to avoid losses they should be antireflection coated. The absorption coefficients of these materials at 10.6 μm are higher than for some of the other materials, so they are unsuitable if a high-power beam must be transmitted.

A number of semiconductor materials, most notably germanium, gallium arsenide, cadmium telluride and zinc selenide have been developed. All of these materials are available with low absorption coefficients at 10.6 μm. All of them are opaque in the visible except for zinc selenide, which does offer visible transmission. All these materials have high cost with germanium currently being lowest and gallium arsenide the highest. They all have a high index of refraction and require coating to avoid excessive reflection loss. All the semiconductor materials are susceptible to a phenomenon called thermal runaway, in which the absorption increases with increasing temperature. Thus, if a high-power beam is transmitted and there is some absorption, heating and further increased absorption can lead to catastrophic overheating of the element. Germanium is particularly susceptible to thermal runaway. Germanium windows, mirrors, and lenses are often water cooled. The other semiconductor materials are less susceptible to thermal runaway. Of these four semiconductor materials, zinc selenide probably offers the best combination of absorption coefficient and thermal conductivity for use as a transmitting element for high-power CO_2 laser beams. Cadmium telluride provides a reasonably close second choice, at somewhat lower cost.

In summary, there are a variety of infrared transmitting materials suitable for use as optics for CO_2 lasers. One can make a choice based on cost and on the requirements of the particular system.

D. Detectors

Detectors suitable for measuring laser power are commonly employed along with lasers. For an application such as a laser communication system, a detector is necessary as the receiver. For applications involving interferometry, optical detectors are used to measure the position and motion of fringes of high light intensity. In applications such as machining or metalworking, a detector is necessary to monitor the output of the laser to provide reproducable conditions. In all applications, one usually desires to measure the level at which the laser is operating. Therefore, suitable detectors for measuring laser power and energy are essential.

The detection and measurement of optical and infrared radiation has been a long-established field of technology. In recent years, however, this technology has been applied more specifically to lasers, and detectors particularly suitable for use with lasers have been developed. Commercial developments have also kept pace. Detectors have been specially packaged and are marketed by many laser manufacturers. Such packaged detectors are commonly called laser power meters. They generally include a complete system for measuring the output of a specific class of lasers, and will include a detector, a housing, amplification if necessary, and a readout, often a meter.

In this section we shall discuss some of the types of detectors that are available for use with lasers. We shall not attempt to cover the entire field of radiation detection, which is very large. Instead, we shall emphasize those detectors which have found most practical application in conjunction with lasers.

There are two broad classes of detectors that we shall describe. These are photon detectors and thermal detectors. Photon detectors rely on the action of quanta of light to interact with electrons in the detector material. Thus, photon detectors rely on generating free electrons. In order to produce such photoeffects, the quantum of light must have sufficient energy to free an electron. The wavelength response of photon detectors will then show a long wavelength cutoff. When the wavelength becomes very long, the photon energy will have too low a value to liberate electrons and the response of the detector will drop to zero.

Thermal detectors respond to the heat energy delivered by the light. They all rely on some type of heating effect. The response of such detectors will be independent of wavelength.

The performance of radiation detectors is commonly described by a number of figures of merit. These figures of merit are in widespread use throughout the radiation detection field. They were developed primarily to describe the capabilities of a detector in responding to small signals in the presence of noise. As such, they are not always pertinent to the case of lasers. In detection of laser radiation, for example, monitoring of a high-power beam for metalworking purposes, there is no question of detection of a small signal in a background of noise. The laser signal far outweighs any other source that is present. However, in applications such as laser communications and detection of backscattering from air pollutants, noise considerations do become important. It is worthwhile to discuss the figures of merit that are commonly used, because the manufacturers of detectors usually describe the performance of their detectors in these terms.

The first term that is commonly used is the responsivity. This essentially defines how much output one obtains from the detector per unit input. The unit of responsivity will be either volts per watt or amperes per watt, depending on whether the signal from the detector is a voltage or an electric current. This will depend on the particular type of detector and how it is used.

The responsivity is an important characteristic and is commonly specified by the manufacturer, at least as a nominal value. Knowledge of the responsivity allows the user to measure the power output of the laser directly. However, absolute calibrations of the responsivity are difficult to make with accuracy. One may also characterize the spectral responsivity, which is the responsivity as a function of wavelength.

A second figure of merit is the noise equivalent power (NEP). This is defined as the radiant power which produces a signal voltage equal to the noise voltage from the detector. Since the noise will be dependent on the bandwidth of the measurement, that bandwidth must be specified. The equation defining noise equivalent power is

$$\text{NEP} = HAV_{\text{N}}/V_{\text{s}}(\Delta f)^{1/2} \tag{7.2}$$

where H is the power density incident on the detector of area A, V_{N} is the root mean square noise voltage within the measurement bandwidth Δf, and V_{s} is the root mean square signal voltage. The noise equivalent power has units of watts per (hertz to the half power). From the definition, it is apparent that the lower the value of noise equivalent power, the better are the characteristics of the detector for detecting a small signal in the presence of noise.

The measurement of noise equivalent power yields a result which is dependent on the area of the detector and also on the bandwidth of the measurement. In order to provide a figure of merit under standardized conditions, a term called detectivity is defined. Detectivity is commonly represented by the symbol D^*. (This figure of merit is pronounced as D-star.) It is defined by the equation

$$D^* = A^{1/2}/\text{NEP} \tag{7.3}$$

Since many detectors have noise equivalent power proportional to the square root of the detector area, D^* is independent of the area of the detector and yields a measure of the quality of the detector material itself, independent of the area with which the detector happens to be made. The dependence of D^* on the wavelength λ, the frequency f at which the measurement is performed, and the bandwidth Δf are specified in the following notation: $D^*(\lambda, f, \Delta f)$. The reference bandwidth is generally taken as 1 Hz. The units of D^* are centimeters (square root hertz) per watt. A high value of D^* means that the detector is suitable for detecting weak signals in the presence of noise.

All the detectors that we shall consider in this section are so-called square law detectors. They respond to the power in the beam, which is proportional to the square of the electric field. Microwave detectors, in contrast, can measure the field intensity directly. However, all the types of detectors that we shall consider here exhibit square law response. This is also true of such common detectors as the human eye and photographic film.

We shall now describe the operation of photon detectors. As mentioned previously, photon detectors rely on liberating an electron and require the light to have a sufficient photon energy in order to be able to exceed some threshold energy. There are three types of photoeffects which we shall consider, which have been used for photon detectors. These include the photovoltaic effect, the photoemissive effect, and the photoconductive effect.

1. Photovoltaic Effect and Photodiodes

The photovoltaic effect and the operation of photodiodes both rely on the presence of a p–n junction in a semiconductor. When a junction in a semiconductor is in the dark, an electric field is present internally in the junction. The change in the level of the valence and conduction bands at the junction leads to the familiar rectification effect.

When light is incident on the junction, it is absorbed and produces free electron–hole pairs. This assumes that the wavelength of the light is short enough, i.e., that the photon energy is high enough to produce a free electron–hole pair. The electric field at the junction separates the pair and moves the electron into the *n*-type region and the hole into the *p*-type region. This leads to the production of a voltage which may be measured externally. This is the so-called photovoltaic effect. We note explicitly that the photovoltage generated may be detected directly, and that no bias voltage nor ballast resistor is required.

It is also possible to use a *p–n* junction to detect the radiation by applying a bias voltage to the junction in the reverse direction. By reverse direction we mean the direction of low current flow, that is, with the positive side of the voltage applied to the *n*-type material. A *p–n* junction operated with bias voltage is termed a photodiode. The characteristics of a photodiode are shown in Fig. 7-5. The curve labeled dark represents conditions in the absence of light; it displays the familiar rectification characteristics. The current–voltage characteristics when the sample is illuminated at different light levels are shown by the other curves. The current that may be drawn through an external load resistor increases with increasing light level. In practice, one measures the voltage drop appearing across the resistor.

In order to increase the frequency response of photodiodes, a type called the PIN photodiode has been developed. There is a layer of nearly intrinsic material bounded on one side by a relatively thin layer of highly doped *p*-type semiconductor, and on the other side, by a relatively thick layer of heavily doped *n*-type semiconductor. A sufficiently high reverse bias is applied so that the depletion layer, from which carriers are swept out, spreads to occupy the entire intrinsic region. This volume then becomes a region of high and nearly constant electric field. It is called the depletion region because all mobile charges have been swept out. Radiation which

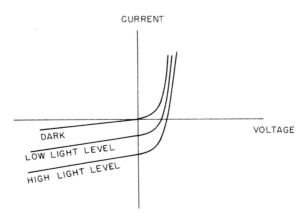

Fig. 7-5   Current–voltage characteristics for photovoltaic detectors.

is absorbed in the intrinsic region will produce electron–hole pairs, provided the photon energy is high enough. Carriers generated by this means are swept across the region with high velocity and are collected at the heavily doped regions. The frequency response of such PIN diodes can be very high, of the order of 10^{10} Hz. This is higher than the frequency responses of which p–n junctions are capable.

The production techniques for photodiodes have been the subject of considerable development. A variety of photodiode structures have been employed. No single photodiode structure can best meet all system requirements. Therefore, a variety of types are available. These include the planar diffused photodiode, which is shown in Fig. 7-6a, and the Schottky photodiode, which is shown in Fig. 7-6b. The planar diffused photodiode is formed by taking a slice of high-resistivity silicon, growing an oxide layer over it, etching a hole in the oxide and allowing boron to diffuse into the silicon through the hole. High voltage breakdown and low leakage current can be obtained with this structure. Devices with area in excess of 1 cm² have been fabricated successfully on silicon of resistivity 4000 Ω cm.

The Schottky barrier photodiode uses a junction between a metal layer and a semiconductor. If the metal and semiconductor have work

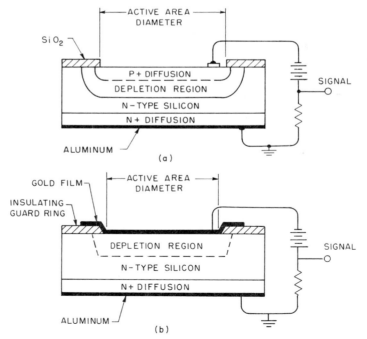

Fig. 7-6   Photodiode structures. (a) Planar diffused photodiode. (b) Schottky photodiode.

functions related in the proper way, this can be a rectifying barrier. The junction is formed by oxidation of the silicon surface and then evaporation of thin transparent and conducting gold layers. The insulating guard rings serve to reduce the leakage currents through the devices.

Photodiodes have been fabricated from a number of materials. Different materials are suitable for use in different spectral regions. Figure 7-7 shows the spectral D^* or detectivity of a number of commercially available photovoltaic detectors. The choice of detector will depend on the wavelength region that is desired. For example, for a laser operating at 5 μm, an indium antimonide photovoltaic diode would be a suitable detector.

The shape of the curves in Fig. 7-7 is characteristic of photon detectors. One photon releases one electron in the material, so long as the photon has sufficient energy. Absorption of one photon gives a certain response, independent of the wavelength (so long as the wavelength lies within the range

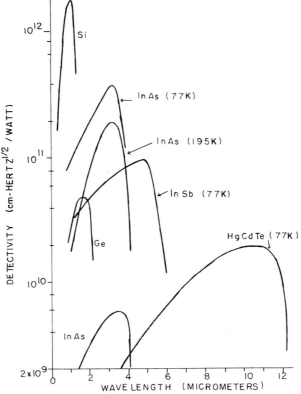

Fig. 7-7 Detectivity as a function of wavelength for a number of different types of photodiode. Where the temperature of operation is not indicated, it is room temperature.

of the spectral sensitivity of the detector). One photon of ultraviolet radiation and one photon of infrared radiation can each produce the same result, even though they deliver much different energies. Therefore, the response depends on the number of photons arriving per unit time. As the wavelength increases, for constant photon arrival rate, the incident power decreases, but the signal remains constant. Therefore, the value of D^* increases. At a wavelength corresponding to approximately the band gap divided by Planck's constant, the value of D^* is a maximum. At longer wavelengths, it drops off rapidly because the photons do not have sufficient energy to excite an electron into the conduction band.

Silicon photodiodes cover the visible and near infrared, and in particular are suitable for detection of many of the common lasers, including argon, helium–neon, ruby, Nd:YAG, and Nd:glass. As a practical matter, silicon photodiodes have become the detector of choice for many laser applications. They have been commercially built into laser power meters, and have become widely available. They probably provide the most common type of laser detector for lasers operating in the visible and near infrared portions of the spectrum.

Another variation of the silicon photodiode is the avalanche photodiode. The avalanche photodiode offers the possibility of internal gain; thus it is sometimes referred to as a "solid state photomultiplier." An avalanche photodiode has a diffused p–n junction in silicon, with surface contouring to permit high reverse bias voltage without surface breakdown. The high internal field strength leads to internal multiplication gain, so that the signal can be perhaps 100–200 times as great as for a nonavalanche device. The signal-to-noise ratio is thus increased greatly, provided that limiting noise is not from background radiation. Drawbacks of avalanche photodiodes include much greater cost than conventional photodiodes and the need for temperature-compensation circuits to maintain the optimum bias. Still, avalanche photodiodes offer an attractive choice of detector when high speed and high performance are important.

2. Photoemissive Detectors

The photoemissive detector employs a cathode coated with a material which emits electrons when radiation of wavelength shorter than a certain value falls on the surface. The electrons emitted from the cathode may be accelerated by an applied voltage to an anode, giving rise to a current in an external circuit. These devices, of course, are operated in vacuum. The time constants are short, often of the order of 10^{-8} sec. These detectors are available commercially from many manufacturers and form a broad class of laser detectors of interest. For photoemissive detectors, some standardized spectral responses are shown in Fig. 7-8. Some common phototubes use the S-1 photosurface, which extends farthest into the infrared, and which is the only practical photoemissive surface for a Nd:glass laser. The S-20 photosurface gives high response for the ruby and He–Ne laser wavelengths. The ERMA (extended-red) surface has good response in the red. The usefulness of these devices ranges from the ultraviolet to the near-infrared. Beyond a wavelength of 1.2 μm, no photoemissive response is available.

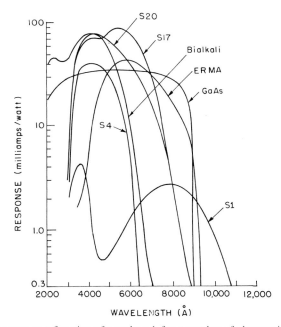

Fig. 7-8   Response as a function of wavelength for a number of photoemissive surfaces.

The photomultiplier is a vacuum photoemissive photodetector containing a number of secondary emitting stages called dynodes. In the secondary emission process, electrons which are emitted from the cathode are accelerated by an applied voltage to the first dynode where their impact causes production of numerous secondary electrons. These proceed to the next dynode and more amplification is obtained. By arranging a number of such surfaces so that electrons from each dynode are delivered to the next dynode, a very large multiplication factor (of the order of 10^4 to 10^5) is obtained. The high gain means that photomultiplier tubes offer the highest available responsivity in the ultraviolet, visible, and near-infrared portions of the spectrum.

3.   Photoconductive Detectors

Other infrared detectors utilize the principle of photoconductivity. A semiconductor in thermal equilibrium contains free electrons and holes. The concentration of the electrons and holes is changed if light is absorbed by the semiconductor. The photon energy must be enough to cause excitation, either raising electrons across the forbidden band gap or activating impurities present in the band gap. When light falls on a photoconductor to which a bias voltage is applied, the increased number of carriers leads to an increase in electrical conductivity, so that the current flowing through the detector will increase. This can be measured in an external circuit (after amplification if necessary). There are many types of infrared photoconductors in use. Some typical values of D^* are shown in Fig. 7-9. The exact value of D^* depends on the temperature of operation, and also on the field of view of the detector. The photoconductive detectors are more widely used in the infrared, at wavelengths where photoemissive detectors and photodiodes no longer respond. The cryogenic requirements of long-wavelength photoconductive detectors may mean inconvenience.

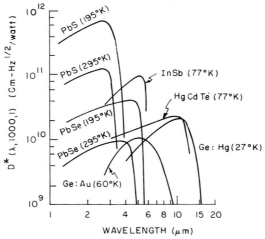

Fig. 7-9   Detectivity (D^*) as a function of wavelength for a number of photoconductive detectors at the indicated temperatures.

4.   Thermal Detectors

The thermal detector responds to the total energy absorbed, regardless of wavelength. Thermal detectors have no long wavelength cutoff, as do photon detectors. The value of D^* for a thermal detector is independent of wavelength. Thermal detectors generally do not offer as rapid response as photon detectors and for laser work are not often used in the wavelength region for which photon detectors are effective. They are often used at longer wavelengths.

Pyroelectric detectors are one type of thermal detector whose use is increasing rapidly. Pyroelectric detectors operate by virtue of the change in electric polarization that occurs in certain ceramic or crystalline materials as their temperature changes. The change in polarization can be measured as an open circuit voltage or as a short circuit current. The time response can be rapid enough to detect short pulses of laser light. This is in contrast to many thermal detectors, where response speed is a problem. Pyroelectric detectors are finding favor for use with CO_2 lasers.

Another example of thermal detectors involves calorimeters. Calorimetric measurements provide simple determination of the total energy in a laser pulse. Calorimeters designed for laser measurements involve blackbody absorbers of low thermal mass with thermocouples or other temperature measuring devices (e.g., bolometers or thermistors) in contact with the absorber to measure the temperature rise. Both bolometers and thermistors involve measurement of the change in electrical resistivity with temperature. A bolometer uses a metal element; a thermistor uses a semiconductor element.

There have been many types of calorimeters developed for measuring the total energy in a laser pulse or for integrating the power from a continuous laser. Since the total energy in a typical laser pulse is not large, the calorimetric techniques must be rather delicate. The absorbing medium must be small enough that the absorbed energy can be rapidly distributed throughout the body.

One form that a calorimeter can take is for the absorber to be a small hollow cone of carbon, shaped so that radiation entering the base of the cone will not be reflected back out of the cone. Such a design makes an efficient absorber. Thermistor beads are placed intimately in contact with the cone. The thermistor beads form an element of a balanced bridge circuit, the output of which is delivered to a recorder. As the cone is heated by a pulse of energy, the resistance of the thermistors changes, resulting in an unbalance of the bridge and a voltage pulse at a recorder. The pulse decays as the cone cools to ambient temperature. The magnitude of the voltage pulse gives a measure of the energy in the light pulse.

One of the first calorimeters specifically designed for lasers used this design. The carbon cone weighed about 0.33 gm. Thermistor beads embedded in slots were held in place by Glyptal. Two identical cones with matching thermistors were connected to form a conjugate pair in the bridge circuit. Ambient temperature variations produced equal changes in both cones, so that the net effect was cancelled out. The output of the bridge circuit could drive a recorder or a microvoltmeter.

It is important to have the thermal relaxation time of the system long enough so that the measuring portion of the circuit can come to equilibrium. The decay time is determined by the rate at which the cone loses heat to its surroundings and should be much longer than the rise time required for the heat to distribute itself in the cone. For power measurements on continuous lasers, the calorimeters can be used to integrate over a known period of time.

A calorimeter employing a carbon cone or a similar design is a simple and useful device for measurement of laser pulse energies. In the range of output energies below 1 J or so, an accuracy of 4% or better should be attainable. The main sources of error in conical calorimeters for pulsed laser energy measurements are loss of some of the energy out of the calorimeter by reflection, loss of heat by cooling of the entire system before the heat is redistributed evenly, and imperfect calibration. Methods of calibrating using a current pulse applied to a heating element can make the last source of error small. With careful technique, the other sources of error can be held to a few percent. A cone with proper apex angle allows multiple reflection and eventual absorption.

Calorimeters can also be employed for laser pulses of energy containing tens to hundreds of joules. For very high power lasers, destructive effects (vaporization of part of the absorbing surface) may limit the usefulness. In the measurement of Q-switched pulses, the high surface temperature may lead to significant reradiation losses. Good calorimeters must be designed to have small surface temperature rise. Since most calorimeters require surface absorption of the beam energy, there are upper limits on the power densities that the calorimeters can withstand without damage.

If the response time of the calorimeter is reasonably fast, it can be used for measurement of power in a continuous laser beam. The temperature

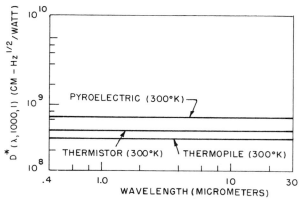

Fig. 7-10 Detectivity (D^*) as a function of wavelength for several typical thermal detectors.

of the absorber will reach an equilibrium value dependent on the input power. Such units have been packaged and are available commercially as laser power meters.

Some values of D^* for thermal detectors are shown in Fig. 7-10. The values are independent of wavelength. In the visible and near infrared, the values of D^* for thermal detectors tend to be lower than those of good photon detectors, but the response does not fall off at long wavelengths.

5. Calibration

The response of any photodetector in voltage (or current) output per unit power input is often taken as the nominal value specified by the manufacturer. For more precise work, the detector may be calibrated separately, for example, by use of a secondary standard tungsten lamp. Absolute measurements of power and energy are difficult. A good calibration requires painstaking work.

Quantitative measurements of the power of a laser can be troublesome. The intense laser output tends to saturate the output of detectors, so that absorbing filters are used to cut down the input to the detector. Suitable filters avoid saturation of the detector, keep the detector in the linear portion of its operating characteristics, and make it blind to background radiation. They also protect detectors from damage. Photoemissive surfaces in particular are degraded by exposure to multiple high-power laser pulses. Many types of attenuators have been used, for example, neutral density filters, silicon wafers, and liquid filters. Gelatin or glass neutral density filters are easily damaged by high power laser beams, but such filters as solutions of copper sulfate may be useful for high-power ruby laser beams. These liquid filters are not too susceptible to permanent damage.

The calibration of filters is a very difficult task, because filters can saturate in the presence of a high light flux. If a certain attenuation is measured for a filter when a relatively low light intensity is incident on the filter, the attenuation may be less for an intense laser beam. Filters may be calibrated by measuring the power incident on the filter and the power transmitted through the filter, but the measurements must be done in the same power range that the filter will be used.

A useful and widespread method of attenuating the beam before detection is to allow it to fall on a diffusely reflecting surface such as a magnesium oxide block. The goniometric distribution of the radiation reflected by such a surface is independent of the azimuth angle and depends on the angle θ from the normal to the surface in the following simple manner:

$$P_\omega \, d\omega = P_{tot} \cos \theta \, d\omega/\pi \qquad (7.4)$$

where P_ω is the power reflected into the solid angle $d\omega$ at angle θ from the normal, and P_{tot} the total power. This relation is called Lambert's cosine law, and a surface which follows such a law is called a Lambertian surface. There are many practical reflecting surfaces that follow this relation approximately. The power that reaches the detector after reflection from such a surface is given by

$$P_{detector} = P_{tot} \cos \theta \, A_d/\pi D^2 \qquad (7.5)$$

where A_d is the area of the photodetector (or its projection on the plane perpendicular to the line from the target to the detector), and D the distance from the reflector to the detector. This approximation is valid when D is much larger than the transverse dimension of the laser beam and the detector dimensions. With a Lambertian reflector, the power incident on the photosurface may simply be adjusted in a known way by changing the distance D. The beam may be spread over a large enough area on the Lambertian reflector so that the reflector surface is not damaged. The measurement of the power received by the detector, plus measurement of a few simple geometric parameters, gives the total beam power.

A thin membrane beam splitter called a pellicle is also used for attenuating laser beams. A pellicle inserted in a laser beam at an angle about 45° to the beam reflects approximately 8 % and is not easily damaged. Using several pellicles in series, each one reflecting the light from the previous pellicle in the series, one can easily reduce the intensity to a tolerable level.

One of the most widely used calibration methods involves measurement of total energy in the laser beam at the same time that the detector response is determined. The time history of the energy delivery is known from the shape of the detector output. Since the power integrated over the known pulse shape equals the total energy, the calibration is obtained in terms of laser power per unit response.

In many practical applications, one uses a calorimeter to calibrate a detector which is ultimately used as a monitor of laser output from one pulse to another. A small portion of the laser beam is picked off and sent to a detector and the remainder is delivered to a calibrated calorimeter. The total energy arriving at the calorimeter is measured. The output of the detector gives the pulse shape. Then, by numerical or graphical integration, the response of the detector can be calibrated relative to the calorimeter.

Electrical calibration of laser power meters is becoming fairly common. The absorbing element can be heated by an electrical resistance heater, and the power dissipation determined from electrical measurements. The measured response of the instrument to the known electrical input provides the calibration. It is implicitly assumed that deposition of a given amount of energy provides the same response, independent of whether the energy was radiant or electrical.

The difficulty of accurate measurement of radiant power on an absolute basis is well known. Different workers attempting the same measurement often obtain widely different results. This fact emphasizes the necessity for care in the calibration of laser detectors.

E. Modulators

Amplitude modulation of laser output is especially important for applications involving information storage and processing. Much effort has been devoted to development of methods for producing high frequency modulation. The most successful wide-bandwidth systems have involved electrooptical effects and acoustooptical effects.

1. Electrooptic Modulation

The operation of an electrooptic element is schematically shown in Fig. 7-11. Polarized light is incident on the modulator. The light may be polarized originally or a polarizer may be inserted as shown. The analyzer, oriented at 90° to the polarizer, will prevent light from being transmitted, when no voltage is applied. When the correct voltage is applied to the electrooptic switch, the plane of polarization of light transmitted through the switch can be rotated by 90°. Then light can pass through the analyzer. In practice, the electrooptic effects that have been employed are the Kerr electrooptic effect, which is shown by liquids such as nitrobenzene, and the Pockel's electrooptic effect, which is shown by crystalline materials such as

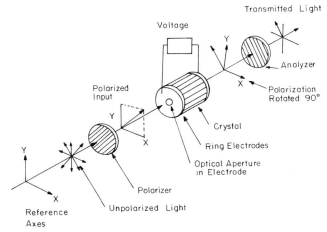

Fig. 7-11 Schematic diagram of the operation of an electrooptic modulator, in which the voltage is applied parallel to the direction of light propagation.

ammonium dihydrogen phosphate or potassium dihydrogen phosphate. The earliest electrooptic devices usually employed nitrobenzene. However, the nitrobenzene tends to polymerize under the action of intense light. Later models of electrooptic modulators usually use Pockel's elements.

In Fig. 7-11 the direction of polarization of the beam is shown as being at 45° to the vertical direction. The polarization can be considered to consist of two perpendicular components of equal intensity, one vertical and one horizontal. The crystal element is oriented with its crystalline axes in specified directions in relation to the applied voltage, with the exact orientation being dependent on the material. When the birefringence is induced by applying voltage through the electrodes, the two components of polarization have different velocities inside the crystal.

They travel in the same direction through the crystal and will not become physically separated. However the two components, originally in the same phase as they enter the crystal, will emerge with different phases. As the two beams traverse the sample, a phase difference accumulates between these beams, which is proportional to the distance travelled and which depends on the applied voltage. When the two beams emerge, they recombine to form a beam which will have a polarization depending on the relative phase. If the phase difference is one-half wavelength, it is equivalent to rotation of the direction of polarization by 90°. This does not by itself lead to any intensity change in the beam. However, with the analyzer, the transmission of the entire system can vary depending on the relation between the pass direction of the analyzer and the direction of the polarization of light emerging from the electrooptic cell. The transmission is given by

$$T = T_0 \sin^2(\pi \, \Delta n L / \lambda) \qquad (7.6)$$

where T is the transmission, T_0 the intrinsic transmission of the assembly taking into account all the losses, Δn the birefringence in the crystal (i.e., the difference in the index of refraction for the two different polarization states), L the length of the crystal, and λ the wavelength of the light. The birefringence Δn is a function of the applied voltage. The maximum transmission thus occurs when

$$\Delta n = \lambda / 2L \qquad (7.7)$$

This occurs at a value of voltage called the half-wave voltage, which is usually denoted $V_{1/2}$. The transmission of the electrooptic device will be an oscillatory function of applied voltage. The half-wave voltage for a particular electrooptic material increases with wavelength. This means that in the infrared, electrooptic modulators require higher operating voltages than in the visible. This limits the application of electrooptic modulators in the infrared.

A performance parameter for electrooptic modulators is the contrast ratio, defined as the ratio of the transmission when the device is fully open, to the transmission when it is fully closed. In practice, the transmission never becomes zero because of inevitable imperfections and light leakage. High contrast ratio is desirable because it defines the maximum signal-to-noise ratio that may be obtained in a system using the modulator. It also defines the minimum light leakage in the fully closed condition. Polarizers that are used with the modulator should not limit the performance, so high quality prism polarizers are usually recommended.

Electrooptic light modulators can be devised with a number of different physical forms. In one form, voltage may be applied parallel to the direction of light propagation, by using transparent electrodes on the ends of the crystal, or by using electrodes with central apertures. Such devices are called longitudinal modulators and are illustrated in Fig. 7-11.

In another type of construction, metallic electrodes are applied to the top and bottom of the crystal. The voltage is then perpendicular to the direction of light propagation. Such a configuration is called a transverse modulator. In a variant construction, a transverse modulator may be constructed in two sections, with the voltage applied in perpendicular directions in the two sections. This construction compensates for temperature changes of the birefringence.

Let us compare transverse and longitudinal modulators. For longitudinal modulators, the entrance aperture can be large, but the half-wave voltage will fall in the kilovolt range. Longitudinal modulators are often used with conventional light sources. The disadvantage of longitudinal modulators is the high voltage required. Because of the difficulties of changing kilovolt power supplies rapidly, the frequency response of longitudinal modulators tends to be low. Some applications in which longitudinal modulators might be used are impressing relatively low frequency (e.g., audio) information on laser beams, or relatively slow switching of the intensity or polarization of a beam of laser light.

In transverse modulators, where the electric field is applied at right angles to the direction of light propagation, one may increase the amount of phase retardation for a given applied voltage simply by making the crystal longer, without increasing the width. The length and width of the crystal may be completely independent. The crystal elements used in transverse modulators may be long thin parallelepipeds. The half-wave voltage may be much lower for such transverse modulators. It follows that the frequency response can be much better. Transverse modulators may often be used for fast broadband applications. A problem with transverse modulators is the relatively small angular aperture they permit. The low half-wave voltage and flat frequency response characteristics of transverse modulators make them ideal for fast, broadband applications but several restrictions must be observed.

The physical configurations of the crystal elements require use with well-defined laser beams only. Contrast ratios achievable with transverse devices are normally an order of magnitude below those of longitudinal modulators and external optical techniques must be used to produce comparable results.

Applications of transverse field modulators include broadband optical communication, display and printing systems and fast image and signal recorders. Table 7-2 lists some properties of commercially available electro-optic light beam modulators.

TABLE 7-2

Electrooptic Modulators

Material	Spectral range (μm)	Bandwidth (MHz)	Typical voltage requirements (V)
Potassium dihydrogen phosphate	0.3–1.1	~100	~7000
Deuterated potassium dihydrogen phosphate	0.3–1.1	~100	~3000
Ammonium dihydrogen phosphate	0.3–1.2	~100	300
Deuterated ammonium dihydrogen phosphate	0.3–1.2	~100	200
Lithium niobate	0.5–2	150	50
Lithium tantalate	0.4–1.1	400	35
Gallium arsenide	10.6	5	
Cadmium telluride	10.6	100	5000

2. Acoustooptic Modulators

A different thrust in modulator technology has been the utilization of interactions between light and sound waves to produce changes in optical intensity, phase (frequency), and propagation direction. Operation of acoustooptic modulators is based on the diffraction of a light beam by a sound column in a suitable interaction medium. A piezoelectric transducer is attached to the medium and excited by an rf driver to launch a traveling acoustic wave. The elastooptic properties of the medium result in a spatially periodic disturbance of the refractive index. A light beam incident on this spatial array at the proper angle is partially deflected in much the same way that a diffraction grating deflects light. The operation is shown in Fig. 7-12. The alternate rarefactions and compressions associated with the sound wave form a grating which diffracts the incident light beam. No light can be deflected out of its path unless the sound wave is present in the acoustooptic material.

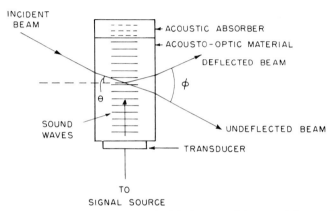

Fig. 7-12 Operation of an acoustooptic light beam modulator or deflector, defining the angles θ and ϕ used in the text.

For a material with a fixed acoustic velocity, the acoustic wavelength or grating spacing is a function of the rf drive frequency; the acoustic wavelength controls the deflection angle. The amplitude of the disturbance, a function of the rf power applied to the transducer, dictates the fractional quantity of optical energy deflected and may be used as an intensity control. Modulation of the light beam is then achieved by maintaining a constant rf frequency, allowing only the deflected beam to emerge from the assembly, and modulating the rf power applied to the transducer.

The transmission characteristic of an acoustic modulator is given as

$$T = T_0 \sin^2(\beta P_x^{1/2}/\lambda_0) \tag{7.8}$$

where β is a material and configurational parameter and P_x the acoustic power supplied to the medium. The inherent transmission, T_0, is a function of reflective and absorptive losses in the device.

The design and performance of acoustooptic light beam modulators have several restrictions. The transducer and the acoustooptic medium must be carefully designed to provide maximum light intensity in a single diffracted beam, when the modulator is in the open condition. The transit time of the acoustic wave across the diameter of the light beam imposes a limitation on the rise time of the switching and therefore limits the modulation bandwidth. The acoustic wave travels with a finite velocity and the light beam cannot be switched fully on or fully off until the acoustic wave has traveled all the way across the diameter of the light beam. Therefore one focuses the light beam to as small diameter as possible to achieve minimum transit time and maximum bandwidth. Frequently the diameter to which the beam can be focused is the ultimate limitation on the bandwidth. If the laser beam has relatively high power, it cannot be focused in the acoustooptic medium without causing optical damage.

Acoustooptic light beam modulators have a number of important advantages for use in systems. The electrical power required to excite the acoustooptic wave is relatively small. Often only a few watts of electrical drive power are required. Very high contrast ratios may be obtained easily, because when the electrical power is turned off, no light will emerge in the direction of the diffracted beam. Acoustooptic devices may be compact and offer an advantage for systems in which size and weight are important. The characteristics of some acoustooptic light beam modulators are presented in Table 7-3.

TABLE 7-3
Acoustooptic Modulators

Material	Spectral range (μm)	Bandwidth (MHz)	Typical drive power (W)
Lead molybdate	0.4–1.2	23	2.25
Quartz	0.4–1.5	8	30
Germanium	10.6	5	30.5

F. Deflectors

Light beam deflectors, sometimes called scanners, form an important class of accessory which is required for applications such as display, laser printers, and optical data storage. Three main methods have been employed for light beam deflection: mechanical, electrooptical, and acoustooptical. The capabilities of present and projected systems are summarized in Fig. 7-13 [1]. Light beam deflection is measured in terms of "resolvable spots," rather than in terms of absolute angle of deflection. One resolvable spot means deflecting the beam by an amount equal to its own angular spread.

The access time is the random access time to acquire one spot or resolution element.

We shall now describe these methods in more detail individually.

1. *Mechanical Deflectors*

Typical mechanical deflectors consist of a mirror or prism which is driven by galvanometric or piezoelectric means. The advantages of the mechanical system are

(1) a large number of resolvable spots or large deflection angle,
(2) low loss of deflected light beam,
(3) relatively low driving power, and
(4) wide variety in scanning mode.

In general, the deflection accuracy and scanning speed are both low in the mechanical deflectors because of the inertia of a moving mass. However, recent advances in fabrication techniques are gradually overcoming these shortcomings. Figure 7-13 shows that mechanical techniques such as rotating or vibrating mirrors are good for applications where many resolvable spots are required and that where higher speed of deflection is needed, acoustooptic or electrooptic techniques are needed.

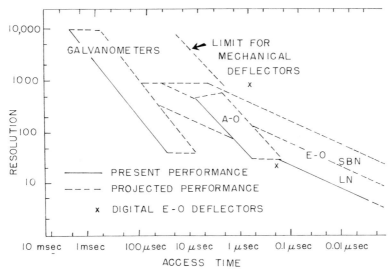

Fig. 7-13 Present and projected status of various types of light beam deflectors. The notation SBN stands for strontium barium niobate and LN for lithium niobate. Acoustooptic and electrooptic deflectors are denoted AO and EO respectively. (From J. D. Zook, *Appl. Opt.* **13**, 875 (1974).)

2. Electrooptic Deflectors

Electrooptic techniques involve application of a voltage to certain crystals (e.g., potassium dihydrogen phosphate, lithium niobate) to change the index of refraction and hence the direction of the beam traveling through a prism of the material. Figure 7-14 shows a typical prism setup for an electrooptical light beam deflector element.

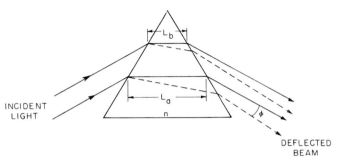

Fig. 7-14 Prism setup for an electrooptical light beam deflector element, where *n* is the index of refraction of the material and φ the angle of deflection. The figure defines the widths L_a and L_b used in the text.

The deflection angle of a plane wave passing through such a prism is obtained by applying Snell's law at the boundaries of the prism. The number *N* of resolvable spots is given by

$$N = \Delta n(L_a - L_b)/\varepsilon\lambda \tag{7.9}$$

In this equation, Δn is the induced change in index of refraction, which is a function of the applied voltage, L_a and L_b are the widths defined in Fig. 7-14, ε is a factor of the order of unity and λ is the wavelength.

A common construction is shown in Fig. 7-15. This uses two prisms, one of which has a positive value of Δn and one of which has a negative value of Δn, when voltage is applied to the electrodes as shown.

The electrooptic deflectors were extensively investigated in the early stages of the laser age. These deflectors inherently exhibit fast response and may be considered likely to survive long in the future. However, suitable electrooptic materials, with high optical quality throughout a sufficiently large area, are not available at present. This obstacle of material unavailability limits the use of the electrooptic deflector for the present. Also, requirements on operating power are uncomfortably high.

Figure 7-13 shows projected performance limits for electrooptic deflectors. The curve marked LN refers to lithium niobate and the curve marked SBN refers to strontium barium niobate. Electrooptic deflectors

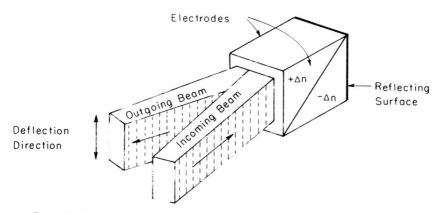

Fig. 7-15   A possible construction of an electrooptic light beam deflector using two prisms, one of which has a positive value of Δn and the other of which has a negative Δn.

appear most useful for deflectors which require fast response, but for which the number of resolvable spots can be small.

So far we have described analog light beam deflectors, in which the deflection will change continuously with the applied voltage. One may also construct digital electrooptic light beam deflectors for which there may be a certain number of preset beam positions. A basic building block in a digital light beam deflector will be an electrooptic polarization switch followed by a birefringent crystal. The polarization switch will switch the polarization of the light beam between two mutually perpendicular states, depending on whether voltage is on or off. The two different polarizations will emerge from the birefringent crystal in either of two output locations, depending on the polarization. Thus a single such unit will provide two positions, between which the beam may be switched back and forth at will. If one arranges N such units in a series, one will have 2^N separate locations for the emerging beam. Such a digital deflector eases the requirements on voltage control, as compared to the analog deflector. If one applied voltage is not exactly correct in an analog deflector, the beam will go to a position which is not right. For a digital deflector, if the voltage is slightly wrong, the beam will still go to the correct position, only with somewhat reduced contrast ratio. Some models of digital light beam deflectors that have been developed are indicated by the crosses in Fig. 7-13.

3.   *Acoustooptic Deflectors*

Acoustic waves propagating from a flat, thickness-mode piezo-electric transducer into a crystal form almost planar wave fronts traveling in the crystal. Light rays passing through the crystal approximately parallel to the acoustic wave fronts are diffracted by the phase grating formed by the acoustic waves. If the light strikes the acoustic wave fronts at the proper angle, the light appears to be reflected from these fronts. This diffraction satisfies the same relationship as in the x-ray case and thus is known as Bragg reflection. The Bragg angle is defined by the equation

$$\sin \theta = \lambda/2n\Lambda \qquad (7.10)$$

where λ is the light wavelength, Λ the acoustic wavelength, and n the index of refraction of the crystal.

The operation has already been illustrated in Fig. 7-12 in connection with acoustooptic modulators. When an acoustooptic device is used as a deflector, the acoustic wavelength is varied in order to change θ. One then has for ϕ, the angle formed by the undeflected and deflected beams,

$$\phi \simeq \sin \phi = \lambda/\Lambda \qquad (7.11)$$

The bandwidth is limited by the time it takes for the acoustic wave front to propagate across the diameter of the light beam.

Materials which have been employed for acoustooptic deflectors include fused silica with a quartz acoustic transducer or lead molybdate with a lithium niobate acoustic transducer.

The acoustooptic deflector has several advantages:

(1) Capability of high capacity.
(2) Ease in control of deflection modes and position.
(3) Simple structure of deflector element.
(4) Wide variety of usage from a single deflector.

The present and projected status of acoustooptic deflectors is shown in Fig. 7-13. Commercial models offering 100–1000 resolvable spots are becoming available, with deflection times of the order of a few microseconds. Two-dimensional deflectors can be obtained by placing two deflectors in series, with their directions of deflection at right angles.

G. Q-switches

We have already described the operation of a Q-switch in Chapter 2, using a mechanical type of Q-switch to illustrate the operation. There are a variety of different types of Q-switch which have been employed. In describing the methods other than the mechanical method, we shall use a solid state laser as an illustration, but the basic methods are applicable to other types of lasers.

The other methods involve a variable absorber between the laser rod and the mirrors. During the time while the flashtube pumps the rod, the absorber is opaque and laser action is held off. When the absorber is suddenly switched to a transparent condition, the laser rod can see the output mirror. The energy stored in the rod can be extracted by the stimulated emission process and emitted through the output mirror in a short pulse.

The various types of absorber that can be used define three other classes of Q-switch. The three types of variable absorber that are commonly used are electrooptic elements, acoustooptic elements, and bleachable dyes.

The use of the electrooptic and acoustooptic effects in light modulators has been described in earlier sections of this chapter. It is obvious that a Q-switch is a specialized type of modulator. The operation of an electro-optic Q-switch is schematically shown in Fig. 7-16 with reference to a ruby laser. Crystalline anisotropy of the ruby polarizes the light in a horizontal direction.† The analyzer will prevent this light from reaching the output mirror. When a voltage is applied to the electrooptic switch, the plane of polarization of light transmitted through the switch is rotated by 90°. Then light from the ruby can pass through the analyzer and reach the output mirror. In practice, the electrooptic effects that have been employed are the Kerr electrooptic effect, which is shown by liquids such as nitrobenzene, and the Pockel's electrooptic effect, which is shown by crystalline materials such as ammonium dihydrogen phosphate or potassium dihydrogen phosphate. The earliest electrooptic Q-switches usually employed nitrobenzene. Later models of electrooptic Q-switches usually use Pockel's cells.

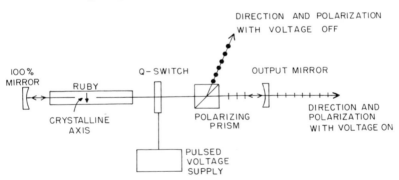

Fig. 7-16 Application of an electrooptic Q-switch in a ruby laser.

Acoustooptic Q-switches rely on the diffraction grating set up by sound waves in certain materials, as discussed above. The light beam can be deflected out of its path when a sound wave is present in the acoustooptic material. The operation of an acoustooptic Q-switch is shown in Fig. 7-17. When the acoustic wave is present, some of the light is diffracted and misses the mirror.

† See the discussion in Chapter 3 about crystalline properties of ruby.

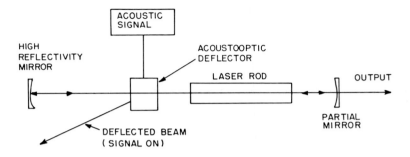

Fig. 7-17 Application of an acoustooptic Q-switch in a solid state laser.

The angle of diffraction, usually around 0.5°, is exaggerated in the figure. The energy loss due to the diffraction is enough to drive the laser below threshold. When the pumping of the laser is complete, the acoustic wave is suddenly switched off. Then all the light reaches the mirror and a laser pulse builds up rapidly. Materials which have been employed for acoustooptic Q-switches are fused silica with a quartz acoustic transducer or lead molybdate with a lithium niobate acoustic transducer.

Acoustooptic Q-switches are commonly used with Nd:YAG lasers operating at 1.06 μm. The voltage required for operation of an electrooptic Q-switch increases with wavelength, and bleachable dyes that are available for Q-switching in the infrared have less desirable properties than those in the visible; the acoustooptic Q-switching has proved to be more useful in the infrared.

Bleachable dye Q-switches initially absorb light at the laser wavelength, but under the action of light, the absorption saturates. The transmission of a particular bleachable dye material at 0.6943 μm is shown in Fig. 7-18 [2] as a function of incident light intensity. This dye, vanadium phthalocyanine in nitrobenzene solution, has low transmission at low light intensity. Thus a cell filled with this material can isolate a ruby laser rod from the mirror. As pumping proceeds, the fluorescent light intensity from the ruby builds up and the dye bleaches slightly, allowing some light to reach the mirror. This provides further light amplification and hence further bleaching. At high light intensities the solution has become almost 100% transparent.

The physical basis for this behavior arises from a saturation of all the available absorbing dye molecules. The dye molecules are absorbing, and are raised by the absorption of light from the ground state, to an excited state, which has no absorption. The molecules should remain in the excited state for a time longer than the laser pulse duration, before decaying back to the

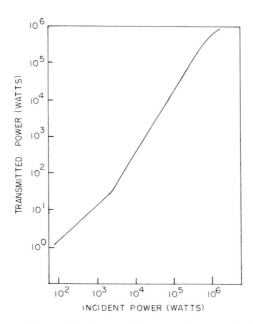

Fig. 7-18   Transmission of the dye vanadium phthalocyanine in nitrobenzene solution as a function of incident light intensity. This indicates the capabilities of this dye for bleaching at high light intensities. (From J. A. Armstrong, *J. Appl. Phys.* **36**, 471 (1965).)

ground state. Since the number of dye molecules is limited, they can all be raised by intense light to the excited state; the dye has then bleached. Between laser pulses, the molecules decay back to the absorptive state. Thus, the switch can operate for many pulses without degradation.

Perhaps the most common dye is cryptocyanine in solution in methyl alcohol, used for ruby lasers. Other dye materials [examples are chloro-aluminum phthalocyanine and DDI (1,1'-diethyl-2,2'-dicarbocyanine iodide)] and other solvents (examples are acetone, nitrobenzene, aceto-nitrile, and chloronaphthalene) have been used for ruby lasers. For neody-mium: glass lasers, the so-called 9740 and 9860 dye solutions are often used.

The bleachable dyes are the simplest and least expensive type of *Q*-switch, but they offer somewhat less desirable properties than electrooptic and acoustooptic devices. If synchronization with external events is impor-tant, bleachable dyes are not suitable.

Table 7-4 compares properties of the various types of *Q*-switch. Bleach-able dyes offer the simplest and lowest cost systems, but the properties of the output pulse are less good than when the more expensive electrooptic or acoustooptic elements are used.

TABLE 7-4
Comparison of Q-switching elements

Type of switch	Cost	Pulse to pulse stability	Switching speed	Beam profile	Comments
Rotating mirror	Medium	Good	Slow	Good	Used in early devices, less common now
Electrooptic element	High	Good	Fast	Good	Often used with ruby
Acoustooptic element	High	Good	Medium	Good	Usually used with Nd:YAG
Bleachable dye	Low	Fair	Medium	Poor	Often used with ruby and Nd:glass

H. Nonlinear Optical Elements

The electromagnetic field associated with the passage of a light wave through a material produces a polarization of the material. For simplicity we will consider only the electric polarization P. In the classical discussion of wave propagation, the polarization is treated as being proportional to the electric field E. The polarization of a dielectric medium by an applied electric field may result from orientation of molecules under the action of the field. For isotropic materials, the vector P is parallel to the vector E. Thus, for simplicity we will drop the vector notation. If E varies sinusoidally, the polarization will vary sinusoidally also. When the intensity of the light propagating through the material is low, the linear view of polarization being proportional to the electric field is valid. However, the intensity available with lasers may exceed the limits of the linear theory. In a one-dimensional form, the polarization P may be written

$$P = XE(1 + a_2 E + a_3 E^2 + \cdots) \tag{7.12}$$

Here X is the linear polarizability and the a's are nonlinear coefficients, which are generally much smaller than unity. If the electric field is suitably small, the variation of P with an applied electric field of the form $E_0 \cos \omega t$ will be of the form $P_0 \cos \omega t$. However, when the term $a_2 E$ becomes comparable to unity, the second term will contribute a polarization

$$P_2 = a_2 XE_0^2 \cos^2 \omega t = \tfrac{1}{2} a_2 XE_0^2 (1 + \cos 2\omega t) \tag{7.13}$$

Consider the form $(1 + \cos 2\omega t)$. The first term is a dc component. The second term, which oscillates at frequency 2ω, results in generation of a second harmonic. Higher harmonics can also arise from the higher-order terms. They are generally smaller because the coefficients decrease rapidly in the series represented by Eq. (7.12).

In many materials the coefficient a_2 will be identically zero. For a_2 to be other than zero, one requires a crystalline material in which the crystalline cell does not have a center of symmetry. For such materials, the passage of a laser wave at frequency ω and with suitably high intensity will cause a polarization with a component at frequency 2ω, i.e., the second harmonic of the incident light. This polarization then can radiate second harmonic radiation so that one has the possibility of producing radiation at a frequency twice that of the incident frequency.

The intensity $I_{2\omega}$ of the frequency-doubled radiation after the light has traversed a thickness L of the material varies as

$$I_{2\omega} = \sin^2[(2\pi/\lambda_1)(n_1 - n_2)L]/(n_1 - n_2)^2 \qquad (7.14)$$

where λ_1 is the incident wavelength, and n_1 and n_2 are the indices of refraction for the original and frequency-doubled wavelengths, respectively. If $n_1 = n_2$, the intensity will increase as the square of crystal thickness, at least until the power in the incident beam becomes depleted. However, the index of refraction is a function of wavelength, and generally $n_1 \neq n_2$. Then the original and frequency-doubled waves will get out of phase and I_2 will reach a maximum value when the crystal thickness is

$$L_c = \tfrac{1}{4}(\lambda_1/|n_1 - n_2|) \qquad (7.15)$$

This thickness, called the coherence length, will be around 10 μm for typical crystals. This would severely limit the power that could be converted into the second harmonic radiation. This limitation can be avoided by the technique of phase-matching. In some birefringent crystals, one can choose combinations of direction of propagation and directions of the polarization of the two beams so that $n_1 = n_2$. Then the intensity $I_{2\omega}$ can build up without the two waves getting out of phase. Exact matching is achieved by varying the angular orientation of the crystal or by varying its temperature or by both. With phase matching, conversion of the incident power into second harmonic power is possible with high conversion efficiency.

In this brief discussion, which greatly abbreviates the very extensive subject of nonlinear optics, we have emphasized the production of second harmonic radiation. Many other phenomena have been investigated. One can mix two beams of different frequencies ω_1 and ω_2 and obtain outputs at the sum frequency $\omega_1 + \omega_2$ or at the difference frequency $\omega_1 - \omega_2$. In this way, many different combinations of wavelengths can be derived, starting with a few basic laser wavelengths. Commercial models of such devices are starting to become available, and a brief sampling of some experimental results has been presented in Chapter 4. Frequency-doubled Nd:YAG lasers are available with output at 0.53 μm and frequency-doubled tunable lasers with the original output in the visible can provide tunable output in the ultraviolet.

REFERENCES

[1]   J. D. Zook, *Appl. Opt.* **13**, 875 (1974).
[2]   J. A. Armstrong, *J. Appl. Phys.* **36**, 471 (1965).

SELECTED ADDITIONAL REFERENCES

A. Mirrors

V. R. Costich, Multilayer Dielectric Coatings, *in* "Handbook of Lasers" (R. J. Pressley, ed.),
Chemical Rubber Co., Cleveland, Ohio, 1971.
O. S. Heavens, "Thin Film Physics," Methuen, London, 1970.
Z. Knittl, "Optics of Thin Films," Wiley, New York, 1976.
T. T. Saito *et al.*, 10.6 μm Absorption Dependence on Roughness of UHV-Coated Supersmooth
Mirrors, *Appl. Opt.* **14**, 266 (1975).
T. T. Saito and J. R. Kurdock, Diamond Turning and Polishing of Infrared Optical Compo-
nents, *Appl. Opt.* **15**, 28 (1976).
W. J. Spawr and R. L. Pierce, Guide to Metal Mirrors, *Laser Focus*, p. 37 (March 1975).

B. Polarizers

M. Born and E. Wolf, "Principles of Optics," Pergamon, Oxford, 1975, Chapter 14.
E. S. Hass, Polarization and Lasers, Laser Tech. Bull. No. 7, Spectra-Physics, Inc., Mountain
View, California (1975).
F. A. Jenkins and H. E. White, "Fundamentals of Optics," McGraw-Hill, New York, 1957.

C. Infrared Materials

R. H. Anderson *et al.*, Preparation of High-Strength KCl by Hot-Pressing, *J. Am. Ceram. Soc.*
56, 287 (1973).
C. R. Andrews and C. L. Strecker (eds.), *Proc. Annu. Conf. Infrared Laser Window Mater.*, *4th
Tucson, Arizona* November 18–20, 1974, Sponsored by Advanced Research Projects
Agency, Arlington, Virginia.
E. Bernal G., Heat Flow Analysis of Laser Absorption Calorimetry, *Appl. Opt.* **14**, 314 (1975).
A. Feldman, Measuring the Optical Properties of Infrared Laser Windows, *Electro-Opt.
Systems Design*, p. 36 (December 1976).
D. T. Gillespie, A. L. Olsen, and L. W. Nichols, Transmittance of Optical Materials at High
Temperatures in the 1 μ to 12 μ Range, *Appl. Opt.* **4**, 1489 (1965).
J. A. Harrington, D. A. Gregory, and W. F. Otto, Infrared Absorption in Chemical Laser
Window Materials, *Appl. Opt.* **15**, 1953 (1976).
M. Hass *et al.*, Infrared Absorption Limits of HF and DF Laser Windows, *Appl. Phys. Lett.*
25, 610 (1976).
P. W. Kruse, L. D. McGlauchlin, and R. B. McQuistan, "Elements of Infrared Technology,"
Wiley, New York, 1962, Chapter 4.
F. M. Lussier, Guide to IR-Transmissive Materials, *Laser Focus*, p. 47 (December 1976).
D. E. McCarthy, The Reflection and Transmission of Infrared Materials. I, Spectra from 2–50
Microns, *Appl. Opt.* **2**, 591 (1963).
D. E. McCarthy, The Reflection and Transmission of Infrared Materials. VI: Bibliography,
Appl. Opt. **7**, 2221 (1968).
P. Miles, High Transparency Infrared Materials, *Opt. Eng.* **15**, 451 (1976).
A. V. Nurmikko, Fundamental Absorption Edge in Laser Window CdTe, *Appl. Opt.* **14**, 2663
(1975).

C. A. Pitha, H. Posen, and A. Armington (eds.), *Proc. Conf. High Power Infrared Laser Window Mat., 3rd, Hyannis, Massachusetts* November 12–14, 1973, Sponsored by Air Force Cambridge Res. Lab., Bedford, Massachusetts.

M. Sparks, Optical Distortion by Heated Windows in High Power Laser Systems, *J. Appl. Phys.* **42**, 5029 (1971).

D. Detectors

L. K. Anderson and B. J. McMurtry, High-Speed Photodetectors, *Appl. Opt.* **10**, 1573 (1966).

H. P. Beerman, Advances in Pyroelectric Infrared Detectors, *Proc. Tech. Program, Electro-Opt. Syst. Design Conf., New York* (September 12–14, 1972).

D. E. Bode, Optical Detectors, *in* " Handbook of Lasers " (R. J. Pressley, ed.), Chemical Rubber Co., Cleveland, Ohio 1971.

E. L. Danahy, The Real World of Silicon Photo-Diodes, *Electro-Opt. Systems Design*, p. 36 (May 1970).

W. M. Doyle, The New Frontier in Thermal Detection, *Opt. Spectra*, p. 21 (November 1972).

J. Geist, L. B. Schmidt, and W. E. Case, Comparison of the Laser Power and Total Irradiance Scales Maintained by the National Bureau of Standards, *Appl. Opt.* **12**, 2773 (1973).

S. M. Johnson, Photomultiplier Tubes: Surviving the Rugged Life, *Electro-Opt. Systems Design*, p. 20 (March 1972).

R. J. Keyes and R. H. Kingston, A Look at Photon Detectors, *Phys. Today*, p. 48 (March 1972).

T. Koehler, HgCdTe Photodiodes for CO_2 Laser Systems, *Electro-Opt. Systems Design*, p. 24 (October 1974).

P. W. Kruse, L. D. McGlauchlin, and R. B. McQuistan, "Elements of Infrared Technology," Wiley, New York, 1962, Chapter 10.

R. J. Locker and G. C. Huth, A New Ionizing Radiation Detection Concept Which Employs Semiconductor Avalanche Amplification and the Tunnel Diode Element, *Appl. Phys. Lett.* **9**, 227 (1966).

V. J. Mazurczyk, R. N. Graney, and J. B. McCullough, High Performance, Wide Bandwidth (Hg, Cd) Te Detectors, *Opt. Eng.* **13**, 307 (1974).

B. C. McIntosh and D. W. Sypek, Pyroelectric Detector Arrays, *Laser Focus*, p. 38 (December 1972).

D. B. Medved, Photodiodes for Fast Receivers, *Laser Focus*, p. 45 (January 1974).

National Materials Advisory Board, Materials for Radiation Detection, Publ. NMAB 287, Nat. Acad. of Sci., Washington, D.C. (1974).

R. P. Riesz, High Speed Semiconductor Photodiodes, *Rev. Sci. Instrum.* **33**, 994 (1962).

T. W. Russell, A Procedure for Intercomparing Laser Power Meters, *Electro-Opt. Systems Design*, p. 28 (August 1974).

W. B. Tiffany, Detectors—You have to Know Them, *Opt. Spectra*, p. 22 (February 1975).

P. H. Wendland, Silicon Photodiodes Revisited, *Electro-Opt. Systems Design*, p. 48 (August 1970).

P. H. Wendland, Silicon Photodiodes Come into Their Own, *Opt. Spectra*, p. 33 (October 1973).

E. Modulators

F. S. Chen, Modulators for Optical Communications, *Proc. IEEE* **58**, 1440 (1970).

D. E. Flinchbaugh, Acousto-Optic Laser Modulation Today, *Electro-Opt. Systems Design*, p. 24 (January 1974).

R. Goldstein, Pockels Cell Primer, *Laser Focus*, p. 21 (February 1968).

I. P. Kaminow and E. H. Turner, Electrooptic Light Modulators, *Appl. Opt.* **5**, 1612 (1966).

I. P. Kaminow, "An Introduction to Electrooptic Devices," Academic Press, New York, 1974.

T. Nowicki, A–O and E–O Modulators, Basics and Comparisons, *Electro-Opt. Systems Design*, p. 23 (February 1974).

M. T. V. Scibor-Rylski, Calling Out Light Modulators, *Opt. Spectra*, p. 30 (February 1976).

F. Deflectors

H. J. Aronson, Acousto-Optic Scanning, *Laser Focus*, p. 36 (December 1976).

L. Beiser, Laser Scanning Systems, *in* "Laser Applications," Vol. 2 (M. Ross, ed.), Academic Press, New York, 1974.

P. J. Brosens, Fast Retrace Optical Scanning, *Electro-Opt. Systems Design*, p. 21 (April, 1971).

R. Compton, Optical Scanners; Comparisons and Application, *Electro-Opt. Systems Design*, p. 16 (February 1976).

V. J. Fowler and J. Schlafer, A. Survey of Laser Beam Deflection Techniques, *Appl. Opt.* **5**, 1675 (1966).

E. I. Gordon, A Review of Acoustooptical Deflection and Modulation Devices, *Appl. Opt.* **5**, 1629 (1966).

E. P. Grenda and P. J. Brosens, Closing the Loop on Galvo Scanners, *Electro-Opt. Systems Design*, p. 32 (April 1974).

U. Kruger, R. Pepperl, and U. J. Schmidt, Electrooptic Materials for Digital Light Beam Deflectors, *Proc. IEEE* **61**, 992 (1973).

W. Kulcke *et al.*, Digital Light Deflectors, *Appl. Opt.* **5**, 1657 (1966).

T. C. Lee and J. D. Zook, Light Beam Deflection with Electrooptic Prisms, *IEEE J. Quantum Electron.* **QE-4**, 442 (1968).

J. F. Lotspeich, Electrooptic Light-Beam Deflection, *IEEE Spectrum*, p. 45 (February 1968).

H. M. Tenney and J. C. Purcupile, Laser/Galvo Scanner Design, *Electro-Opt. Systems Design*, p. 40 (October 1975).

N. Uchida and N. Niizeki, Acoustooptic Deflection Materials and Techniques, *Proc. IEEE* **61**, 1073 (1973).

J. D. Zook, Light Beam Deflector Performance: A Comparative Analysis, *Appl. Opt.* **13**, 875 (1974).

G. Q-Switches

J. A. Armstrong, Saturable Optical Absorption in Phthalocyanine Dyes, *J. Appl. Phys.* **36**, 471 (1965).

R. C. Benson and M. R. Mirarchi, The Spinning Reflector Technique for Ruby Laser Pulse Control, *IEEE Trans. Military Electron.*, p. 13 (January 1964).

R. B. Chesler, M. A. Karr, and J. E. Geusic, An Experimental and Theoretical Study of High Repetition Rate Q-Switched Nd:YAlG Lasers, *Proc. IEEE* **58**, 1899 (1970).

T. F. Deutsch and M. J. Weber, Optical Properties of Cryptocyanine, *J. Appl. Phys.* **37**, 3629 (1966).

J. E. Geusic, M. L. Hensel, and R. G. Smith, A Repetitively Q-Switched, Continuously Pumped YAG:Nd Laser, *Appl. Phys. Lett.* **6**, 175 (1965).

R. W. Hellwarth, Q Modulation of Lasers, *in* "Lasers, A Series of Advances," Vol. 1 (A. K. Levine, ed.), Dekker, New York, 1966.

D. Hull, Combination Laser Q-Switch Using a Spinning Mirror and Saturable Dye, *Appl. Opt.* **5**, 1342 (1966).

D. Maydan and R. B. Chesler, Q-Switching and Cavity Dumping of Nd:YAlG Lasers, *J. Appl. Phys.* **42**, 1031 (1971).

F. J. McClung and R. W. Hellwarth, Giant Optical Pulsations from Ruby, *J. Appl. Phys.* **33**, 828 (1962).

F. J. McClung and R. W. Hellwarth, Characteristics of Giant Optical Pulsations from Ruby, *Proc. IEEE* **51**, 46 (1963).

M. Seiden, Mechanical Q-Switch Revisited, *Laser Focus*, p. 40 (July 1971).

C. H. Skeen and C. M. York, The Operation of a Neodymium Glass Laser Using a Saturable Liquid Q-Switch, *Appl. Opt.* **5**, 1463 (1966).

R. G. Smith and M. F. Galvin, Operation of the Continuously Pumped, Repetitively Q-Switched YAlG:Nd Laser, *IEEE J. Quantum Electron.* **QE-3**, 406 (1967).

B. H. Soffer, Giant Pulse Laser Operation by a Passive, Reversibly Bleachable Absorber, *J. Appl. Phys.* **35**, 2551 (1964).

P. P. Sorokin *et al.*, Ruby Laser Q-Switching Elements Using Phthalocyanine Molecules in Solution, *IBM J. Res. Develop.*, p. 182 (April 1964).

J. L. Wentz, Novel Laser Q-Switching Mechanism, *Proc. IEEE* **52**, 717 (1964).

E. J. Woodbury, Five-Kilohertz Repetition-Rate Pulsed YAG:Nd Laser, *IEEE J. Quantum Electron.* **QE-3**, 509 (1967).

H.   Nonlinear Optical Elements

R. S. Adhav and M. Orszag, Frequency Doubling Crystals—Unscrambling the Acronyms, *Electro-Opt. Systems Design*, p. 20 (December 1974).

R. W. Minck, R. W. Terhune, and C. C. Wang, Nonlinear Optics, *Appl. Opt.* **5**, 1595 (1966).

S. Singh, Non-Linear Optical Materials, *in* "Handbook of Lasers" (R. J. Pressley, ed.), Chemical Rubber Co., Cleveland, Ohio, 1971.

R. W. Terhune and P. D. Maker, Nonlinear Optics, *in* "Lasers, A Series of Advances," Vol. 2 (A. K. Levine, ed.), Dekker, New York, 1968.

A. Yariv, "Quantum Electronics," Wiley, New York, 1975, Chapter 16.

F. Zernike and J. E. Midwinter, "Applied Nonlinear Optics," Wiley, New York, 1973.

CHAPTER 8

ALIGNMENT AND TOOLING

One of the most common applications of lasers in industry arises in the solution of alignment problems for which knowledge of displacement from a reference line is needed. Use of laser alignment techniques has replaced older and more tedious methods of performing such tasks as machine tool alignment. Such applications are possible because of the specialized properties of laser radiation. The properties that are of particular interest for alignment are the directionality of the beam and the brightness.

A collimated laser beam has an irradiance distribution given by the Gaussian distribution

$$I(r) = I_0 \exp -(2r^2/r_0{}^2) \tag{8.1}$$

where r is the radial distance from the beam center, and r_0 the radius at which the power density I is $1/e^2$ of the value I_0 at the center of the beam. This equation holds for lasers operating in the TEM_{00} mode. The intensity profile will remain Gaussian at all distances from the laser. For a Gaussian beam the divergence angle θ is

$$\theta = 2\lambda/\pi R_0 \tag{8.2}$$

where λ is the wavelength and R_0 the radius at the exit aperture of the laser. This defines the full angle of the cone into which the light beam diverges at distances far from the laser. For a typical helium–neon laser with an exit diameter around 1 mm, the beam divergence angle will typically be of the order of 10^{-3} rad. With external optics to collimate the beam, the divergence angle can be reduced even further.

The brightness is important because the laser beam will be easily visible even in an ambient background of room light or day light. Virtually all practical laser systems for alignment have employed the helium–neon gas laser, typically at power levels around 1–5 mW. The orange–red beam from such a laser is clearly visible, even in a sunlit background at fairly large distances from the laser. It is unnecessary to employ lasers capable of emitting higher powers. The helium–neon laser, moreover, is the cheapest, most reliable, and most durable of currently available lasers emitting visible light. This makes it an ideal choice for applications requiring a visible beam where the power requirements are not high.

A.   Laser Tooling

Consider the problem of positioning objects along a straight line. An obvious method is to project a beam of light and measure the deviation of the objects from the center of the beam. Use of a beam of light for alignment can offer many advantages, particularly if the alignment is to be carried out over great distances. Mechanical methods of alignment are difficult and cumbersome. Laser alignment systems have become available for aligning large structures and large machines to greater accuracy and at lower cost than is possible with mechanical methods.

The practice of optical alignment has come to be called optical tooling and has been in use for a number of years. Optical tooling predates the invention of lasers. The term "optical tooling" originated in the aircraft industry. Optical methods for alignment of components of an aircraft (such as wing or fuselage sections) have been in use for many years. Optical tooling at first employed instruments such as autocollimators to measure rotation of mirror targets and precision alignment telescopes, capable of measuring the displacement of objects from the optical axis of the telescope.

The development of the laser led to significant advances in optical tooling. Lasers have been specifically developed for optical tooling applications, and as such are sometimes called "tooling lasers." Laser tooling involves the use of the unidirectional output beam in such tasks as determining displacement of objects from a line, determining angular alignment, establishing planes, performing leveling, and establishing right angles.

A typical small helium–neon laser has an angular divergence angle of the order of 10^{-3} rad. The accuracy of laser tooling is limited by the angular divergence of the beam. For example, for an accuracy of 1 mm at a distance of 10 m, an angular spread of 0.1 mrad is required, about equal to the diffraction limit of a 1 cm aperture. A helium–neon laser beam can easily be expanded and collimated to 0.1 mrad with a small telescope. Low-power helium–neon lasers provide considerably higher brightness for tooling applications than conventional light sources. They are easily visible in ambient lighting conditions even at distances of hundreds of feet from the laser.

Many practical alignment systems employ a laser and a centering detector which automatically determines the exact center of the beam. One type of centering detector uses four silicon photodiodes. The arrangement of the diodes in the centering detector is shown schematically in Fig. 8-1. The photocells are electronically connected in quadrature to compare the intensity in each half of the beam, both horizontally and vertically. If the beam accurately strikes the center of the detector, each quadrant receives the same amount of light. One then measures a null between the pairs of detectors electronically. If the beam does not accurately strike the center of the detector, there will be a different amount of light incident on each quadrant. The output is presented on a meter. A null on the meter shows that the center of the beam has accurately been located. Deviation from a null indicates the amount of misalignment quantitatively.

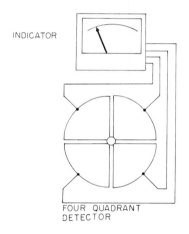

Fig. 8-1   Schematic diagram of centering detector consisting of four split quadrant detectors.

We note that the laser should operate in the TEM_{00} Gaussian mode. If admixtures of higher-order modes are present, it will complicate the measurement. Quoted accuracies for the determination of the beam center by this method are of the order of 10 μin./ft. Thus, over distances of the order of 100 ft, alignment can be obtained to tolerances of the order of 0.001 in. using a centering detector.

Another newer type of position-sensitive detector is the lateral-effect photodiode. The lateral-effect photodiode is a PIN photodiode (see Chapter 7) with metallic contacts separated by a distance L. As a spot of light is scanned across the interelectrode spacing, the fraction of the current carried by each of the electrodes varies. The current I carried by one of the electrodes when the spot of light is at a distance x from that electrode is

$$I = I_0 \sinh \alpha(L\text{-}x)/\sinh \alpha L \qquad (8.3)$$

where I_0 is the total photocurrent and α is a material parameter. When $x = 0$, $I = I_0$, and all the current is carried by the specified electrode. When $x = L$, $I = 0$, and all the current is carried by the other electrode. Thus, monitoring the current carried by each electrode can provide a direct readout of the displacement of the spot of light across the active area of the sensor.

Lasers are specially constructed for alignment applications. On a tooling laser, the laser is constructed so that the beam emerges from the center of the laser to within 0.001 in. The angle at which the beam emerges is controlled to within a few arc seconds. The laser must be specially designed to provide directional stabilization of the beam direction.

Without stabilization, the direction of the beam can drift approximately 10 seconds of arc for a temperature change of 1°C. This value is representative for many popular inexpensive lasers. The direction of travel of the beam can actually change by this amount. Even after a warm-up period, fluctuations of laser temperature around 1°C/hr may occur, even if the ambient temperature remains constant. Such instabilities can introduce an error of the order of 10 sec of arc into the measurement. The stabilization techniques for high quality alignment lasers reduce the drift to less than 1 sec of arc for a temperature change of 1°C.

Table 8-1 presents some typical quoted specifications for commercial alignment lasers. We note that the power level is low enough so that the applications are in compliance with safety standards. Collimation of the beam leads to a beam cross section around 10 mm, which further alleviates safety problems.

TABLE 8-1
Typical Specifications of Commercial Alignment Systems

Laser power	0.7 mW
Beam diameter	10 mm
Beam centering on instrument	±0.001 in.
Beam alignment to instrument axis	±4 arc sec
Alignment stability (after warmup)	±0.0005 in./hr
Readout accuracy	0.001 in.

As compared to conventional optical tooling, the laser offers several advantages. Laser tooling requires only one man to set up and operate. The measurement is made by reading linear displacement directly from a meter.

One great advantage of the laser alignment is that readings compare well between different operators. This is in contrast to the case of conventional optical tooling, where a single operator may be able to reproduce his readings, but different operators may arrive at different results.

The limitation on the use of laser alignment involves stability of the atmosphere. Beam wandering in a turbulent atmosphere will reduce the accuracy. The index of refraction of air fluctuates in a random manner because of temperature changes and air turbulence. These inhomogeneities cause random movement of a laser beam which is propagated for a distance through the atmosphere. Figure 8-2 [1] shows some experimental results on the wandering of a laser beam due to turbulence. These results are applicable to a propagation distance of 100 m along a line of sight 30 cm above the ground. The measurements were taken on a windy evening at a temperature of 68°F. Beam excursions of a few centimeters are observed even for fairly short times. As the beam propagation distance increases, the amplitude of the excursions will also increase.

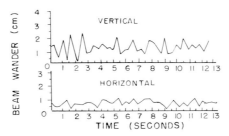

Fig. 8-2 Wandering of a laser beam over a 100-m path length, 0.3 m above the ground on a windy evening, with a temperature of 68 F. (From A. Chrzanowski, *XIII Intern. Congr. Surveyors, Wiesbaden*, September 1–10, 1971).)

A simple technique to reduce beam wander, which is often employed in enclosed areas, involves an axial flow of air along the direction of alignment. A 20-in. window fan blowing air along the direction of the laser beam will reduce the noise due to air turbulence.

A more sophisticated method of compensating for atmospheric turbulence involves use of centering detectors which integrate over a period of time. The distribution of light energy averaged over a suitable time interval is symmetric about the beam center. With the integrating feature, measurements of the beam center can be obtained with great accuracy, even in the presence of turbulence. Devices have been constructed which allow alignment of laser beams in a turbulent atmosphere to an accuracy of 0.2 arc sec over distances of a few kilometers [2].

Some results showing the capabilities of angular measurements using an integrating centering detector at ranges out to 1500 m are shown in

TABLE 8-2
Integrating Centering Detector Measurements[a]

Distance (m)	Movement of laser spot (mm)	Standard deviation of angular measurement (arc sec)
100	2	0.28
200	5	0.37
300	5	0.29
400	3	0.20
500	3	0.16
600	10	0.21
800	40	0.09
1000	50	0.16
1500	40	0.20

[a] From A. Chrzanowski, *XIII Intern. Congr. Surveyors, Wiesbaden,* September 1–10. 1971.

Table 8-2. The measurements were taken under field conditions with low temperatures and high winds. The integration time was 5 sec. The movement of the laser spot due to beam wander was somewhat erratic, but generally the amplitude of the movement increased with increasing distance. The angular measurements were accurate to around 0.2 arc sec, a much better value than might be expected under the circumstances.

Another fact that must be considered for alignment over long paths is the change in air pressure (and hence in index of refraction) with height. This effect can cause light traversing a nominally horizontal path to bend downward. For typical conditions of air pressure and temperature, this effect can lead to a downward displacement of the beam by around 100 cm over a length of 10 km [3].

Let us consider now the accuracy that is attainable with laser alignment. The values that have been quoted vary from one observer to another, and there is no complete consensus on the best values that are attainable. Part of the problem is the lack of other means to check the precision of laser alignment. The laser provides the best alignment over large distances, and it is very difficult to check independently. In a practical sense, for alignment in a shop, where doors may be opened and people are moving about, the best that appears reasonable is about 0.0002 in. over distances of the order of 12–15 ft. The accuracy decreases as the distance increases. For distances of the order of 20 ft, perhaps 0.0005 in. is a reasonable number; at 50 ft, 0.001 in.; and perhaps 0.002 in. at distances of the order of 100–200 ft. These numbers represent reasonable values that should be attainable in shop conditions.

Commercial versions of laser alignment systems are available and are being employed in practical applications in a number of industries, including the aircraft industry, the construction industry, and the machine tool industry. The line or bright spot produced by the laser is used to check locations, position equipment, establish accurate blasting patterns, and guide automated equipment such as automatic boring machines. Lasers are used to align components in building and construction jobs, to determine line and grade in excavations, tunnels and piping, check pipes, position floating marine equipment, and provide accurate alignment in the building of bridges, sewer pipes and waterways. In the control of construction operations, the laser beam is used as a reference line to guide equipment alignment or location. For example, in overwater construction, equipment operators could line up a barge with reference to a beam reflected from a retroreflector mounted on a far shore from a laser. The retroreflector sends a beam back in the same direction from which it came. Such a system can lead to savings in construction costs, because the operator can make quick intermediate alignment control checks.

The ability of the laser to generate an accurate straight line has especially gained widespread acceptance in the construction industry. Systems have been developed for determining line and grade in sewers, for example, and these have reached the status of common field use.

Complete laser alignment systems designed for field use in pipeline construction have been introduced. They eliminate the need for batterboards (pairs of horizontal boards nailed to posts used to indicate a desired level), line and grade strings, and the associated optical equipment normally used for this type of work. As a result, the setup costs are reduced and construction is speeded. Pipe laying accuracy is also increased. The laser produces more accurate references than conventional systems so that pipe can be laid to a more precise line and grade. Such systems are being used in a practical sense in sewer installations where a precise grade must be maintained.

The lasers developed for this use are reliable, durable, lightweight, battery-operated and otherwise compatible with use in the field. They have led to significant savings in construction costs. Such lasers also found applicability in tunneling, where the tunneling machines are set to follow the straight line projected by the laser beam, with appropriate centering detectors and servomechanism control to keep the tunneling machine accurately on the laser beam.

For use in the alignment of large aircraft fuselages, lasers are superior to previous methods. Tooling holes may be machined in the sections. The laser and centering detector are used to line up the sections, one by one, over distances of the order of 100 ft to an accuracy that was not previously possible. This type of alignment is simpler to use than the telescopes that were previously employed. Alignment is directly read from the detector, so that operator judgment is not required. The straight line is available for tapping at any distance from the source. It has no sag, as any wire inevitably does.

As specific examples of practical use, we mention the following:

(1)   Laser tooling has been used for alignment of assembly tooling jigs for construction of large jet aircraft. According to the users of the system, the accuracy was ten times greater than that of conventional optical systems and provided tolerances within ±0.002 in. at a distance of 200 ft. Figure 8-3 shows use of an alignment laser for alignment of wing jigs on a jumbo jet aircraft [4]. The centering detector is at the left of the workman in the background. The laser and the control box for readout of offsets in two transverse dimensions are visible in the foreground.

Fig. 8-3   Alignment laser being used on wing jigs of large jet aircraft. Centering detector is in background. (From S. Minkowitz, Electro-Opt. Systems Design, p. 22 (April 1970).)

(2)   As an example of application in the machine-tool industry for checking machine ways, the top surfaces of the ways of a custom-built ECM machine were maintained flat and parallel within 0.001 in. over a $10\frac{1}{2}$-ft length. The lead screw was aligned to one of the ways to within 0.0005 in. This technique represented a considerable advance for the machine-tool industry.

(3)   The straightness of machine motion may be checked. Figure 8-4 represents measurements of the straightness of machine motion versus vertical and horizontal displacement of a boring mill, with bed length $22\frac{1}{2}$ ft [5]. For each component of displacement, the two curves represent data obtained at two different times, three months apart. Checking such motions by conventional techniques is extremely difficult. The figure shows that the data are reproducible to within approximately 0.0005 in Figure 8-5 shows a typical setup for determination of straightness of travel of a large machine tool. Centering detector and readout are located on the movable machine bed [5].

(4)   Use in guiding tunnelers in laying out drill-hole patterns [6]. A helium–neon laser was employed in a 3100 ft tunnel under the floor of Lake

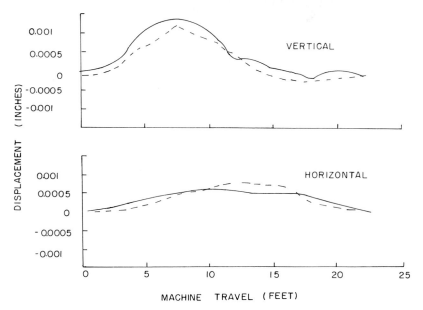

Fig. 8-4   Straightness of travel of boring mill. Horizontal and vertical displacements are shown separately over a run-out distance of 22.5 ft. The solid and dashed curves represent two different sets of data taken 3 months apart. (From C. E. Enderby, "A Practical Laser Tooling System," Electro Optics Assoc., Palo Alto, California.)

Fig. 8-5 Laser determination of the straightness of travel of a large planing machine. (From C. E. Enderby, "A Practical Laser Tooling System," Electro Optics Assoc., Palo Alto, California.)

Ontario. The floor grade was checked by allowing the light beam to intersect target rods. Offset measurements from the beam were used to spot drill holes on fresh faces by swinging radius rods with the beam as the center.

(5) Alignment of the Stanford two-mile-long linear accelerator [7]. This application employed the use of diffraction gratings to increase the precision of the optical alignment. The diffraction gratings led to a more compact intensity pattern at the measurement point. A laser-illuminated rectangular Fresnel zone plate was used to align to an accuracy of 0.5 mm at a distance of 3 km. Such tolerances are beyond the capabilities of conventional methods.

B. Related Uses

In addition to simple alignment along the beam, the laser may be employed in a variety of other tasks, such as establishing right angles, establishing planes, determining the vertical, and measuring angular misalignment.

In order to provide a right angle, the laser beam may be projected into a pentaprism. The beam emerging from the pentaprism will emerge at an angle accurately 90° from the original beam direction. Thus, a right angle may easily be established. The pentaprism is illustrated in Fig. 8-6. The incoming beam makes two reflections off aluminized surfaces and emerges in a direction 90° to its original direction. The pentaprism also has the property that it does not reverse or invert an image.

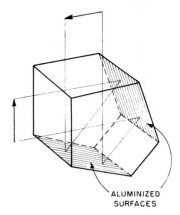

ALUMINIZED
SURFACES

Fig. 8-6   Drawing of pentaprism.

The pentaprism shown deviates the beam by 90°, regardless of its orientation to the line of sight. The angular tolerance of inexpensive pentaprisms is approximately 5 arc sec. This means that regardless of the orientation of the pentaprism to the laser beam, it will always deviate the beam through the same angle, i.e., 90° ±5 seconds of arc. This makes the prism desirable for applications requiring right angles, without the necessity for precise orientation of the prism.

A plane may be established by rotation of the pentaprism about the axis of the original incoming beam. The 90° reflected beam will then trace out a plane. This is a very simple way of generating a plane reference using a laser. It is illustrated in Fig. 8-7. Objects may easily be aligned in the same plane.

The laser may be used to establish a vertical direction by means of apparatus as shown in Fig. 8-8 [8]. The front surface mirror is adjusted to be perpendicular to the axis of the autocollimator telescope. Then the laser

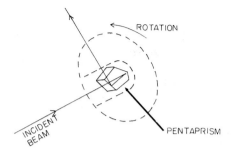

Fig. 8-7 Schematic diagram showing generation of a plane using a rotating pentaprism.

is adjusted until the reflection of the laser beam and the reflection from the pool of mercury both coincide with the center of the reticle of the telescope. This method allows establishment of a vertical direction to an accuracy of 0.5 arc sec. The alignment may be made in a much shorter time than is possible with other techniques.

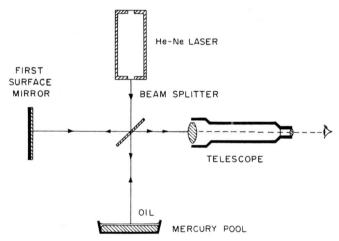

Fig. 8-8 Schematic diagram of optical plummet. (From A. Chrzanowski, F. Ahmed, and B. Kurz, *Appl. Opt.* **11**, 319 (1972).)

Autocollimating lasers have been constructed with a centering detector built into the laser head itself. Such a device is capable of fast accurate angular alignment, when used with a conventional autocollimating target mirror.

The laser projects a collimated beam to the flat target mirror. If the mirror is accurately aligned, the reflected beam will reenter the front aperture of the autoreflector and the beam will return directly superimposed on the projected beam. The centering detector will indicate a null. Misalignment of the target mirror causes angular deviation of the return beam. The centering detector readout will indicate relative angular position of the target mirror with respect to the laser beam. The centering detector can sense this offset in two axes simultaneously and display the angular deviation on two meters. Commercial models of this instrument are available, with a quoted resolution of 0.2 arc sec with typical air noise and an operating range from 0 to 250 ft.

Such an autocollimating laser may be used for measurement of angular deviation and for angular alignment. The conventional method involves use of a measuring autocollimator and requires a skilled operator. The use of an autoreflecting laser eliminates the need for a skilled operator and reduces set-up time to minutes. Angular displacement may be read directly from the output of the centering detector.

As an example of the applicability of this type of measurement, straightness of runout of machine tools has been checked by mounting a reflector on the moving part, and monitoring its angular changes during the tool motion. Installation of ceiling tile using a plane swept out by a laser and a rotating pentaprism has become common. The laser beam is accurately aligned with the vertical so that the scanned plane of light is horizontal. Workmen move the supporting grid for the tile up or down until it is in line with the plane of light, and then tie the grid at that position. It has been estimated that this method saves 25–50% of the needed manpower, and that it is more accurate than conventional methods based on strings or wires.

In a similar type of system, lasers have been used for control of road building equipment such as graders. One can set exact grade levels. A number of different types of systems have been devised. For simplicity we shall describe one specific example. In this example, the system consists of a tripod-mounted laser with a rotating prism, a sensor which mounts on the equipment and a readout instrument on the operator's control panel. The transmitter and rotating prism project a level plane. In addition, if the work calls for a slope or grade, it can be tilted at the desired angle. The sensor is mounted on a mast high enough above the other portions of the machinery that the beam can be detected, no matter what the relative orientation of the machinery and the laser. The sensor detects whether the position of the plane of light striking it is at the desired level or high or low. Lights and audible signals then inform the operator of the condition. The operator holds the reading on center by adjusting the blade level or depth of cut of his machinery. The specifications of this system allow holding blade levels to accuracy within ±0.25 in. over distances up to 1500 ft from the laser. This particular system involves operator control to complete the feedback loop. Obviously, more sophisticated systems with servocontrol can provide automatic adjustment.

A similar development involves projection of a laser beam as an elevation reference for automatic depth control of pipelaying machinery. Control of the digging depth for a pipelaying machine to install pipe at the prescribed depth and grade, particularly for drainpipe, has been fraught with problems of accuracy and reliability. The depth and grade of drains installed with many trencher-type pipelaying machines are controlled manually by an operator using a line-of-sight established by grade targets for reference. This technique of grade control has several disadvantages; among these are the excessive amount of time required to preset the targets and the stress on the machine operator in manually controlling the machine to achieve a prescribed degree of accuracy.

A stretched wire grading reference could be used as a reference for controlling the plow-type installation equipment, but the time and labor required to preset the wire to the desired grade are excessive. Moreover, with the stretched wire as a reference line, rather sophisticated hydraulic components would still be needed for control. It is much easier to project a collimated laser beam across the field as an elevation reference than it is to stretch a wire or set up grade targets.

In summary, laser methods of alignment have gained considerable acceptance in highly competitive construction applications.

REFERENCES

[1] A. Chrzanowski, *XIII Intern. Congr. Surveyors, Wiesbaden*, September 1–10, 1971.
[2] A. Chrzanowski, F. Ahmad, and B. Kurz, *Appl. Opt.* **11**, 319 (1972).
[3] J. C. Owens, "Laser Applications in Metrology and Geodesy, Laser Applications 1" (M. Ross, ed.). Academic Press, New York, 1971.
[4] S. Minkowitz, Electro-Opt. Systems Design, p. 22 (April 1970).
[5] C. E. Enderby, "A Practical Laser Tooling System," Electro Optics Assoc., Palo Alto, California.
[6] Construction Methods and Equipment, July 1967.
[7] W. B. Herrmannsfeldt, *et al.*, *Appl. Opt.* **7**, 995 (1968).
[8] A. Chrzanowski, F. Ahmed, and B. Kurz, *Appl. Opt.* **11**, 319 (1972).

SELECTED ADDITIONAL REFERENCES

A. Laser Tooling

B. Feinberg, Laser Tooling Goes to Work, *The Tool and Manufacturing Engineer* (October 1967).
P. A. Hickman, Optical Tooling Viewed in a New Light, *Laser Focus*, p. 23 (March 1968).
B. O. Kelly, Lateral-Effect Photodiodes, *Laser Focus*, p. 38 (March 1976).
Lasers—The Thin Red Line Tells It Like It Is, *The Excavating Contractor* (January 1974).
S. Minkowitz, The Laser as a Measuring Device, *Electro-Opt. Systems Design*, p. 22 (April 1970).

E. J. Schneider, Optical Tooling, *Opt. Spectra*, p. 19 (April–May–June, 1967).

J. A. Strasser, Laser-Assisted Optical Tooling Making Big Gains in Aerospace Industry, *Aerospace Technol.*, p. 24 (April 8, 1968).

B.   Related Uses

J. Cornillault, Using Gas Lasers in Road Works, *Appl. Opt.* **11**, 327 (1972).

J. L. Fouss and R. C. Reeve, The Laser in Construction: "Lite-Line" Guides a Pipeline, *Laser Focus*, p. 31 (December 1968).

M. R. Hamar, Laser-Alignment-Current Uses and Applications, Soc. of Manufacturing Eng. Tech. Paper MR74-967 (1974).

CHAPTER 9

PRINCIPLES USED IN MEASUREMENT

Some of the most important applications of lasers involve their use for measurement. Optical methods of measurement have been known for many years. Previously they have been limited by the available light sources, which traditionally have lacked sufficient coherence or sufficient brightness. The development of the laser has removed these limitations. The availability of lasers has led to improvements on conventional optical measurement techniques and the introduction of many new techniques.

Some of the measurement applications for lasers offer the best prospects for large scale use of lasers in industrial applications. The laser provides a very versatile tool capable of being employed for a diversity of different measurements. In this chapter, we shall describe principles that are common to several types of measurements. In the following chapters, we shall describe the specific applications including distance measurement, velocity measurement, angular rotation rate measurement, fiber or wire diameter measurement, and measurement of thickness and surface contours.

The principles described in this chapter have been utilized in many of these measurement applications. These same principles recur in different situations. It will assist understanding laser measurements if we first review some of the phenomena that are used for a variety of applications. It happens that some of the most common measurements employ interference phenomena which arise when two beams of light are combined. Our discussion will center on interference.

We shall first describe use of the Michelson interferometer to produce an interference pattern between two beams of light that traverse different paths. We then describe production of a beat which involves interference by two light beams of slightly different frequency. We then describe the production of a frequency shift using the Doppler effect and the role of coherence of the laser light in providing the ability to produce usable interference patterns.

A.   The Michelson Interferometer

The Michelson interferometer consists of two mirrors, M_1 and M_2, arranged as shown in Fig. 9-1, with a beamsplitter inclined at 45° to the mirrors. The collimated beam of laser light is incident on the beamsplitter B and is divided into two beams when it strikes the semireflecting surface of the beamsplitter. Part of the light follows the path to mirror M_1, is reflected by mirror M_1 and retraces its path to the beamsplitter. Some of this light will pass through the partially transparent beamsplitter. Another light ray originally passes through the beamsplitter to mirror M_2, is reflected back to the beamsplitter, is reflected again by the semireflecting surface of the beamsplitter, and is superimposed on the first ray. We note that these two superimposed rays contain only part of the original light, the rest having been lost by reflection or transmission at the beamsplitter and returned back toward the laser. A compensating plate C, which has the same thickness and is made of the same glass as the beamsplitter, is sometimes placed in

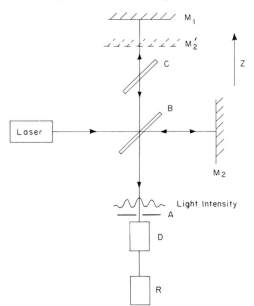

Fig. 9-1   Schematic diagram of Michelson interferometer. In this diagram M_1 and M_2 are the two mirrors, M_2' is the image of mirror M_2 observed when looking in the Z direction, B is a beamsplitter, C a compensator, A an aperture, D a detector, and R a recorder.

the path of the first ray. The first ray traverses the thickness of the beam-splitter only once, whereas the second ray traverses the beamsplitter three times. Insertion of the compensator provides for equal path length in glass.

The mirror M_2 has an image parallel to mirror M_1 because of reflection at the semireflecting surface of the beamsplitter. This image is indicated by the dotted surface M_2'. The Michelson interferometer produces interference that is identical to the interference produced by the wedge of air contained between the mirror M_1 and the image M_2'. An observer views the light emerging from the beamsplitter containing the superimposed reflections from M_1 and M_2'. These reflections are seen by the observer in the same direction. The path difference between the two beams is $2S \cos \theta$, where S is the distance between M_1 and M_2' and θ is the angle between the axis of the system and the angle at which the rays are observed. The phase difference δ between the two beams is

$$\delta = 2\pi(2S \cos \theta/\lambda) \tag{9.1}$$

The formation of fringes is a result of the wave nature of light. A light wave moving in the negative z direction may be described by the function

$$E(z, t) = E_0 \exp i(kz + \omega t + \phi) \tag{9.2}$$

where E_0 represents the amplitude of the wave, ω the angular frequency, and $k = 2\pi/\lambda$, with λ the wavelength. The function ϕ represents the phase of the wave, and is generally a function of position and time. For monochromatic and coherent laser light, we may consider ϕ as a constant.

The two beams from M_1 and M_2' combine to yield an intensity I given by

$$I = |E_1 \exp i(kz + \omega t + \phi_1) + E_2 \exp i(kz + \omega t + \phi_2)|^2 \tag{9.3}$$

where E_1 and E_2 are the amplitudes of the two beams, and ϕ_1 and ϕ_2 are their phases.

Equation (9.3) represents the so-called "square law." When two light beams are combined, the resulting intensity is given by summing the amplitudes of the two beams and then taking the square of the absolute value. Common detectors of light—phototubes, photographic film, the human eye—all respond to the intensity as given by the square of the amplitude of the light wave. This is in contrast to the microwave case, where linear detectors which respond directly to the amplitude are available. If we carry out the operations indicated by Eq. (9.3), we obtain

$$\begin{aligned}
I &= [E_1 \cos(kz + \omega t + \phi_1) + E_2 \cos(kz + \omega t + \phi_2)]^2 \\
&\quad + [E_1 \sin(kz + \omega t + \phi_1) + E_2 \sin(kz + \omega t + \phi_2)]^2 \\
&= E_1^2 + E_2^2 + 2E_1 E_2 \cos(\phi_1 - \phi_2) \\
&= E_1^2 + E_2^2 + 2E_1 E_2 \cos \delta
\end{aligned} \tag{9.4}$$

where δ is given by Eq. (9.1). This result is not simply an exercise in trigonometry, but really represents the pattern of light intensity seen by a detector. The result expressed by Eq. (9.4) corresponds to a general background intensity $E_1^2 + E_2^2$ modulated by fringes expressed by the cosine term. If the amplitudes are adjusted so that $E_1 = E_2$, then

$$I = 2E_1^2(1 + \cos \delta) \tag{9.5}$$

and the light intensity is zero when

$$\delta = (2M + 1)\pi \tag{9.6}$$

where M is an integer. This corresponds to formation of a dark fringe. The light intensity is maximum at

$$\delta = 2M\pi \tag{9.7}$$

This corresponds to formation of a bright fringe in the interference pattern. The fringe will be circular, because a given value of phase difference corresponds to a circle about the line from the eye normal to the mirror. At the center of the fringe pattern, where $\theta = 0$, we have $2S = M\lambda$ (with M an integer) as the condition for a bright spot.

Consider the condition where M_1 is more distant from the beamsplitter than M_2'. A pattern of circular fringes is seen. The fringes are packed closer together as their radius increases. As M_1 is moved away from M_2', the fringes move out from the center and the radii increase. A new fringe appears at the center of the pattern for each movement of the mirror through a distance $\lambda/2$. In other words, as the path difference between M_1 and M_2' changes by half a wavelength, the light intensity varies from maximum to minimum and back to maximum. This variation along the axis ($\theta = 0$) can be detected by insertion of an aperture A as shown in Fig. 9-1 with a photodetector D behind the aperture.

B. Beat Production (Heterodyne)

Another type of interference effect that is often employed in measurement is the phenomenon of a beat note. If two light beams of slightly different frequency interfere, the interference pattern will not remain fixed in time, but will be in motion. At a given position in space, the motion of the interference pattern will produce a varying intensity.

The phenomenon is exactly analogous to the familiar case in acoustics where two tuning forks of slightly different frequency are struck simultaneously. Interference of the two sound waves produces a sound of rising and falling amplitude. This phenomenon is referred to as beating, and the frequency of the variation of the sound is called the beat frequency. As utilized in optics, this is sometimes called optical heterodyne, or photo-mixing.

We shall treat this case analogously to the discussion centering around Eqs. (9.3) to (9.5). As before, we represent the light wave traveling in the negative z direction by the functional form given in Eq. (9.2). We note again that photodetectors respond to the square of the amplitude of the light field. Then, the intensity I for the two combined beams is

$$
\begin{aligned}
I &= |E_1 \exp i(k_1 z + \omega_1 t + \phi_1) + E_2 \exp i(k_2 z + \omega_2 t + \phi_2)|^2 \\
&= [E_1 \cos(k_1 z + \omega_1 t + \phi_1) + E_2 \cos(k_2 z + \omega_2 t + \phi_2)]^2 \\
&\quad + [E_1 \sin(k_1 z + \omega_1 t + \phi_1) + E_2 \sin(k_2 z + \omega_2 t + \phi_2)]^2 \\
&= E_1^2 + E_2^2 + 2E_1 E_2 \cos[(k_1 - k_2)z + (\omega_1 - \omega_2)t + \phi_1 - \phi_2] \\
&= E_1^2 + E_2^2 + 2E_1 E_2 \cos[(k_1 - k_2)z + (\omega_1 - \omega_2)t + \delta]
\end{aligned}
\tag{9.8}
$$

where δ is again the difference in phase between the two beams, and where now we consider the beams to have different angular frequencies, ω_1 and ω_2.

For a fixed position, corresponding to given values of z and of δ, the light intensity viewed by a detector will contain a time-varying component of angular frequency $\omega_1 - \omega_2$.

Derivation of Eq. (9.8) in this way shows how production of the beat frequency is related to production of fringes as expressed by Eq. (9.4). Indeed, this is a generalization of the earlier result, and contains the result that if the two beams are of different frequency, the fringe pattern will move with time. For a constant position on the z axis, the bright fringes will satisfy

$$
(\omega_1 - \omega_2)t + \delta = \text{const}
\tag{9.9}
$$

where δ may be considered as a function of the viewing angle θ. The optical frequency of the light beam is of the order of 10^{15} Hz. This variation is too fast for photodetectors to follow. If the two angular frequencies ω_1 and ω_2 are reasonably close together, the difference $\omega_1 - \omega_2$ will be small enough for the detector to respond and the output of the detector will contain an oscillating term at the difference frequency. The output of a photo-tube viewing two combined laser beams of slightly different frequencies does in fact contain this time-varying component.

Many measurement applications involve a shift in frequency of part of the beam from a laser. Subsequent recombination of the original beam and the frequency-shifted beam yields the beat note, the frequency of which can be measured. This technique proves useful for a varied range of measurements.

C. The Doppler Effect

A common method of providing the frequency difference that can be employed to produce a beat pattern involves the Doppler effect. When light of frequency f is reflected from the surface moving at velocity v, there is a change in frequency because of the Doppler effect. The Doppler effect is exactly analogous to the familiar frequency shift which is observed from moving sources of sound waves. The magnitude Δf of the Doppler shift is given by the equation:

$$\Delta f/f = 2v/c \qquad (9.10)$$

where c is the velocity of light.

Since v is much smaller than c, the frequency shift will be a small fraction of the original frequency. For example, a velocity v of 1500 cm/sec (a little more than 30 miles/hr) yields $\Delta f/f = 10^{-7}$, or $\Delta f \approx 10^8$ Hz. Two conclusions are apparent:

(1)   In order to be able to detect the frequency shift, the laser must be well stabilized.

(2)   Frequency shifts arising by Doppler shifting light from surfaces moving at reasonable velocities can be in the range where measurements can be made.

Thus, the Doppler effect can provide a convenient source of two light beams at slightly different frequencies. The beam from a laser is split by a beamsplitter; one of the beams is Doppler-shifted, and then the beams are recombined at a detector in order to detect the beat frequency arising because of the frequency shift. Because both beams originate from the same laser, the small difference in frequency may be carefully controlled, to a degree that would not be possible if two independent lasers were used.

D. Coherence Requirements

The laser is useful for the measurements described in this chapter because the properties of the wavefront do not change erratically with time and with position, i.e., the laser light is coherent, as we have described previously. The electric field associated with the light wave may be written in the form

$$E(\mathbf{r}, t) = A(\mathbf{r}, t) \exp i[\mathbf{k} \cdot \mathbf{r} - \omega t + \phi(\mathbf{r}, t)] \qquad (9.11)$$

where A, the amplitude function, and ϕ, the phase function, are both functions of position $\mathbf{r}$ and time t. Consider first the time variation. If A and ϕ

are slowly varying compared with the function $e^{i\omega t}$, then during one period of the light wave, the functions A and ϕ can change only slightly. In our previous analysis, we assumed that for the time periods of interest, A and ϕ are constant. However, in reality, these functions will fluctuate to some extent and reduce the coherence of the system.

The effect of the variations in the phase function ϕ will be to wash out interference fringes. If the time of measurement is long compared to the characteristic time over which ϕ varies, the intensity of the light beam at any position in the light field will fluctuate during the time of measurement. The averaged light intensity in the field will then reach some common average value and the fringes will disappear.

In order to preserve the structure of the fringes, it is necessary that the coherence time of the laser be long enough for the two beams traversing the different arms of the interferometer to remain mutually coherent when they are recombined. In other words, the coherence length of the laser must be longer than the path differences in the two arms. This leads to the equation

$$S < \Delta L = c\,\Delta t \cong c/\Delta v \qquad (9.12)$$

where ΔL is the coherence length of the laser, Δt the coherence time, and Δv the spectral width of the laser line.

Some results for coherence lengths are shown in Table 9-1. For the high-pressure mercury arc, the coherence length is too short to make worthwhile

TABLE 9-1
Coherence Properties of Light Sources

Source	Line width Δv (Hz)	Coherence length ΔL (cm)
High-pressure mercury arc	5×10^{12}	0.006
Low-pressure arc lamps	10^9	30
Multimode He–Ne laser	up to 1.5×10^9	20
Single-mode He–Ne laser	10^6	3×10^4

measurements. For some lower-pressure arc lamps, the coherence length may be longer, but the intensity is very much reduced. The fluorescent line of neon (which defines the envelope under which the laser line can occur) would by itself lead to a coherence length around 20 cm. For single-mode helium–neon lasers, the coherence length can be long enough to make measurements over path differences which are significant.

Similar remarks apply to the phenomenon of beat production. The coherence time must be long compared to the period of the beat. This leads to the requirement

$$1/\Delta v > 1/(f_1 - f_2) = 1/\Delta f \qquad (9.13)$$

where Δf is the beat frequency. For shifts that occur by virtue of the Doppler effect, we have

$$1/\Delta v > 1/(2vf/c) \qquad (9.14)$$

or

$$\Delta v < 2vf/c \qquad (9.15)$$

This sets a lower limit to the velocity that can be detected by a frequency shift using the Doppler effect.

The spatial coherence properties of the laser are also important. Spatial coherence is a measure of how the phase function varies across the wavefront at different positions at one instant of time. Conventional nonlaser sources have poor spatial coherence properties. To increase the spatial coherence of nonlaser sources, the area of the source can be masked off, so that only a small fraction of the source can contribute. This increases the spatial coherence sufficiently so that fringes can be formed. However, this reduces the amount of light available.

The spatial coherence properties of a laser beam are related to the mode structure. Any one transverse mode of a laser will be spatially coherent. However, the lowest-order mode, TEM_{00}, will provide more uniform illumination. Virtually all interferometry uses lasers operating in the TEM_{00} spatial mode.

The high brightness of the laser is related to these considerations. All the light from the laser may be employed in forming the fringe pattern, in contrast to the situation for nonlaser sources. The laser beam is well collimated, so that all the light is easily collected. Thus, fringes formed by a laser are easily visible under conditions where the interference pattern formed by a nonlaser source will not be visible.

SELECTED REFERENCES

A.   The Michelson Interferometer

K. M. Baird and G. R. Haines, *in* "Applied Optics and Optical Engineering," Vol. IV, Part I (R. Kingslake, ed.), Academic Press, New York, 1967.
M. Born and E. Wolf, "Principles of Optics," 5th ed., Pergamon Press, New York, 1975, Chapter 7.

M. Françon, "Optical Interferometry," Academic Press, New York, 1966, Chapter IV.
F. A. Jenkins and H. E. White, "Fundamentals of Optics," McGraw-Hill, New York, 1957, Chapter 13.
S. Tolansky, "An Introduction to Interferometry," Longmans, Green, London, 1955, Chapter 8.

B. Beat Production (Heterodyne)

E. B. Brown, "Modern Optics," Van Nostrand–Reinhold, Princeton, New Jersey, 1965, Chapter 13.
B. M. Oliver, Signal-to-Noise Ratios in Photoelectric Mixing, *Proc. IRE* **49**, 1960 (1961).
M. Ross, "Laser Receivers," Wiley, New York, 1966, Chapter 4.
A. E. Siegman, S. E. Harris, and B. J. McMurtry, Optical Heterodyning and Optical Demodulation at Microwave Frequencies, *in* "Optical Masers" (J. Fox, ed.), Polytechnic Press, Brooklyn, New York, 1963.

C. The Doppler Effect

J. C. Angus *et al.*, Motion Measurement by Laser Doppler Techniques, *Ind. Eng. Chem.* **61**, 8 (1969).
E. B. Brown, "Modern Optics," Van Nostrand–Reinhold, Princeton, New Jersey, 1965, Chapter 1.
F. A. Jenkins and H. E. White, "Fundamentals of Optics," McGraw-Hill, New York, 1957, Chapter 11.
L. Levi, "Applied Optics, A Guide to Optical System Design," Vol. I, Wiley, New York, 1968, Chapter 2.
A. Sommerfeld, "Optics, Lectures on Theoretical Physics," Vol. IV, Academic Press, New York, 1964, Chapter II.

D. Coherence Requirements

R. J. Collier, C. B. Burckhardt, and L. H. Lin, "Optical Holography," Academic Press, New York, 1971, Chapter 7.
M. Garbuny, "Optical Physics," Academic Press, New York, 1965, Chapter 6.
M. V. Klien, "Optics," Wiley, New York, 1970, Chapter 6.
H. M. Smith, "Principles of Holography," Wiley (Interscience), New York, 1969, Chapter 6.

CHAPTER 10

DISTANCE MEASUREMENT
AND DIMENSIONAL CONTROL

Lasers can be applied in a variety of ways to solve practical problems in distance measurement. The main methods which have been employed are summarized in Table 10-1, along with some areas in which these methods have found application. Interferometric systems using helium–neon lasers offer precise distance measurement over a scale of distances less than 100 m and in an indoor environment. Such devices are suitable for dimensional control of machine tools. For longer distances and for measurements in the field, interferometric methods are less appropriate. An often employed method is beam modulation telemetry. In this method, the beam is modulated, transmitted to a distant object, and a reflected return signal is detected. The phase of the modulation of the return signal is compared to the phase of the transmitted signal. This method is useful for distances of hundreds of meters, and is used in surveying.

TABLE 10-1
Laser-Based Distance Measurement

Method	Typical laser	Range	Typical accuracy	Typical applications
Interferometric	He–Ne	Meters (indoors)	Microinch	Machine tool control, seismic and geodetic uses, length standard calibration
Beam modulation telemetry	He–Ne, GaAs	Up to kilometers	1 part in 10^6	Surveying
Pulse time of flight	Q-switched solid state	Kilometers	1 part in 10^5	Military range-finding, satellite ranging

For other applications, where very rapid measurements are desired, such as in military range finding and where somewhat less precision can be tolerated, the roundtrip transit time for a very short high-power pulse of light can be measured.

A. Interferometric Distance Measurement

Lasers offers two main advantages for interferometric measurement of distances:

(1) High intensity.
(2) High temporal coherence.

These properties allow measurements to be made over considerable distances, greater than would be possible with conventional light sources.

Most interferometric systems use a helium–neon laser operating at a wavelength of 0.6328 μm (corresponding to a frequency around 5×10^{14} Hz). For an unstabilized laser, there is a possible variation in frequency of 10^9 Hz, the width of the fluorescent line in neon. This would mean an uncertainty of one part in 5×10^5 in the measurement, since the laser could operate on more than one cavity mode that falls within this line width. Thus, for interferometric applications, a laser must be stabilized to keep its frequency at the center of the neon fluorescent line. A typical design for frequency stability involves construction of the laser with the mirrors rigidly mounted on a single piece of invar, which has low thermal expansion. The laser is contained in a temperature-controlled oven to provide additional thermal stability. Then the length of the resonator is servocontrolled to the position of a small dip in power which appears when the laser frequency lies at the center of the neon emission line. As the length of a laser operating in a single longitudinal mode shifts slightly, the power output varies. When the cavity length is adjusted to the exact center of the emission line, there is a slight decrease in the laser power. This decrease is called the Lamb dip. Servocontrol of the cavity length using piezoelectric transducers allows the laser to be stabilized at the frequency corresponding to the center of the emission line. The frequency stability obtained in this manner is approximately 1 part in 10^9.

The laser is constrained to operate in a single longitudinal mode by making the cavity short. The spacing between longitudinal modes is $c/2L$, where c is the velocity of light and L the length of the laser cavity. Thus, shortening the laser cavity will increase the mode separation, until the separation becomes comparable to the width of the fluorescent line of neon. This means that there will be only one resonant cavity mode within the fluorescent line width. This leads to operation at a single frequency. Single-mode operation is achieved at an expense in the power output of the laser, because of the decreased length of the laser tube. Such mode-stabilized helium–neon lasers typically operate at power outputs of a fraction of a milliwatt.

We describe first an interferometric method based on the Michelson interferometer, because the basic operation of interferometric laser distance measuring systems can be understood easily with reference to the Michelson interferometer systems. Later we shall describe variations which provide better stability under conditions of atmospheric turbulence and attenuation. We have already discussed operation of a Michelson interferometer in Chapter 9. A schematic diagram is shown in Fig. 10-1. The laser beam is split into two parts: a measurement beam and a reference beam. Each of these beams traverses one leg of the interferometer. The measurement beam travels to a

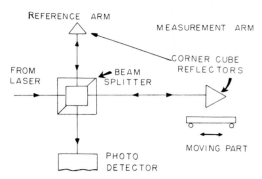

Fig. 10-1 Schematic diagram of application of Michelson interferometry to measurement of distance traversed by moving parts.

mirror mounted on the moving part, whose distance is to be measured. The mirror is usually a cube corner reflector, which provides accurate return of the beam. The use of the cube corner reflector is a valuable feature. A cube corner reflector is essentially a trihedral prism. The light makes one bounce off each face of the prism and returns in a direction almost exactly opposite to that from which it originally came (see Fig. 10-2). This feature is not critically dependent on the exact angular alignment of the prism, so that restrictions on the alignment are considerably less than if a conventional mirror were employed.

The two beams are recombined by a beamsplitter and travel to a detector. The two combined beams form the interference pattern. The amplitude of the light at the detector depends on the phase between the reference beam and the measurement beam, which in turn depends on the

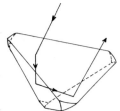

Fig. 10-2 Schematic diagram of cube corner reflector.

difference of optical path that the two beams have traveled. The phase difference δ is given by

$$\delta = 2ks \cos \theta \qquad (10.1)$$

where $k = 2\pi/\lambda$ with λ the wavelength, s the path difference, and θ the angle between the common axes of the beams and the direction of observation. Usually one chooses $\theta = 0$. When the moving part travels one-half wavelength of light, the total difference in optical path goes through one wavelength of light. This gives rise to the factor 2 in Eq. (10.1). The phase of light at the position of the detector will change by 2π, i.e., the fringe system will go through one period. This will correspond to a change from light to dark to light at the position of the detector. In other words, the motion of the moving part leads to amplitude modulation of light which is sensed by the detector. Electronic circuitry can then count the periods of the amplitude modulation and feed the information to a computing element which calculates the distance through which the moving part has moved.

It is important to note that laser interferometers provide measurement of displacement from an arbitrary zero, rather than absolute measurement of distance. The instrument reading is set to zero at the initial position of the workpiece. Then motion is measured relative to this preset zero.

The distance measured in interferometric measurements is the optical path length, which differs from the physical path length by a factor equal to the index of refraction of the air. Corrections for the index of refraction of the air are necessary to obtain good accuracy. Variations in index of refraction often are the limiting factor in the overall accuracy of length measurements. The index of refraction for dry air at 15°C and at a pressure of 760 Torr with 0.03% CO_2 is $n_{dry} = 1.0002765$ at the helium–neon laser wavelength, which in vacuum is 0.63299138 μm. Corrections for humidity, temperature, and pressure variations are given by the equation [1]

$$(n - 1) = (n_{dry} - 1)P/720.775$$

$$\times [1 + P(0.817 - 0.0133T) \times 10^{-6}/(1 + 0.003662T)]$$

$$- 5.6079 \times 10^{-8}f \qquad (10.2)$$

where f is the partial pressure of water vapor in an atmosphere of total pressure P Torr at temperature T C. Under most conditions the effects of a varying concentration of CO_2 are small, but there may be some conditions where a high content of CO_2 could cause noticeable errors.

To summarize the implications of the above formula, an increase in pressure of 1 Torr increases the index of refraction by 0.36 ppm near standard conditions. A temperature increase of 1°C decreases the index of refraction by 0.96 ppm and an increase of 1 Torr in the partial pressure of water vapor decreases the index of refraction by 0.06 ppm, near standard conditions. To a reasonable approximation, these corrections may be added independently.

In order to convert the measured optical path length to physical path length, these corrections must be taken into account. Systems that automatically sense the variations in ambient air pressure, temperature and humidity, calculate the appropriate correction and display the corrected result have been developed. In addition, the temperature of the workpiece can be monitored, and length measurements normalized to a reference temperature.

A schematic diagram of such a complete apparatus appears in Fig. 10-3. The system shown in Fig. 10-3 uses two photodetectors so as to determine the direction of motion. The two detectors collect light from regions of the

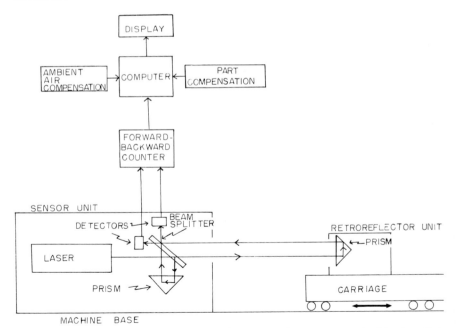

Fig. 10-3 Schematic diagram of complete system for measurement of tool motion, including compensation for ambient air parameters and part temperature, and sense of motion.

fringe pattern where the phase difference of the interfering beams differs by
$\lambda/2$. Then the relative phase of the amplitude modulation viewed by the two
detectors will be different depending on the direction of remote mirror
motion. A logical circuit processes the signals so that a count is added for an
increase of $\lambda/2$ or subtracted for a decrease of $\lambda/2$ in the optical path. A fast
reversible counter then gives the correct displacement.

More sophisticated arrangements for measuring direction of motion
have also been developed. One system is shown in Fig. 10-4. The beam from
the laser passes through a quarter-wave plate and becomes circularly polarized (see Chapter 1). The beam is split at a beamsplitter to form the reference
beam and the measuring beam. Since the reference beam has been reflected,
the handedness of its circular polarization is opposite to that of the measuring
beam. For specificity, consider the measuring beam as being left-hand circularly polarized and the reference beam as being right-handed. The measuring and reference beams recombine at the upper beamsplitter surface. The
measuring beam, having undergone three reflections at the trihedral reflector,
is now right-handed. The measuring beam, passing through to the left of
the beamsplitter, remains right-handed. The reference beam to the left of the
beamsplitter has undergone a second reflection and reverts to left-handed.
The opposite occurs above the beamsplitter where the reference beam passes
through, remaining right-handed, and the measuring beam reflects, reverting
to left-hand circular polarization. In both cases, the recombination is
of two beams with opposite hands of circular polarization. When two
beams of opposing circular polarization are combined, the resulting polarization is linear. The orientation of the linear polarization vector will depend
on the relative phase of the two circularly polarized beams. Thus, the plane
of polarization of the combined beam rotates as a function of target position.
This rotation is equivalent to 180° for a target movement of half a wavelength. Polarizers in front of the detectors are oriented at 45° with respect
to each other. Since one fringe passes for each half-wavelength of target

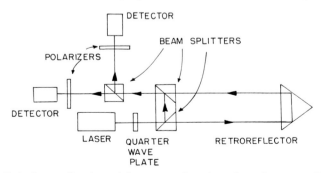

Fig. 10-4 System for determining sense of motion of moving retroreflector.

movement, the fringes as observed through the polarizing screen oriented at 45°, are in quadrature with those observed through the polarizing screen oriented at 0°. Thus, the outputs of the photodetectors are in phase quadrature. The outputs are amplified and combined to give a square wave output which provides sensitivity to the direction of target motion.

Other variations have involved multiple passes of the measuring beam along the path to the remote retroreflector. Such systems increase the sensitivity, so that in some cases one count on the counter has corresponded to 1/32 of a wavelength.

The Michelson interferometric technique described above was developed early in the history of lasers and descriptions of it are plentiful in the technical literature extending through the 1960s. This Michelson interferometric method of distance measurements is sensitive to variations in the intensity of the laser light. If the light beam intensity changes because of air turbulence, shifts in laser output, or air turbidity, improper fringe counting can result and the measurement will be in error. These shortcomings are overcome by development of a two-frequency laser system which mixes beams of two different frequencies and measures Doppler shift of the beam reflected from a moving retroreflector [2]. The schematic diagram of the operation is shown in Fig. 10-5. The helium–neon laser emits light of two slightly different frequencies f_1 and f_2, with different polarization properties.

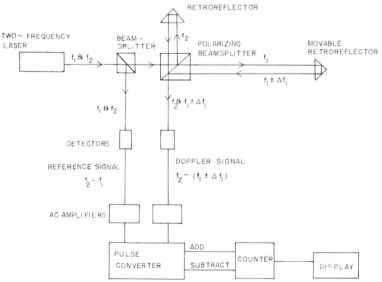

Fig. 10-5 Schematic diagram of two-frequency laser distance-measuring system. In this figure, f_1 and f_2 are the two laser frequencies with transverse polarizations and Δf_1 is the Doppler shift produced by the motion of the retroreflector.

This is achieved by applying an axial magnetic field which splits the fluorescent line of the neon into two components. The frequency stabilized laser then emits two frequency components separated by about 2×10^6 Hz. In this system one measures the velocity of the moving reflector. Light reflected from the moving reflector has its frequency shifted by an amount Δf:

$$\Delta f/f = v/c \tag{10.3}$$

where v is the velocity of motion. This shift is essentially a Doppler shift, analogous to the familiar Doppler effect in acoustics, and is described in Chapter 9. The velocity v is then integrated to obtain linear displacement.

A polarization-sensitive beamsplitter separates the light of the two different frequencies so that the beams of different frequencies travel different paths. Frequency f_2 is sent to a fixed reflector and frequency f_1 travels to the movable reflector on the part whose distance is to be measured. The two beams are recombined at the beamsplitter and produce a beating which manifests itself as an amplitude modulation of the light, according to the discussion in Chapter 9. If the movable reflector moves, the returning beam will be frequently shifted by Δf, so that the beat frequency will be $f_2 - (f_1 \pm \Delta f_1)$. The frequency difference signal is amplified by ac amplifiers. A portion of each of the two output frequencies is picked off near the end of the laser and is used to generate a reference signal $f_2 - f_1$. The two signals are fed into the pulse converter which extracts Δf_1 and produces one pulse per quarter wavelength of motion of the movable mirror.

The two-frequency systems can be adapted to measure pitch and yaw components of motion and straightness of travel. The operation as a straightness sensor is indicated in Fig. 10-6. This adaptation allows both frequencies f_1 and f_2 to be sent to the remote mirror which is configured as illustrated. A Wollaston prism, constructed of birefringent materials, separates the two frequencies, by virtue of their different polarizations. The interferometer measures relative lateral displacement between the interferometer and the reflectiong mirror axis. Relative lateral displacement affects the difference

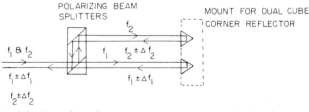

Fig. 10-6   Adaption of two frequency system to measure pitch and yaw components of motion. In this figure, f_1 and f_2 are the two incident laser frequencies and Δf_1 and Δf_2 are the respective Doppler shifts.

in optical path lengths between the two beams. This causes a difference in accumulated fringe counts. For relative lateral translation x, the fringe counts accumulated will be $2x \sin(\theta/2)$ where θ is the angle between the two beams. With this adaptation, straightness of travel of tools can be measured.

Operation as an angle sensor is illustrated in Fig. 10-7. Beams of the two frequencies f_1 and f_2 are sent out parallel to the remote reflector. Angular rotation of the mount which contains the two trihedral prism retroreflectors causes a Doppler shift between the return frequencies, $f_1 \pm \Delta f_1$ and $f_2 \pm \Delta f_2$, which is not affected by axial displacement. The accumulated number of fringe counts is proportional to the sine of the angular rotation.

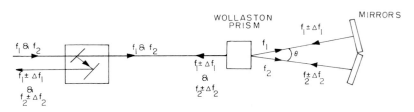

Fig. 10-7 Operation of a two frequency laser system as an angle sensor. In this figure f_1 and f_2 are the two laser frequencies and Δf_1 and Δf_2 are their respective Doppler shifts.

This adaptation can be used to measure pitch and yaw components of motion, which can potentially cause errors greater than lead errors. It can also be employed to measure flatness of surfaces, for example surface plates or machine tool beds.

Now let us consider some of the sources of error in laser interferometric distance measurements. These will apply to both types of interferometers described above.

We consider first the error due to instrumental accuracy. Interferometric distance measuring systems similar to those described above are commercially available and can in principle measure to a fraction of the wavelength of light. The accuracy quoted for some instruments is 1 ppm of the distance measured, plus or minus one count. The 1 ppm is due to the effect of the transducers and calculations employed to correct for the ambient environment. The error of one count represents the intrinsic resolution of the fringe counting. For many instruments, one count is 1/4 wavelength ($\approx 1.6 \times 10^{-5}$ cm). For some multiple path instruments, one count has represented a value as small as 1/32 wavelength ($\approx 2 \times 10^{-6}$ cm). Commercial instruments have been stated to be useful at distances up to 200 ft and are usually used indoors or in an evacuated environment.

The errors due to the atmospheric index of refraction and to material temperature effects are summarized in Table 10-2. Often these errors are

TABLE 10-2
Atmospheric Errors and Material Temperature Errors in an Uncompensated Interferometer

Variable	Variation from standard conditions	Error[a]
Atmospheric pressure	1 Torr	0.36 ppm
Atmospheric temperature	1 C	0.96 ppm
Relative humidity	1 Torr partial pressure	0.06 ppm
Workpiece temperature	1 C	$D_m K_m$

[a] D_m = measured distance, K_m = coefficient of expansion of material being measured (units of cm/cm °C).

compensated automatically and will be included in the stated instrumental accuracy. Other errors arise from the particular mechanical setup of the measurement.

The interferometric system measures travel in the direction of the laser beam. If the direction of the beam does not coincide with the direction of travel, an error called cosine error is introduced. If the axis of the beam is misaligned from the axis of travel by an angle α, the measured distance D_m differs from the actual distance traveled D_a by

$$D_a = D_m \cos \alpha \approx d^2/2D_m \qquad (10.4)$$

where d is the lateral displacement at the distance D_m. As an example, for a measurement at a distance of 10 ft, with an angle of 0.1° between the beam and the machine travel, the cosine error would be about 0.2 mils.

When the travel is longer than 1 m, alignment of the laser for maximum signal over the distance of the travel minimizes cosine error. Alternatively, use of a centering detector to align the beam to the direction of travel of the machine, as described in Chapter 8, can be helpful.

Offset error is an error that arises when the beam is not in line with the tool bit on all axes. Offset error is illustrated in Fig. 10-8. The error arises because of any rotation about an axis normal to the plane of the offset.

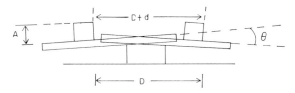

Fig. 10-8   Illustration of offset error. In this figure, A is the distance from the instrument to the tool, θ is the angle at which the carriage tilts, D is the motion and d is the offset error.

Rotations arise because of crooked ways, thermal gradients, dirt, wear, etc. The magnitude d of the offset error is

$$d = A \sin \theta \tag{10.5}$$

where A is the distance from the instrument to the tool, and θ the angle at which the carriage tilts. We note that offset error is independent of the distance being measured. If the tilt angle θ is $0.1°$ and the displacement A is 10 in., the offset error d is about 1.7 mils. This error is reduced by eliminating the offset between the object being measured and the laser beam, if possible, and by assuring that the object being measured is parallel to the measuring system. Use of centering detectors, which give direct indications of deviations from the laser beam, can be useful.

An error called preset (or deadpath) error arises because the laser interferometer measures the distance moved from an arbitrary zero. If the interferometer is zeroed with the target a distance D_1 from the interferometer, and then the target is moved a distance D_2, the compensation for atmospheric index of refraction is applied over distance D_2. However, the beam actually travels distance $D_1 + D_2$ and is affected by the changes in index of refraction over the entire path. Preset error is proportional to the preset distance and to the magnitude of the change in index of refraction. Similar considerations arise relative to compensation for material temperature.

The preset error E_p can be represented by the equation

$$E_p = D_1[0.96 \times 10^{-6} \Delta T - 0.36 \times 10^{-6} \Delta P + K_m \Delta T_m] \tag{10.6}$$

where D_1 is the preset distance (between the interferometer to the zero reference position), ΔT the change in air temperature (°C) during the measurement, ΔP the change in air pressure (torr) during the measurement, K_m the coefficient of thermal expansion of the workpiece material (cm/cm °C) and ΔT_m the change in material temperature (°C) during the measurement. We note explicitly that preset error is present in interferometers which are compensated for part temperature and for atmospheric index of refraction. This error can be compensated by allowing the interferometer to accept preset data and correct for it. It can also be minimized by positioning the interferometer optics as close as possible to the physical starting point for the motion to be measured.

As an example, at a preset distance of 30 cm, for a carbon steel workpiece with a thermal expansion coefficient $K_m \approx 10 \times 10^{-6}$ cm/cm °C, and a change of 1°C in the workpiece temperature, with constant atmospheric temperature and pressure, the preset error would be 3×10^{-4} cm.

An estimate of the accuracy of the measurement must include all of the sources of error in Table 10-3. It is usually helpful to compile an error budget

TABLE 10-3
Summary of Errors

Error	Typical estimate[a]
Instrument error	1 ppm ± 1 count
Atmospheric error	Included in instrument error in compensated instruments
Material temperature error	Included in instrument error in compensated instruments
Cosine error	$d^2/2D_m$
Offset error	$A \sin \theta$
Preset error	$D_1[0.96 \times 10^{-6} \Delta T - 0.36 \times 10^{-6} \Delta P + K_m \Delta T_m]$

[a] See text for definition of symbols.

and calculate each of the errors and then add the individual errors to obtain a total error estimate. In different measurement situations, any of the individual errors listed in Table 10-3 could be the dominant error source. In careful work, the entire error budget must be calculated in order to understand the accuracy of the measurement.

Laser interferometry may be considered in the following cases:

(a)   When the dimensional tolerances on the work approach the tolerances of the machine tool.

(b)   When the machine tool must be calibrated or checked.

(c)   When systematic deviations in the machine tool must be compensated.

(d)   When machine tool travel covers relatively large distances.

As an example of large distance requirements, spar milling machines with carriage travel of 140 ft have been employed in the aircraft industry.

The ultimate limiting distance will be set by the coherence properties of the laser. The fringe pattern will be washed out as one approaches a path difference of the order of the coherence length of the laser. Thus one has

$$s < c/\Delta v \tag{10.7}$$

where Δv is the linewidth for the stabilized single frequency laser. In practice the frequency stability for commercially available stabilized lasers is of the order of 10^5–10^6 Hz. Hence, one is restricted to path differences of the order of 3×10^4–3×10^5 cm, or 300–3000 m. Practical limitations will be somewhat less.

The use of the laser in industrial environments will depend on the control of the environment. In practice, variations in air stability will tend to wash out the interference fringes for distances much over 100–200 ft in practical machine-shop environments. For geodetic and seismic measurements, operation at kilometer distances has been demonstrated with evacuated optical paths [3]. A second practical limitation is the ability to determine part temperature and ambient air conditions and correct the readings to the desired reference conditions.

Laser interferometric distance measurements are useful in many industrial applications because size and accuracy requirements for machines and machine tools have increased to the point that conventional linear positioning devices no longer provide adequate performance. The conventional devices include engraved line scales, magnetic pickoffs, and linear encoders. In order to check machine positioning, one should use a measurement device that has accuracy much better than the accuracy of the device being measured. The usual methods mentioned above do not offer this degree of accuracy. It is sometimes doubtful that the checkout method is better than the machine being checked. The accuracy requirements for machine parts now lead to requirements to check the accuracy of the machine tool itself to within a few parts per million. In particular, where the machine tool travel extends over long distances, the measurement errors become great. The errors become cumulative when the conventional devices are employed over long distances. Laser interferometry offers improved accuracy of measurement in many practical cases.

The laser interferometer has been used in many industrial applications. Examples have included:

(1) Accurate setup of work-holding fixtures for the production of aircraft engine components. This application involves a considerable reduction in time.

(2) Checkout of the motion of machine tools, leading to automatic compensation for errors in machine tool runout. An example of the use of a laser interferometer to check the movement of a machine carriage appears in Fig. 10-9 [4]. This shows checkout of the motion of a jig bore machine.

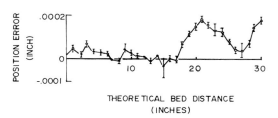

Fig. 10-9 Example of the results of a laser interferometer checking the motion of a jig-bore machine, yielding position error as a function of machine runout. (Data presented by E. A. Haley at the Engineering Seminar on New Industrial Technology, Pennsylvania State University, July 7–9, 1969.)

The cumulative position error is shown as a function of the machine runout. The known error can be used to compensate and correct the runout. The measurements shown in Fig. 10-9 took a total time of 1 hr, including setup and averaging of three runs. The actual position of the carriage is determined to a high degree of accuracy. Such measurements were previously very difficult to make over large distances of travel. The errors shown in Fig. 10-9 are the result of the limited capability of conventional systems. It is apparent that the laser can offer considerable improvement.

(3) Speedup of positioning operations. In one application, laser interferometry assisted a manual positioning operation in a machining process that was otherwise numerically controlled. Aluminum plates 4 ft long required 40-in.-spaced hole locations to a close tolerance, at an ambient temperature of 86°F. Automatic sensing of the part temperature and coupling this with the expansion coefficients of the part and the machine gave 0.0005-in. accuracy on the 40-in. spacing. The numerical control program allowed manual operation for this particular set of operations. The operator located one hole with the tool, zeroed the interferometer, retracted the tool, and manually positioned the table to the position of the next hole. The desired accuracy could be achieved by taking the reading for hole position from the interferometer.

(4) Building vibration measurement. In one example, workers obtained a time record of north–south displacement of a large building in San Francisco. A laser interferometer on the ground was aimed at a cube corner reflector on the 41st floor. From such measurements, the power spectrum of the building vibration could be extracted.

(5) Measurement of strains in the earth. As an example, 25-m-long interferometers near a fault in Bakersfield, California measured shifts of 1 part in 10^7 per day because of motion of the fault [5]. Other seismic and geodetic measurements involving measurement of earth strains due to tides, seismic waves, and continental drift have been carried out.

Figure 10-10 shows a photograph of an application involving interferometric measurement in an industrial environment. The figure illustrates setup of sixty-six 8-in. racks on a 44-ft horizontal boring mill. The previous method used a precision peg bar and electronic indicator, and required a total time of 11 hr. Use of the laser interferometer reduced the total time to 22 min.

The examples above are not an exhaustive list of possible applications for laser interferometric distance measurement, but are intended to give some ideas of representative situations in which laser interferometers may be used to advantage.

Fig. 10-10   Laser interferometry in an industrial environment, for rack setup on a 44-ft horizontal boring mill. (Photograph courtesy of R. E. Stark.)

B.   Beam Modulation Telemetry

For measurements of large distances in the field, interferometric methods with microinch sensitivity are no longer applicable. Fluctuations in the density of the atmosphere over paths exceeding a few hundred feet make interferometry in the field impractical.

A common method of distance measurement over long distance involves amplitude modulation of the laser beam and projection of the modulated beam at a target. The lasers most often used have been He–Ne lasers and GaAs lasers. The target may be a retroreflector† which is positioned at the spot whose distance is desired. Alternatively the target may be a diffusely reflecting surface whose distance is to be measured. Measurements have been made using many different types of surfaces as targets (water, grass, buildings), but the reflected signal will be enhanced if a retroreflector can be used.

The reflected return from the target is received by a telescope and sent to a detector. The phase of the amplitude modulation of the returning light is compared to the phase of the emitted light. A difference in phase occurs

---

† For example, a corner cube mirror that reflects the beam of light back along the direction from which it came.

because of the finite time for the light to travel to the object and return to the telescope. The phase shift ϕ is related to the total path length L by the equation:

$$\phi = 2\pi(2n_g L/\lambda_v) \tag{10.8}$$

where λ_v is the vacuum wavelength of the laser and n_g the group index of refraction.

The index of refraction to be used in correction for atmospheric effects is the group index of refraction, which is defined by

$$n_g = n + \sigma(dn/d\sigma) \tag{10.9}$$

with σ the reciprocal of the vacuum wavelength, and n the phase index of refraction, which we have used previously. The difference arises because in dispersive media (i.e., media in which the index of refraction is a function of wavelength), the group velocity (with which the energy associated with the wave propagates) differs from the phase velocity (at which the phase of the wave propagates). In interferometric measurements, as described above, the phase velocity (and phase index) are to be used, but in beam modulation telemetry, the group velocity (and group index) are the appropriate parameters. The value of the group index at the He–Ne laser wavelength (0.63299138 μm in vacuum) is

$$n_g = 1.0002845073$$

for dry air at 15°C, 760 Torr, and 0.03% CO_2. Corrections for the group index as a function of varying atmospheric temperature and pressure are available in the literature [6].

Determination of the proper corrections to be applied to n_g is difficult for field measurements. One must actually use a value for n_g averaged over the entire path traversed by the light. If the terrain is uniform and meteorological conditions are stable, the situation is not too bad. Correction for the atmospheric index of refraction can be made if the temperature and pressure are known along the path of the light beam. Measurements could be made at several points along the path and averaged. However, this would still not lead to complete accuracy and would involve much additional labor. Also, in many cases, the beam may travel through regions where the temperature and pressure may be different from the values at the end points, and where it is not practical to measure along the beam path. An example would be in mountainous terrain. A method for averaging the index of refraction has been suggested by making measurements at two different optical wavelengths, and a system to demonstrate this method has been constructed at the Environmental Science Services Agency Research Laboratories in Boulder, Colorado [7]. In tests of this instrument using a 1-mile path at the Stanford linear accelerator, a reproducibility of 1 part in 10^7 was found.

In order to measure distance, one measures the phase shift of the collected light relative to the outgoing light, and uses Eq. 10.8. Figure 10-11 shows a schematic diagram of a system for such a measurement. The light is amplitude modulated at a given frequency, collimated, and transmitted

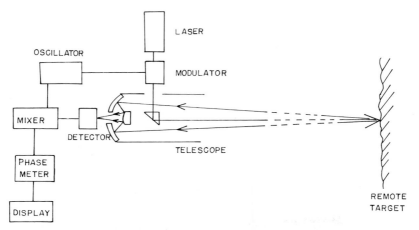

Fig. 10-11   Schematic diagram of distance measurement in beam modulation telemetry.

to a distant reflector. Reflected light is collected by a telescope and focused on a photomultiplier. A phase detector compares the relative phase of the amplitude modulation on the return beam with that originally present.

The phase difference can be represented as

$$\phi = (N + f)2\pi \qquad (10.10)$$

where N is an integer and f a fractional part less than unity. The original measurement gives f, but no information about N.

Thus, it is necessary to make measurements at several values of the modulating frequency. One may begin at a low value of frequency, where $\phi < 2\pi$, (i.e., $N = 0$) and obtain a course measurement. Then the frequency is increased, f remeasured more accurately, with N known from the earlier coarse measurement. In several steps, one may obtain ϕ with high accuracy.

Another variation allows varying the frequency so as to obtain a null output when the original signal and the return signal are 180° out of phase. If the frequency of modulation is varied, nulls will be obtained for variations in the optical path corresponding to one-half wavelength of the modulation frequency. Measurements of several nulls will provide enough information to obtain the distance. Because the frequency at which a null occurs can be found accurately, this method offers greater accuracy than direct measurement of the phase shift.

Beam modulation telemetry units have been developed for which the claimed accuracy is ± 1 mm at distances up to approximately 1000 meters and one part in 10^6 for greater distances. Such devices are finding application for a variety of uses, including land profiling and geodetic survey work. They have been used for measurement of motion of large structures, such as bridges and dams. The system can be used as an altimeter, and can be used for purposes such as slope surveys and gravimeter calibration.

With further improvements in accuracy and sensitivity, promising geophysical applications involving measurement of the motions in the earth's crust should become possible. For example, applications such as accurate measurement of the motions of the blocks making up the opposite sides of the San Andreas fault in California and measurements of the widening of the Red Sea due to motion of the Arabian peninsula away from Africa have been suggested. For geodetic measurements, the common method might require a second operator to position the retroreflector. When the device is used as an altimeter, it can be used as a single-ended unit, provided the reflection from the target surface is sufficiently high. Such a system can be operated by a single person.

Competing methods for measurement of large distances include the use of microwave radar and the use of geodetic measuring tapes. Microwave radar measurements are limited in accuracy by high sensitivity to atmospheric water vapor and also by problems arising from multiple returns caused by reflection of the microwaves along multiple paths. The use of geodetic tapes requires a large amount of time for accurate measurements.

A rather dramatic example of the use of a laser beam modulation distance measuring device occurred in mapping of the Grand Canyon in 1972 [8]. The mapping provided accurate profiles for a large-scale map of a 310 square mile area in Grand Canyon National Park. The survey was completed in three days with two men and provided improved maps of the park area. It was estimated that conventional taping methods would have taken 1 yr and would have required 100 men.

Fig. 10-12 Ocean wave profile obtained using beam modulation telemetry with a laser flown in an airplane at an altitude of 500 ft.

Other more routine surveying tasks are also being speeded by the use of such laser units. As one specific and perhaps unusual example of the capabilities of beam modulation measurements, we show ocean wave profiles obtained by using the device as an altimeter. Figure 10-12 shows the record of an ocean wave profile obtained 20 miles off the New Jersey coast using a laser device flown in an airplane at an altitude of 500 ft. The laser beam diameter at the water surface was 4 in. Thus, the laser could provide resolution of a few inches on the waves. This figure shows very graphically the capabilities of laser distance measuring systems for making difficult types of measurements.

C. Pulsed Laser Range Finders

Another type of distance measurement involves measurement of the round trip transit time for a very short pulse. Such devices have commonly used high peak power Q-switched lasers. The most common uses of this technique have been in military range finding systems and in satellite tracking. These applications are outside the scope of this book, but this technique represents a different method of distance measurement, and so is included for purposes of completeness.

Pulsed laser range finders operate on the principle of using a very short laser pulse and measuring the transit time for the laser pulse to reach an optically visible target and for the reflected pulse to return to a receiver, essentially an optical radar. Thus, the system consists of a laser, a telescope to collect the reflected return, a photodetector, and a precision clock to measure the difference in time between the outgoing pulse and the return pulse. The narrowly collimated beam of the laser makes it possible to measure the range of specific targets and to receive a return free from ground clutter or interference with other objects near the target. Because the target must be visible, the range is limited by atmospheric conditions.

For military uses, beam modulation telemetry is not desirable because of the need to measure at several different frequencies. This takes time and usually for military purposes, the result is desired very quickly. The nature of the application requires a short high-power pulse, especially a pulse with a fast rise time. Therefore, Q-switched lasers have usually been employed in military range finders. Devices have been developed for use with aircraft, tanks, and artillery. Some of these units have reached production status.

The characteristics of one particular range-finding device, the so-called model M-70B, are shown in Table 10-4 [9]. The characteristics represent the capabilities of laser range-finding devices for a typical device employing a Nd:YAG laser. The device has been designed to be self-contained in a package of reasonable size and weight.

TABLE 10-4
M70B Range Finder

Laser type	Nd:YAG
Pulse length	≤ 20 nsec
Energy per pulse	150 mJ
Pulse repetition rate	1 pps or 10 pps
Range resolution	± 5 m
Maximum range	10 km (at 20 km meteorological visibility)
Target discrimination	20 m (can discriminate between targets at least 20 m apart)
Weight	23.6 lbs.
Dimensions	$6 \times 6 \times 17$ in.

This same technique has also been employed for measurement of the distance to the moon. The round-trip transit time to retroreflectors left by Apollo astronauts has been measured. Panels containing a number of cube corner reflectors were deployed on the Apollo 11, 14, and 15 missions and also on the unmanned Soviet Luna 17 and 21 missions. At the McDonald Observatory of the University of Texas, a Q-switched ruby laser pulse with a duration around 4 nanoseconds is transmitted through a 2.7 m telescope [10]. The round-trip transit time for the reflected return signal is measured. The lunar distance has been measured to an accuracy of 15 cm in the one-way distance [11]. It is anticipated that the use of subnanosecond pulses can reduce this to the 2–3 cm range. Data obtained in these experiments has been used to generate a new lunar emphemeris and to check scientific theories about the lunar orbit and lunar mass distribution.

As an example of the use of lasers for satellite tracking, we describe the NASA Laser Geodetic Satellite (Lageos), launched in 1976. Lageos is a passive satellite, whose surface is studded with 426 fused silica retroreflectors. Lageos is illuminated by Nd:YAG laser pulses with duration around 200 psec. Range accuracies around 2 cm are expected. Range data taken over a number of years on stations mounted on different tectonic plates will provide sensitive data on movements in the earth's crust, and could lead to prediction of global earthquake patterns.

D. A Laser Interferometer Application in Mask Production: A Specific Example of Distance Measurement and Dimensional Control

In this section, we present a specific example of the application of a laser interferometer for use in distance measurement and dimensional control in a photolithographic mask making process. The system is shown in Fig. 10-13. The laser is a two-frequency helium–neon laser located near

Fig. 10-13 Laser system for dimensional control in a photolithographic maskmaking process. (Photograph courtesy of Electromask, Inc.)

the right edge of the photograph. The helium–neon laser power is approximately 1 mW. The small rectangular units directly in front of the laser are beam splitters. The system employs the Doppler shift type of measurement described in Section 10.A.

The system has two applications, one as a pattern generator and the second as a step and repeat camera for production of copies of a pattern. In the pattern generation mode, the workpiece is moved to a previously determined position. Movable apertures are positioned to form a rectangular area which is to be exposed to light. The workpiece has photosensitive material which is exposed by pulsing an incoherent light source, visible at the top of the figure. Five degrees of freedom are possible, two dimensional coordinates for the center of the rectangle, the width and length of the rectangle as defined by the movable apertures, and the angular orientation of the rectangle. The entire process is under the control of a minicomputer. The rectangles are automatically exposed sequentially in order to form the desired pattern, which is then processed photolithographically.

In the step and repeat mode, a previously generated pattern is used as a reticle. The system steps and repeats exposures with a factor-of-ten reduction in size. In this mode one controls the two positional coordinates only.

The laser system has also been used to measure the straightness of the runout of the mechanical system, the pitch and yaw of the mechanical motion and the flatness of the surface in which the positional measurements occur. All these parameters must be controlled to high accuracy in order to achieve acceptable results. The accuracy attainable with this system is $\pm 0.25\,\mu$m in the positional coordinates and 0.1 arc sec in the rotation coordinate.

The system is employed for production in an enclosed room in which the temperature and relative humidity are controlled. The system is mounted on a granite slab with large thermal mass, which controls the temperature of the machine. Changes in ambient air pressure are monitored and compensated separately.

The system is employed because it offers greater accuracy and better control than competing techniques, such as mechanical indexing systems, stepping motors, and grid-type glass scales which generate a Moiré pattern. The system is routinely used under the control of a minicomputer for highly accurate generation of mask patterns.

REFERENCES

[1]   J. C. Owens, *Appl. Opt.* **6**, 51 (1967).
[2]   J. N. Dukes and G. B. Gordon, *Hewlett-Packard J.*, p. 2 (August 1970).
[3]   V. Vali and R. C. Bostrom, *Rev. Sci. Instrum.* **39**, 1304 (1968).
[4]   E. A. Haley, Engineering Seminar on New Industrial Technology, Pennsylvania State University, July 7–9, 1969.
[5]   P. L. Bender, *Proc. IEEE* **55**, 1039 (1967).
[6]   J. C. Owens, *Appl. Optics* **6**, 51 (1967).
[7]   J. C. Owens, "Laser Applications in Metrology and Geodesy, Laser Applications 1," (M. Ross, ed.). Academic Press, New York, 1971.
[8]   *Opt. Spectra*, p. 32 (May 1973).
[9]   P. Hermet, *Appl. Opt.* **11**, 273 (1972).
[10] P. L. Bender *et al.*, *Science* **182**, 229 (1973).
[11] E. C. Silverberg, *Appl. Opt.* **13**, 565 (1974).

SELECTED ADDITIONAL REFERENCES

A.   Interferometric Distance Measurement

R. R. Baldwin, G. B. Gordon, and A. F. Rudé, Remote Laser Interferometry, *Hewlett-Packard J.* (December 1971).
R. R. Baldwin, B. E. Grote, and D. A. Harland, A Laser Interferometer that Measures Straightness of Travel, *Hewlett-Packard J.* (January 1974).

R. Baldwin, Inspecting Machine Tool Geometry Interferometrically, *Electro-Opt. Systems Design*, p. 36 (October 1976).

R. H. Burns, Optics in Machine Tools, *Laser Focus*, p. 43 (April 1973).

G. D. Chapman, Interferometric Angle Measurement, *Appl. Opt.* **13**, 1646 (1974).

J. Dyson, "Interferometry as a Measuring Tool," Machinery Publ., Brighton, England, 1970.

J. D. Garman and J. J. Corcoran, Measuring the Variable Speed of Light Improves Laser Distance Measurement, *Electronics*, p. 91 (April 24, 1972).

J. Koch, Laser Interferometers for Position Feedback, Machine Design (February 20, 1975).

S. Minkowitz, Machine Design on the Fringe of Change, *Laser Focus*, p. 43 (October 1971).

J. C. Owens, Laser Applications in Metrology and Geodesy, *in* "Laser Applications" 1, (M. Ross, ed.), Academic Press, New York, 1971.

G. Roth, Measurement Error Considerations for Users of Laser Interferometers, Quality Management and Engineering, Part I, p. 14 (October 1972), Part II, p. 28 (November 1972).

A. Sona, Lasers in Metrology: Distance Measurements, *in* "Laser Handbook," Vol. 2 (F. T. Arecchi and E. O. Schulz-Dubois, ed.), North-Holland Publ., Amsterdam, 1972.

B.   Beam Modulation Telemetry

P. L. Bender, Laser Measurements of Long Distances, *Proc. IEEE* **55**, 1039 (1967).

K. B. Earnshaw and J. C. Owens, A Dual Wavelength Optical Distance Measuring Instrument Which Corrects for Air Density, *IEEE J. Quantum Electron.* **QE-3**, 544 (1967).

K. D. Robertson, Laser by a Dam Site, *Laser Focus*, p. 45 (October 1971).

A. Sona, Lasers in Metrology: Distance Measurements, *in* "Laser Handbook," Vol. 2 (F. T. Arecchi and E. O. Schulz-Dubois, eds.), North-Holland Publ., Amsterdam, 1972.

C.   Pulsed Laser Rangefinders

D. W. Coffey and V. J. Norris, YAG:Nd^{3+} Laser Target Designators and Range Finders, *Appl. Opt.* **11**, 1013 (1972).

A. M. Johnson, J. Nunez, and L. Bushor, Laser Rangefinders—Today's Military and Industrial Systems, *Electro-Opt. Systems Design*, p. 34 (March 1976).

M. L. Stitch, Laser Rangefinding, *in* "Laser Handbook," Vol. 2 (F. T. Arecchi and E. O. Schulz-Dubois, eds.), North-Holland Publ., Amsterdam, 1972.

D.   A Laser Interferometer Application in Mask Production: A Specific Example of Distance Measurement and Dimensional Control.

E. A. Hilton and D. M. Cross, Laser Brightens the Picture for IC Mask-making Camera, *Electronics*, p. 119 (August 7, 1967).

CHAPTER 11

ENVIRONMENTAL MEASUREMENTS

A.   Introduction to Laser Monitoring of the Environment

There is a definite need for improved accurate sensitive methods for measurement of concentration of pollutants in the atmosphere. There are many different types of impurities in the atmosphere. Some of the major pollutants of interest include oxides of nitrogen, carbon monoxide, sulfur dioxide, and ozone. In addition, there are various types of particulate materials, such as ash, dust, and soot. There have been maximum pollutant levels established, along with plans for reduction of pollutant levels.

In order to carry out the desired reductions, there is a need for improved measurement techniques. There are many problems involved with measurement of pollutants in the atmosphere. Conventional methods of measurement are difficult. For example, if one desires to measure gases from a smokestack, there are problems of the inhospitable environment inside the smokestack, problems with accessibility, and problems of perturbing the quantities that are to be measured by the measurement technique. Traditionally, measurements have involved collecting a sample of gas from the smokestack and making a spectroscopic or chemical analysis. Usually such analysis was conducted remotely, and results were not immediately available. Stratification of gases and particulate materials in the smokestack led to error. Instruments that were immersed in the smoke had problems with maintenance. Thus the monitoring of the gaseous composition in a smokestack has been a problem. There have been similar problems with the monitoring of other sources of air pollution.

Thus there is a need for improved measurement techniques to ensure air quality. In addition, there is a need for instrumentation to determine compliance with standards that have been adopted. Lasers offer an attractive possibility for remote monitoring of air quality. It is possible to carry out measurements at a large distance from the sample that is to be monitored, without the necessity of sample collection or of any chemical processing. The results of the measurement can be available immediately and there is no distortion of the quantities being measured. For the example of smoke-stack monitoring, there is no contact with the hot corrosive gas so that maintenance is reduced.

The basic method involves sending a beam of laser radiation through the sample of atmosphere to be investigated and monitoring either the transmission or the scattering of the light. There are a variety of physical processes that can occur. The different processes lead to a variety of different techniques for monitoring the air quality. The transmission of the light through the atmosphere can be expressed by the equation

$$I(x) = I_0 e^{-\alpha x} \qquad (11.1)$$

where I is the intensity transmitted through a distance x when the incident intensity is I_0, and α is the extinction coefficient that arises from a variety of terms. We can express α as the sum of an absorption term and scattering terms,

$$\alpha = \alpha_{ABS} + \alpha_{MIE} + \alpha_{RAY} \qquad (11.2)$$

where α_{ABS} is the extinction coefficient arising from absorption and the other terms are due to scattering. The absorption coefficient is strongly wavelength dependent and depends on the gases that are present in the atmosphere. This is particularly true in the infrared. Figure 11-1 shows the infrared transmission of the atmosphere. It is apparent that there are a

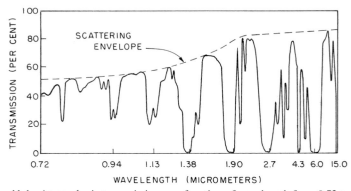

Fig. 11-1 Atmospheric transmission as a function of wavelength from 0.72 to 15.0 μm. The dashed line shows the envelope due to scattering. The solid curve indicates molecular absorption bands. (From R. M. Langer, Report on Signal Corps Contract No. DA-36-039-SC-72351 (1957).)

number of regions of strong absorption. These correspond to the presence of various gases in the atmosphere. In particular, carbon dioxide and water vapor dominate the infrared absorption spectrum. But there are a number of other lines characteristic of other gases of interest. The presence of these gases can easily be monitored by measuring the transmission of the atmosphere at a wavelength corresponding to the absorption line of the particular gas of interest.

There are a number of different scattering processes. In a scattering process, the light is redirected from its original direction of propagation. In order to use scattering to monitor atmospheric quality, one generally measures the light scattered in some particular direction, rather than measuring directly the attenuation due to the scattering. Two important types of scattering are Mie scattering and Rayleigh scattering, whose contributions to the extinction coefficient are denoted as α_{MIE} and α_{RAY} in Eq. (11.2).

Rayleigh scattering is due to molecules of gases. The scattering coefficient α_{RAY} for Rayleigh scattering is given by the equation

$$\alpha_{\text{RAY}} = 32\pi^3(n - 1)^2/3N\lambda^4 \tag{11.3}$$

where N is the density of scattering molecules, n is the index of refraction of the air, and λ is the wavelength. Thus Rayleigh scattering increases as the wavelength decreases and becomes particularly strong in the ultraviolet.

Mie scattering is due to particulate material and aerosols. It depends in a complicated fashion on the particle size, the refractive index, the absorption, the particle size distribution, the wavelength, and the scattering angle. One may define a size parameter b according to the equation

$$b = 2\pi R/\lambda \tag{11.4}$$

where R is the particle radius. If we define a backscatter coefficient β as the power scattered per unit incident irradiance, one has

$$\beta = S(\lambda)\pi R^2 \tag{11.5}$$

where S gives the variation of the backscattering as a function of wavelength. For a single value of R, one can obtain the backscattering coefficient through an infinite series involving orthogonal polynomials. An example is given in Fig. 11-2, which shows the function S as a function of the size parameter for particles having an index of refraction equal to 1.33, which is typical of water droplets. We see that Mie scattering is large when the particle size is approximately equal to the wavelength. Generally, one has a distribution of sizes of particles rather than a single size. Then, it is necessary to integrate over the particle size distribution in order to determine the backscattering coefficient. It is perhaps more useful to employ an

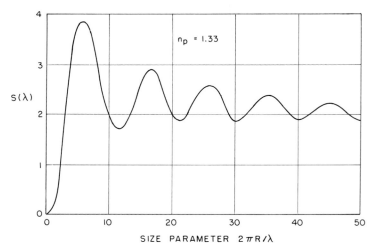

Fig. 11-2. Scattering function $S(\lambda)$, defined in test as a function of scattering parameter $b = 2\pi R/\lambda$, with λ the wavelength and R the particle radius for a particle index of refraction $n_p = 1.33$. (From H. G. Houghton and W. R. Chalker, *J. Opt. Soc. Am.* **39**, 955 (1949).)

empirical equation that relates the backscatter coefficient to the visibility. One such empirical expression [1] is

$$\beta = (3.91/V)(0.55/\lambda)^{(0.585V^{1/3})} \quad km^{-1} \tag{11.6}$$

where λ is the wavelength in micrometers and V is the visibility in kilometers. This expression is an approximation and does not necessarily hold under all conditions. Moreover there are measurements that indicate that this equation applies less well to laser sources than it does to white light sources [2]. Still, this empirical relation does provide an approximate measure of the Mie backscattering coefficient that will hold under a variety of conditions. It also relates the scattering coefficient to an easily measured quantity. More detailed treatment of Mie scattering is extremely difficult, and it is not often used for practical measurements.

Laser monitoring systems can make use of any one of a variety of different types of physical phenomena, including several methods for absorption (as we shall discuss later) and several different types of scattering. The general method involves use of a laser source that illuminates the sample of air, a receiving telescope that collects light transmitted through the sample or scattered by the sample, a filter to select a desired range of wavelengths, a detector that detects the collected light, and some sort of processing system to process the detector output and to obtain a measure of the impurity. In what follows, we shall describe a variety of the physical methods that can use lasers for measurement of air pollution and air quality.

The methods which we shall discuss are not the only possible methods, but they appear to be among the leading candidates. Each of the methods we shall describe leads to a somewhat different system, with different capabilities.

B. Optical Radar

We shall first discuss optical radar techniques, sometimes called lidar, which stands for light detection and ranging. Lidar operates in a manner similar to radar. In its basic form, lidar employs a pulsed laser simply as a source of pulsed energy. Energy is backscattered by the atmosphere. Light that is backscattered to the position of the detector is detected by the photodetector. The resulting signal can be processed as a function of time from the transmission of the short pulse, in a manner similar to that of a radar ranging system. Generally lidar systems have relied on Mie scattering to provide the source of the backscattered radiation. Thus they are most useful in determining concentrations of particulate matter and usually give no information on the nature of the scattering particles. Lidar systems have been used most often for applications in which the total concentration of particulate material is desired. They have the capability of determining this concentration as a function of distance from the measuring station.

The basic apparatus for carrying out air-pollution measurements is sketched in Fig. 11-3, which shows some of the essential components. First one uses a laser, most often a pulsed laser, with a short pulse duration to obtain good range resolution. A telescope collects the radiation returned

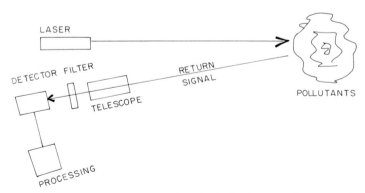

ELEMENTS OF MONITORING SYSTEM

Fig. 11-3 Generalized diagram showing the basic elements of a remote laser-based environment monitoring system.

from the sample volume. One also requires some frequency-discriminating element to discriminate the radiation of the proper wavelength from background radiation. This could be a narrow band interference filter with high rejection performance, or in some instances a monochromator. One also needs a photodetector and a data-processing and display system.

Often an oscilloscope is used in simple optical radar systems to display the data. A typical oscilloscope trace from a clean atmosphere is shown in Fig. 11-4a. This shows the return as a function of time, which is equivalent to the return signal as a function of range from the laser. The signal first increases because near the laser the volume irradiated by the laser does not overlap well with the volume viewed by the collecting telescope. At the point A the volumes overlap well, and the signal decreases as the square of the range. This is the signal that would be obtained as a result of Rayleigh scattering from the clean atmosphere. Figure 11-4b shows a typical result when backscattering is obtained from a layer in the atmosphere. The signal is enhanced between points B and C as compared to the signal from the clean atmosphere. This means that there is a layer of scattering particles lying between the ranges corresponding to points B and C. It is apparent that there is a considerable amount of processing required to obtain good data from

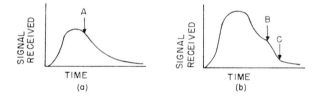

Fig. 11-4   Drawing of typical oscilloscope traces of returns from an optical radar system. Trace (a) shows the return from clean air. At point A the volume irradiated by the laser overlaps well with the volume viewed by the collecting telescope. Trace (b) shows a result when backscattering is obtained from a layer located between points B and C.

an optical radar system. The equation that describes the power P_r received as a function of range R is

$$P_r(R) = P_0 KT(R)A Y(R)N(R)\sigma \, \Delta R/4\pi R^2 \qquad (11.7)$$

In this equation, P_0 is the transmitted power, K the efficiency of the optical system, T the transmission of the atmosphere, A the area of the receiver, $Y(R)$ a factor that accounts for the overlap of the transmitted and received beam paths, $N(R)$ the concentration of the scatterers as a function of range, σ the backscattering cross section, and ΔR the range sampled, that is, the range over which it is possible to obtain a return at a particular value of time. The sampled depth ΔR is

$$R = c\tau/2 \tag{11.8}$$

where τ is the laser pulse duration. The cross section σ is the cross section relevant for Mie scattering, which has been discussed earlier. The transmission is given by the equation

$$T(R) = \exp\left[-2\int_0^R \alpha(r)\,dr\right] \tag{11.9}$$

where $\alpha(r)$ is the extinction coefficient as a function of range. The extinction coefficient will contain terms due to absorption, Mie scattering, and Rayleigh scattering. If the wavelength of the laser is chosen in a transparent atmospheric window, the contribution due to absorption can be small and α will be related to the Mie backscattering cross section. In real polluted atmospheres, which have absorptive particles with irregular shapes and sizes that are distributed according to some distribution law, the relation between the volume backscattering coefficient and the attenuation coefficient will be somewhat uncertain. This implies that it is difficult to obtain an absolute calibration for particle concentration. Various power law relations have been used for particle size distributions to derive at least approximate values. Alternatively, the relation between Mie scattering and visibility can be used.

Figure 11-5 shows results that are typical of the resolution of laser radar. These show contours of particulate concentration, derived from

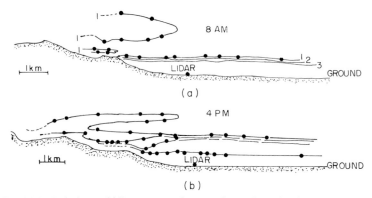

8 AM

1km

LIDAR

GROUND

(a)

4 PM

1km

LIDAR

GROUND

(b)

Fig. 11-5 Relative turbidity contours in a north–south section in a canyon near Pasadena, California, as obtained with an optical radar system. The contours in (a) were obtained at 8:00 A.M. and the contours in (b) at 4:00 P.M. (From F. F. Hall, in "Laser Applications 2" (M. Ross, ed.). Academic Press, New York, 1974.)

data obtained with a 20-MW pulse from a ruby laser, 50-nsec long, with a 20-cm-diameter receiving telescope. The system was used to probe the marine layer structure in the Los Angeles basin. Some stratified profiles of the marine layer structure are shown in the figure. In Fig. 11-5a, the contours are shown early in the morning and indicate a relatively horizontal top to the marine layer, which is less than 1 km thick. In the afternoon, as indicated in Fig. 11-5b, the marine layer has thickened and slopes up the mountainsides. This shows the flow of aerosols in the channel formed by solar heating of the mountain slopes. These data have been used as a tool for determining the effectiveness of the pumping of pollutants from the Los Angeles basin.

Figure 11-6 shows another example, which is a drawing of contours in a cross section of a plume from a high smokestack. Contours of relative particulate concentration are defined. The example shows clearly how laser radar can perform valuable monitoring of stratification, flow, and the changing profiles of turbid layers in the atmosphere.

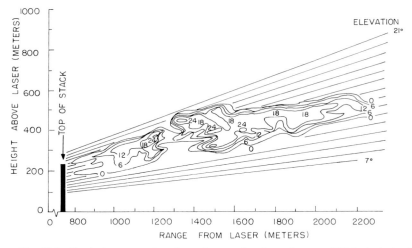

Fig. 11-6 Vertical cross section through a smoke plume from an 800-foot stack of a coal-fired power plant, as observed by optical radar. The contours represent the range-corrected signal in decibels relative to that from the ambient background aerosol. They can be interpreted as representing relative particulate concentrations. (From W. B. Johnson, *J. Air Pollution Contr. Assoc.* **19**, 176 (1969).)

In order to correct the returns from lidar systems for such factors as crossover of the transmitted and received beams and subtraction of the Rayleigh scattering, more sophisticated systems than oscilloscope displays have come into use. Sometimes an amplification stage proportional to the square of the range is used to compensate for the R^2 dependence of received signal in Eq. (11.7).

Lidar techniques for remote observation of particulate concentrations in the atmosphere have been developed and are in use in research. For operational use, e.g., in routine monitoring of emission sources, further work on development of inexpensive eye-safe instrumentation is required.

C. Raman Backscattering

The Raman effect involves scattering of light by molecules of gases, accompanied by a shift in the wavelength of the light. This is in contrast to Rayleigh and Mie scattering, in which the scattered light has the same wavelength as the incident light. For a particular gas, the Raman effect involves the appearance of additional spectral lines at a wavelength near the original wavelength. The Raman spectrum is characteristic of the scattering material. Figure 11-7 shows the shifts characteristic of certain gases. The shifts

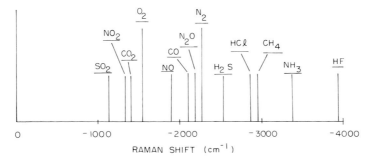

Fig. 11-7   Raman shifts characteristic of a number of gases.

involve a constant change of wave number, so that the change in wavelength is dependent on the incident wavelength. The shift in wavelength $\Delta\lambda$ is related to the shift in wave number Δk by the equation

$$\Delta\lambda = \lambda^2 \, \Delta k \qquad (11.10)$$

Table 11-1 gives the wavelengths of the Raman scattered radiation for some different incident wavelengths.

Since the values of the Raman shift are characteristic of individual molecules, examination of the spectrum of light scattered from gases in the atmosphere can provide an analytical tool that is sensitive to the presence of certain gases. Thus Raman spectroscopy of backscatterd light can provide identification of various pollutant gases such as SO_2. Figure 11-8 shows a schematic diagram of apparatus that has been used for detection of Raman-shifted lines from pollutant gases in the atmosphere. This is a system that has been built into a mobile van and used for monitoring specific sources

TABLE 11-1
Wavelengths of Raman-Shifted Laser Lines

| | | Wavelength of Raman-scattered light (μm) | | |
Molecule	Raman shift (cm^{-1})	Nitrogen laser (0.3371 μm)	Argon laser (0.5145 μm)	Ruby laser (0.6943 μm)
CO	−2145	0.3615	0.5713	0.7977
NO	−1877	0.3584	0.5642	0.7848
N_2O	−1290	0.3518	0.5486	0.7565
SO_2	−1151	0.3502	0.5450	0.7498
CH_4	−2914	0.3702	0.5916	0.8348
H_2S	−2611	0.3668	0.5836	0.8202
NH_3	−3340	0.3751	0.6029	0.8553

of pollutants in Japan, such as smokestacks. The laser is a frequency-doubled Nd:YAG laser that emits short pulses of high peak power at a wavelength of 530 nm. The return signal arises from Raman scattering from various Raman-active gases. The monochromator is set at a particular wavelength, corresponding to the wavelength of the Raman-shifted laser line for a particular gas of interest. The photodiode monitors the strength of the output signal. The Raman-shifted signal can be displayed on an

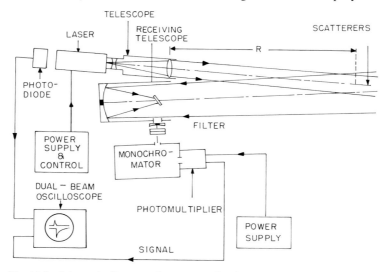

Fig. 11-8 Schematic diagram of apparatus for detection of Raman-shifted lines from pollutant gases in the atmosphere. (From T. Kobayasi and H. Inaba, Proc. IEEE **58**, 1568 (1970).)

oscilloscope to give the Raman signal as a function of range, or in more elaborate systems can be integrated over a number of pulses.

The Raman return signal as a function of range can be described with reference to Eq. (11.7). For Raman scattering, one interprets the equation in the following manner. The scattering cross section σ is the Raman scattering cross section, which is known for specific gases but is small, typically of the order of 10^{-30} cm^2. The density $N(r)$ is the density of the specific Raman-active gas. Because the Raman cross section is small, Raman systems have most utility for relatively short ranges, and most of the work done has involved investigation of specific point sources of pollution that can be identified and investigated at a range of some tens of meters. Thus this technique may have applicability to remote monitoring of smoke-stack emissions. Because the Raman scattering cross section is small, the loss out of the beam is also smaller than for the Rayleigh and Mie scattering mechanisms, so that the Raman attenuation has not been included in Eq. (11.2). In addition, because the Raman system is likely to be used at close range, the factor TY in Eq. (11.7) may be close to unity.

Figure 11-9 gives a specific example of the return from a smokestack at a distance of approximately 30 m. This shows the strength of the return signal over a range of wavelengths. A number of specific pollutants, such as SO_2 and H_2S, are identified. In addition, the major constituents of the atmosphere, such as oxygen and nitrogen, show up as relatively strong signals.

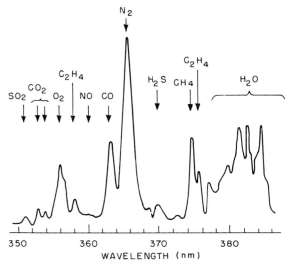

Fig. 11-9 Spectral distribution of Raman-shifted components from a variety of molecular gases in an oil–smoke plume, remotely analyzed by the laser Raman method. (From H. Inaba and T. Kobayasi, *Opto-Electron.*, **4**, 101 (1972).)

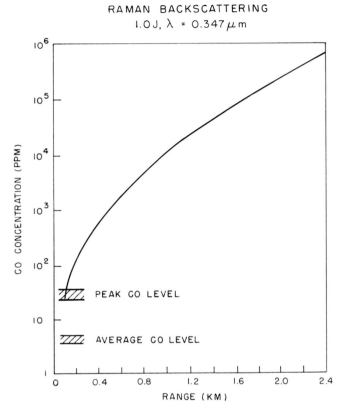

Fig. 11-10   Minimum carbon monoxide concentration detectable by the Raman back-scattering method, as a function of range. Peak and average carbon monoxide levels in ambient air are indicated. This curve was calculated assuming a 1-J pulse with 100-nsec duration at a wavelength of 0.347 μm. (From R. L. Byer, *Opt. and Quantum Electron.* **7**, 147 (1975).)

Figure 11-10 shows some calculations of the minimum detectable concentration of CO as a function of range, for a Raman backscattering system operating with a 1-J pulse at a wavelength of 347 nm. Values are indicated for peak and average levels of CO in the ambient atmosphere. It is apparent that the Raman scattering techniques will not be sensitive enough to detect ambient levels of pollutants at great ranges. So far, Raman techniques have been used mainly with laser sources of fixed frequency. Development of higher power tunable lasers may offer a redeeming feature that will allow Raman scattering to be used as a probe at longer distances. Tuning of the source wavelength would permit use of the Raman effect at a wavelength near a resonant absorption of the molecule. Near a resonance, the Raman effect can be much enhanced, varying as $[(\omega_L - \omega_0)^2 + \gamma^2]^{-1}$, where ω_L is the angular frequency of the laser, ω_0 the resonant frequency, and γ the pressure-broadened line width of the absorption. If the laser is tuned so that $\omega_L \approx \omega_0$, the Raman cross section can increase very much above its nonresonant value. Relative increases of several orders of magnitude are possible. Thus, when high-power tunable pulsed lasers become available, much smaller pollutant concentrations can be investigated. Figure 11-11 shows some calculated results for resonant Raman scattering at 0.05 mJ/pulse with integration over 10^4 pulses. The minimum detectable pollutant concentration is shown as a function of range. The laser is assumed to be tuned to the resonant frequency separately for each of the different molecules. It is apparent that detection capabilities with the Raman effect would be enhanced considerably. Monitoring over considerably greater distances would be possible. There has been little experimental work regarding the use of the resonant Raman effect. Real applications of Raman backscattering thus await advances in lasers, such as higher power tunable pulsed dye lasers.

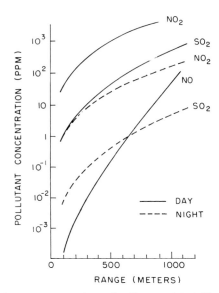

RESONANCE RAMAN SCATTERING
0.05 mJ / PULSE ; 10^4 PULSES

Fig. 11-11 Minimum concentrations of pollutants detectable with a resonant Raman laser system as a function of range. These curves were calculated assuming integration over 10^4 pulses of energy 0.05 mJ/pulse. (From H. Rosen, P. Robrish, and O. Chamberlain, *Appl. Opt.* **14**, 2703 (1975).)

The Raman effect, of course, has range discrimination, so that pollutant concentration as a function of distance can be obtained. The Raman effect suffers from problems with laser safety because the peak powers required are high. The wavelengths that would be employed are in the visible or near ultraviolet portions of the spectrum, where the hazard of laser damage is high and where laser safety standards will be stringent.

D. Resonance Fluorescence

Another possibility is the use of resonance fluorescence. The laser beam will be directed to a gaseous mass, which is assumed to contain molecules that can fluoresce upon excitation with the proper wavelength. One such example would be NO_2, which fluoresces when excited with blue light, such as 0.4880-μm light from an argon laser. The fluorescent light is then collected by a receiving telescope. Equation (11.7) can also be interpreted to cover this situation. In this case, the backscattering cross section σ can be interpreted as the cross section for fluorescence. Typical values may be of the order of 10^{-13}–10^{-16} cm^2/sr. Thus the sensitivity is considerably higher than for Raman backscattering. The density $N(r)$ is the density of fluorescent gases as a function of range. This system also has range resolution, but the resolution is degraded because of the finite fluorescent lifetime. The range resolution ΔR must be interpreted as $c(\tau + \tau_F)$ where τ_F is the fluorescent lifetime. This will reduce the range resolution considerably because the fluorescent lifetime can typically be as long as 50 μ sec.

Figure 11-12 shows calculated results for resonant fluorescent backscattering from CO, assuming a laser wavelength of 4.6 μm and a pulse energy of 100 mJ. The average CO level is shown for comparison. Results are shown for optically thin layers and optically thick layers of CO. "Optically thin" means no depletion of the 4.6-μm light in passing through the layer, and "optically thick" means depletion.

Some experimental results on laser detection of NO_2 levels, using an argon laser on samples collected in ambient air in Los Angeles on a smoggy day, are shown in Fig. 11-13. The concentration of NO_2 varies as a function of the time of day, with peaks appearing nearing the time of peak automobile traffic. The rapidity of the changes shows the desirability of real-time monitoring systems with fast response.

A slightly different technique detects the fluorescence naturally emitted by the pollutant, without excitation of the fluorescence by a laser. Because the amount of such fluorescent light that can be collected at a distance is small, very sensitive detection must be available. Optical heterodyne techniques have been employed, using tunable lasers or fixed-frequency laser lines

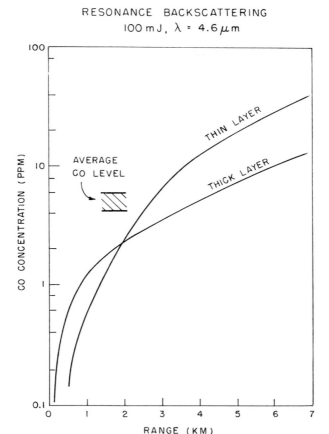

RESONANCE BACKSCATTERING
100 mJ, $\lambda = 4.6 \mu m$

Fig. 11-12 Minimum detectable concentration of carbon monoxide as a function of range for a resonance fluorescence backscatter technique. The average carbon monoxide level is indicated. These curves are calculated for optically thin (1-m) and optically thick pollutant layers, assuming a 100-mJ pulse with 100-nsec duration at a wavelength of 4.6 μm. (From R. L. Byer, *Opt. Quantum Electron.* **7**, 147 (1975).)

that fall very near the wavelength of the fluorescence. Optical heterodyne techniques, described in Chapter 9, offer extremely sensitive detection of small signals in the presence of noise. Semiconductor lasers can be produced to match any desired wavelength over a broad range. First, a laser of suitable composition is selected so that the output falls near the desired wavelength. The spectral ranges that can be spanned by semiconductor lasers of various composition are shown in Fig. 3-26. Then the laser is fine tuned to match the fluorescent wavelength. Tuning methods include variation of

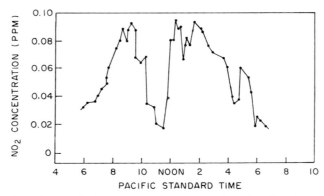

Fig. 11-13 Nitrogen dioxide concentrations measured as a function of local time during a smoggy period in the Los Angeles basin. These data were obtained using a resonance fluorescence technique. (From J. A. Gelbwachs *et al., Opto-Electron.* **4**, 155 (1972).)

current in the diode, temperature, and magnetic field. In modern devices, tuning over a range of a few micrometers is possible through temperature control, and over a range around 1 μm through current control [3]. Such tuning allows one to investigate a number of different pollutants with a single diode laser. The narrow line widths available from such lasers make them highly suitable for use as the local oscillator in a heterodyne.

In such an application, the laser light is not sent out to the position of the pollutant gases. Rather, a telescope collects the fluorescent radiation emitted naturally by the pollutant gases some distance away. The laser is part of the detection system, acting as the local oscillator. If the fluorescent light has angular frequency ω_1 and the laser frequency is ω_2, the difference frequency $\omega_1 - \omega_2$ is produced when the two frequencies are mixed, as described in Chapter 9. The signal-to-noise ratio can be much higher for detection of the difference frequency than for direct detection of the signal at frequency ω_1.

Detection of pollutant gases passively, using heterodyne detection of fluorescence, has been demonstrated using tunable semiconductor lasers [4] and using individual fixed-frequency lines from CO_2 lasers [5].

E.   Absorption Methods

Absorption methods provide an extremely sensitive method of detection of specific gases. The usual method of absorption is to transmit the beam through the sample. Thus one needs a detector on the opposite side of the sample from the transmitter. This may be inconvenient for remote monitoring systems because it involves a double-ended system with installation at two different points. The absorption system has great utility when specific samples of gas can be collected. One possibility for removing the difficulty for remote measurement is the use of a remote topographic reflector. One may use reflection from some topographic feature present in the countryside, such as a large building or a hill. This can provide a single-ended system in which the transmitter and receiver are located at the same position. It does, however, require a suitable reflecting feature to be present.

Figure 11-14 shows calculated results for minimum detectable pollutant concentration from long-path absorption methods for SO_2, NO_2, and CO. The average SO_2 level is indicated for comparison. The average CO level is several parts per million. Thus, long-path absorption methods provide extremely high sensitivity for monitoring common pollutants over relatively long ranges. We see that, in contrast to the backscattering methods, the sensitivity of the measurement increases with increasing path length.

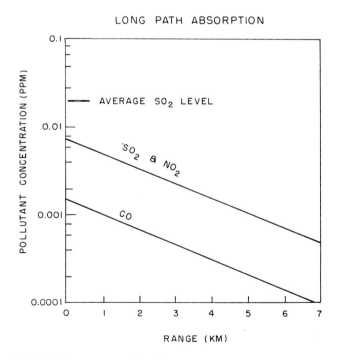

Fig. 11-14 Minimum measurable pollutant concentration for several pollutants using a long-path absorption technique, assuming 1-μW laser power. The average SO_2 level is indicated. (From R. L. Byer, *Opt. and Quantum Electron.* **7**, 147 (1975).)

The equations relative to the power received in long-path absorption methods are

$$P_{OFF} = KP_0 A / R^2 \tag{11.11}$$

$$P_{ON} = KP_0 A \exp\left[-\int_0^R N(r)\sigma \, dr\right] \Big/ R^2 \tag{11.12}$$

where P_{OFF} and P_{ON} are the collected powers when the laser wavelength is off and on the absorption line, respectively, P_0 the transmitted power, A the area of the collecting optics, K the efficiency of the optical system, R the range, $N(r)$ the density of the absorbing molecules as a function of distance, and σ the absorption cross section.

The absorption measurement is carried out by changing the frequency of the laser so that it is first on and then off the position of the absorption line. The pollutant concentration is then given by

$$\int_0^R N(r) \, dr = (1/\sigma)\ln[P_{OFF}/P_{ON}] \tag{11.13}$$

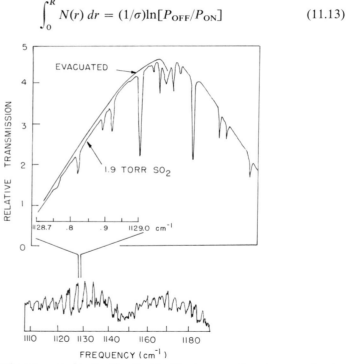

Fig. 11-15 Infrared spectrum of sulfur dioxide. The lower portion shows results of a scan with a good grating spectrometer. The upper portion shows an expanded portion of the scan made with a laser diode, indicating the greatly enhanced resolution obtainable with tunable laser diodes. (From G. A. Antcliffe and J. S. Wrobel, *Appl. Opt.* **11**, 1548 (1972).)

We see that this method has no range discrimination, but only gives the integrated concentration over the entire path.

The extreme selectivity of absorption methods is illustrated by Fig. 11-15. This shows an expanded portion of a trace near the wavelength 8.8 μm of the absorption spectrum of SO_2, taken using a tunable semiconductor diode laser. For comparison, the trace over the similar region with a good monochromator is shown. The resolution with the tunable diode laser is much improved. Thus lasers offer the possibility of extremely sensitive detection of a particular pollutant without interference from other possible constituents of the sample. An example is shown in Fig. 11-16 for the absorption spectrum of ethylene in automobile exhaust. The top trace is a calibration scan and shows the signature of ethylene. This unique signature allows ethylene to be distinguished from other constituents of the exhaust. The bottom three traces show scans for three different automobiles.

Absorption methods have been shown to offer great sensitivity and specificity, but have drawbacks of requiring either a double-ended system or a remote reflector and of offering only range-integrated gas concentrations.

The method of differential absorption backscatter eliminates some of the weaknesses of the other long-path absorption methods, while at the same

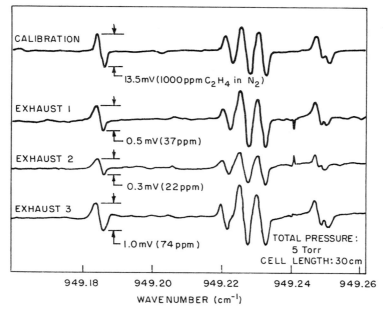

Fig. 11-16 Measurements of ethylene in three automobile exhausts using a scan with a tunable diode laser. The gain is switched between scans. (From E. D. Hinkley, *Opto-Electron.* **4**, 69 (1972).)

time preserving high sensitivity and specificity for particular gases. This method uses ambient atmospheric particulates and molecules as a distributed reflector to provide the return signal. The laser is tuned both on and off a characteristic molecular absorption peak to give sensitivity for a particular pollutant. The pollutant concentration is determined from the difference in return signal strength for on-line and off-line operation. The essential feature is that one monitors radiation backscattered by Rayleigh and Mie scattering so that no physical reflector is required. This removes some of the objections to either double-ended absorption or topographic absorption methods. There is enough Rayleigh and Mie backscattering to provide a reasonable return signal. In addition, this method provides range resolution, which other absorption methods do not.

In order to analyze the differential absorption backscatter technique, we can again refer to Eq. (11.7). In this application, we interpret the backscattering as the sum of the contribution of Rayleigh and Mie backscattering, namely

$$N\sigma = \alpha_{RAY} + \alpha_{MIE} \tag{11.14}$$

The transmission factors are interpreted as

$$T_{ON} = \exp\left[-2\int_0^R N(r)\sigma_{ABS}\,dr\right]\exp\left[-2\int_0^R \alpha(R)\,dr\right] \tag{11.15}$$

for the on-line case and

$$T_{OFF} = \exp\left[-2\int_0^R \alpha(r)\,dr\right] \tag{11.16}$$

for the off-line case. If one takes the ratio of the return signal in the off-line case to the on-line case and differentiates the logarithm, one obtains

$$N(R) = (1/2\sigma_{ABS})\,d/dR\,\ln[P_{OFF}/P_{ON}] \tag{11.17}$$

where σ_{ABS} is the absorption cross section at the particular wavelength for the particular gas, $N(r)$ is the concentration of the gaseous species as a function of range, and $\alpha = \alpha_{RAY} + \alpha_{MIE}$. It is apparent that the method does provide range sensitivity.

An example of some calculated results is shown in Fig. 11-17 for backscattered power as a function of range for a particular assumed concentration of carbon monoxide, as indicated in the lower part of the figure. The effect of the 300-m-thick layer of CO assumed in the lower part of the figure in reducing the on-line signal is apparent. The off-line and on-line curves are the returned powers when the laser is tuned off and on the CO absorption line respectively. The energy requirement for the differential

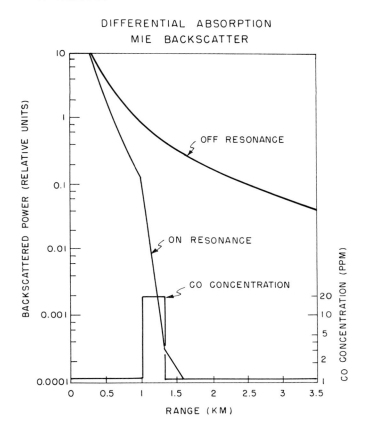

Fig. 11-17   Relative backscattered power as a function of range for detection of carbon
monoxide by differential absorption. The curves are calculated for the laser tuned both on and
off the carbon monoxide absorption line, for an assumed carbon monoxide concentration as
indicated. (From R. L. Byer, *Opt. and Quantum Electron.* **7**, 147 (1975).)

absorption method is higher than that for topographic absorption, but the
advantages outweigh this increased energy requirement.

Figure 11-18 shows some experimental results obtained using the differential absorption backscatter method, which is shown in Fig. 11-18a,
to measure NO_2 concentration. The laser was a dye laser that could be tuned
over a range of approximately 440–450 nm. It emitted from 4 to 8 mJ in
a 200-nsec-duration pulse. This laser was used to monitor NO_2 concentration in a sample chamber 365 m from the laser. The atmosphere behind the
chamber acted as a distributed reflector. Fig. 11-18b shows a comparison
of the NO_2 concentrations measured with this technique, as compared
with those obtained with a transmissometer.

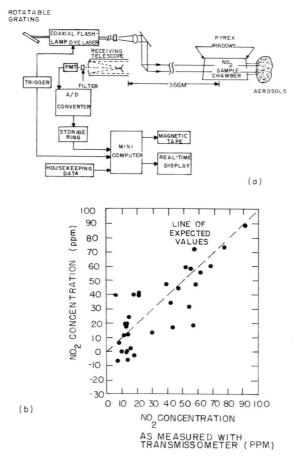

(a)

(b)

Fig. 11-18 Remote measurement of nitrogen dioxide using differential absorption backscatter. (a) Block diagram of the apparatus. (b) Values of nitrogen dioxide concentration in the sample chamber as measured by differential absorption and compared with transmissometer measurements. (From W. B. Grant *et al., Appl. Phys. Lett.* **24**, 550 (1974).)

F.   Summary and Comparison of Methods

In summary, there are a number of promising methods that involve application of lasers to monitoring of ambient air pollution. Table 11-2 shows a compilation of representative results obtained by a number of research workers using some different techniques. This is not an exhaustive list, but it serves to illustrate the range of techniques that have been applied

TABLE 11-2
Selected Examples of Remote Detection of Pollutants

Pollutant	Method	Laser	Wavelength	Sensitivity	Range	Ref.
SO_2	Differential absorption	Frequency-doubled dye	2923, 2933 Å	0.1 ppm over 100 m	306 m	a
O_3	Differential absorption	Frequency-doubled dye	2923, 2940 Å	1.2 ppm over 100 m	306 m	a
SO_2	Differential absorption	Frequency-doubled dye	2960–3000 Å	0.01 ppm over 800 m	800 m	b
NO_2	Differential absorption	Tunable dye	4550–4700 Å	0.2 ppm over 100 m	4000 m	c
CH_4	Long-path absorption (topographic reflector)	DF	3.715 μm	6 ppm over 1 km	300 m	d
N_2O	Long-path absorption (topographic reflector)	DF	3.890 μm	0.24 ppm over 1 km	300 m	d
O_3	Long-path absorption (with retroreflector)	CO_2	9.5 μm	0.002 ppm over 3.75 km	1875 m	e
CO	Long-path absorption (with retroreflector)	$PbS_{0.82}Se_{0.18}$	4.7 μm	0.005 ppm over 0.61 km	305 m	f
NO_2	Fluorescence	HeCd	4420 Å	0.0006 ppm	Short[k]	g
NO_2	Fluorescence	HeCd	4416 Å	0.001 ppm	Short[k]	h
SO_2	Raman scattering	Frequency-doubled ruby	3471 Å	30 ppm over 10 m[l]	200 m	i
CH_4	Raman scattering	Frequency-doubled ruby	3471 Å	28 ppm over 10 m	200 m	j

[a] W. B. Grant and R. D. Hake, *J. Appl. Phys.* **46**, 3019 (1975).
[b] J. M. Hoell, W. R. Wade, and R. T. Thompson, *Proc. Int. Conf. Environmental Sensing and Assessment, Las Vegas, September 14–19, 1975.*
[c] K. M. Rothe, U. Brinkmann, and H. Walther, *Appl. Phys.* **3**, 115 (1974).
[d] E. R. Murray, J. E. van der Laan, and J. G. Hawley, *Appl. Opt.* **15**, 3140 (1976).
[e] R. T. Menzies and M. S. Shumate, *Appl. Opt.* **15**, 2080 (1976).
[f] R. T. Ku, E. D. Hinkley, and J. O. Sample, *Appl. Opt.* **14**, 854 (1975).
[g] A. W. Tucker, M. Birnbaum, and C. L. Fincher, *Appl. Opt.* **14**, 1418 (1975).
[h] J. A. Gelbwachs *et al.*, *Opt.-Electron.* **4**, 155 (1972).
[i] T. Hirschfeld *et al.*, *Appl. Phys. Lett.* **22**, 38 (1973).
[j] H. P. DeLong, *Opt. Eng.* **13**, 5 (1974).
[k] Laboratory measurement of sample of ambient gas.
[l] Amount detected; not necessarily limiting sensitivity.

and the state of the art of detection. Raman scattering provides good specific identification of the particular gases, but only at relatively high concentrations and at relatively short ranges. Absorption methods provide identification at lower concentrations and to greater ranges. Fluorescence methods have mostly been applied to samples of collected ambient air, for the purpose of monitoring concentrations of a pollutant in real-time at a specific location.

Table 11-3 compares the features of the different methods described in the preceding. All the methods except optical radar, which is useful for monitoring particulate materials, provide the capability for identification of gases. All methods except the long-range, double-ended transmission and topographic absorption methods provide range resolution. Similarly, the methods are all capable of operation from a single station, except for absorption methods, which can be single-ended if a suitable reflecting topographic feature is available. The absorption methods provide highest sensitivity, the Raman method has the lowest, and optical radar and resonance fluorescence are intermediate in sensitivity.

TABLE 11-3
Comparison of Methods

	Identification of gases	Range resolution	Ease	Sensitivity
Optical radar	No	Yes	Single-ended	Moderate
Raman scattering	Yes	Yes	Single-ended	Low
Resonance fluorescence	Yes	Yes	Single-ended	Moderate
Absorption				
Transmission	Yes	No	Double-ended	High
Topographic	Yes	No	Single-ended but needs reflecting feature	High
Differential	Yes	Yes	Single-ended	High

Although many workers have been active in this area for a number of years and have shown the capability for obtaining remote monitoring of ambient air pollutants, the status of the technique remains experimental, and no commercial laser instrumentation has been produced. The intensity of the work in this area, the development of ideas and techniques, and the need for improved instrumentation for monitoring of air quality indicate that this may become a viable technique. It has already become useful as a research tool; concentration of particular gases around discrete sources such as chemical factories have been derived. The Raman method offers the capability of monitoring discrete sources such as smokestacks at relatively short ranges. The other methods offer the capability of monitoring gaseous concentrations over larger areas.

The rapidity with which these techniques reach practical application will depend on the development of lasers. The techniques depend critically on the availability of narrow-band, high-energy tunable lasers in spectral regions of interest. In addition, eye safety requirements will be an important factor. From this point of view, the optimum spectral range is in the infrared, at wavelengths greater than about 2 μm. Tunable lasers with adequate pulse energy are just beginning to be developed in this spectral region.

Laser safety standards set relatively low maximum permissible exposures in the visible and near ultraviolet. These very low values for maximum permissible exposures may rule out high-energy pulsed laser sources for remote measurement. Public safety and eye safety standards will be extremely important considerations in the application of remote laser sensing methods as they move from the research stage to practical application in the environment.

REFERENCES

[1] P. W. Kruse, L. D. McGlauchlin, and R. B. McQuistan, "Elements of Infrared Technology." Wiley, New York, 1962.
[2] L. W. Carrier and L. J. Nugent, *Appl. Opt.* **4**, 1457 (1965).
[3] E. D. Hinkley, *et al.*, *Appl. Opt.* **15**, 1653 (1976).
[4] E. D. Hinkley, *Opt. Electron.* **4**, 69 (1972).
[5] R. T. Menzies, *Appl. Phys. Lett.* **22**, 592 (1973).

SELECTED ADDITIONAL REFERENCES

A. Introduction to Laser Monitoring of the Environment

R. L. Byer, Review, Remote Air Pollution Measurements, *Opt. Quantum Electron.* **7**, 147 (1975).
H. Kildal and R. L. Byer, Comparison of Laser Methods for the Remote Detection of Atmospheric Pollutants, *Proc. IEEE* **59**, 1644 (1971).
P. W. Kruse, L. D. McGlauchlin, and R. B. McQuistan, "Elements of Infrared Technology," Wiley, New York, 1962, Chapter 5.
F. F. Hall, Laser Systems for Monitoring the Environment, *in* "Laser Applications," Vol. 2 (M. Ross, ed.), Academic Press, New York, 1974.
H. Tannenbaum *et al.*, Long-Path Monitoring of Atmospheric Pollutant Gases, *Opt. Quantum Electron.* **8**, 194 (1976).

B. Optical Radar

R. J. Allen and W. E. Evans, Laser Radar (LIDAR) for Mapping Aerosol Structure, *Rev. Sci. Instrum.* **43**, 1422 (1972).

R. T. H. Collis, Lidar, *Appl. Opt.* **9**, 1782 (1970).

R. T. H. Collis and E. E. Uthe, Mie Scattering Techniques for Air Pollution Measurement with Lasers, *Opto-Electron.* **4**, 87 (1972).

G. W. Grams, E. M. Patterson, and C. M. Wyman, Airborne Laser Radar for Mapping Two-Dimensional Contours of Aerosol Concentration, *Opt. Quantum Electron.* **7**, 187 (1975).

F. F. Hall, Laser Measurements of Turbidity in the Atmosphere, *Opt. Spectra*, p. 67 (July/August 1970).

F. F. Hall, Laser Systems for Monitoring the Environment, *in* "Laser Applications," Vol. 2 (M. Ross, ed.), Academic Press, New York, 1974.

W. B. Johnson, Lidar Applications in Air Pollution Research and Control, *J. Air Pollut. Contr. Assoc.* **19**, 176 (1969).

C. A. Northend, R. C. Honey, and W. E. Evans, Laser Radar (Lidar) for Meteorological Observations, *Rev. Sci. Instrum.* **37**, 393 (1966).

C. Raman Backscattering

R. L. Byer, Review, Remote Air Pollution Measurement, *Opt. Quantum Electron.* **7**, 147 (1975).

R. K. Chang and D. G. Fouche, Gains in Detecting Pollution, *Laser Focus*, p. 43 (December 1972).

J. A. Cooney, Measurements of the Raman Component of Laser Atmospheric Backscatter, *Appl. Phys. Lett.* **12**, 40 (1968).

H. P. DeLong, Air Pollution Field Studies with a Raman Lidar, *Opt. Eng.* **13**, 5 (1974).

T. Hirschfeld *et al.*, Remote Spectroscopic Analysis of PPM-Level Air Pollutants by Raman Spectroscopy, *Appl. Phys. Lett.* **22**, 38 (1973).

H. Inaba and T. Kobayasi, Laser Raman Radar for Chemical Analysis of Polluted Air, *Nature (London)* **224**, 170 (1969).

H. Inaba and T. Kobayasi, Laser–Raman Radar, *Opto-Electron.* **4**, 101 (1972).

H. Kildal and R. L. Byer, Comparison of Laser Methods for the Remote Detection of Atmospheric Pollutants, *Proc. IEEE* **59**, 1644 (1971).

T. Kobayasi and H. Inaba, Spectroscopic Detection of SO_2 and CO_2 Molecules in Polluted Atmosphere by Laser–Raman Radar Technique, *Appl. Phys. Lett.* **17**, 139 (1970).

T. Kobayasi and H. Inaba, Laser–Raman Radar for Air Pollution Probe, *Proc. IEEE* **58**, 1568 (1970).

D. A. Leonard, Observation of Raman Scattering from the Atmosphere Using a Pulsed Nitrogen Ultraviolet Laser, *Nature (London)* **216**, 142 (1967).

D. A. Leonard, Raman and Fluorescence Lidar Measurements of Aircraft Engine Emissions, *Opt. Quantum Electron.* **7**, 197 (1975).

S. H. Melfi, J. D. Lawrence and M. P. McCormick, Observation of Raman Scattering by Water Vapor in the Atmosphere, *Appl. Phys. Lett.* **15**, 295 (1969).

S. H. Melfi, Remote Measurements of the Atmosphere Using Raman Scattering, *Appl. Opt.* **11**, 1605 (1972).

H. Rosen, P. Robrish, and O. Chamberlain, Remote Direction of Pollutants Using Resonance Raman Scattering, *Appl. Opt.* **14**, 2703 (1975).

D. Resonance Fluorescence

R. L. Byer, Review, Remote Air Pollution Measurement, *Opt. Quantum Electron.* **7**, 147 (1975).

J. A. Gelbwachs *et al.*, Fluorescence Determination of Atmospheric NO_2, *Opto-Electron.* **4**, 155 (1972).

E. D. Hinkley and P. L. Kelley, Detection of Air Pollutants with Tunable Diode Lasers, *Science* **171**, 635 (1971).

E. D. Hinkley, Tunable Infrared Lasers and Their Applications to Air Pollution Measurements, *Opto-Electron.* **4**, 69 (1972).

H. Kildal and R. L. Byer, Comparison of Laser Methods for the Remote Detection of Atmospheric Pollutants, *Proc. IEEE* **59**, 1644 (1971).

I. Melngailis, The Use of Lasers in Pollution Monitoring, *IEEE Trans. Geosci. Electron.* **GE-10**, 7 (1972).

R. T. Menzies, Use of CO and CO_2 Lasers to Detect Pollutants in the Atmosphere, *Appl. Opt.* **10**, 1532 (1971).

R. T. Menzies, Remote Detection of SO_2 and CO_2 with a Heterodyne Radiometer, *Appl. Phys. Lett.* **22**, 592 (1973).

R. T. Menzies and M. S. Shumate, Air Pollution: Remote Detection of Several Pollutant Gases with a Laser Heterodyne Radiometer, *Science* **184**, 570 (1974).

A. W. Tucker, M. Birnbaum, and C. L. Fincher, Atmospheric NO_2 Determination by 442-nm Laser Induced Fluorescence, *Appl. Opt.* **14**, 1418 (1975).

E. Absorption Methods

G. A. Antcliffe and J. F. Wrobel, Detection of the Gaseous Pollutant Sulfur Dioxide Using Current Tunable $Pb_{1-x}Sn_x$ Te Diode Lasers, *Appl. Opt.* **11**, 1548 (1972).

R. L. Byer, Review, Remote Air Pollution Measurement, *Opt. Quantum Electron.* **7**, 147 (1975).

W. B. Grant *et al.*, Calibrated Remote Measurement of NO_2 Using the Differential-Absorption Backscatter Technique, *Appl. Phys. Lett.* **24**, 550 (1974).

W. B. Grant and R. D. Hake, Calibrated Remote Measurements of SO_2 and O_3 Using Atmospheric Backscatter, *J. Appl. Phys.* **46**, 3019 (1975).

P. L. Hanst, Optical Measurement of Atmospheric Pollutants: Accomplishments and Problems, *Opt. Quantum Electron.* **8**, 87 (1976).

T. Henningsen, M. Garbuny, and R. L. Byer, Remote Detection of CO by Parametric Tunable Lasers, *Appl. Phys. Lett.* **24**, 242 (1974).

E. D. Hinkley and P. L. Kelley, Detection of Air Pollutants with Tunable Diode Lasers, *Science* **171**, 635 (1971).

E. D. Hinkley, Tunable Infrared Lasers and Their Applications to Air Pollution Measurements, *Opto-Electron.* **4**, 69 (1972).

E. D. Hinkley *et al.*, Long Path Monitoring: Advanced Instrumentation with a Tunable Diode Laser, *Appl. Opt.* **15**, 1653 (1976).

T. B. Hirschfeld, Long Path Infrared Spectroscopic Detection of Atmospheric Pollutants, *Opt. Eng.* **13**, 15 (1974).

H. Kildal and R. L. Byer, Comparison of Laser Methods for the Remote Detection of Atmospheric Pollutants, *Proc. IEEE* **59**, 1644 (1971).

L. B. Kreuzer, N. D. Kenyon, and C. K. N. Patel, Air Pollution: Sensitive Detection of Ten Pollutant Gases by Carbon Monoxide and Carbon Dioxide Lasers, *Science* **177**, 347 (1972).

R. T. Ku, E. D. Hinkley, and J. O. Sample, Long-Path Monitoring of Atmospheric Carbon Monoxide with a Tunable Diode Laser System, *Appl. Opt.* **14**, 854 (1974).

F. L. McNamara, A Tunable Semiconductor Laser for Pollution Studies, *Opt. Eng.* **11**, 9 (1972).

I. Melngailis, The Use of Lasers in Pollution Monitoring, *IEEE Trans. Geosci. Electron.* **GE-10**, 7 (1972).

R. T. Menzies and M. S. Shumate, Remote Measurements of Ambient Air Pollutants with a Bistatic Laser System, *Appl. Opt.* **15**, 2080 (1976).

CHAPTER 12

LASER INSTRUMENTATION AND MEASUREMENT

The unusual properties of laser light can be used in a variety of important measurements for industrial applications. We have already discussed the application of lasers for interferometric distance measurement in Chapter 10 and have considered how those techniques can be adapted to determine such parameters as straightness and flatness. In this chapter we discuss other techniques.

A wide variety of inspection and measurement procedures can be accomplished easily with lasers. Some such devices are being developed as commercially available instrumentation. For other applications, the instrumentation and procedures are developed by the individual user for his own needs. The variety of different approaches is so large that it is impossible to give a complete summary. The selection in this chapter will show the great adaptability of lasers for providing unusual types of instrumentation and their capabilities for making measurements that might be difficult with conventional techniques. In particular, we include methods for measuring velocity (both of fluids and of solid surfaces), for measuring angular rotation rate, for measuring the diameter of fibers and wires, for determining surface profile and position, and for determining dimensions of finished products. The variety of measurement capabilities illustrated here can serve to stimulate the imagination for applying similar techniques to one's own measurement problems.

A. Velocity Measurement

1. *Fluid Flow Measurements*

Measurements of the velocity of fluid flow can be made by scattering a laser beam from a liquid or gas. The laser beam interacts with small particles carried along by the flow stream. The particles scatter some of the light in all directions. Many fluids contain enough scattering particles so that the scattered light can be easily detected.

The frequency of the scattered light is slightly Doppler-shifted, with a frequency shift proportional to the velocity of the fluid.† Measurement of the frequency shift directly gives the flow velocity. Laser Doppler velocimeters have been developed at a number of laboratories and have been employed in specialized engineering applications to measure flow. These velocimeters have been used in situations where other measurements would have been difficult.

Laser Doppler velocimetry offers the advantage of being a noncontact method. It is not necessary to introduce any equipment into the flow, so that there are no problems with disturbance of the flow. Hot or corrosive fluids can be probed. Accurate velocity measurements can be made because of the narrow line width of the laser. The Doppler shift of light scattered from the moving fluid can be greater than the line width of the laser, even for relatively slow flow velocity. This fact gives the laser velocimeter a potentially wide dynamic range.

Velocities measured by experimental models cover the range from centimeters per second to hundreds of meters per second.

Frequency response is high, in contrast to most conventional devices, which have a more sluggish response. Thus, it becomes possible to obtain data in transient conditions and to investigate turbulent flow characteristics.

The main disadvantage of laser Doppler velocimetry is the necessity for scattering particles to be entrained within the flow. Particles are needed to provide sufficient scattered light to yield a detectable signal. In many practical circumstances, the fluid has sufficient scattering centers to provide a measurable output. In cases where the fluid is very transparent, with no scattering centers, it may be necessary to introduce scattering particles artificially. Particles seeded into the flow must have small diameter in order to follow the flow faithfully.

Laser Doppler velocimetry is based on the Doppler shift. The frequency of light scattered by an object moving relative to a radiating source is changed by an amount that depends on the velocity and the scattering geometry. The technique basically consists of focusing laser light at a point within the flowing fluid. Light scattered from the fluid or from particles entrained within the fluid flow is collected and focused on a detector. Signal processing of the detector output yields the magnitude of the Doppler frequency shift and hence the velocity of the flow.

Various optical arrangements have been used. The one most commonly

† See Chapter 9 for a discussion of the Doppler shift.

used in modern instruments is illustrated in Fig. 12-1a. This arrangement is called a dual-beam mode. Light from a continuous laser is split into two equal parts by a beam splitter as shown. The focusing lens is inserted in front of the beam splitter. The operation of this dual-beam system is possible to visualize in terms of fringes. Where the two beams cross, the light waves interfere to form alternate regions of high and low intensity. The fringe pattern is illustrated in Fig. 12-1b. If a particle traverses the fringe pattern, it will scatter more light when it is passing through regions of high intensity.

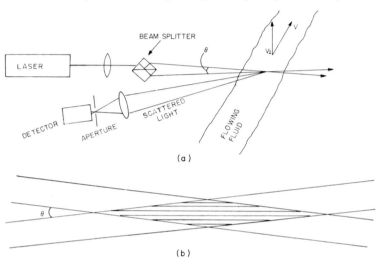

Fig. 12-1 (a) Schematic diagram for laser fluid velocimetry. This arrangement is the so-called dual-beam mode. (b) shows the interference pattern generated by the superposition of the two beams at a small angle θ. The fluid flows with velocity v, with a component $v_\perp$ perpendicular to the direction of the laser beam.

Thus the light received at the detector will show a varying electrical signal, the frequency of which is proportional to the rate at which the particle crosses the interference fringes. The distance s between fringes is given by the expression

$$s = \lambda/2 \sin(\theta/2) \tag{12.1}$$

where θ is the angle between the two converging beams. If a particle passes across the fringes with a component of velocity v in the direction perpendicular to the fringes, the signal will be modulated at the frequency

$$\Delta f = 2v \sin(\theta/2)/\lambda \tag{12.2}$$

The wavelength λ is the wavelength in the fluid, which may be different from the vacuum wavelength by a factor equal to the index of refraction n. Thus Eq. (12.2) can be rewritten

$$\Delta f = 2nv \sin(\theta/2)/\lambda_0 \tag{12.3}$$

where λ_0 is the wavelength of the laser in vacuum.

The collecting optics are arranged to view the point of intersection of the two beams in the fluid. In the arrangement in Fig. 12-1, the detector receives backscattered light. This allows the system to be single-ended, i.e., the laser and collector are near the same position. This is the most convenient arrangement when remote measurements are to be made at a distance. When the sample is nearby and accessible, the collector may be located on the opposite the side of the flowing fluid, and forward scattered light is collected. The use of forward scattering increases the signal and reduces requirements on laser power.

Laser power requirements depend on the geometry and the distance at which the measurement is to be made. For measurements made at distances of 10 or 20 cm, helium–neon lasers with output powers in the range from 5 to 15 mW are generally sufficient. When measurements are to be made at greater distances (remote anemometry), argon lasers with output 1 W or greater are generally used.

The dual-beam mode shown in Fig. 12-1 forms the basis for operation of commercial laser Doppler velocimeters and laser Doppler anemometers that are becoming available. Many variations on this basic scheme are possible, including the simultaneous use of two laser colors or two laser polarizations to determine two components of the velocity in the flow field.

Earlier instrumentation, which served for many experimental measurements and which is described in many articles in the technical literature, used the so-called reference-beam mode. In the reference-beam mode, the scattered light is collected from particles going through a single incident beam. Scattered light is collected and mixed at the phototube with unscattered light directly from the laser. The scattered light is frequency shifted by the amount

$$\Delta f = 2nv \sin(\theta/2)/\lambda_0$$

where θ is the angle of scattering. This is exactly the same as Eq. (12.3). The frequency shift is detected as a beat note in the output of the photodetector. This point of view yields exactly the same result as consideration of a particle traversing a fringe pattern in the dual-beam mode.

The reference-beam mode requires that the aperture in which the scattered light is collected be small. This means that the signal is smaller than that available from the dual-beam mode, and that therefore laser power requirements are higher. In comparison, the dual-beam mode is preferable when the number of scattering particles is relatively small. If the number of scattering particles becomes large, so that many particles are traversing the measurement volume at the same time and the signal is large, the reference-beam mode may be preferable.

A variety of different types of measurement can be made. For example, one can measure the beam velocity in a given direction, i.e., one component of the velocity at a point. Alternatively, with added complexity, one can measure two or three components of the velocity at a point. If the velocity is varying with time, due to turbulence, the frequency will be a rapidly varying function centered about a mean value. Measurement of the frequency spectrum can provide information about the spectrum of the turbulence. For example, one can use a spectrum analyzer with frequency-tracking capability to provide instantaneous information about the turbulence at the point of measurement in the fluid.

Applications to date include study of the flow field in a variety of situations (including laminar flow and turbulent flow), investigation of boundary layer and shock phenomena, determination of three-dimensional velocity data in vortices near the tips of aircraft wings, measurement of flow between the blades of turbines, gas velocity measurements within compressor rotor passages, *in vivo* measurement of blood flow, and remote measurement of wind velocity.

In one application, an air-speed indicator has been developed that offers greater precision than conventional devices. The system uses a CO_2 laser beam focused to a point about 20 m ahead of the aircraft, where the air is undisturbed. Some of the radiation is backscattered by aerosols. The Doppler shift of the backscattered light is measured and recorded by a digital system. The usual method of measuring air speed uses Pitot-tube air-pressure sensors. It is difficult to obtain accurate measurements using such sensors because of turbulence introduced by the sensor and because of boundary layer effects in the air surrounding the aircraft.

In other applications, laser Doppler velocimeters have been used to measure velocity fluctuations of gas flows. For measurements of turbulence, a spectrum analyzer can be used to display the frequency content of the scattered light. The spectrum analyzer output yields the velocity probability distribution averaged over a period of time. To obtain instantaneous information about the turbulent velocity, automatic frequency tracking capability can be provided.

To demonstrate the capabilities of laser velocimetry, we present two specific examples. Figure 12-2 shows the velocity profile of laminar flow of water through a circular pipe, in which a constant flow velocity of 5.35 cm/sec was maintained at the center of the tube [1]. From elementary considerations, one expects the velocity distribution of a viscous fluid undergoing laminar

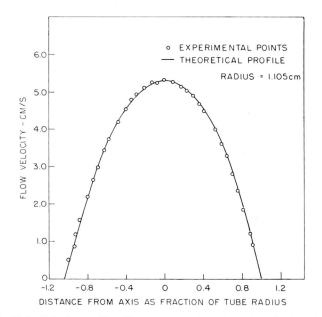

Fig. 12-2   Velocity profile of laminar water flow through a circular pipe. (From J. W. Foreman *et al.*, *IEEE. Quantum Electron.* **QE-2**, 260 (1966).)

flow in a cylindrical pipe to be parabolic, with the velocity besng maximum at the center and zero at the walls. The measurements shown in Fig. 12-2 demonstrate the expected type of behavior and show the versatility of laser Doppler velocimetry in probing a flow profile at various points.

Figure 12-3 shows a longitudinal velocity profile of a shock wave [2]. This date relates to a supersonic jet of nitrogen ejected under pressure into ambient air through a supersonic nozzle. A wedge was set into the flow stream to produce a shock wave. The longitudinal velocity scan shown in Fig. 12-3 was made 0.15 in. above the center line of the flow. High spatial resolution is demonstrated. The figure illustrates direct measurement of the flow velocity throughout a shock wave without introducing any perturbations into the flow. This is a measurement difficult to obtain with other techniques. The two examples together serve to show the wide range of fluid velocities that can be dealt with by this technique.

The status of development of laser velocimeters is compatible with use in engineering studies of flow fields. Commercial instruments using laser Doppler velocimetry have been developed.

Commercial instruments offering accuracy of 0.5% at 40 MHz are available. Such instrumentation provides the capability for making unusual

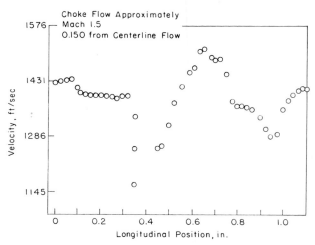

Fig. 12-3 Longitudinal velocity profile of a shock wave for a supersonic jet of nitrogen ejected into ambient air through a supersonic nozzle. (From A. E. Lennert *et al.*, presented at National Laser Industry Association Meeting, 2nd, Los Angeles, October 20–22, 1969.)

types of measurements that would be difficult or impossible with conventional techniques. It can also be applicable for process control in production applications.

2. *Velocity of Solid Surfaces*

The methods described in the preceding can also be used to measure the velocity of solid surfaces using the Doppler shift of laser radiation reflected from the surfaces. Such measurements have been used to study vibrations in large structures and to measure velocity of moving semimolten fibers in the polymer industry. Other, similar applications of laser Doppler velocimetry applied to motion of solid surfaces will be obvious.

In addition, other laser velocimeters for motion of solids have been developed. One example is a system that has been used to determine the motion of metallic plates that have been shocked by high-velocity projectiles [3]. The laser beam is reflected from the surface of the metallic specimen, which can be in rapid motion when the shock front reaches it. The arrangement is shown in Fig. 12-4. The surface motion imparts a Doppler frequency shift to the reflected light beam. Part of the reflected light is picked off at a beam splitter and measured at photomultiplier 1. This provides a reference level for the intensity of the light reaching the detection apparatus and allows compensation of such factors as change in the surface reflection properties of the target. The light that passes through the beam

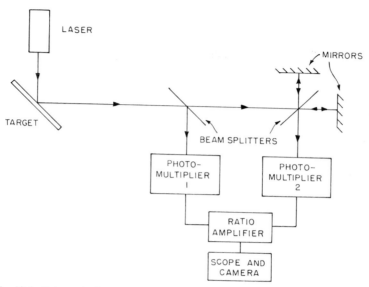

Fig. 12-4   Schematic diagram of a laser velocimeter for determination of motion of solid-surface velocity.

splitter enters an interferometer. The position of the mirrors of the interferometer is adjusted so that the light reaching photomultiplier 2 will be a maximum when the target is stationary (i.e., when there is no frequency shift). When the target moves, the wavelength of the light and the path difference in the arms of the interferometer (measured in optical wavelengths) change. This change is a sensitive function of the velocity of the target. Therefore, when the target is moving, the fringes will move across the photomultiplier, and the light intensity detected by photomultiplier 2 will change. The signals from the two photomultipliers are fed into a ratio amplifier, the output signal of which is proportional to surface velocity. This yields a direct reading of the velocity of the surface as a function of time on an oscilloscope display.

This device has been used to measure the velocity of the free surface of metal plates that have been shocked. Internal stress waves propagating through metal plates that were struck by high-velocity projectiles or by electron beams have been measured. The measurements would have been extremely difficult to make with conventional techniques.

A type of laser velocimeter has been developed that operates on an entirely different principle [4]. This velocimeter is well suited for measuring velocity of solid materials that have relatively rough surfaces. The previously described velocimeter was better adapted to a solid with a relatively shiny surface, so that a reasonably well collimated beam would be reflected from the surface.

This second velocimeter uses the so-called "speckle pattern" that occurs in laser light reflected from a diffusely reflecting surface. When an observer views such a reflection, he sees an illuminated area covered by tiny specks of light. This appearance is produced by diffraction effects in the coherent light. The light reflected from the small irregularities present on the surface can interfere to produce areas with a local high light intensity. The observer sees these areas as bright specks. Such constructive interference effects are also present transiently when a surface is illuminated by conventional incoherent light sources, but the pattern changes randomly in a short time compared to the resolving time of the human eye. Thus, a surface illuminated by a conventional light source appears uniformly lighted, whereas a surface illuminated by coherent laser light appears to have a granulated, speckled texture.

When there is a relative motion between the surface and the observer, the speckle pattern appears to move. The entire pattern of bright spots moves as a whole. The apparent velocity of movement of the specks is twice the velocity of the relative motion.

If the eye of the observer is replaced by a photodetector with a small aperture, movement of one speck of light over the aperture will result in a pulse from the photodetector. This gives a method of measurement for the surface velocity. To provide quantitative measurements, an apparatus is set up as shown in Fig. 12-5. The speckle pattern is shown schematically as an irregular pattern of reflected light with bright lobes. A grating is placed in front of the photodetector. As one bright spot in the speckle pattern moves

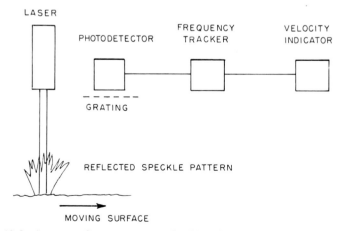

Fig. 12-5 Apparatus for measurement of solid-surface velocity using speckle pattern of reflected laser light.

over the grating, it will produce a series of pulses. The frequency f of this series of pulses will be

$$f = 2V/d \qquad (12.4)$$

where V is the velocity of the surface and d is the grating spacing. There will be a large number of spots crosssng the grating. Each will give a series of pulses. The pulse trains from each spot will have random phases, but they will all have the same frequency. Thus it is only necessary to measure the center frequency of the output of the photodetector to determine the surface velocity.

Instruments based on this principle of optical diffraction velocimetry can provide a noncontact sensing element suitable for use on surfaces with any color or texture. The accuracy of the measurement is stated to be 0.1% over the range of velocities from 50 to 1000 ft/min. This type of velocimeter has been tested successfully on a wide variety of materials including aluminum, black and shiny steel, copper wire, newsprint, clear and opaque plastics, leaves, sand, and dirt. It can be used to measure velocity of vehicles with greater accuracy than a fifth wheel. It appears especially suited for use in automated processes involving continuous measurement of speed or length of product. It is finding acceptance for monitoring speed of strip metal in steel and aluminum mills.

A few precautions are necessary in a mill environment. The optical path must be kept relatively clear. The surface must be free of excessive amounts of liquid. This requirement may be satisfied by having an air jet that blows surface liquid away from the spot that is being illuminated. Finally, if the surface undergoes accelerations, so that the velocity of the spots changes substantially during the time of their passage over the grating, the accuracy may be reduced.

B. Angular Rotation Rate

The use of lasers offers the possibility for angular rotation rate sensing, or in other words a laser gyroscope. The so-called laser gyro has a ring configuration, usually a triangle, in which two counterrotating laser beams can travel in opposite directions around the ring. The wavelength of operation of the ring laser adjusts itself so that the distance around the ring is an integral number of wavelengths. A change in the length of the ring causes a change in the wavelength of operation. If the ring rotates about a direction perpendicular to the plane of the ring, there will be a change in length between the paths seen by two counterrotating beams. The beam traveling in the same direction as the sense of rotation will have to travel slightly farther on each traversal of the ring. The system is shown schematically in Fig. 12-6, which shows a triangular arrangement. The active laser medium

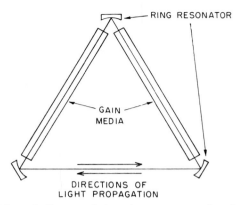

Fig. 12-6   Schematic diagram of triangular arrangement for a laser gyro.

is in two legs of the triangle. Reflectors at the end of each leg define the path of the beam.

Although the velocity of light is very high, it is not infinite. During the time it takes the light wave to travel once around the ring, the ring will have moved slightly, so that one beam must travel slightly farther to traverse the ring. The light beam traveling in the direction opposite to the sense of rotation will have a slightly shorter distance to travel. This is the same as changing the effective distance that the beam travels around the ring in each of the two directions. The frequencies at which the two light beams oscillate are determined by the optical length of the path. The condition is that the number of wavelengths in the cavity must be an exact integer. The path difference in the two different directions of rotation causes the two light beams to have slightly different frequencies in order to satisfy the condition for an integral number of wavelengths.

To understand the operation of the laser gyro, it is instructive to work through the mathematical conditions. Consider light leaving point Q in Fig. 12-7 and making one transit around the equilateral triangle. The point Q will move a small distance during the time that the light is moving around the path.

The time T for light to make one round trip around the ring is

$$T = P/c = 3L/c \qquad (12.5)$$

where P is the total perimeter of the triangle and c is the velocity of light.

The distance moved by point Q in this time is

$$D = \omega T x \qquad (12.6)$$

where ω is the angular rotation rate and x is the distance from Q to the

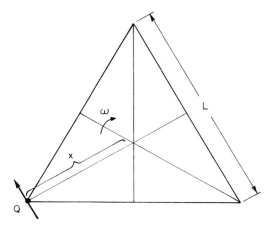

Fig. 12-7   Diagram for analysis of laser gyro operation. In this figure, ω is the angular rotation rate.

intersection of the perpendicular bisectors of the three sides of the equilateral triangle, as shown in Fig. 12-7. The change in path ΔP is the component of D along the direction of light propagation,

$$\Delta P = D \cos 60° = D/2 \qquad (12.7)$$

We have

$$\Delta P = \omega Tx/2 = \sqrt{3}\,\omega L^2/2c \qquad (12.8)$$

where we have used $x = \sqrt{3}\,L/3$. Since the area A of the equilateral triangle is

$$A = \sqrt{3}\,L^2/4 \qquad (12.9)$$

we may rewrite

$$\Delta P = 2\omega A/c \qquad (12.10)$$

Since for a standing light wave

$$N\lambda = P \qquad (12.11)$$

where N is an integer, when the path length shifts by ΔP the wavelength shifts by $\Delta \lambda$:

$$\Delta \lambda = \Delta P/N = \lambda\,\Delta P/P \qquad (12.12)$$

and the operating frequency shifts by Δf:

$$\Delta f/f = \Delta \lambda/\lambda = \Delta P/P \qquad (12.13)$$

Each of the two counterrotating beams will see the same shift, although in opposite senses.

Hence, the difference Δ in frequency between the two counterrotating beams is

$$\Delta = 2\,\Delta f = 2f\,\Delta P/P = 4\omega A/\lambda P \tag{12.14}$$

This equation is the basic equation that is used for angular rotation rate measurement. Although it has been derived here for the case of an equilateral triangle, it turns out to be valid for any path with enclosed area A and perimeter P. It is valid even when general relativity is taken into account. This frequency difference can be detected as a beat note according to methods described in Chapter 9. Equation (12.14) shows that the beat frequency is proportional to angular rotation rate, which can be measured for a known geometry simply by measuring the beat frequency.

Let us consider a specific numerical example. For a helium–neon laser ($\lambda = 0.633\ \mu$m), a rotation rate of 0.1 rad/hr, and a triangle with 10-cm legs, this equation gives a beat frequency around 2.5 Hz. This frequency is easily detectable, even though it is only about 2×10^{-14} of the laser frequency.

For readout, part of each counterrotating beam is fed out through a semitransparent mirror, and the two beams are combined at a detector. One method for carrying this out is shown in Fig. 12-8.

If the apex angle of the prism in Fig. 12-8 is exactly 90°, θ will be 0 and the two beams will emerge colinearly. They will have slightly different frequencies, and the frequency difference will be detectable as a beat note in the detector. One measures the frequency difference Δ characteristic of the beat note and applies Eq. (12.14) to determine the angular rotation rate.

The arrangement is not critically dependent on the apex angle of the prism being exactly 90°. If the apex angle differs from 90° by a small angle θ, the two beams will emerge with an angular separation equal to $2\theta n$, where n is the index of refraction of the prism. This means that a fringe pattern

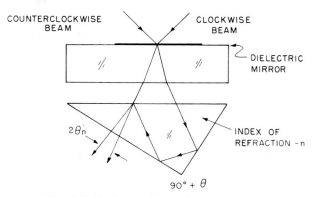

Fig. 12-8   Readout technique for laser gyro.

will be in motion. The detector determines the motion of the fringes from the amplitude modulation of light reaching the detector. The higher the rotation rate, the greater is the frequency difference, and the faster the rate of fringe motion.

In principle, laser gyros should be very sensitive. Analysis of the fundamental considerations limiting the accuracy of the laser gyro indicates a limit of less than 10^{-6} degrees/hr. In practice, the limits of accuracy are set by construction, cleanliness, and other factors. A phenomenon called "lock-in" has also proved troublesome. The frequency difference between the two counterrotating beams falls to zero for small values of the input, i.e., of the rotation rate. This is the same as the locking together of two coupled oscillators operating at slightly different frequencies. Thus there is a threshold value of the angular rotation rate below which the laser gyro gives no output signal. This is shown schematically in Fig. 12-9. Various techniques have been devised to reduce the lock-in problem. One technique involves dithering

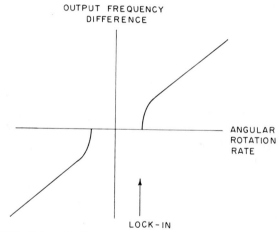

Fig. 12-9   Schematic diagram of lock-in at low angular rotation rates.

the gyro, i.e., rotating it back and forth at a known rate so that the total rotation rate lies outside the threshold for lock-in. When one integrates over a number of periods of the oscillatory dither, the contribution of the total fringe shift from the opposite motions of the dither will cancel. The remaining signal is due to the rotation that is to be measured. With this technique, one can measure rotation rates that would be below the threshold for lock-in if the dithering were not used.

The construction of the laser gyro is very simple, allowing low-cost fabrication. A quartz block can be drilled out in a triangular geometry as shown in Fig. 12-10. Corner mirrors are attached to the block, anodes and

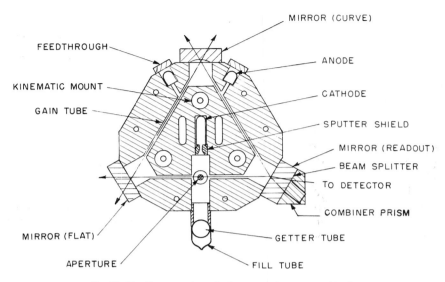

Fig. 12-10   Construction of a laser gyro in a quartz block.

Fig. 12-11   Three-axis laser gyro in a single quartz sphere. (Photograph courtesy of F. Aronowitz.)

cathodes are attached to the housing, and the gyro is filled with helium and neon and then sealed. Three-axis systems can be constructed within a single quartz block. Figure 12-11 shows the construction detail of a solid-block three-axis laser gyro. Such devices are rugged and low in cost. They involve short run-up time and provide an integrating rate gyro with potentially high precision.

At present, the laser gyro is in the status of advanced research and development. It can potentially be an accurate low-cost instrument which has no sensitivity to acceleration. Performance figures that have been published to date indicate that laser gyros are capable of measurement down to rates about 0.003 degrees/hr. For navigation, one integrates the rotation rate output of the device. Laser gyros can operate in a strapped-down mode (i.e., rigidly attached to the vehicle). Thus, no gimbal system is needed.

Current limitations on the laser gyro include the limitation on rotation rate measurements because of lock-in and the fact that very long lifetimes have been difficult to achieve.

C. Wire Diameter

Measurement of small dimensions can be achieved with high accuracy by use of the diffraction pattern produced when a laser beam bends around the small object. A specific example of this process is the measurement of wire diameters. The diffraction pattern produced by a wire inserted in the path of a laser beam of larger diameter than the wire is an array of dots extending in a line perpendicular to the wire. Figure 12-12 represents this situation. A central undiffracted spot appears at the center of the pattern projected on the screen. In the direction perpendicular to the wire are the diffracted spots. The separation of the spots varies inversely with the wire diameter D according to the following equation:

$$\sin \phi_n = n\lambda/D \qquad (12.15)$$

where ϕ_n is the angle of the nth spot from the direction of the laser beam measured from the position of the wire and λ is the laser wavelength. This equation may be used to obtain D from geometrical measurements of the position where the diffracted spot appears. Figure 12-12A shows the pattern for a wire of given diameter. Figure 12-12B shows how the pattern changes as the wire diameter decreases. The diffracted spots spread farther apart.

In practical terms, this method can measure wire or fibers with a diameter of 0.00025 cm to an accuracy of 0.5%.

This type of measurement has attractive features. There is no contact with the wire. Movement of the wire along the direction of its length does

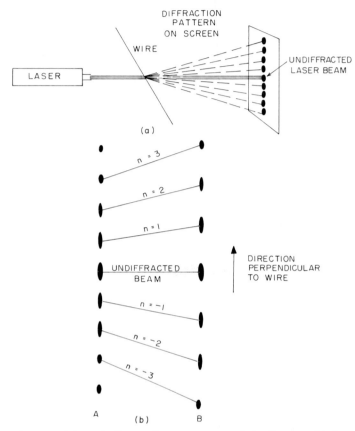

Fig. 12-12 (a) Schematic diagram for measurement of wire diameter. (b) shows how the pattern changes with wire diameter, where A indicates the pattern for a wire diameter larger than B and n is the order of the interference.

not affect the pattern. Thus, wires can be measured while moving, e.g., during the extrusion process.

One can make a continuous real-time measurement on the wire as it is extruded or drawn. Motion of the wire perpendicular to its length, so long as it remains within the laser beam, does not affect the separation of the spots. The method is particularly applicable to small wires.

This is one potential application which makes use of the brightness of the laser. Diffraction phenomena have long been known, but available previous sources were not bright enough to make this type of measurement attractive. An automated system has been developed [5] in which an array of photodiodes or a TV vidicon detects the diffraction pattern. The signal from the detector is analyzed to measure the separation of the maxima in the diffraction pattern and thus to give direct readout of the wire diameter.

Diffraction measurements can also be used in a variety of ingeneous ways to determine dimensions of other types of products besides fibers and wires. For example, one of the edges of a part could be used as one side of a narrow aperture. A stationary flat edge would form the other side of the aperture. If the dimensions of the product change, the width of the aperture would change. The spatial pattern of light diffracted through the narrow aperture would then change. Photodetectors sensing a change in the pattern could be used to determine the changing dimension of the part. Alternatively, if the part is stationary and the laser beam is scanned along the length of the aperture formed by the part and the reference surface, changes in the diffraction pattern as a function of the beam position could be used to determine parameters such as flatness or waviness of the surface. Such measurements can be made with high accuracy and can detect small changes in the dimensions or surface profile of the object. The method can provide solutions to measurement problems in which conventional techniques are difficult to apply.

Some functions that can be performed by diffraction measurements include diameter or width of extruded and drawn products or manufactured parts, thickness of items produced in sheet form such as plastic, rubber, or paper, flatness or waviness of surfaces of items such as seals or piston rings, concentricity of manufactured parts, proximity of items such as turbine blades or compressor parts, and angular orientation of piece parts. Commercial systems aimed at solving a variety of commonly encountered measurement, gauging, and inspection problems in industry are becoming available.

D. Profile and Surface Position Measurement

Contouring measurements are commonly made with the use of a sharp probe. The probe is in mechanical contact with the surface. Changes in position of the surface induce motion in the probe that can be measured. In this way, a profile of the surface can be mapped out.

Such mechanical devices have been used in a variety of applications. Examples include examination of parts for surface finish or dimensional tolerances, continuous monitoring of thickness or surface parameters in automated process control (with feedback servocontrol of the machining process), and accurate determination of dimensions of a model. Profileometry of models in the automobile industry is performed to obtain accurate dimension for full-scale devices.

Use of a light beam as an optical probe in a profileometer offers a number of obvious advantages. Surface finish is not damaged, and there is no wear on the measuring tool. Measurements can be made on soft materials, e.g., clay models.

A variety of techniques for noncontact laser profileometry have been suggested, and prototype models have been developed. With these models, the microscopic texture and surface finish are resolved by a focused laser beam. The devices that have been developed are capable of measuring a wide variety of materials with surface finishes from polished metals to dull plastics. It is claimed that higher accuracy can be obtained than with conventional stylus profileometers. Area scan mapping and profiling can be done much faster.

A variety of approaches to this type of measurement have been suggested.

The basic ingredient is the optical measurement of the distance from the measuring head (located at a fixed position) to some surface. If the measurement is performed with relatively high resolution over the surface of an object, it is basically a profiling measurement and will generate a contour of the object. If the measurement is performed with relatively coarser resolution, it can be considered more as a thickness measurement. Such a measurement could be performed on a product that is continually being moved under the measuring head.

The following two examples illustrate the types of devices that have been developed to perform these two types of gauging functions.

(1) *Profile measurement system* In one system the laser light reflected from the surface being inspected is imaged onto a vibrating pinhole [6]. The experimental arrangement shown in Fig. 12-13 uses a lens to focus the laser beam onto a small spot on the surface of the object. The same lens collects light scattered from the surface. Another lens then images the light on a pinhole that is driven by a tuning fork oscillator. Light passing through the pinhole reaches a photodetector. The ac signal from the photodetector is amplified and then rectified by a lock-in detector synchronized with the signal from the tuning fork oscillator. Whether or not the laser beam is focused on the surface can be determined from the output signal of the lock-in detector. The output voltage of the lock-in detector, as a function of displacement of the surface of the sample in the x direction, is shown in Fig. 12-14. These results were obtained under conditions of oscillation of the pinhole at 525 Hz with a pinhole diameter of 250 μm. The sign of the output signal from the lock-in detector corresponds to the direction of displacement of the surface from the focus of the laser beam.

Contour maps can be obtained in such a way that two-dimensional

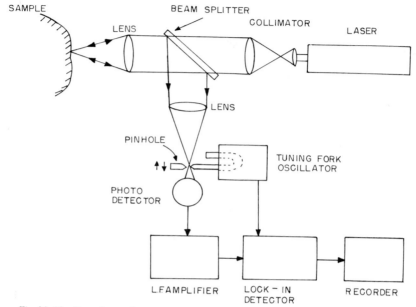

Fig. 12-13 Experimental arrangement for profile measurement system. (From S. Ando *et al., Appl. Opt.* **5**, 1961 (1966).)

coordinates corresponding to displacement of the sample surface in the x and y directions are obtained. Contour maps were obtained from plaster, clay, and brass surfaces with this system. Reproducibility in indicated distances was within 2 μm for surfaces at rest. The errors were estimated to

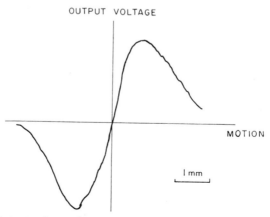

Fig. 12-14 Output voltage of system shown in Fig. 12-13 for the indicated surface motion. (From S. Ando *et al., Appl. Opt.* **5**, 1961 (1966).)

be less than 40 μm for the plaster surfaces and in the range of 50–80 μm for a rough brass surface. The resolution of the system corresponded to the diameter of the focused laser beam, which could be as small as 5 μm.

Such an optical profiling system could be employed for model scanning in an automated system for functions such as die cutting. It could also serve to measure surface roughness on machined parts.

(2) *Thickness measurements* can be performed by a system that generates two beams that strike the target surface at a known angle, as shown in Fig. 12-15. The two beams will produce two spots of light on the surface. The separation of the spots of light depends on the distance from the sensor head to the surface of the object. The spot spacing s is given by the equation

$$s = 2D/\tan\theta \qquad (12.16)$$

where D is the displacement of the surface from its nominal position and θ is the angle between the beam and the plane of the surface. As the target surface changes position, the spots move closer together or farther apart, depending on whether the surface moves farther away from the sensor head or closer.

The sensor head contains a lens assembly and an optical scanning system that scans the surface and generates two pulses when the positions of

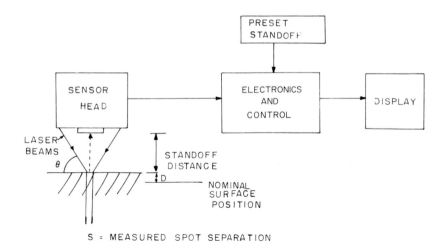

S = MEASURED SPOT SEPARATION

S = 2D / tan θ

Fig. 12-15 System for thickness measurements using two beams that strike the target surface at an angle θ. (From D. L. Cullen, reprinted from the *Proceedings of the Technical Program, Electro-Optical Systems Design Conference 1970*, page 783. Copyright © 1970 by Industrial & Scientific Conference Management, Inc.)

the spots are scanned over. The time separation of the two pulses is proportional to the separation between the two spots of lights. Thus, the time between pulses depends on the position of the target surface. Specifications of this system are quoted as follows: accuracy ±0.0003 in., repeatability 0.0001 in., dynamic range 0.2 in., and response speed 1 msec.

A second, similar system has been developed that involves a laser beam striking the target surface, as shown in Fig. 12-16. At a specified angle, backscattered light is focused on a split photocell. The system seeks a null output. In the null position, both parts of the photocell are illuminated equally. Unbalance of the photocell can be detected and fed to a servo-amplifier that repositions the surface. Alternatively, the amount of movement of the surface can be displayed on a meter or recorded. Specifications of this system are as follows: accuracy 0.0005 in., repeatability 0.0003 in., dynamic range 0.05 in., and response speed 1 in./sec.

To make thickness measurements, the material to be measured is compared to a preset dimension that is acquired from a standard. In flow processes such as making sheet food, strip metals, plastic extrusions, or rubber, the material passes under the noncontact laser probe. The position of the surface is automatically compared to the established standard position, and a digital readout of the dimension variation from the standard is

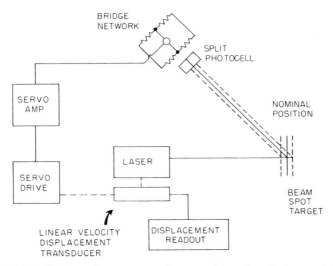

Fig. 12-16 Split photocell arrangement for measuring surface displacement through the motion of the backscattered light pattern on the photocell. (From D. L. Cullen, reprinted from the *Proceedings of the Technical Program, Electro-Optical Systems Design Conference 1970*, page 783. Copyright © 1970 by Industrial & Scientific Conference Management, Inc.)

obtained. This mode requires a stable zero reference such as a roller or surface.

An example is measurement of thickness of food materials on a continuous basis. Conventional gauging techniques have not been accurate or fast enough to be used in food-processing industries. A laser surface position monitor can measure thickness of food materials, e.g., dough sheets. The information can be used to adjust roll openings to control the thickness.

By simultaneous use of two surface position devices, one can measure the thickness of a sheet that has no backing surface. A specific example of such a system has been developed for measurement of thickness of plywood. The position of each side of the sheet of plywood in a continuous processing system is measured. A system of beam splitters allows measurement at ten different points over an 8-ft span normal to the direction of travel. The position of each side is measured independently, and the results are fed into logic circuits that compute thickness information at each of the ten measurement points. The output is continuous information about thickness at ten points across the sheet as it moves.

As another example, the profile of turbine blades has been measured with a two-beam profiler, as described in the preceding. The system uses two profiling heads to measure both sides of a blade in one scan. For a scan at a particular height along the axis of the blade, the blade must be rotated to a particular angular orientation. The scan across the blade is then performed, and the thickness of the blade and the contour of each surface is obtained. This information can be displayed in a variety of forms, such as on a chart recorder. Scans are repeated at a number of positions along the blade axis, with the blade being rotated between scans. The result completely defines the turbine blade profile, which can then be compared to a standard.

This method allows inspection of turbine blades of any size, with speed capable of yielding complete blade inspection in less than 1 min and with an accuracy of 0.0001 in. in profile measurement. It is compatible with complete automation, so that 100% inspection is possible.

The profile of turbine blades has been measured in the past by a gauging system that has two knife blades that contact the edge of the turbine blade. The knife blades ride across the contour of the turbine blade and actuate position sensors. These in turn plot the turbine blade profile at one particular position along the axis of the turbine blade. This measurement is carried out at several positions along the height of the turbine blade. The resulting profile can be compared to the standard profile.

Although this measurement technique has been in general use, it suffers from several drawbacks including speed of inspection, the ability to deal with a variety of blades of different sizes, and limited accuracy. The use of a laser profiling system can reduce these problems.

E.   Measurements of Product Dimension

Lasers can be used in a variety of ways for measuring the dimensions of manufactured parts or products. The discussion in the previous sections has described some of the possible procedures. In this section we shall summarize some of the available procedures for measuring product dimensions and compare the approaches. Use of laser light for measuring product dimension has taken a wide variety of forms. Because of the enormous diversity of manufactured products and the variety of requirements for gauging such products, it is not possible to describe a generalized system. Lasers have been suggested for measurement of a wide range of products including sheet thickness for plastics, extruded rubber and metals, dimensions of piston rings, cylinder heads and spark plugs, fuel injector components and choke springs in the automotive industry, the diameter of glass tubing, fuel rods, and munitions, the shape of items such as roller bearings and valve lifters, and the amount of materials used in food products, such as cookies and butter. This is not meant to be a complete list, but merely an indication of the diversity of the capabilities of laser measuring systems. The specific method in which a laser system can be applied to a particular product necessarily depends on the manufacturing process and on the requirements of the particular product.

Basically, there have been three fundamental techniques by which dimensional measurements have been made:

(1)   measurement of the obscuration of a laser beam that passes across the item,

(2)   dimensional comparison measurements in which the position of a surface of the product is determined relative to the sensor head, and

(3)   diffraction-type measurements, which have already been described in relation to measurement of fiber and wire diameter.

The first type of measurement involves interposing the object to be measured in the path of a beam which is scanned across the object. The beam is detected by a photodetector. An object in the path of the beam interrupts the detector output for a time. In one arrangement, the interruption produces a timing signal. A digital readout that is generated from the timing signals as the beam crosses the edges of the object provides a direct display of the dimensions. Such devices can measure dimensions of fast-moving parts or continuously produced products without precise positioning. The measurement result can be used for closed-loop control of the process. Small modifications can be effected to measure profiles (moving the scan down the length of the object) or roundness (rotating the object in the beam).

This method is perhaps the simplest to implement and to interpret. It requires calibration because of the nonuniform spatial profile of the laser beam. The edge of the beam is not infinitely sharp, so that the detector signal decreases as a function of time while the beam sweeps across the edge.

Dimensional comparison measurements make use of principles such as discussed in Section 12.D, in which the position of an edge of the object in relation to the sensor is measured. With two sensors, each measuring the position of a different side of a flat sheet of extruded material, it is possible to determine the thickness of the material. Alternatively, if parts are moving on a belt or roller, the position of the belt can serve as a zero reference. Then, as the parts move under the beam, the thickness is determined by comparison with the reference surface. Such measurements are useful when the dimension that is desired can be determined from measurement of surface position.

The diffraction type of measurement offers great versatility and the greatest accuracy. It is not universally applicable to all types of products. A system must be devised wherein a small dimension such as a slit can be measured. Large parts are less suitable than small parts, and the total range over which measurements can be made is limited.

Some stated characteristics of commercial models of each of the three approaches described in the preceding are listed in Table 12-1.

TABLE 12-1
Approximate Specifications of Laser Gauging Systems

Measurement type	Minimum resolution	Accuracy	Maximum range	Typical use
Scanned beam	5 μm	0.12% of measurement	12 cm	Diameter of rods, profiles of extruded parts
Dimensional comparison	12 μm	1% of measurement	25 cm	Thickness of sheet products
Diffraction	0.025 μm	0.12 μm	0.25 cm	Wire diameter, width of small gaps

Although the specific implementation of dimensional measurement for process control or inspection for a particular part or product varies according to the shape, size, and requirements of an individual operation, the general principles described in this section can be used for a broad class of automated inspection processes in industrial applications. Laser gauging as described here offers the advantage of being a remote noncontact gauging process, with capability for high speed and good accuracy.

F.   Measurement of Surface Finish

The profiling system described in Section 12.D can be used to inspect surface flatness and roughness, at least to within the tolerances set by their specifications. In this section, we shall describe an alternative method of surface inspection that aims more at detecting imperfections on a surface. The method relies on the fact that surface imperfections scatter light.

Various surface defects such as scratches, pits, small raised areas, etc., reflect light differently than perfect surface areas. Scattered light from the defects can be detected simply by a photodetector. The technique is complicated by background reflection and scattering from the other areas on the surface. In practice, it is difficult to distinguish between light scattered from defects and light scattered in all directions by the general roughness of the surface.

One technique for resolving this problem was illumination of the surface at a grazing angle coupled with spatial filtering to eliminate the specular component of reflected light. A typical arrangement is shown in Fig. 12-17. The spatial filter consisted of a central stop which passed only scattered light and eliminated the central diffraction peak due to specularly reflected light.

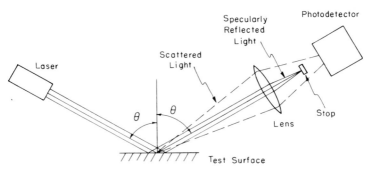

Fig. 12-17   Arrangement for measurement of surface finish. (From T. Sawatari, *Appl. Opt.* **11**, 1337 (1972).)

Some experimental results are shown in Fig. 12-18 for a steel plate with a surface finish of 0.2 μm. Scratches about 10 μm wide and 5 mm long were made on the plate with a diamond scriber. The scratches were perpendicular to the plane of incidence, but the change in scattered light was not a strong function of orientation. The results are shown for an angle of incidence of 88°, which maximizes the signal under these conditions. The system clearly distinguishes between a lightly scratched and a heavily scratched surface.

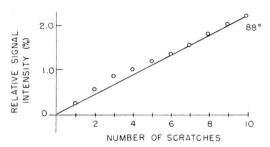

Fig. 12-18 Experimental results from the system shown in Fig. 12-17 for an incidence angle of 88°, which maximizes the signal. (From T. Sawatari, *Appl. Opt.* **11**, 1337 (1972).)

An alternative arrangement can provide scanning of a surface to detect the location of individual defects. An oscillating mirror was used to sweep the helium–neon laser beam in a line across the surface, which was moving underneath the beam. As an example, the system was used for detection of nodules of gold on a ceramic surface. The evaporated gold nodules scattered more light at large angles than did background areas of the surface. A photodetector positioned to view light scattered at large angles could effectively map out the number and location of the nodules.

These examples show how lasers can be employed to inspect surfaces for defects. In applications where high surface perfection is important, laser systems offer a significant advantage in cost and speed as compared to visual inspection under a microscope.

G. Surface Flaw Inspection Monitor—A Specific Example

As a specific example, we shall discuss an optical surface flaw inspection that has been developed for 100% inspection of small caliber ammunition [7]. This example is chosen because it demonstrates the versatility of laser inspection in a situation for which it has been difficult to provide automatic inspection by conventional techniques. It can also serve as a prototype for the considerations that enter into choice of laser systems. The technique can be applied to inspection of surface quality for other mass-produced parts when the surface must be inspected rapidly. The technique has application for inspection of continuous webs or of any small precise part, e.g., automotive parts.

The surface flaw monitor uses scattered light and electrooptical instrumentation to detect the presence of flaws on 100% of the cartridge case surface, at throughput rates exceeding 1200 cases/min. The inspection of surfaces of cartridge cases has been difficult to implement with automatic flaw inspection because the surface has several discontinuities in profile, a rough surface finish, and acceptable variations in surface coloration. A number of different types of flaws must be inspected for including dents, scratches, scaly metal, split cases, folds, wrinkles, buckles, and variations in mouth and primer pocket shape.

The complex case surface and the high speed requirements make optical inspection desirable. There are many types of surface flaws, but all can be determined from deviations from a normal configuration of reflected light. A helium–neon laser beam is spread into a line with cylindrical optics and automatically tracks a spinning cartridge on the perimeter of a wheel-type mechanical handling system. A number of fiber-optic elements located at different case positions collect scattered light from surface zones of the case and deliver the light to photomultiplier detectors. The detected signals are frequency processed with the frequency pattern of the detected scattered light relating directly to the type of surface flaw, such as dents or scratches. Signals are then fed to a minicomputer for acceptance or rejection by mechanical handling hardware.

At each test station, a spindle holds and spins the cartridge case. As the case spins, 100% of its surface is illuminated. Scattered light is picked up by the fibre-optic bundle that is located off-axis from specularly reflected light. The frequency spectrum of the reflected light is compared to the signature of an acceptable case. A large signal-to-background ratio for flaws allows the use of simple threshold-level signal processing to produce a digital accept or reject signal. The reject signal initiates a reject sequence that removes the cartridge from the line. The system can achieve a throughput rate up to 1200 cartridge cases/min.

H. Summary

Lasers have many potential applications in measurement and instrumentation. We have described some of them in the preceding. This discussion is not a complete listing, but rather a collection of examples that may stimulate the imagination of the reader to apply lasers to his own particular measurement problem. In addition to the type of measurements described, we shall later discuss measurement of vibration and strain in connection with holography.

The phenomena used for measurement have been well known, typically involving interferometry, diffraction, etc. The basic methods are not new. What is new is the light source. The laser has intensity and coherence properties that make many types of optical measurement much easier. In particular, the field of interferometry has been revitalized by the laser. In many cases, laser-based methods can make certain types of measurement more easily and economically than any other method.

At present lasers can often be used to advantage for engineering measurements that were previously difficult. They have so far been used to make specialized measurements of phenomena that were difficult to measure by other means. A specific example of this is the use of a velocimeter to measure spallation properties of solids. There have been relatively few measurements that have made the transition to continuous use by unskilled or semiskilled personnel on a day-to-day basis. A few applications involving measurement of product dimension are in use, but many applications have remained specialized engineering measurements made under the control of skilled personnel. We can reasonably expect routine day-to-day uses of lasers to increase for many industrial measuring applications as commercial instrumentation for specific measurement tasks continues to develop.

REFERENCES

[1] J. W. Foreman et al., IEEE J. Quantum Electron. **QE-2**, 260 (1966).
[2] A. E. Lennert, D. B. Brayton, and F. L. Crosswy, AEDC Rep. TR-70-101 (1970).
[3] F. J. Lavoie, Mach. Des., p. 212 (March 20, 1969).
[4] G. Stavis, Instrum. and Contr. Syst., p. 99 (February 1966).
[5] F. P. Gagliano, R. M. Lumley, and L. S. Watkins, Proc. IEEE **57**, 114 (1969).
[6] S. Ando et al., Appl. Opt. **5**, 1961 (1966).
[7] F. Reich and W. J. Coleman, Opt. Eng. **15**, 48 (1976).

SELECTED ADDITIONAL REFERENCES

A.   Velocity Measurement

J. B. Abbiss, T. W. Chubb, and E. R. Pike, Laser Doppler Anemometry, Opt. Laser Technol. **6**, 249 (1974).

J. C. Angus et al., Motion Measurement by Laser Doppler Techniques, Ind. Eng. Chem. **61**, No. 2, p. 9 (1969).

F. Durst, A. Melling, and J. H. Whitelaw, "Principles and Practice of Laser–Doppler Anemometry," Academic Press, New York, 1976.

W. M. Farmer and J. O. Hornkohl, A Two-Component Self-Aligning Laser Vector Velocimeter, Appl. Opt. **12**, 2636 (1973).

C. E. Fuller and C. E. Craven, Three Dimensional Gas Velocity Measurements with a Laser Doppler Velocimeter System, Proc. Tech. Program, Electro-Opt. Syst. Design Conf. East, New York, (September 14–16, 1971).

G. R. Grant and K. L. Orloff, Two Color Dual-Beam Backscatter Laser Doppler Velocimeter, Appl. Opt. **12**, 2913 (1973).

A. Hauer and J. W. Alwang, Design of Laser Velocimeter for Use in Turbo-Machinery, Proc. Tech. Program. Electro-Opt. Syst. Design Conf. East, New York (September 14–16, 1971).

A. J. Hughes and E. R. Pike, Remote Measurement of Wind Speed by Laser–Doppler Systems, Appl. Opt. **12**, 597 (1973).

S. J. Ippolito, S. Rosenberg, and M. C. Teich, Velocity Measurement of Slowly Moving Surfaces Using a He–Ne Laser Heterodyne System, *Rev. Sci. Instrum.* **41**, 331 (1970).

H. J. Pfeifer and H. D. vom Stein, Application of the Laser Velocimeter in Supersonic Wind Tunnel, *Opto–Electronics* **5**, 53 (1973).

M. J. Rudd, The Laser Anemometer—A Review, *Opt. Laser Technol.* **3**, 200 (1971).

C. P. Wang and D. Snyder, Laser–Doppler Velocimetry: Experimental Study, *Appl. Opt.* **13**, 98 (1974).

B. Angular Rotation Rate

F. Aronowitz, The Laser Gyro, *in* "Laser Applications" 1. (M. Ross, ed.), Academic Press, New York, 1971.

F. Aronowitz, J. E. Killpatrick, and S. P. Callaghan, Power-Dependent Correction to the Scale Factor in the Laser Gyro, *IEEE J. Quantum Electron.* **QE-10**, 201 (1974).

J. E. Killpatrick, The Laser Gyro, *IEEE Spectrum*, p. 44 (October 1967).

C. Wire Diameter

M. Koedam, Determination of Small Dimensions by Diffraction of a Laser Beam, *Philips Tech. Rev.* **27**, 208 (1966).

D. Profile and Surface Position Measurement

S. Ando *et al.*, A New Method of Contour Measurement by Gas Lasers, *IEEE J. Quantum. Electron.* **QE-3**, 576 (1967).

D. L. Cullen, Miscellaneous Laser Measurements, including Strain and Product Dimension, Soc. of Manufacturing Eng. Tech. Paper MR74-970 (1974).

D. L. Cullen, Optical Thickness Measurements. *Electro-Opt. Systems Design*, p. 52 (July 1975).

J. R. Kerr, A Laser-Thickness Monitor, *IEEE J. Quantum Electron.* **QE-5**, 338 (1969).

E. Measurements of Product Dimension

T. R. Pryor and J. C. Cruz, Laser-Based Gauging and Inspection, *Electro-Opt. Systems Design*, p. 26 (May 1975).

F. Measurement of Surface Finish

S. S. Charschan (ed.), "Laser in Industry," pp. 335–340. Van Nostrand–Reinhold, Princeton, New Jersey, 1972.

G. Surface Flaw Inspection Monitor—A Specific Example

S. M. Ward and P. H. Brew, Design and Analysis of a Laser-Oriented Automatic Dimensional Inspection System for High-Speed Process Control, *J. Vacuum Sci. Technol.* **10**, 1044 (1973).

CHAPTER 13

INTERACTION OF HIGH-POWER
LASER RADIATION WITH MATERIALS

The ability of lasers to produce intense pulses of radiation means that there are many potential applications involving heating, melting, and vaporization. The feature of the laser that allows it to be used in metalworking is, of course, its ability to deliver very high power per unit area to localized regions on a workpiece. A conventional thermal source such as a welding torch delivers much lower power per unit area, and it cannot be localized so well. Only an electron beam can compare with a laser in this respect. The total power is not necessarily so important as the ability to focus to a small spot, producing a high power per unit area. A 200-W light bulb does not melt metal; a 200-W continuous laser can do so.

In this chapter, we shall discuss the physical processes that occur in the interaction of high-power laser radiation with materials. Knowledge of these processes is important in understanding capabilities and limitations of laser processing. These interactions are the basis for laser applications in materials processing. We shall emphasize metallic targets, but much of what we say is applicable to other absorbing materials.

When laser radiation falls on a target surface, part of it is absorbed and part is reflected. The energy that is absorbed begins to heat the surface. There are several regimes that should be considered, depending on the time scale and the laser power per unit area. For example, losses due to thermal conduction are small if the pulse duration is very short, but they can be important for longer pulses. There can be important effects due to absorption of the radiation in the plasma formed by vaporized material above the target surface. We note that losses due to reradiation from the target surface are usually insignificant.

The heating effects due to absorption of high-power beams can occur very rapidly. The surface quickly rises to its melting temperature. Laser-induced melting is of interest because of welding applications. One often desires maximum melting under conditions where surface vaporization does not occur. Melting without vaporization is produced only within a narrow range of laser parameters. If the laser power per unit area is too high, the surface begins to vaporize before a significant depth of molten material is produced. This means that there is a maximum power per unit area suitable for melting applications. Alternatively, for a given total energy in the laser pulse, it is often desirable to stretch the pulse length.

Melting of a material by laser radiation depends on heat flow in the material. Heat flow is dependent on the thermal conductivity K. But thermal conductivity is not the only factor that influences the heat flow, since the rate of change of temperature also depends on the specific heat c of the material. In fact, the heating rate is inversely proportional to the specific heat per unit volume, which is equal to ρc, where ρ is the material density. The important factor for heat flow is $K/\rho c$.

This factor has the dimensions of cm^2/sec, characteristic of a diffusion coefficient, and has therefore been given the descriptive term "thermal diffusivity" (to recognize that it represents the diffusion coefficient of temperature or, more properly, heat).

For all unsteady-state heat-flow problems, the term $K/\rho c$ is involved. the significance of this term is that it determines how rapidly a material will accept and conduct thermal energy. Thus, for welding, high thermal diffusivity normally allows larger penetration of the fusion front with no thermal shock or cracking. Table 13-1 lists the thermal diffusivity of several metals and alloys. The thermal diffusivity of alloys is generally lower than that of the pure metal that is the major component of the alloy. Stainless steel and some nickel alloys have especially low values. A low value of thermal diffusivity limits the penetration of heat and may reduce the laser weldability.

The depth of penetration of heat in a time t is given approximately by the equation

$$D = (4kt)^{1/2} \tag{13.1}$$

where D is the depth of penetration of the heat and k is the thermal diffusivity. Let us consider an example. For a metal with thermal diffusivity 0.25 cm^2/sec, heat can penetrate only about 3×10^{-4} cm during a pulse of 90-nsec duration (typical of a Q-switched laser). During a pulse of 100-μsec duration (typical of a normal-pulse laser), heat can penetrate about 0.01 cm into the same metal.

These ideas lead to the concept of a thermal time constant for a metal plate of thickness x. The thermal time constant is equal to $x^2/4k$. The

TABLE 13-1
Thermal Diffusivity and Thermal Time Constants

Metal	Thermal diffusivity (cm²/sec)	Thermal time constants (msec)			
		0.01-cm thick	0.02-cm thick	0.05-cm thick	0.1-cm thick
Silver	1.70	0.015	0.059	0.368	1.47
Aluminum alloys					
Commercially pure	0.850	0.029	0.118	0.74	2.94
2024 alloy	0.706	0.035	0.142	0.89	3.54
A13 casting alloy	0.474	0.053	0.2111	1.32	5.27
Copper alloys					
Electrolytic (99.95 %)	1.14	0.022	0.088	0.55	2.19
Cartridge brass	0.378	0.066	0.265	1.65	6.61
Phosphor bronze	0.213	0.117	0.470	2.93	11.74
Iron alloys					
Commercially pure	0.202	0.124	0.495	3.09	12.38
303 stainless steel	0.056	0.446	1.786	11.16	44.64
Carbon steel (1.22 C, 0.35 Mn)	0.119	0.210	0.840	5.25	21.01
Nickel alloys					
Commercially pure	0.220	0.114	0.454	2.84	11.36
Monel	0.055	0.455	1.818	11.36	45.46
Inconel	0.039	0.641	2.564	16.03	64.10

thermal time constant represents the pulse duration required to give a specified penetration depth.†

For effective melting, the laser pulse could be approximately equal to the thermal time constant for the given metallic specimen. Table 13-1 shows some thermal time constants. For thin material (0.01–0.02 cm), the thermal time constants are shorter than (or at least comparable to) pulse durations from normal-pulse lasers (msec durations). The thermal time constants are all much longer than pulse durations characteristic of Q-switched lasers ($\sim 10^{-7}$ sec). This means that effective melting with Q-switched lasers is not practical. Lasers suitable for welding are either continuous (e.g., CO_2 or Nd:YAG) or have pulse durations of the order of milliseconds (e.g., ruby laser pulses).

For somewhat thicker metals (~ 0.1 cm), the thermal time constant becomes too long for heat to penetrate through metals with low thermal

† Strictly, this thermal time constant is the time required for the temperature of the back surface of the plate to rise to 37% of the temperature of the front surface when heat is absorbed in a very short pulse at the front surface. This time constant gives a convenient order-of-magnitude estimate of the time required for heat flow through the plate.

diffusivity, but metals with high values have thermal time constants around a few milliseconds. Heat can penetrate through these metals (e.g., silver, pure aluminum, and copper) during the period of a normal laser pulse.

For thicker metals (>0.1-cm thick), the thermal time constants become long, of the order of milliseconds, even for metals with the highest values of thermal diffusivity. The thermal time constants are much longer than the laser pulse durations that are available. This means that it becomes difficult to weld metals with thickness of 0.1 cm or greater. One could, of course, allow a continuous laser to dwell on a spot for a long time, but this technique is usually not employed because the heat spreads over too large an area and because the welding rate becomes unacceptably slow.

Low values of thermal diffusivity mean that heat does not penetrate well into the material. A high value of thermal diffusivity can also cause problems by allowing rapid removal of heat from the surface. This can reduce the amount of melting that is produced. To compensate for these effects, one should vary the laser parameters for optimum welding of different metals. To weld copper, for example, one should use a relatively higher laser power and shorter pulse lengths to overcome losses due to the high thermal diffusivity. To weld stainless steel, one should use relatively lower laser power and longer pulse lengths to achieve good penetration.

Effective melting (and hence welding) with lasers depends on propagation of a fusion front through the sample during the time of interaction, at the same time avoiding vaporization of the surface. Figure 13-1 shows

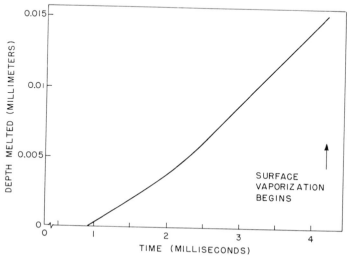

Fig. 13-1 Calculated depth melted in nickel as a function of time for an absorbed laser power density of 10^5 W/cm^2.

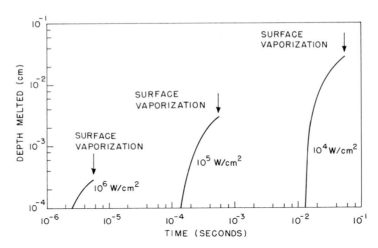

Fig. 13-2 Calculated depth melted in stainless steel as a function of time for several different absorbed laser power densities. The point at which surface vaporization begins is indicated for each curve.

the time dependence of the penetration of the molten front into a massive nickel sample for an absorbed power density of 10^5 W/cm^2. After about 4 msec, the surface begins to vaporize. The depth of penetration without surface vaporization is limited.† To obtain greater depth, one can tailor the laser parameters to some extent. Generally, one lowers the power density and increases the pulse duration. The control is very sensitive. One must make careful adjustments to achieve a balance between optimum melting depth and avoidance of vaporization.

One is mainly interested in welding in a regime where surface vaporization does not occur. Melting without vaporization is produced only within a narrow range of laser parameters. If the laser power per unit area is too high, the surface begins to vaporize before a significant depth of molten material is produced. This means that there is a maximum power per unit area suitable for melting applications. Alternatively, for a given total energy in the laser pulse, it is often desirable to stretch the pulse length to allow time for penetration of the fusion front through the workpiece. Figure 13-2 shows depth of melting in stainless steel as a function of time.† Good fusion can be achieved over a range of pulse lengths if the laser output energy is carefully controlled. For pulse lengths shorter than 1 msec, surface vaporization cannot easily be avoided.

† The results in Figs. 13-1 and 13-2 are calculated using an analog computer routine developed by M. I. Cohen [1].

In fairly recent work, deep-penetration welding has been achieved with continuous multikilowatt CO_2 lasers. Samples of stainless steel up to 2-cm thick have been welded. The welding probably occurs because of "keyholing," in which a hole is opened up so that the laser energy can penetrate deep into the workpiece and be deposited there. Flow of molten material recloses the hole after passage of the beam to a new area. This is similar to electron-beam welding with deep penetration.

In this regime of continuous laser power, one is no longer limited by considerations of heat flow such as were outlined in the preceding, and impressive seam-welding rates in thicker metals have been demonstrated.

As a specific example, $\frac{1}{4}$-in.-thick stainless steel has been seam-welded with 3.6 kW of laser power at a speed of 50 in./min. The weld was of good quality, with no damage outside the fusion zone. Deep-penetration laser welding has been demonstrated in the laboratory and may be nearing production status, particularly in the automotive industry. One specific application in production will be described in Chapter 15.

The reflectivity of a metal surface is another important parameter. It defines how much of the light that falls on a surface is actually absorbed and can be used for melting. The reflectivity is defined as the ratio of the radiant power reflected from the surface to the radiant power incident on the surface. Thus the reflectivity is a dimensionless number between zero and unity.

The reflectivity of several metals as a function of wavelength is shown in Fig. 13-3. These curves represent typical smooth surfaces of the metals. The exact value of reflectivity is a function of variable conditions including surface finish and state of oxidation of the surface. Thus, the values in

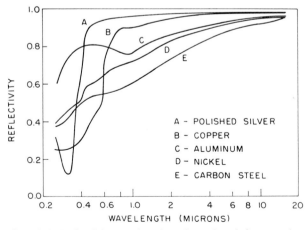

Fig. 13-3   Reflectivity as a function of wavelength for several metals.

Fig. 13-3 cannot be interpreted as being exact values for a specific metal. The figure does show several important general features. A metal such as gold has reflectivity that is low in the blue portion of the visible spectrum and that increases at the red end of the spectrum. This accounts for the color of gold metal. Metals such as aluminum and silver have high reflectivity that is fairly uniform through the visible spectrum. This accounts for the whitish appearance of these metals. Ferrous metals (steels and nickel alloys) have typically lower reflectivity throughout the entire spectrum. Thus, they usually appear more dull than metals such as silver. This is represented by the curve for carbon steel.

The reflectivity of all metals becomes high as one goes to long infrared wavelengths. For wavelengths greater than 5 μm, almost all metals have reflectivity greater than 90%. At wavelengths greater than 5 μm, the reflectivity is dependent on the electrical conductivity. Metals with high electrical conductivity have the highest values of infrared reflectivity. Thus the reflectivity of gold is higher than that of aluminum, which in turn is higher than that of steel. This ordering is the same as for the electrical conductivities of these metals.

The amount of light absorbed by a metallic surface is proportional to $1 - R$, where R is the reflectivity. At the CO_2 laser wavelength of 10.6 μm, where R is close to unity, $1 - R$ becomes small. This means that only a small fraction of the light incident on the surface is absorbed and is available for melting of metal.

The difference in values of $1 - R$ becomes important in the infrared. For silver or copper at 10.6 μm, $1 - R$ is about 0.02, whereas for steels $1 - R$ is about 0.05. Thus steel absorbs about 2.5 times as much of the incident light as silver or copper. In practice, this means that steels are easier to weld with a CO_2 laser than are metals such as copper or aluminum.

The wavelength variation is also important. At shorter wavelengths, the factor $1 - R$ is much higher than at the CO_2 laser wavelength. For example, the value $1 - R$ for steel is near 0.35 at 1.06 μm, about 7 times as high as at 10.6 μm. This means that, at least initially, 7 times as much light is absorbed from a Nd:YAG laser as from a CO_2 laser if equal intensities of light from the two lasers strike a steel surface. This makes it easier in some cases to carry out welding operations with a shorter wavelength laser.

There is evidence that the reflectivity can decrease during a laser pulse. For example, in one experiment [2], an aluminum surface exposed to a Nd:glass laser pulse with a power per unit area of 10^7 W/cm^2 had an original reflectivity of approximately 70%. When the laser pulse began, the reflectivity dropped steadily to 20% after a time of 200 μsec and then remained fairly constant for the remainder of the 1-msec-duration laser pulse. Therefore, the laser energy was considered to be coupled effectively into the material, since the average reflectivity during the pulse was low.

Reflectivity of the surface is especially important for CO_2 lasers. One difficulty in welding metals with a continuous CO_2 laser has been the high reflectivity at 10.6 μm. It is difficult to couple the energy from the optical beam into the workpiece. Coating of the metallic surfaces with paints or other materials is not always effective in starting the heating because of poor thermal coupling between the coating and the underlying metal. The coating is effectively evaporated from the surface. The high-reflectivity problem has in fact been a barrier to the application of CO_2 lasers to welding of metals such as gold. Ferrous metals, e.g., steels, have somewhat lower reflectivity at 10.6 μm and are better candidates for welding with CO_2 lasers. The reduction of surface reflectivity that can occur during the laser pulse is especially helpful for welding with CO_2 lasers. Because of this reduction in reflectivity which occurs when a high-power CO_2 laser beam is focused on a surface, CO_2 lasers do indeed have many practical uses for welding.

Figure 13-4 shows some data on the reflectivity of a stainless steel surface struck by a 100-nsec-duration pulse from a CO_2 TEA laser, which delivered a power density of 2×10^8 W/cm^2 to the target. The reflectivity dropped rapidly over a period of a few hundred nanoseconds. These data show how the reflectivity can decrease during the time when the surface is being illuminated so as to increase the effective absorption of the surface.

We have so far considered surface vaporization as undesirable. However, in many cases one does desire vaporization, as in cutting and hole

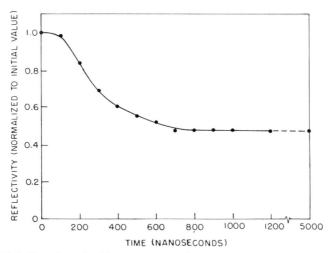

Fig. 13-4   Specular reflectivity at 10 μm as a function of time for a stainless steel surface struck by a CO_2 TEA laser delivering 1.5×10^8 W/cm^2 in a pulse 200-nsec wide.

drilling. When the laser delivers high power density to a surface, the metallic surface quickly reaches its boiling temperature and a hole is vaporized in the surface. The lasers most useful for hole drilling have pulse lengths in the 100-μsec to 1-msec regime.

When a normal-pulse laser beam (duration about 1 msec) interacts with a surface, the process of material removal involves conventional melting and vaporization. The time scale is long enough so that the vaporized material flows away from the area of interaction. The material is removed without further interaction with the laser beam. Vaporization occurs at a continually retreating surface. Under typical conditions, the boiling temperature is reached in a very short time. The time t_B to reach the boiling temperature is given by the following equation:

$$t_B = (\pi/4)(K\rho c/F^2)(T_B - T_0)^2 \tag{13.2}$$

where K, ρ, and c, respectively, are the thermal conductivity, density and specific heat of the material, T_B is its boiling temperature, T_0 the ambient temperature, and F the absorbed laser power density.

Table 13-2 shows values calculated using the above equation for absorbed laser power densities of 10^5–10^7 W/cm^2. These are typical values for the power density obtainable at target surfaces using commonly available ruby or Nd:glass lasers and simple focusing optics. Because of surface reflection, the absorbed power density may be lower than the incident power density. The values in Table 13-2 can be very short. For high values of absorbed laser power density, the surface begins vaporizing very rapidly.

TABLE 13-2

Time to Reach Vaporization Temperature

Metal	Absorbed laser power density (W/cm^2)		
	10^5	10^6	10^7
Lead	118 μsec	1.18 μsec	12 nsec
Zinc	128 μsec	1.28 μsec	13 nsec
Magnesium	245 μsec	2.45 μsec	24.5 nsec
Titanium	319 μsec	3.19 μsec	31.9 nsec
Chromium	1.54 msec	15.4 μsec	154 nsec
Nickel	1.84 msec	18.4 μsec	184 nsec
Iron	1.86 msec	18.6 μsec	186 nsec
Aluminum	3.67 msec	36.7 μsec	367 nsec
Molybdenum	5.56 msec	55.6 μsec	556 nsec
Copper	8.26 msec	82.6 μsec	826 nsec
Tungsten	10.46 msec	104.6 μsec	1046 nsec

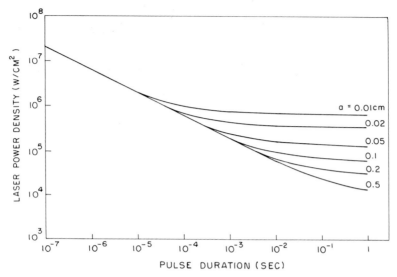

Fig. 13-5 Laser power density required to raise a massive aluminum surface to its vaporization temperature as a function of the pulse duration and the Gaussian radius *a* of the laser beam.

The values in Table 13-2 are calculated on the assumption that uniform laser power is absorbed at an infinite plane surface. Figure 13-5 shows how the finite spot size influences this result. The figure shows the absorbed laser power density required for surface vaporization to begin as a function of laser pulse duration and the radius of the laser beam on an aluminum target. For short pulse durations, there is no time for transverse thermal conduction, and the result is independent of spot size. It also corresponds with the results of Eq. (13.2). For longer pulse durations, transverse thermal conduction becomes important, and the spot size is important. For small spot size, the transverse thermal gradients are higher, and heat is conducted out of the focal area more rapidly. Thus, a small focal spot requires higher power densities to produce vaporization.

Before boiling begins, the surfaces must first start to melt. Because of the great rapidity with which boiling begins, there is not time for much material to melt. Thus, at high laser power densities ($\gtrsim 10^6$ W/cm^2), the dominant physical process is vaporization, and the role of melting tends to be less significant. After the surface reaches its vaporization temperature, the laser continues to deliver more energy to the surface. This energy supplies the latent heat of vaporization. Material is then removed from the target as a vapor. The result is to produce a hole in the target.

The time to reach the boiling temperature is often a short fraction of the pulse length. An equilibrium is set up, and the vaporizing surface retreats at a steady rate V_B equal to

$$V_B = F/\{\rho[L + c(T_B - T_0)]\} \tag{13.3}$$

where L is the specific latent heat of vaporization. For reasonable conditions, this velocity may persist throughout most of the duration of a normal laser pulse. For a typical metal and 1-msec pulse, one derives a depth vaporized around 1 mm. This is the scale that is usually attained with normal ruby pulse lasers.

Additional results for the depth vaporized under various conditions will be presented in Chapter 16. We will mention here that one important factor is the latent heat of vaporization. Metals with low latent heat of vaporization are vaporized to a larger extent then metals with a high latent heat of vaporization, such as tungsten. The amount of material vaporized depends on the exact conditions of the laser irradiation. Thus, quoted results from different workers may disagree. It is important to emphasize that the depths of holes produced by pulsed lasers are limited. Typically, an ordinary laser can drill holes only one or a few millimeters deep in metals.

In hole drilling with lasers, not all the ejected material is vaporized. When a hole begins to be produced in the metallic target, the vapor builds up a pressure that causes a flow toward the exit aperture of the crater. The flow can carry along some of the molten material formed along the face of the crater. This flushing process removes some of the mass as unvaporized droplets of the material. It results in a mass removal larger than if only vaporization occurred. Molten material on the surface of the hole may be ejected in globules. Typically, in many metals there is a shower of glowing sparks of hot particles ejected along with the vapor.

A scanning electron microscope photograph of an area impacted by several successive CO_2 laser pulses of 100-nsec duration is shown in Fig. 13- 6. It is apparent that even for this short pulse duration, molten material flowed within the crater and was splattered around the crater edges. Under some conditions, most of the mass removed is carried away as liquid. Figure 13-7 shows data on the fraction of the material ejected as liquid by a Nd:glass laser pulse of 30-kW power. Early in the pulse, most of the metal was removed as vapor, but after a few hundred microseconds about 90% of the material removal occurred as liquid droplets.

One might think that to increase material removal one should use lasers that deliver very high peak power. Paradoxically, it is not the highest peak powers that are best for material removal. The high powers from a Q-switched laser vaporize a small amount of material and heat it to a high temperature. Early in the laser pulse, some material is vaporized from the

Fig. 13-6 Scanning electron microscope photograph of crater produced by the impact of several successive TEA laser pulses delivering 3×10^8 W/cm^2 to a stainless steel surface. The width of the area shown in the photograph is 1.8 mm.

surface. This material is slightly thermally ionized and produces an opaque, high-temperature plasma. The laser light is effectively cut off from the surface. The material removed from the surface continues to interact with the incident laser beam. This means that new physical processes have become important as the power density becomes higher. This is shown schematically in Fig. 13-8, which shows depth vaporized as a function of time. The laser pulse shape is also shown for comparison. This is a typical shape for Q-switched lasers. Early in the pulse, the surface starts to vaporize. Then a plasma is formed, and the incoming laser light is absorbed. The flat portion of the curve is the region where light is cut off. Late in the pulse,

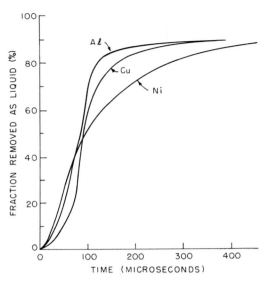

Fig. 13-7   Fraction of material removed in liquid form. Results are shown as a function of time for several metals struck by an Nd:glass laser pulse. (From M. K. Chun and K. Rose, *J. Appl. Phys.* **41**, 614 (1970).)

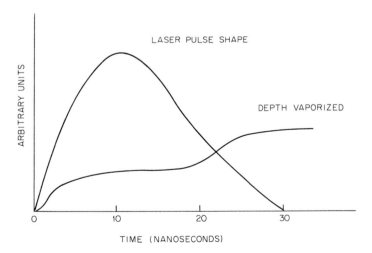

Fig. 13-8   Schematic representation of the depth vaporized in a metallic target as a function of time by a 30-nsec-duration laser pulse with the indicated shape. The effect of shielding of the target surface by blowoff material produced early in the laser pulse is apparent. (From J. F. Ready, "Effects of High-Power Laser Radiation." Academic Press, New York, 1971.)

the plasma expands and becomes transparent so that more material is removed. The amounts of material that can be removed by a Q-switched laser are strictly limited. Such lasers are not well adapted for hole drilling or for cutting.

The shielding of the target by high-temperature opaque plasma leads to the so-called laser-supported absorption (LSA) wave. The LSA wave is a plasma that is generated at the target surface and propagates back toward the laser. While it is present, it effectively shields the target surface.

Some experiments [3] directly demonstrating the LSA wave were carried out using a gas dynamic laser that could deliver pulses containing 1–2 MW/cm² to a target surface. The pulse duration was about 5 msec.

When 2024 aluminum was irradiated, the beam coupled to the specimen surface. This generated a plume of incandescent aluminum particles that extended above the specimen surface toward the laser. The form of the plume, as photographed with a high-speed framing camera, is shown schematically in Fig. 13-9. By approximately 1.71 msec, the plume had

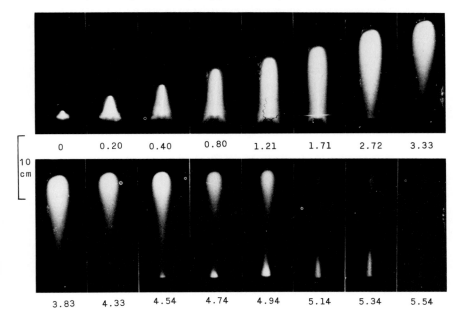

Fig. 13-9 Vaporized material emitted from an aluminum surface struck by a CO_2 laser pulse delivering 1.5×10^6 W/cm². The beam was incident from the top. Selected frames of high-speed photography of the plume show decoupling of the laser radiation from the surface after 2.72 msec. Each frame is identified by the time in milliseconds after the start of the laser pulse. (From R. L. Stegman, J. T. Schriempf, and L. R. Hettche, *J. Appl. Phys.* **44**, 3675 (1973). Photograph courtesy of J. T. Schriempf).

started to decouple from the surface because of attenuation caused by the plume. Approximately 1 msec later, the plume was completely decoupled from the specimen, at which time little laser energy was being delivered to the specimen surface.

When this occurred, the specimen no longer supplied incandescent aluminum particles, and so the plume began to dissipate and its attenuation effect was diminished. In addition, the plume had propagated back toward the laser to a position where the laser power density was lower. At 4.54 msec, the laser beam once again coupled to the specimen surface and remained coupled until 5.14 msec, when the laser pulse ended. The entire plume then quickly dissipated until another laser pulse impinged on the specimen.

Different target materials gave much different responses. An MgO target gave approximately four cycles of plume generation, decoupling, plume dissipation, and recoupling during the 5-msec pulse. A pyroceram target went through about eight cycles. On the other hand, graphite and fiberglass–epoxy targets remained continuously coupled through the pulse. Thus, the optimum pulse duration for a laser depends on the target material. If the pulse duration exceeds the optimum value for the particular target material, laser energy is wasted in the LSA wave.

The effect of the shielding by the LSA wave is demonstrated in Fig. 13-10, which shows relative amounts of material removal from an alumina target by a CO_2 laser with a 1-μsec pulse duration. This figure shows specific mass

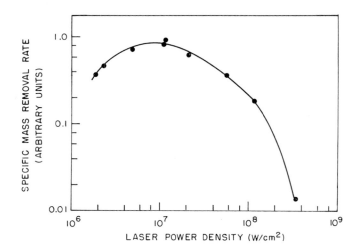

Fig. 13-10 Specific mass removal from an alumina surface struck by a CO_2 laser pulse with 1-μsec duration as a function of laser power density. The specific mass removal rate is the amount of material removed per unit incident energy. It gives a measure of the efficiency of the utilization of the incident energy for material removal.

removal, i.e., mass removal divided by laser energy, which is a measure of the efficiency of the material removal process. As power density increases, the specific material removal increases up to about 10^7 W/cm². Above 10^7 W/cm², the laser-supported absorption wave develops and shields the target so that the material removal becomes inefficient. The specific material removal then decreases with further increases in laser power density.

The absorption of laser radiation at a surface can also produce large pressure waves in the target material. One mechanism by which pressure pulses could be produced is evaporation of material from the surface, with recoil of the heated material against the surface. This would cause motion of the target as a whole. A considerable amount of experimental work on measuring the impulse transmitted to the target has been done. This work has been carried out mainly at high laser irradiance, where a considerable amount of material is removed from the surface.

Variation of momentum transfer with light intensity has been investigated. The momentum delivered by a focused ruby laser beam to a simple pendulum consisting of a sphere of target material suspended by a thread was measured through a window to the vacuum chamber with a calibrated microscope. The specific momentum transfer† as a function of laser irradiance for a number of materials is shown in Fig. 13-11. The specific impulse is much higher than the impulse that would be transferred through photon

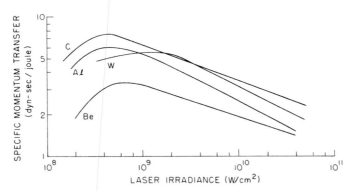

Fig. 13-11 Specific momentum transfer to several metals from a Q-switched ruby laser pulse. (Adapted from D. W. Gregg and S. J. Thomas, *J. Appl. Phys.* **37**, 2787 (1966).)

reflection. The results show that there is an optimum intensity for each material that gives the maximum momentum transfer per joule of laser energy. At values below the optimum, some of the energy is lost by thermal

---

† Specific momentum transfer (also called specific impulse) is the momentum transferred to the target per unit incident energy. It has units of dyn-sec/J.

conduction. Above the maximum, some of the energy goes into increasing the degree of ionization and the temperature of the vaporized material. This is less efficient in imparting momentum than if the same amount of energy were used to vaporize a larger number of atoms.

Peak pressures of the order of tens of kilobars were obtained from metallic materials irradiated by high-power Q-switched Nd:glass lasers. The pressure pulse can be large enough to cause back surface spallation in a thin metallic plate.

At very high values of power density, plasma processes dominate. The solid target surface disappears rapidly, and the phenomena are almost entirely plasma phenomena. Between 10^9 and 10^{12} W/cm^2 the physical mechanisms are apparently dominated by a process called inverse Bremsstrahlung. In this process (see Fig. 13-12a), a photon of the laser light is absorbed by a free electron in the plasma. The electron is raised to a higher-lying energy state in the continuum of states available to it. At the same time, an ion recoils in order to take up the residual momentum. This process has a cross section that increases as the square of the laser wavelength, and hence it proceeds more easily with CO_2 lasers than with visible or near-infrared lasers.

At still higher levels, $\gtrsim 10^{13}$ W/cm^2, collective plasma efforts become important. In these effects, light is absorbed by interaction with a number of plasma particles collectively, instead of by a single electron. This leads to

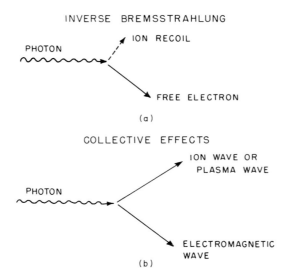

Fig. 13-12 Schematic diagrams of (a) absorption through inverse Bremsstrahlung and (b) through collective effects in which ion waves or plasma waves are excited.

generation of ion waves in the plasma, collective oscillations of charged particles in the plasma, or other electromagnetic modes (i.e., reradiation of light at frequencies different from the incident laser frequency). (See Fig. 13-12b.) The theoretical understanding of these collective effects is still in an early state, but they appear to have cross sections that increase as wavelength decreases. Hence, workers in the area of laser-assisted thermonuclear fusion are interested in development of short-wavelength, high-power lasers. Thermomonuclear fusion work is proceeding with lasers that can deliver up to 10^{16} W/cm^2 to a target. This work is strongly influenced by the collective plasma effects.

We have now described the relevant physical phenomena and are in a position to summarize. The physical phenomena involved in the interaction are sketched in Fig. 13-13. Figure 13-13a indicates absorption of the incident laser light according to the exponential absorption law

$$I(x) = I_0 e^{-\alpha x} \tag{13.4}$$

where $I(x)$ is the light intensity penetrating to depth x. The incident light intensity is I_0; for the purposes of this figure, we neglect the fraction of the light that is reflected. For metals, the absorption coefficient α is of the order of 10^5 cm^{-1}; thus the absorbed energy is deposited in a layer about 10^{-5} cm thick. It penetrates into the material by thermal conduction; thus the top part of the figure indicates heating of the material. When the surface reaches the melting temperature, a liquid interface propagates into the material is indicated in Fig. 13-13b. With continued irradiation, the surface begins to vaporize as shown in Fig. 13-13c, and a hole begins to be drilled. If the laser light is intense enough, absorption in the blowoff material leads to a high-temperature opaque plasma. The plasma can grow back along the beam toward the laser as an LSA wave. The plasma absorbs the incident light and shields the surface as shown in Fig. 13-13d.

The ranges of laser power density for which individual processes dominate the interaction are given in Table 13-3. Values are stated for two wavelength regimes: the visible and near-infrared region and the CO_2 laser region near 10 μm. The values in Table 13-3 are to be interpreted as approximate and will vary depending on the exact circumstances of pulse duration, target properties, etc. At relatively low power densities, melting is the main effect. At somewhat higher power densities, vaporization becomes the most important effect. This is conventional vaporization without much interaction between the incident light and the vaporized material.

At still higher values, LSA waves are kindled and dominate the physical processes, whereas vaporization is diminished. The threshold for kindling the LSA wave is about an order of magnitude lower for CO_2 lasers than for

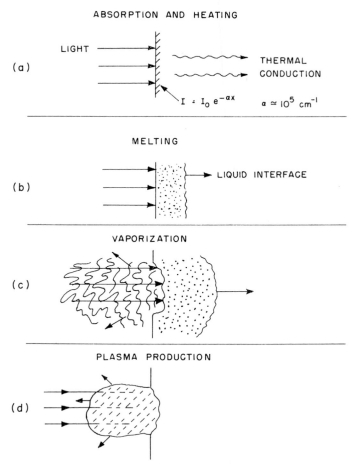

Fig. 13-13 Schematic diagrams of physical processes occurring when a high-power laser beam strikes an absorbing surface.

shorter wavelength lasers. The values given in Table 13-3 for the threshold for LSA waves are those appropriate to one specific case, i.e., a titanium target with a laser pulse duration in the microsecond regime. The threshold will vary as the circumstances change. However, the numbers in Table 13-3 serve to identify an order of magnitude for the threshold at which such interactions occur.

Finally, when the power density becomes very high, additional absorption mechanisms may become operative. This includes absorption of the laser light in the laser-produced plasma through inverse Bremsstrahlung or

TABLE 13-3
Approximate Ranges of Laser Power Density at Which Different
Processes Dominate

Process	Range for visible and near-infrared laser (W/cm^2)	Range for CO_2 laser (W/cm^2)
Melting	$\sim 10^5$	$\sim 10^5$
Vaporization	$10^6 - 1.5 \times 10^8$	$10^6 - 2.5 \times 10^7$
Laser-supported absorption wave	$>1.5 \times 10^8$	$>2.5 \times 10^7$
Plasma-inverse Bremsstrahlung	$\lesssim 10^{12}$	?
Plasma-collective effects	$\gtrsim 10^{13}$	?

through collective plasma effects. These mechanisms have been investigated theoretically, and experimental verification of the theory is proceeding. In this region, coupling of the radiation to the plasma can become large. This absorption is more pronounced for shorter wavelength lasers. Not much work has yet been done using CO_2 lasers in this regime. Investigation of this region is of particular interest for workers in the field of laser-assisted thermonuclear fusion.

We note that the region of greatest interest for material-processing applications (i.e., welding, drilling, etc.) lies below the threshold for ignition of LSA waves. Below this threshold, the laser energy goes into changing the target, whereas above the threshold the energy goes mainly into the LSA wave or into plasma effects. An LSA wave can appear spectacular: there is a loud noise and a bright flash of light, but the solid target surface can be very little affected. Thus, for the practical applications that we will describe in the next chapters, the production engineer will generally want to work at laser power densities below the threshold for LSA waves.

We summarize these phenomena in Fig. 13-14, which identifies various regions of interaction and their potential applications. In the region marked "welding," one obtains reasonable depths of melted material. Below the line marked "no melting," the surface is not heated to the melting point. Above the line marked "surface vaporization," the surface begins to vaporize, and welding is less desirable in this region. To the left of the welding region heat penetration is too small, and to the right it spreads over too great an area. Similarly, regions useful for cutting, hole drilling, and material removal (i.e., small amounts of material such as cutting of thin films for resistor

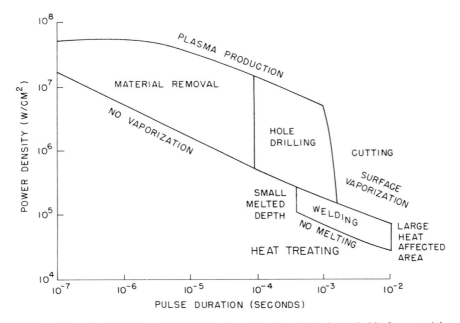

Fig. 13-14 Regimes of laser power density and pulse duration suitable for material-processing applications.

trimming) are identified. Above the line marked "plasma production," the LSA wave develops.

We emphasize that the regions identified in Fig. 13-14 are not exact, but will vary with target material, laser wavelength, etc. Nevertheless, they define approximate regions where certain applications are most productive. In the next three chapters, we shall discuss some of these applications in more detail.

REFERENCES

[1] M. I. Cohen, *J. Franklin Inst.* **283**, 271 (1967).
[2] M. K. Chun and K. Rose, *J. Appl. Phys.* **41**, 614 (1970).
[3] J. T. Schriempf, R. L. Stegman, and G. E. Nash, *IEEE J. Quantum Electron.* **QE-9**, 648 (1973).

SELECTED ADDITIONAL REFERENCES

F. J. Allen, Surface Temperature and Disposition of Beam Energy for a Laser-Heated Target, *J. Appl. Phys.* **42**, 3145 (1971).
W. D. Brewer, Ablative Material Response to CO_2 Laser Radiation, *J. Spacecr.* **7**, 1449 (1970).

S. S. Charschan (ed.), "Lasers in Industry," Van Nostrand–Reinhold, Princeton, New Jersey, 1972, Chapter 3.

M. K. Chun and K. Rose, Interaction of High-Intensity Laser Beams with Metals, *J. Appl. Phys.* **41**, 614 (1970).

C. DeMichelis, Laser Interaction with Solids—A Bibliographical Review, *IEEE J. Quantum Electron.* **QE-6**, 630 (1970).

W. W. Duley, "CO_2 Lasers, Effects and Applications," Academic Press, New York, 1976, Chapters 4 and 5.

B. P. Fairand *et al.*, Quantitative Assessment of Laser-Induced Stress Waves Generated at Confined Surfaces, *Appl. Phys. Lett.* **25**, 431 (1974).

J. A. Fox and D. N. Barr, Laser-Induced Shock Effects in Plexiglas and 6061-T6 Aluminum, *Appl. Phys. Lett.* **22**, 594 (1973).

F. P. Gagliano, R. M. Lumley, and L. S. Watkins, Lasers in Industry, *Proc. IEEE* **57**, 114 (1969).

F. P. Gagliano and U. C. Paek, Observation of Laser-Induced Explosion of Solid Materials and Correlation with Theory, *Appl. Opt.* **13**, 274 (1974).

J. N. Gonsalves and W. W. Duley, Interaction of CO_2 Laser Radiation with Solids, I. Drilling of Thin Metallic Sheets, *Can. J. Phys.* **49**, 1708 (1971).

J. E. Lowder and L. C. Pettingill, Measurement of CO_2-Laser-Generated Impulse and Pressure, *Appl. Phys. Lett.* **24**, 204 (1974).

T. J. Magee, R. A. Armistead, and P. Krehl, Laser-Induced Stresses in Coated and Uncoated Targets, *J. Phys. D: Appl. Phys.* **8**, 498 (1975).

S. A. Metz *et al.*, Effect of Beam Intensity on Target Response to High-Intensity Pulsed CO_2 Laser Radiation, *J. Appl. Phys.* **46**, 1634 (1975).

A. M. Prokhorov *et al.*, Metal Evaporation under Powerful Optical Radiation, *IEEE J. Quantum Electron.* **QE-9**, 503 (1973).

J. F. Ready, Effects due to Absorption of Laser Radiation, *J. Appl. Phys.* **36**, 462 (1965).

J. F. Ready, "Effects of High-Power Laser Radiation," Academic Press, New York, 1971, Chapter 3.

J. F. Ready, Impulse Produced by the Interaction of CO_2 TEA Laser Pulses, *Appl. Phys. Lett.* **25**, 558 (1974).

J. F. Ready, Change of Reflectivity of Metallic Surfaces during Irradiation by CO_2-TEA Laser Pulses, *IEEE J. Quantum Electron.* **QE-12**, 137 (1976).

J. E. Robin and P. Nordin, Effects of Gravitationally Induced Melt Removal on CW Laser Melt-Through of Opaque Solids, *Appl. Phys. Lett.* **27**, 593 (1975).

B. A. Sanders and V. G. Gregson, Optical Reflectivity of Some Metals Using a CO_2 TEA Laser, *in Proc. Tech. Program, Electro-Opt. Systems Design Conf., New York* (September 18–20, 1973).

I. P. Shkarofsky, Review on Industrial Applications of High-Power Laser Beams III, *RCA Rev.* **36**, 336 (1975).

M. Sparks, Theory of Laser Heating of Solids: Metals, *J. Appl. Phys.* **47**, 837 (1976).

R. L. Stegman, J. T. Schriempf, and L. R. Hettche, Experimental Studies of Laser-Supported Absorption Waves with 5-ms Pulses of 10.6 μ Radiation, *J. Appl. Phys.* **44**, 3675 (1973).

M. von Allmen, Laser Drilling Velocity in Metals, *J. Appl. Phys.* **47**, 5460 (1976).

R. E. Wagner, Laser Drilling Mechanics, *J. Appl. Phys.* **45**, 4631 (1974).

T. E. Zavecz, M. A. Saifi, and M. Notis, Metal Reflectivity under High-Intensity Optical Radiation, *Appl. Phys. Lett.* **26**, 165 (1975).

CHAPTER 14

LASER APPLICATIONS IN MATERIAL PROCESSING

Material processing refers to a variety of industrial operations in which the laser operates on a workpiece to melt it or remove material from it. Some of the possible applications include welding, hole drilling, cutting, trimming of electronic components, and heat treating.

The properties of a laser that are important are its collimation and its coherence. Because of these properties, the laser is a source of energy that can be concentrated by a lens to achieve extremely high power density at the focal spot.

Let us compare an ordinary CO_2 laser and a large mercury arc lamp. If the CO_2 laser emits 100 W of power in a beam with a divergence angle of 10^{-2} rad, it can be focused by a lens with a focal length of 1 cm to a spot 0.01 cm in diameter. This diameter is calculated using the equation

$$D = f\theta \tag{14.1}$$

where D is the diameter of the focal spot, f the focal length of the lens, and θ the beam divergence angle of the laser. All the laser light can be collected and delivered to this small spot. The power density D in this focal spot is approximately

$$P \simeq 10^2 \text{ W}/(0.01)^2 \text{ cm}^2 = 10^6 \text{ W/cm}^2 \tag{14.2}$$

(This approximation neglects a factor of $\pi/4$, which is of the order of magnitude of unity.)

Consider, by comparison, a mercury arc lamp which emits 1000 W of power uniformly into a solid angle of 4π sr. This is a large lamp which requires a big power supply. Because not all the light from this lamp can be collected by the focusing optics, and because the large divergence angle of the light leads to a much larger focal area, the power density from this lamp is much lower. It can be shown (although we shall not go through the details here) that the same lens (with a focal length of 1 cm) will deliver a power density of 100 W/cm^2 in the focal area. Thus, the total power emitted by the lamp is 10 times larger than for the laser, but the power density delivered to a workpiece by the laser is 10,000 times larger.

Pulsed lasers, of course, can deliver much higher values of power density in a short pulse. For example, small, easily available ruby lasers can produce 1-msec-duration pulses with a power density of 10^9 W/cm^2 at a target. No other source (except an electron beam) can produce such high values of power density.

When the high-power beam interacts with the surface of a workpiece, it can cause melting or vaporization, as we have discussed in Chapter 13. Because of these effects, the laser can be a useful tool in material processing. Lasers still remain somewhat specialized tools; they will not replace conventional methods on a wholesale basis. The cost of operating lasers remains relatively high, so that applications must be carefully selected. If conventional methods are economical and work reasonably well, lasers often cannot compete with them. Lasers offer most promise for situations in which there is a problem with conventional methods, but continuing development has brought lasers close to a competitive position even for some routine applications.

Laser development has helped make lasers more useful and economical. Early investigations used ruby and Nd:glass lasers; some applications were shown to be feasible, but costs were often too high. Development of the repetitively pulsed CO_2 laser, the high-power continuous CO_2 and Nd:YAG lasers, and the repetitively Q-switched Nd:YAG laser has changed this situation. These lasers offer possibilities for economical operation. The lasers of choice now for most materials working applications are the CO_2 and Nd:YAG lasers.

Of the many different types of lasers that have been developed, only a few are useful for materials processing applications. These are listed in Table 14-1. The properties listed are an indication of commercially available levels, not necessarily the highest values ever achieved. We enter the CO_2 laser and the Nd:YAG laser more than once, in each useful type of operation. The mechanisms of interaction and the applications will vary according to the type of operation. We note that not all indicated properties may be achieved simultaneously. Operation at the highest peak power may involve a sacrifice of pulse repetition rate or of pulse length.

The ruby and Nd:glass lasers appear to be most useful where single pulses of high energy are desired, for example, in spot welding and in large hole drilling. The CO_2 lasers and Nd:YAG lasers, in their various modes of operation, appear to offer the best economy and the greatest versatility. Both the Nd:YAG and the CO_2 lasers have been carefully engineered and are durable enough for production use.

TABLE 14-1
Commercial Lasers Suitable for Materials Processing

Laser	Wavelength (μm)	Type of operation	Typical power (W)	Pulse rep. rate (pps)	Pulse length	Typical use
CO_2	10.6	Continuous	100–several thousand	—	—	Seam welds
CO_2	10.6	Repetitively pulsed	100 (average)	100	100 μs	Seam welds, hole drilling
CO_2	10.6	TEA	10^7	Up to 100	10 μsec	Hole drilling, marking
Nd:YAG	1.06	Continuous	Up to 1000	—	—	Seam welds
Nd:YAG	1.06	Repetitively Q-switched; continuously pumped	10^4 (peak) 10 (average)	≥25,000	200 nsec	Component fabrication and trimming
Nd:YAG	1.06	Pulse pumped	10^4 (peak)	100	1–10 msec	Spot and seam welds, marking
Ruby	0.6943	Normal pulse	10^5 (peak)	Single pulse	0.2–5 msec	Spot welds and single holes
Nd:Glass	1.06	Normal pulse	10^6 (peak)	Single pulse	0.5–10 msec	Spot welds and single holes

An important development is high-voltage pulsing of the carbon dioxide laser. The output occurs in the form of repetitive pulses with typically about 1 J of energy and pulse lengths in the 10 to 100 μsec range. The pulse repetition rate can be 100 pulses per second for an average power of 100 W. This repetitively pulsed mode of operation yields high peak powers that break down a metallic surface within a single pulse, even though a substantial fraction of the energy is reflected. Once the surface begins to vaporize, further energy absorption occurs efficiently. With this type of operation, carbon dioxide lasers can work metals more effectively than a continuous carbon dioxide laser of equal average power. The operation of repetitively pulsed carbon dioxide lasers can be economical. The cost of the gas supply, electricity, and routine maintenance is estimated as $1.00 per hour (or perhaps less if the helium from the gas mixture is reclaimed). Within an hour, one might process several hundred parts, so that the cost per part can be a small fraction of one cent.

A continuous CO_2 laser operating at several hundred watts has metalworking capabilities and can perform seam welding in thin metal sheet. The CO_2 laser has developed from a laboratory instrument to a rugged industrial-rated tool. Continuous CO_2 lasers have been constructed with outputs ranging up to many kilowatts. Although a small number of multikilowatt CO_2 lasers have been delivered to industry, they are not (as of 1977) really "commercially available" in the same sense as CO_2 lasers of somewhat lower power. A level around 1000 W appears to represent the status of "off-the-shelf" lasers. Higher power levels are subject to special order and high expense.

The CO_2 TEA laser is listed in Table 14-1 with a pulse duration of 10 μsec. A TEA laser operated with relatively small amounts of CO_2 and enriched with nitrogen can provide pulse durations in that regime, with somewhat reduced peak power. Such TEA lasers appear suitable for some applications in hole drilling. In its more usual type of operation, with pulses $\sim$100 nsec long and very high peak power, the TEA laser is not very suitable for materials processing, because of the laser-supported absorption wave which is produced and which can shield the target from the incoming radiation.

The wavelength of operation of the CO_2 laser is not an obstacle to its utilization. With presently available infrared materials, there is no particular difficulty in working with 10.6 μm radiation. Available lens and window materials have been described in Chapter 7. It is sometimes stated that the carbon dioxide laser is less useful than the Nd:YAG laser because its beam cannot be focused to a small spot. This consideration is usually not important, because thermal conductivity of the work piece will spread the heat over a broader area than that in which it is delivered. The width of a weld or cut often turns out to be comparable for carbon dioxide lasers and for shorter wavelength lasers, although if an extremely small focal diameter is required, the Nd:YAG laser will be better suited.

Developments in Nd:YAG lasers have included continuously pumped, repetitively Q-switched operation and continuous operation at levels around 1000 W. The high-power continuous operation is suitable for melting and vaporization of metals, because at the 1.06 μm wavelength of the neodymium laser, metallic surfaces are not too highly reflecting. However, at the highest power level, the krypton lamps required for the high output are relatively short-lived, so that the operating cost of a Nd:YAG laser at several hundred watts is high. At lower levels (at perhaps 10 W), the continuous Nd:YAG laser can be very economical.

In the repetitively Q-switched mode of the Nd:YAG laser, the pumping source remains on all the time, and an acoustooptic switch periodically opens the channel between the laser rod and the mirror. When a laser pulse is desired, sonic waves from the transducer diffract the light beam so that it reaches the mirror, and laser action can proceed. The Q-switched pulses have durations of the order of a few hundred nanoseconds and pulse repetition rates up to several kilohertz. Peak powers may reach tens of kilowatts. In this mode of operation, the Nd:YAG laser can be economial. Operating costs can be less than $0.50 per hour, if the output requirements are not too high.

The ruby and Nd:glass lasers have been employed in industrial applications where pulse repetition rate is not too important. They are useful for items such as spot welding, but for seam welding, the pulse repetition rate is often too low to make these lasers attractive. We enter only the normal pulse type of operation, i.e., pulse duration around 1 msec. The Q-switched type of operation is usually not suitable because of target shielding by the blowoff material. Q-switched ruby and Nd:glass lasers do not often seem to be used in industrial material processing.

The unfocused, "raw" output of the laser generally does not provide sufficient power density to raise the temperature of most materials to their melting points. Also, the diameter of the raw beam, typically at least a few millimeters, is too large for those applications requiring that heat-affected zones be small. For these reasons, focusing or concentrating the output beam of the laser is required for the majority of material processing applications.

As compared to conventional processes, lasers offer some advantages in materials processing:

(1) No vacuum is required. For most applications the work can be done in any atmosphere, although for some reactive metals a shielding atmosphere may be desirable.

(2)   There is no contact of any material with the workpiece; contamination problems are reduced.

(3)   The heat-affected zone surrounding the work area is small.

(4)   The laser offers advantages for some "difficult" materials. The laser works well with hard, brittle materials or with refractory materials, and it can sometimes work for joining dissimilar metals which are difficult to weld by conventional techniques.

(5)   Small hole diameters can be achieved.

(6)   The operation is very fast, occurring in approximately 1 msec.

(7)   The processing is readily adapted to automation.

(8)   No welding electrodes are required.

(9)   Extremely small welds may be accomplished on delicate materials.

(10)   Inaccessible areas or even encapsulated materials can easily be reached with the laser beam.

(11)   Higher power per unit area can be delivered than with any other thermal source.

There are also some disadvantages which must be remembered:

(1)   The depth of penetration for laser-produced holes is limited, although repeated shots can increase the depth.

(2)   Recondensation of vaporized material occurs on the walls of the hole and on the lip of the hole, forming a raised rim around the entrance.

(3)   The walls of the holes are generally rough.

(4)   The cross sections of laser-produced holes are not completely round, and the holes taper from the entrance to the exit sides.

(5)   The control of size and tolerances on laser-produced holes is not perfect.

(6)   In laser welding, careful control of the pulse parameters is required to prevent vaporization of the surfaces.

(7)   The sizes of the pieces than can be welded are relatively small and the depth of penetration is limited, except for multikilowatt lasers.

(8)   The costs often are high.

The most serious drawback can be the cost. Indeed for many applications the laser is too costly, even though it can do the job well.

Some samples of considerations which might suggest use of lasers include:

(1)   The necessity of producing a small heat-affected zone.

(2)   The necessity of avoiding deposition of any electrical charge in the work piece.

(3)   The use of difficult materials, e.g., hole drilling in ceramics.

In the early days of laser technology, lasers were delicate instruments. Although the technical feasibility of an operation could be demonstrated, lasers were not suited for continuous use on a production line. As time went on, there were considerable advances in the design and engineering of lasers. Commercial models are durable enough to operate in a factory environment and have acceptable maintenance costs. Lasers have now been used in a routine fashion in production in a variety of areas. Lasers will certainly not take over all the manifold welding, cutting, and trimming operations of industry. The emphasis will probably remain on relatively small pieces and on areas where conventional methods have problems.

A laser is not used all by itself but rather is used in connection with many different pieces of equipment. All the different pieces of equipment must work together properly to achieve the end result, which may be a desirable application of lasers. However, the laser is only one element in the system and of itself could not carry out the application. The system will include the laser, focusing optics, a power monitoring device, fixturing for holding and moving the workpiece, the workpiece itself, appropriate safety devices, possibly some method for observing the effect of the beam on the workpiece, and often numerically controlled drives for automatically positioning the workpiece. All these parts are interdependent. For example, the output of the laser is measured by the power monitor; the measured values are in turn used to control the laser output so as to produce the desired effect. The parts also interact in the sense that they must be chosen to fit the laser properties—for example, spectral transmission characteristics and lens diameter must be chosen to be appropriate for the specific laser. To produce a successful application, the user must carefully design and select all parts of the system.

Two other considerations are worth mentioning:

(1)   Often the optical path can be completely enclosed. A pipe with prisms to provide bending at the corners can be very effective and can enhance safety, especially with infrared beams.

(2)   It may be desirable to provide a sheet of material between workpiece and lens to protect the lens from blowoff material. It will be much cheaper to use such a disposable material (transparent at the laser wavelength) than to replace damaged lenses.

There are a variety of possible laser operations involving both metals and nonmetallic materials, including ceramics, gems, glass, silicon, and organic materials. Many such applications are serious economic contenders for routine production use. The laser parameters such as wavelength, power, and pulse duration must be optimized to fit the particular application.

Figure 14-1 shows a complete laser system employed in production welding. It is based on a continuous 600-W CO_2 laser, in the large cabinet

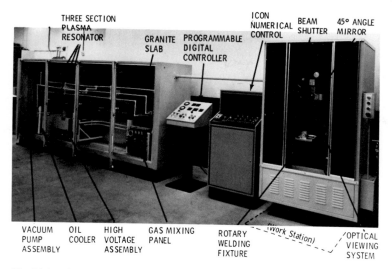

THREE SECTION
PLASMA
RESONATOR

GRANITE
SLAB

PROGRAMMABLE
DIGITAL
CONTROLLER

ICON
NUMERICAL
CONTROL

BEAM
SHUTTER

45° ANGLE
MIRROR

VACUUM
PUMP
ASSEMBLY

OIL
COOLER

HIGH
VOLTAGE
ASSEMBLY

GAS MIXING
PANEL

ROTARY
WELDING
FIXTURE

(Work Station)

OPTICAL
VIEWING
SYSTEM

Fig. 14-1 Photograph of industrial welding system based on a 600-W CO_2 laser.

at the left. The CO_2–N_2–He gas mixture flows through three tubes, each 3 m long, and arranged in a Z shape. The operation of the laser is controlled by a programmable digital controller, which can turn the laser on and off at prescribed times for automated welding. Safety features include transmission of the beam to the work station in a metal pipe, and enclosure of the work station with lucite panels, which are opaque at 10.6 μm. These panels are interlocked with the beam shutter so that the beam cannot enter the work station when the panels are opened. The beam is bent through 90° and focused on the workpiece by a lens. The workpiece is held in a rotary fixture, and rotated under the beam so as to make seam welds. The motion of the workpiece is controlled automatically by the programmable numerical control. This example gives an idea of some of the elements that are employed in a complete system for laser welding.

Specific material processing applications include heat treating, welding, hole drilling, cutting, and scribing. In the next two chapters we shall describe some of these applications in more detail.

SELECTED REFERENCES

S. S. Charschan (ed.), "Lasers In Industry," Van Nostrand–Reinhold, Princeton, New Jersey, 1972, Chapter 3.
M. I. Cohen, Material Processing, *in* "Laser Handbook," Vol. 2 (F. T. Arecchi and E. O. Schulz-Dubois, eds.), North-Holland Publ., Amsterdam, 1972.
M. Eleçcion, Materials Processing with Lasers, IEEE Spectrum, p. 62 (April, 1972).

V. G. Gregson *et al.*, Basis of Laser Material Processing, *Electro-Opt. Syst. Design*, p. 25 (November 1976).

E. Hoag *et al.*, Performance Characteristics of a 10-KW Industrial CO_2 Laser System, *Appl. Opt.* **13**, 1959 (1974).

E. V. Locke, Multikilowatt Industrial CO_2 Lasers, Soc. of Manufacturing Eng. Tech. Paper MR74-952 (1974).

M. Pasturel, CO_2 Industrial Laser Machining, *Electro-Opt. Syst. Design*, p. 23 (May 1976).

J. F. Ready, Selecting a Laser for Material Working, *Laser Focus*, p. 38 (March 1970).

J. F. Ready, "Effects of High Power Laser Radiation," Academic Press, New York, 1971, Chapter 8.

G. Schaffer, Lasers in Metalworking, *Am. Mach.*, p. 41 (July 1, 1975).

CHAPTER 15

APPLICATIONS OF LASER
HEATING: WELDING AND HEAT TREATING

In the preceding two chapters, we have described the physical mechanisms by which high-power laser radiation interacts with materials and the potential applications of these interactions to material processing. In this chapter we shall describe the specific application of welding in greater detail.

Early in the 1960s it was recognized that specialized welding applications could be performed with ruby lasers, and that in certain cases laser welding provided technical advantages, such as a very small heat-affected zone. Most early descriptions of laser welding center on the use of the ruby laser. Spot welds were emphasized in most cases, but seam welding by overlapping pulses was demonstrated, although at slow rates. By the late 1960s, some production applications of spot welding with ruby lasers were announced. By 1970, development of CO_2 and Nd:YAG lasers made them the leading contenders for welding applications. Seam welds could be made more rapidly using continuous or repetitively pulsed CO_2 or Nd:YAG lasers. In 1971, deep penetration welding with multikilowatt CO_2 lasers was announced. By the mid-1970s, laser welding had reached production status for a number of applications.

The parameters of the laser beam and the properties of the workpiece both influence the results of a laser welding operation. The thermal diffusivity of the workpiece is important. High thermal diffusivity allows greater conduction of the heat and in general permits greater depth of welding. High reflectivity of the metal surface can reduce the energy absorbed by the surface. The reflectivity of the surface can drop abruptly during the interaction with a high-power beam, so that most of the energy in the laser pulse is absorbed even if the initial reflectivity is high. Welding in high reflectivity metals does require somewhat more energy than in low reflectivity metals.

The surface may be darkened by a surface coating, but this is not always effective; the coating can be blasted off, leaving the shiny metal untouched. Surface finish can also influence light absorption and hence melting depths. The depth of penetration is decreased by polishing a metallic surface.

The important point here is that one cannot view the laser as an individual component, but must treat the whole system as a single entity to perform the desired job. The following points are important:

(1)   Suitable selection of tasks compatible with lasers;
(2)   economic analysis of laser use;
(3)   choice of the proper type of laser;
(4)   careful design of fixturing for motion of the workpiece;
(5)   design of optics for delivery of the beam to the workpiece;
(6)   measurement of the effect of the beam with feedback control of the laser;
(7)   safety considerations.

There are many properties of the laser that must be properly chosen to fit into a system. These include the following:

(1)   The wavelength should be absorbed well by the workpiece.
(2)   The power level must be high enough to produce melting.
(3)   The pulse duration for pulsed lasers must be long enough to permit penetration of the heat into the material.
(4)   The pulse repetition rate for pulsed lasers must be high enough to weld a seam if seam welding is required.
(5)   The power density and pulse duration should be chosen to fall in a regime where surface vaporization is not excessive.

As we noted before, for laser welding the material should be melted with penetration of the heat through the sample. Vaporization should be avoided and the pulse length or dwell time tailored to allow adequate penetration. With good control of the laser pulse, vaporization can be minimized. Optimization of the welding is largely controlled by the pulse length. For many pulsed welding applications, it is desirable to stretch the pulse length as long as possible.

Continuous lasers should in principle be adaptable to welding of thicker samples, since the interaction time can be made arbitrarily long. Early continuous lasers were incapable of delivering enough power for welding, but present-day continuous CO_2 and Nd:YAG lasers are both capable of welding at outputs of several hundred watts or more.

The most important parameter in laser welding is the power per unit area delivered to the surface. This is determined both by the laser output and by the focusing.

Of the lasers listed in Table 14-1, all except the TEA CO_2 laser and the repetitively Q-switched Nd:YAG laser have been employed for industrial welding applications.

We begin by describing seam welding with lasers of average power below 1000 W, a region in which penetration of the heat by thermal conduction is a dominant factor. Then we discuss welding with multikilowatt lasers, for which the limitations imposed by thermal conduction are relaxed so that deeper penetration is possible. We then describe spot welding applications, for which average power is less a consideration. Heat treating, which may be regarded as frustrated welding, is then considered. Finally, a specific example in which laser welding solved a difficult production problem is presented.

A. Seam Welding—Subkilowatt Levels

We describe first seam welding at levels below about 1000–1500 W. Seam welding requires either a continuous laser or a repetitively pulsed laser with reasonably high pulse repetition rate. Thus, one usually employs either a CO_2 laser or Nd:YAG laser for seam welding. At levels of average power below 1000–1500 W, as we saw in Chapter 13, the laser energy is absorbed at the surface of the workpiece, and penetration is limited by thermal conduction. With several kilowatts average power, other mechanisms begin to have an effect. For this reason, welding with multikilowatt lasers will be treated separately later.

Both the pulsed CO_2 laser and the continuous CO_2 laser operated at several hundred watts average power are well suited for seam welding. The pulsed CO_2 laser makes a seam weld by overlapping spots, but the pulse repetition rate is high enough to form seams rapidly. Similar remarks apply to the repetitively pulsed and continuous Nd:YAG lasers.

At 100 W average power, a repetitively pulsed laser is a welding tool for some metals. A continuous laser must emit several hundred watts to have comparable metalworking capabilities because the high peak power in the pulse can break down the surface reflectivity, allowing more efficient absorption of the laser energy. At powers of hundreds of watts, continuous lasers (both Nd:YAG and CO_2) offer welding capabilities. The potential of the high-power continuous CO_2 laser in particular has been demonstrated. The weld speed and smoothness of the weld bead obtained are encouraging.

The tradeoff between penetration depth and welding speed for seam welding with a 375-W continuous CO_2 laser is shown in Fig. 15-1. Data are shown for welding of two different types of steel. These results are for butt welding, with full penetration of the weld zone through the material. There

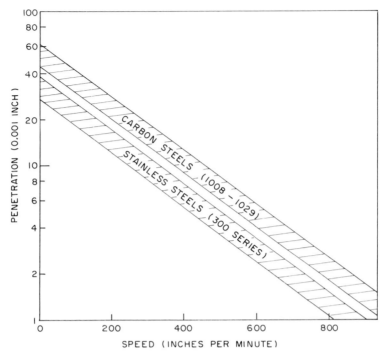

Fig. 15-1 Penetration depth versus speed for some steels welded with a 375 W continuous CO$_2$ laser. (Data from Photon Sources, Inc.)

are many potential applications in welding which can be satisfied by combinations of penetration depth and speed as shown. However, high speeds are possible only with thin material, because of constraints imposed by thermal conduction. Even at very slow rates, penetration is limited to ∼0.040 in. for stainless steel and ∼0.060 in. for carbon steel.

Figure 15-2 shows a similar curve for type 302 stainless steel, welded by a 1500-W continuous CO$_2$ laser [1]. The increased power allows somewhat deeper penetration, but even at slow speeds the maximum penetration is limited to about 0.12 in.

It is worthwhile to compare the results obtained using a CO$_2$ laser and a Nd:YAG laser for welding. Figure 15-3 shows the weld rates obtained with a continuous Nd:YAG laser welding 304 stainless steel [2]. These results indicate that the power required to achieve a given welding speed on a given thickness of material is approximately the same for a Nd:YAG laser as for a CO$_2$ laser at comparable output power. The difference in reflectivity at 1.06 and 10.6 μm is thus not too important for steel. Breakdown of the surface reflectivity by the laser means that the energy is absorbed reasonably well

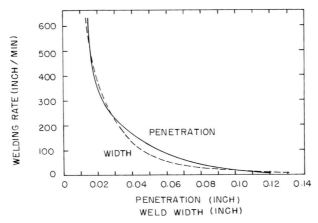

Fig. 15-2 Weld rate versus penetration depth and weld width for welding of 302 stainless steel with a 1500 W continuous CO_2 laser. (From S. L. Engel, *Laser Focus*, p. 44 (February 1976).)

in either case. However, for high conductivity metals (e.g., Cu and Al, see Fig. 13-3), the difference in reflectivity is more significant, and Nd:YAG lasers may be better suited for such welding.

There are many descriptions of laser welding in the literature which date from the 1960s, a time when the lasers available for welding were ruby or Nd:glass with low pulse repetition rates. Seam welding can be carried out with such lasers by overlapping spots, but with low seam speed. The development of pulsed Nd:YAG lasers and CO_2 lasers with high pulse repetition rates has made it possible to produce seam welds at a reasonable rate by overlapping spots.

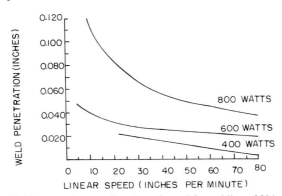

Fig. 15-3 Weld penetration as a function of speed for welding of 304 stainless steel with a continuous Nd:YAG laser at the indicated power levels. (From J. H. Mason and J. H. Wasko, Society of Manufacturing Engineers Technical Paper MR 74-955, 1974.)

    Some data on seam welding with a high-power pulsed Nd:YAG laser
are shown in Fig. 15-4 [3]. The data are relevant to a Nd:YAG laser operating at an average power of 400 W, with a peak power of several kilowatts.
The material is 300-series stainless steel. Weld rates are shown for a variety
of conditions. The curve marked "conduction weld" is for conditions of
focusing such that no substantial amount of vaporization occurs. The
curve marked "penetration weld" is for conditions of focusing to a higher
power density, such that substantial vaporization occurs. A hole is produced
and deeper penetration is achieved. This mode is similar to deep penetration
welding with multikilowatt CO_2 lasers, which will be described in the next
section. In the penetration mode, greater depth and larger aspect ratios are
achieved. Data are shown also for seam welding with a continuous laser of

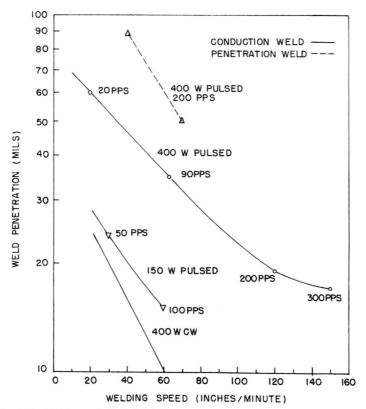

Fig. 15-4   Weld rates for Nd:YAG laser welding of stainless steel. The figure shows weld
penetration under various operating conditions. (From H. L. Marshall, *in* "Industrial Applications of High Power Laser Technology" (J. F. Ready, ed.), Vol. 86. Proc. Soc. Photo-Opt.
Instrum. Engineers, Palos Verdes Estates, California, 1976.)

comparable power. The results for the continuous laser are the same as presented in Fig. 15-3. The repetitively pulsed laser gives deeper penetration because of the higher peak power during the pulse.

The weld depths in Figs. 15-1–15-4 are representative of what can be achieved with lasers emitting a few hundred watts average power. At this level of power, the laser energy is absorbed at the surface of the workpiece, and the penetration of laser energy into the sample is by thermal conduction. This fact limits the depth of penetration, so that the weld depths are relatively small. For such lasers, maximum penetration depths of a few tens of mils are typical. The depth of penetration could, of course, be increased by long dwell times with a continuous laser. This would result in slow weld rates and would lead to large heat-affected zones. The use of a multikilowatt continuous CO_2 lasers relieves these problems and allows welding with much deeper penetration, as we shall describe later.

It is worthwhile to consider the effect of varying the focus of the beam. Fig. 15-5 shows how the penetration depth and bead width vary for welding

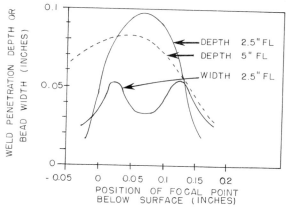

Fig. 15-5   Weld penetration and weld width as a function of focal position for welding of 1018 steel at 50 in./min with 1500 W from a continuous CO_2 laser, with lenses of the indicated focal length (FL). (From S. L. Engel, *Laser Focus*, p. 44 (February 1976).)

of 1018 steel [4]. The welding was done with a 1500-W continuous CO_2 laser at a speed of 50 in./min. The effect of changing the focus is apparent, with maximum penetration occurring when the beam is focused at a point slightly below the surface. When a shorter focal length lens is used, the penetration is deeper, but the half-width of the penetration depth curve is greater. This is because the beam is focused to a smaller minimum beam waist by a short focal length lens (see Chapter 2, Section G).

Examples of the small weld zones produced by a CO_2 laser are shown in Fig. 15-6. Figure 15-6a shows a transverse section of a weld between the base of a cylindrical steel can and a header. Loss of metal through surface vaporation is apparent, but a good hermetic seal was obtained. Complete fusion of the weld bead is apparent. The rapid quenching of the weld bead after passage of the laser beam is demonstrated by the smallness of the grains in the bead. It is also apparent that the heat-affected zone in the material next to the bead is very small, about 0.002 in. thick. Figure 15-6-b shows a longitudinal section along a weld seam run along the top of a lamination stack of high cobalt alloy plates, with one stainless steel end plate. The flow of the molten metal during the welding process is apparent.

Now let us consider some of the practical considerations for laser welding. In order to achieve good welding, the two pieces must be fitted closely together, in order that the material from each part can flow and intermix. Generally, the beam should be delivered approximately equally to each piece.

The workpiece must be positioned and held properly. This operation is called fixturing. The types of fixturing vary widely depending on the exact size and shape of the workpieces. For example, if a top is to be welded to a

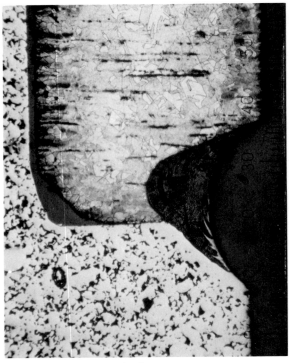

(a)

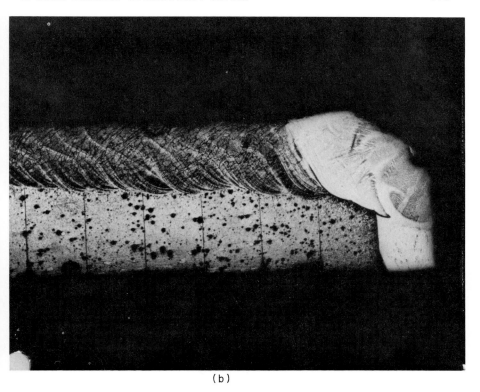

(b)

Fig. 15-6 (a) Cross section of weld made between stainless steel base and header, with a continuous CO_2 laser operating at 360 W. The weld was made at a linear speed of 36 in./min. The width of the area covered by the photograph is 0.047 in. (b) Weld made across the top of a stack of plates of high cabalt alloy by a continuous CO_2 laser operating at 500 W. The right end plate is stainless steel. The width of the area shown in the photograph is 0.1 in. (Photographs courtesy of P. E. Gustafson.)

can, the top must be held in place and the two pieces together must be rotated through the beam, which is focused at the interface between the top and the can. If two flat pieces of metal are to be butt welded, they must be positioned and held tightly together, and then moved so that the interface between the two pieces passes through the focus of the beam. It is impossible to discuss all of the various types of fixturing. We shall simply state that proper fixturing is essential for the success of a laser welding system.

Some rules of thumb for fit-up tolerances for a butt weld are shown in Fig. 15-7a. The tolerances are stated in terms of material thickness t. The parts should be positioned with a gap no larger than $0.15t$ and the misalignment should be less than $0.25t$. The edges should be square and straight.

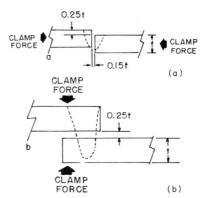

Fig. 15-7   Tolerances for butt welding (a) and lap welding (b) for material of thickness t. (From S. L. Engel, *Laser Focus*, p. 44 (February 1976).)

Compressive clamping as indicated is desirable. Adherence to these restrictions should lead to a fully welded butt joint. Figure 15-7b shows tolerances for a lap joint.

In most laser welding systems, the laser power is monitored continuously to ensure that it remains within proper limits. In setting up a weld schedule, the optimum power output of the laser is determined experimentally. Then, during the welding operation, a portion of the beam may be picked off, for example, by a transparent slide placed in the beam at an angle. Most of the beam passes through the slide and goes on to the workpiece. A small fraction is reflected from the slide and is sent to a detector, which has previously been calibrated to monitor the laser output in this fashion. If the laser output varies from the desired value, the output of the detector may be fed directly back to the laser power supply to restore proper operation.

The size of the heat-affected zone can be small if the material is thin. The narrowness of the heat-affected zone is one attractive feature of laser welding. Since the heat must be conducted through the material, it is difficult to make laser welds that are narrower than the thickness of the workpiece. In thin materials, seams of a few thousandths of an inch wide are attainable.

Successful applications of laser welding with continuous Nd:YAG or CO_2 lasers at power levels of hundreds of watts have been achieved in a variety of diverse areas. The applications have included hermetic sealing of relay cans containing temperature-sensitive elements, welding of caps to vanes in gas turbine engines, formation of contour welds on inconel turbine blades, welding of end caps on zircalloy nuclear fuel rods, and welding of hardened steel edges to steel band saw blades for meat cutting.

An example of a seam weld in a small relay can containing delicate components is shown in Fig. 15-8. The weld, made with a repetitively pulsed CO_2

Fig. 15-8   Photograph of a seam weld of monel to cupronickel in a relay can. The weld was made at 6 in./min with a pulsed CO_2 laser operating at 500 W peak power. (Photograph courtesy of J. M. Webster.)

laser with a peak power of 500 W, is neat. The localization of the heating allows the weld to be made without damage to the internal structures.

B.   Welding with Multikilowatt Continuous Lasers

We have so far emphasized the limited penetration for laser welding in seam welding by lasers with average powers below about 1500 W. With the development of high-power continuous CO_2 laser systems around 1970, deeper penetration welding became possible. So-called deep penetration welding by multikilowatt CO_2 lasers was announced. Figure 15-9 shows penetration depth for welding as a function of laser power [5]. The dashed curve is for welding of stainless steel at levels below 1000 W; as noted before, penetration is limited. Around 1000–1500 W, new physical phenomena became important, and the curve bends upward, as shown by the hatched

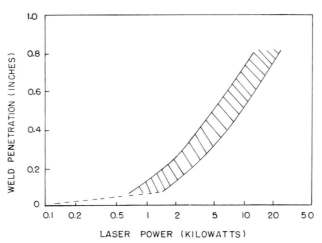

Fig. 15-9 Weld penetration as a function of laser power. Data compiled for a number of metals shows onset of deep penetration welding near 1 kW. (From E. M. Breinan, C. M. Banas, and M. A. Greenfield, United Technologies Research Center Report R75-111087-3, 1975.)

region, which represents data for several metals, including stainless steel, titanium, and aluminum. In this region, the penetration increases approximately as the 0.7 power of the laser power and reaches a depth of 0.8 in. in stainless steel at 20,000 W.

In deep penetration welding, the beam delivers energy to the surface more rapidly than it can be removed by thermal conduction. A hole is then drilled into the material, and laser energy is deposited in the material throughout the depth of the hole. Thus, much deeper penetration of the energy into the workpiece is possible, because one does not rely on thermal conduction of energy deposited at the surface. When the beam is moved across the surface, the hole is translated through the material, with molten material flowing around the hole and filling in behind. Thus, one can form a seam weld with high ratio of depth to width. The process is shown schematically in Fig. 15-10.

Such a multikilowatt laser can produce narrow, deep welds much like electron beam welds. The laser beam can be transmitted through the atmosphere. Tooling can be simplified with a laser because vacuum or magnetic chucks may be utilized, unlike electron beam welding. In addition, a mirror can be used to deflect the laser beam close to the desired area, allowing otherwise inaccessible joints to be welded.

Welds with penetration of 0.8 in. in a single pass have been achieved. Photographs of cross sections of weld beads from bead-on-plate tests on 304 stainless steel at a power of 15,000 W at three different speeds are shown in

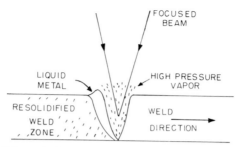

Fig. 15-10   Schematic diagram of the progress of deep penetration welding.

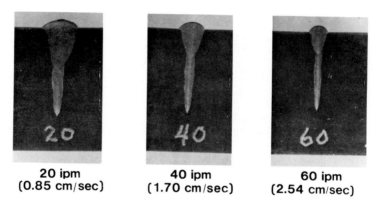

| 20 ipm | 40 ipm | 60 ipm |
| (0.85 cm/sec) | (1.70 cm/sec) | (2.54 cm/sec) |

Fig. 15-11   Weld beads produced in 1-in.-thick plates of 304 stainless steel in bead-on-plate tests with a 15 kW CO_2 laser, at the indicated values of linear transport speed. (From E. M. Breinan, C. M. Banas, and M. A. Greenfield, United Technologies Research Center Report R75-111087-3, 1975.)

Fig. 15-11. At 20 in./min, the penetration depth was about 0.7 in. As speed increased, the weld bead became narrower, but the penetration still exceeded 0.6 in. at a speed of 60 in./min.

The shape of the fusion zone is determined by a number of characteristics. One controllable characteristic is the location of the focal point relative to the surface of the workpiece. In one experiment [6], a 16,000 W beam was moved away from the focal point by 0.060 in. Maximum penetration occurred when the beam was focused slightly inside the surface. Penetration decreased when the beam was focused on the surface or deeper in the workpiece. In some of the cross sections, the fusion zone had an hourglass shape. The position of the waist of the hourglass is controlled by the location of the focal point.

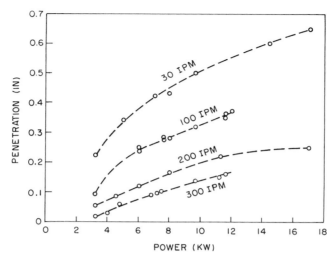

Fig. 15-12 Weld penetration as a function of CO_2 laser power in 304 stainless steel, at the indicated values of weld speed. (From E. V. Locke and R. A. Hella, *IEEE J. Quantum Electron.* **QE-10**, 179 (1974).)

Figure 15-12 summarizes data on the penetration achieved in 304 stainless steel at various rates of beam traverse, for values of laser power up to 17 kW. The penetration depths are large enough to provide many possibilities for practical welding.

Figure 15-13 shows a comparison between hard vacuum electron beam welding data and laser welding data at a 10 kW level in 304 stainless steel. At speeds from 50 to 500 in./min, the laser penetration is approximately 70% of that obtained with the electron beam. The electron beam penetration continually increases at lower speeds, whereas penetration by the laser appears to saturate. This behavior may be associated with production of an absorbing plasma. Multikilowatt CO_2 lasers thus can compete effectively with electron beam welding, without the need for any vacuum system.

Deep penetration welding with highly efficient use of the laser energy is possible. Figure 15-14 shows results for several metals welded with a laser power of 5000 W. The results are expressed in terms of a dimensionless welding speed, Vb/κ, where V is the seam welding rate, b the width of the fusion zone, and κ the thermal diffusivity. The power input is a dimensionless parameter, $P/hK\Delta$, with P the laser power, h the weld penetration, K the thermal conductivity, and

$$\Delta = T_m - T_0 + H/c \qquad (15.1)$$

with T_m the melting temperature of the metal, T_0 the ambient temperature, H the latent heat of fusion, and c the specific heat. The top line in this plot

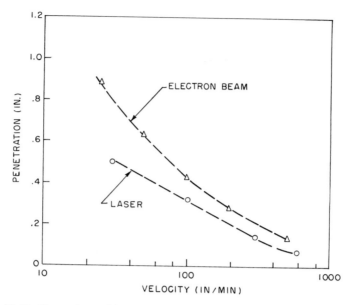

Fig. 15-13 Comparison of laser welding and vacuum electron beam welding in 304 stainless steel at the 10 kW level. The curves show penetration depth as a function of weld velocity. (From E. V. Locke and R. A. Hella, *IEEE J. Quantum Electron.* **QE-10**, 179 (1974).)

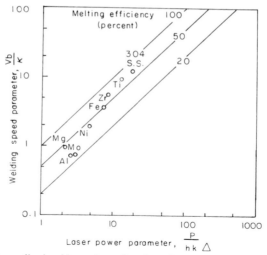

Fig. 15-14 Normalized weld speed as a function of normalized laser power. The straight lines indicate values of efficiency for the use of the incident laser energy in melting. The symbols are defined in the text. (From E. M. Breinan, C. M. Banas, and M. A. Greenfield, United Technologies Research Center Report R75-111087-3, 1975.)

expresses 100% efficient use of the laser energy to heat the material to its melting temperature and to provide the latent heat of fusion. This is true because when

$$Vb/\kappa = P/hK\Delta \tag{15.2}$$

we have

$$
\begin{aligned}
P &= (Vb/\kappa)hK\Delta = (Vb/\kappa)hK(T_m - T_0 + H/c) \\
&= (Vb\rho c/K)hK(T_m - T_0 + H/c) \\
&= (Vbh)\rho[c(T_m - T_0) + H]
\end{aligned}
\tag{15.3}
$$

Since Vbh is the volume of material welded per unit time, and

$$\rho[c(T_m - T_0) + H]$$

is the energy required per unit volume to melt the metal, the above relation states that all the laser power is used to melt metal with no losses. Lines representing several percentage efficiencies are indicated. We see that for several metals, efficiencies of 70% are attainable in practice. Near the left of the figure, metals with high thermal conductivity are welded with somewhat lower efficiency, probably because of losses due to thermal conduction out of the weld zone. The results in Fig. 15-14 are attainable only if initial surface reflectivity is lowered, as discussed in Chapter 13.

The strength of such laser welds can be high, comparable to the strength of the base metal. Some results for the ultimate tensile strength of laser welded nickel and iron-base high-temperature alloys are presented in Table 15-1 (cf. [5]). The ductility of the welds was also satisfactory.

TABLE 15-1
Ultimate Tensile Strength in Laser Welded High-Temperature Alloys

Alloy designation	Base metal	Major alloying elements	Thickness (mm)	Ultimate tensile strength in laser weld (lb/in.^{2})	Ultimate tensile strength in parent metal (lb/in.^{2})
PK33	Ni	Fe, Cr, Ti,	1	142,000	147,000
PK33		Al, Co, Mo	2	155,000	151,000
C263	Ni	Fe, Mn, Cr, Si	1	145,000	141,000
C263			2	141,000	143,000
N75	Ni	Fe, Cr, Ti,	1	104,000	117,000
N75		Co, Mo	2	106,000	117,000
M152	Fe	Ni, Cr, Mo	1	129,000	133,000
M152			2	126,000	126,000

Examples of single- and dual-pass laser welds in arctic pipeline steel are shown in Fig. 15-15. The welds were made with 12 kW of laser power, at 25 in./min for the single pass and at 60 in./min for the dual pass. Cross-weld tensile tests on this material resulted in failure in the base metal.

Currently a number of applications are under study involving multi-kilowatt systems. These include automatic underbody welding in the automotive industry, welding of thin sheet titanium and aluminum for construction of high-speed surface vessels, and welding of gas line for use in arctic environments.

An example in which the multikilowatt lasers have reached production status is in pure lead welding, for production of lead–acid batteries [7]. Positive plates of a cell are cast from pure lead in a circular geometry with six connecting tabs spaced around the diameter. The tabs have dimensions $5.25 \times 80 \times 22.5$ mm. Adjacent tabs are seam welded by a 2 kW CO_2 laser beam moving at 8.5 cm/sec. The presence of the lugs is sensed by means of a pair of semiconductor injection lasers. Light from the injection lasers is reflected from the lugs and detected by photodetectors. The laser welder is in production and is capable of completing about nine batteries per hour, each with 106 welds total. This replaces four conventional bonders, each of which could produce nine or ten batteries per shift.

Experimental welding has been examined at continuous power levels as high as 100 kW [8]. The results indicated that the weld depth scales as the 0.7 power of the laser power. Stainless steel of 2-in. thickness could be welded

Single pass
(12.0 kW−25 ipm) (1.06 cm/sec)

Dual pass
(12.0 kW−60 ipm (2.54cm/sec)
each pass)

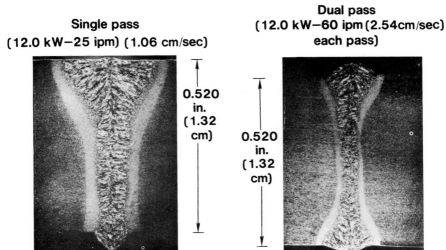

0.520 in. (1.32 cm)

0.520 in. (1.32 cm)

Fig. 15-15   Laser welds in X-80 Arctic pipeline steel. (From E. M. Breinan, C. M. Banas, and M. A. Greenfield, United Technologies Research Center Report R75-111087-3, 1975.)

at 63 in./min, at 80 kW. The use of such high powers for practical industrial welding probably requires additional development.

C. Spot Welding

Laser spot welding has become a versatile tool for welding of small components. It is particularly useful in cases where localized heating is desired, such as welding near glass-to-metal seals or interconnection of leads in delicate heat-sensitive semiconductor circuitry. In such cases, it can be competitive with conventional techniques like resistance welding.

Many of the early investigations of laser welding in the 1960s emphasized spot welding, because the lasers available at the time were ruby or Nd:glass lasers with relatively low pulse repetition rates. Such lasers were best suited for spot welding and there are many papers in the literature describing such applications. Seam welding with such lasers was possible by overlapping spots, but usually was impractical because of the slow speed along the seam, dictated by the relatively low pulse repetition rates available.

When spot welding is required, single pulses from ruby or Nd:glass lasers may be suitable. Spot welding between a nickel tab and a nickel alloy post on a transistor header has reached a routine production status. Good metallurgical joints with small heat affected zones are produced in this welding process.

Figure 15-16 shows data on the required output of a ruby laser to weld copper and nickel wire, as a function of wire diameter [9]. These data are for

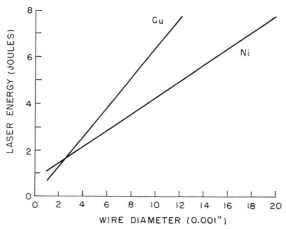

Fig. 15-16   Ruby laser energy required to spot weld wires, as a function of wire diameter, for a 3-msec laser pulse. (Prepared from data presented by W. A. Murray, *Laser Focus*, p. 42 (March 1970).)

ruby laser pulses of duration around 3 msec, focused with simple lenses. As we have mentioned before, Q-switched lasers are not suitable for welding, because the pulse duration (~ 100 nsec) is too short to permit good penetration of the molten zone through the metal. Even lasers with pulse durations around 1 msec are better suited for drilling than welding. The data in Fig. 15-16 represent full strength welds, i.e., the strength of the weld equals the strength of the original wire. If the laser output is too low, the wire will not be completely melted. If the laser output is too high, the surface will be partly vaporized and the strength of the weld will be reduced. The lines in Fig. 15-16 show the laser output required for optimum welding in copper and nickel wires. The reflectivity of copper is relatively high. For wires of metal with lower reflectivity, e.g., nickel or steel, a lower laser output is required for a good weld.

The earliest studies of laser applications in electronics usually involved welding of microcircuits. Ruby lasers welded a variety of electronic components. Specific items included welding of leads to circuit boards and to tabs on silicon chips, welding fine wires to thin films, and welding flatpacks to circuit boards. Laser welds were shown to be true fusion welds, with both elements melted and fused. Technical feasibility was demonstrated for such processes, particularly for interconnecting microelectronic assemblies. Laser welding is readily compatible with present electronic packing and appears particularly applicable for beam-leaded devices. Since such leads are now attached under a microscope, one may boresight the laser beam to the focus of the same microscope, and deliver the laser energy to the center of the microscope field of view.

In order to weld a number of leads simultaneously to integrated circuits or circuit boards, the leads can be positioned automatically and the laser pulse spread by appropriate optics to weld all the leads for a given circuit simultaneously. In one example, four welds are made simultaneously. A ruby laser pulse was spread into a line by a cylindrical lens. Four leads with a Ti–Pt layer facing the laser were welded to a thin gold film on an alumina substrate. The laser light was absorbed well by the beam leads, so that they melted, but the more reflective gold films were not damaged.

One significant investigation of laser spot welding involved use of a high pulse repetition rate Nd:YAG laser to weld insulated copper wires to terminal posts [10]. This investigation had two unusual features which show the capabilities of laser welding for practical production situations. The first is that laser radiation was used both to remove the insulation and to produce the weld. Second, the weld joints were evaluated in real time using stress wave emission techniques.

The welds involved copper wires 312 μm in diameter with polyurethane insulation. The wires were wound on a terminal shoulder which protected the wire from direct irradiation. This geometry eliminated necking of the wire in the welding process. The laser beam was focused on the terminal head, causing the head to flow and weld to the wire. The terminal posts were of monel, which has a higher vaporization temperature than copper. When the monel at the interface between the terminal and the wire reached its melting point, the copper wire, because of its lower melting temperature, melted and developed a fusion zone. The laser pulses were 3.5 msec in duration, contained 8–10 J of energy, and were focused to a spot size of 0.125 cm on the surface of the terminal.

The welds could be evaluated during formation by measurement of the stress wave emission. Stress wave emission arises from the release of energy as a solid material undergoes plastic deformation. The laser spot-welding operation was found to release stress wave emission energy, which propagated as elastic waves through the material. This energy was detected at the material surface using piezoelectric crystals (e.g., lead zirconate titanate) in contact with the surface of the material at some distance from the weld region. An empirical relation was observed between the characteristics of the wave packet of the stress wave emission and the strength of the laser welds. This allowed identification of the weld sequences in which failure mechanisms occurred, such as insulation burning and misalignment of the beam on the terminal. From electronic analysis of the characteristics of the wave packet, a go–no-go signal would be generated by use of a comparator that compared the signature of the envelope to a known characteristic signal. Thus, acceptable welds could be identified in real time.

It is worthwhile to consider two examples of laser spot welding that reached production status. These cases may give an insight into considerations for practical use of lasers. In both cases, the laser was used to solve problems that existed with conventional methods. In the first example, a ruby laser welds a nicket tab to a nickel alloy post on a transistor; this eliminates problems of cracking at a nearby glass-to-metal seal. With conventional tweezer resistance welding, the heat often caused cracking, and the weld often failed in vibration tests. The laser weld showed good intermixing of the constituents, so that the mechanical strength was high. The cracking problem was eliminated because of the small heat-affected zone, and good reliability was achieved in vibration tests. The weld is made with a ruby laser pulse with a 7.5 J energy and 3 msec duration. Production rates of six transistors per minute have been attained [11].

In the second example, precious metal (Paliney 7) wires are welded to phosphor bronze spring members. The weld is difficult to achieve by resistance welding because of the large difference in electrical resistivity of the members. With laser welding, the electrical properties are not significant for the quality of the weld. The weld is made by a 6 J, 3 msec ruby laser pulse, focused by a 43 mm focal length lens. The workpiece was 0.075 in. from the focal plane, allowing enough defocusing so that the laser pulse overlapped both wires and made both welds simultaneously. The heat-affected zone is about 0.002 in. thick, so that the temper of the phosphor bronze spring is not affected. The weld has strength at least as great as that of the original 0.010 in. wires. Figure 15-17 shows a photograph of this weld [11].

These examples are chosen to show some of the considerations involved in successful utilization of lasers. The emphasis is on small parts in areas where technical problems exist; the improved reliability and increased yield available with laser processing can make lasers economical.

Many spot welding operations are performed with resistance welders. This technique is considered to be of relatively low cost. However, one analysis [12] indicated that the cost per weld of an operation using automated laser spot welding can be less than for manual operation of resistance welding equipment, when a large number of welds is required. The initial cost of the automated laser welder is higher than resistance welders needed for the same thruput. However, increased labor costs for the operation of the resistance welders overcame the difference. The analysis assumed a three-year depreciation of the equipment and a thruput of 10^7 spot welds per year. It included

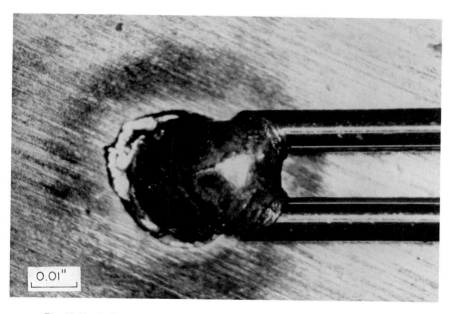

Fig. 15-17 Paliney 7 wires welded to a phosphor bronze spring by a 6-J ruby laser pulse. (From F. P. Gagliano, R. M. Lumley, and L. S. Watkins, *Proc. IEEE* **57**, 114 (1969). Photograph courtesy of F. P. Gagliano.)

allowance for operators' salaries, overhead, and consumable materials. The results of the analysis indicated that automated laser welding under these conditions could be performed for 0.45 cents per weld whereas the resistance welding cost was 0.63 cents per weld. In addition, laser welding offers improved weld quality and elimination of thermal damage and part distortion. This example shows that laser techniques are now capable of competing with conventional techniques, even on an economic basis in some cases.

D. Heat Treatment

A high-power CO_2 laser can be used for heat treating of metals. The laser irradiates the surface and causes very rapid heating of a thin layer of material near the surface. When the laser beam is moved to a different area on the surface, the heat deposited in the thin layer will be quickly conducted away and the heated area will cool rapidly. This can be regarded as a quenching of the surface region. This procedure yields an increase in the hardness of the surface layer.

The basic idea of laser heat treating is to harden the surface of materials that may be hardened through transformation. Lasers are not efficient for heating large volumes of material but they can rapidly raise the surface temperature of a material. When the laser energy is removed, the surface of the material, particularly of a metal, will cool extremely rapidly. In some materials, such quenching can harden the surface through metallurgical transformation. This is equivalent to heat hardening the surface areas which the laser has traversed, and may be selectively used to provide a hard surface in areas subject to wear.

Laser heat treating has sometimes been called "frustrated welding." The beam from a high-power laser is scanned over a surface, similarly to welding, but the motion of the beam is too fast to allow melting to begin. Therefore, control of the scan speed is extremely critical. The depth of the heat-treated region is determined by the thermal conductivity of the material. The surface temperature and the penetration depth can be varied by adjusting the beam power, the focusing, and the speed at which the laser beam is scanned across the surface.

An example of some results is shown in Figs. 15-18 for 1040 carbon steel. (a) shows the change in grain structure near the surface. (b) shows the hardness as a function of depth. In this example, the CO_2 laser beam had dimensions 1.7 cm wide by 1.0 cm in the direction of motion. The beam was moved over the surface at 120 in./min. The change in the material near the surface is apparent. The hardness (Rockwell C) of the base material was

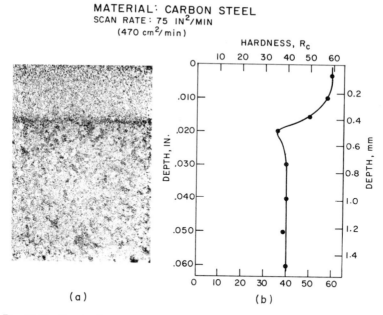

MATERIAL: CARBON STEEL
SCAN RATE: 75 IN²/MIN
(470 cm²/min)

(a)

(b)

Fig. 15-18 Results from heat treating of 1040 carbon steel with a multikilowatt CO_2 laser. (a) shows the grain structure as a function of depth, with the hardened surface layer visible. (b) shows the hardness (Rockwell C) as a function of depth. (From E. V. Locke and R. A. Hella, *IEEE J. Quantum Electron.* **QE-10**, 179 (1974).)

approximately 40, but in the 0.025 cm thick layer near the surface, the hardness reached a value of 57. The rapid heating and quenching by the multikilowatt CO_2 laser beam led to a significant increase in the hardness of the material. Deeper layers of surface hardening could be obtained by sweeping at a slower rate and shallower layers by increasing the sweep rate.

For covering broader areas, overlapping passes of the beam could be used. However, during the overlapping, there is some annealing in the adjacent bands and therefore a decrease of hardness in some areas. Various different scanning methods have been developed to reduce this problem. One may traverse linearly using a defocused beam. Alternatively, one may use a more finely focused beam and oscillate at high frequency in a direction transverse to the direction of traverse. When the laser beam is defocused and traversed linearly, the spatial profile of the beam, which is not uniform, will produce a greater depth of heating in areas where the power density is greatest. This will produce a hardened zone of uneven depth. In order to compensate this, multiple overlapping passes can be used. This still does not provide a completely uniform case depth. Transverse high-frequency oscillation of the laser beam is capable of providing a more even case depth, although it adds more complexity to the system. A uniform case depth may be achieved under conditions when, at some depth in the metal, the oscillating pattern is transformed into an essentially constant heat flow by the time constants associated with thermal conduction.

A solution which is perhaps preferable is to use a uniform beam profile to cover the entire width of the workpiece in one pass. Some high-power lasers have been operated with a "top hat" beam profile, i.e., a relatively flat top with rapid decrease in power at the edges. This beam profile cannot be focused well for applications such as cutting, but it is ideal for covering a broader area with relatively uniform distribution of power.

Some workers in this area analyze beam quality by means of a so-called Harth diagram. The Harth diagram plots curves of constant case depth. The ordinate is the incident energy density delivered by the laser and the abscissa is the feed rate for the workpiece. The resultant curves can be a valuable tool for assessment of beam quality and for quality control of the operation.

All metals that can be transformation hardened can be case hardened with a laser. Some examples are carbon steels, tool steels, and cast irons. In particular, the transformation of austenite steel into martensite steel has been carried out. The response of various steels apparently increases with increasing carbon content. Measured values of hardness (Rockwell C) have exceeded 60 for steels containing 0.75% carbon.

Practical hardening depths are perhaps around 0.030 in. with a multi-kilowatt CO_2 laser, with surface coverages up to 20 in.^{2}/min, depending on the material and the desired case depth. Some tradeoffs between rate of treating and case depth for heat treating of gray cast iron are shown in Fig. 15-19 for a 1500 W CO_2 laser focused to a spot size of 0.05 in. and oscillated transversely at the indicated rates [13]. At the right edge of the figure the

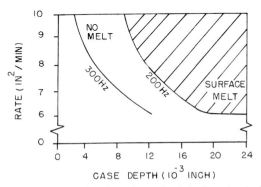

Fig. 15-19   Treating rate as a function of case depth for laser hardening of cast iron with 1500 W from a CO_2 laser. The curves show rates at the indicated frequencies of transverse oscillation of the beam. (From S. L. Engel, *American Machinist*, p. 107 (May 1976).)

traverse is slow enough that surface melting occurs. Thus, the regions of hardening with or without surface melting may be established by the frequency of oscillation, with other considerations being equal.

It is important that the laser energy be absorbed well at the surface. Since the only lasers capable of heat treating at reasonable rates are CO_2 lasers, the high reflectivity of metallic surfaces in the infrared can be a problem. The absorption of the beam can be increased by a thin layer of highly absorbing material on the surface. The coating must be very thin, perhaps around 0.001 in., to allow effective transfer of the energy to the workpiece. If the coating becomes too thick, penetration of the heat is reduced and case depth is lessened.

A variety of materials have been employed as coatings. One source [14] describes the best three coatings, in order, as polycrystalline tungsten, cupric oxide, and graphite. Because of difficulty in applying good coatings of these materials in an industrial environment, perhaps a suitable compromise could be manganese phosphate. This coating can have high absorption, although not quite so high as the other materials named. However, it can be applied with reasonable ease in an industrial plant.

To summarize, good coatings are available to increase the absorption of metallic surfaces to the 80–90% range, allowing good results in surface hardening of transformable metals. In order to make the application really practical in industry, a high-efficiency coating for the surface is needed.

An example of a hardened band produced on the surface of a gray cast-iron sample is shown in Fig. 15-20. The hardened band was produced by a CO_2 laser operating at 1000 W. The beam was scanned across the sample at 10.2 mm/sec and was rapidly oscillated by a mirror in a transverse direction to widen the beam and to achieve a more uniform intensity distribution. The hardened width is 6.8 mm and the hardened depth is 0.53 mm. The hardened zone consists of a matrix of fine martensite containing flake graphite. The hardness in the hardened zone is 58–62 (Rockwell C).

Laser hardening is applicable to geometries with irregular shapes, corners, grooves, lines, and gear teeth. Applications in the automotive industry, where wear on selected surfaces can be a great problem, has been

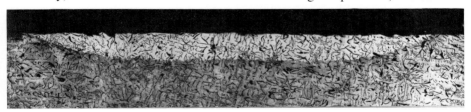

Fig. 15-20   Hardened zone produced on the surface of gray cast iron by a 1000 W CO_2 laser. The width of the hardened zone is 0.53 mm. (Photograph courtesy of M. J. Yessik.)

investigated extensively. The hardening of areas such as engine cylinder bores, valve guides, camshafts, gears, shaft journals, valve seats, and steering gear housings have been investigated. Areas where laser hardening is applicable are in applications which require high hardness and relatively shallow case depths on selected surface areas, with no dimensional distortion.

Competing techniques for hardening metal in a thin surface layer include chemical treating processes. Materials such as carbon or nitrogen are diffused into the surface of the metal in a furnace. Another method is induction heating to produce surface heating in a layer with thickness of the order of a fraction of a millimeter to several millimeters. For applications in which the depth of hardening does not have to be extremely great and where the part geometry is uneven, so that the induction process is not very suitable, laser hardening provides a very good alternative. Laser hardening can generate a substantial increase in surface hardness at a high rate with minimum distortion of the part.

Several other laser techniques of surface treating are under experimental development. They include shock hardening, surface alloying and surface glazing. In shock hardening, a short pulse of very high peak power density ($\gtrsim 10^9$ W/cm^2) is incident on the surface. The rapid surface vaporization and accompanying laser-supported absorption wave drive a shock front into the material. Passage of the shock front produces work hardening. Work with Q-switched Nd:glass lasers has produced hardening in 7075-T73 aluminum alloy [15]. The ultimate tensile strength and the 0.2% yield strength were both increased after laser shock hardening. The low cycle fatigue properties were also improved.

Surface alloying involves spreading of a powder containing alloying elements over the surface to be hardened. Then the laser beam is passed over the surface. The laser typically would be a multikilowatt continuous CO_2 laser. The treating rate would be somewhat slower than for the transformation hardening discussed earlier, because some surface melting is desired. The powder and a thin surface layer on the workpiece melt and intermix. After the passage of the beam, the surface resolidifies. A thin layer containing the alloying elements remains, with hardness greater than the untreated material. In one study [16], the elements carbon, boron, chromium, and silicon were mixed in powder form on the surface of a valve seat. The powder was melted and alloyed into the surface by an 18 kW CO_2 laser beam. The treating and surface hardening took less than 1.5 sec for a valve seat.

Laser glazing [17] also involves some surface melting. As the beam from a multikilowatt continuous CO_2 laser is scanned over the surface, a thin melt layer is produced under proper conditions of laser power density and traverse speed. The interior remains cold. After the beam moves on, resolidification occurs very rapidly. The surface layer is quickly quenched.

The result of the process is surface microstructure with unusual and possibly useful characteristics. The grain size near the surface is very small, because of the high quench rate. The surface structure can often appear glassy; hence the name laser glazing. This technique is also capable of producing surfaces of increased hardness. Laser glazing is applicable both to metals and ceramics. It appears to be controllable and reproducible.

These last three techniques, shock hardening, surface alloying, and laser glazing, are less well developed than the transformation heat treating described earlier in this section. They are processes which offer great promise for future exploitation.

E.   An Example of Laser Welding Capability

We shall now present an example which shows dramatically how the localized heating capability of a laser was used to solve an unusual problem. It involved hermetic sealing of a miniature detonator, with a laser weld made within 0.040 in. of a temperature-sensitive primary explosive. The hermetic sealing of miniature explosive devices such as detonators, small igniters, actuators, etc., has been extremely difficult to achieve on a mass production basis with a high degree of reliability and simplicity.

The particular detonator was in the form of a cylindrical can 0.10 in. in diameter by 0.25 in. long. The hermetic sealing was accomplished using a CO_2 laser to weld a glass-to-metal-sealed header to a can containing the high explosive. The can and header were 320 stainless steel. The laser beam making the weld could be precisely positioned and the welding spot diameter controlled to as little as 0.005 in. in diameter and 0.003 in. in depth. These latter features localized the heat generated by the weld to a tightly controlled area. The weld itself was produced by operation of a CO_2 laser in the repetitively pulsed mode. The pulse energies fell in the 0.1–0.3 J/pulse range and the duration was approximately 40 μsec. The beam produced individual, overlapping spots of metal melt which resolidified to form a homogeneous weld area. A seam weld made by such a laser is shown in Fig. 15-21 [18]. In this photograph, the header is at the top and the seam lies around the periphery of the can, just below the header. The can containing the explosive is at the bottom. The width of this seam is approximately 0.015 in. The marks made by the overlapping of pulses are visible.

The welds are hermetic and enabled the device to withstand such adverse environments as a 28-day temperature and humidity cycling test, 13,000 g shock in any direction, thermal shock alternating continuously between $-65°$F and $+160°$F, and a number of other tests. Other methods of making the weld, including electron beam welding, were not successful, because of the proximity of the weld to the temperature sensitive explosive.

Fig. 15-21   Seam weld around the periphery of a microdetonator can. The weld was made with a repetitively pulsed CO_2 laser. (From J. F. Ready, "Effects of High Power Laser Radiation." Academic Press, New York, 1971.)

A cross section of the laser weld is shown in Fig. 15-22. The weld was etched in equal parts potassium cyanide (10%) and ammonium persulfate (10%), followed by diluted Marble's reagent. The weld nugget has dimensions of 0.010 by 0.003 in. The appearance of the weld nugget indicates the rapidity of the process. The material appears to have been turbently intermixed. The grain size in the weld nugget is very small. The weld cross section does not appear so neat as those of welds made by other techniques. However, the strength of the weld joint is high, and the required hermetic seal is provided.

There have now been many practical applications in which laser welding has reached production status. Laser welding competes with many established techniques, such as arc welding, resistance welding, and electron beam welding. In many cases, the laser process will offer some advantage which is important for the particular application. Among the advantages of laser welding are the following:

(1)   No material contacts the workpiece, so that there is no contamination.

(2)   Laser welding may be performed in the atmosphere. This is in contrast to electron beam welding, which must be done in vacuum.

(3)   The heat-affected zone is very small. This is especially important in cases where a weld must be made near a heat-sensitive element, such as a glass-to-metal seal.

(4)   Welding can be performed in otherwise inaccessible areas, for example for repairs inside an enclosed vacuum tube.

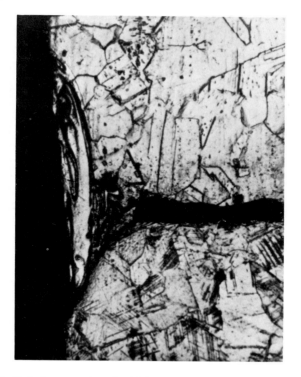

Fig. 15-22 Etched cross section of weld shown in Fig. 15-21. The width of the area shown in the photograph is 0.015 in.

Because of these advantages, lasers are gaining more and more acceptance for routine production welding tasks. Much practical welding still involves small parts because of the limited penetration of heat for all lasers except the expensive multikilowatt CO_2 lasers.

REFERENCES

[1] S. L. Engel, *Laser Focus*, p. 44 (February 1976).
[2] J. H. Mason and J. H. Wasko, Society of Manufacturing Engineers Technical Paper MR74–955, 1974.
[3] H. L. Marshall, *in* "Industrial Applications of High Power Laser Technology" (J. F. Ready, ed.), Vol. 86. Proc. Soc. Photo-Opt. Instrum. Eng., Palos Verdes Estates, California, 1976.
[4] S. L. Engel, Society of Manufacturing Engineers Technical Paper MR75-579, 1975.
[5] E. M. Breinan, C. M. Banas, and M. A. Greenfield, Laser Welding—The Present State of the Art, United Technologies Research Center Report R75-111087-3, June 1975.
[6] E. V. Locke and R. A. Hella, *IEEE J. Quantum Electron.* **QE-10**, 179 (1974).
[7] F. P. Gagliano, Society of Manufacturing Engineers Technical Paper MR74-954, 1974.

[8]   C. M. Banas, at the Society of Manufacturing Engineers Conference on Lasers in Modern Industry, Chicago, January 25–27, 1977.
[9]   W. A. Murray, *Laser Focus*, p. 42 (March 1970).
[10] M. A. Saifi and S. J. Vahaviolos, *IEEE J. Quantum Electron.* **QE-12**, 129 (1976).
[11] F. P. Gagliano, R. M. Lumley, and L. S. Watkins, *Proc. IEEE* **57**, 114 (1969).
[12] S. R. Bolin, Proc. Tech. Program Electro-Opt. Syst. Design Conf., Anaheim, November 11–13, 1975.
[13] S. L. Engel, *American Machinist*, p. 107 (May 1976).
[14] S. L. Engel, at the Society of Manufacturing Engineers Conference on Lasers in Modern Industry, Chicago, January 25–27, 1977.
[15] B. P. Fairand *et al.*, *J. Appl. Phys.* **43**, 3893 (1972).
[16] R. Hella, at the 1975 IEEE/OSA Conference on Laser Engineering and Applications, Washington, D.C., May 28–30, 1975.
[17] E. M. Breinan, B. H. Kear, and C. M. Banas, *Phys. Today*, p. 44 (November 1976).
[18] J. F. Ready, "Effects of High Power Laser Radiation." Academic Press, New York, 1971.

SELECTED ADDITIONAL REFERENCES

A. Seam Welding—Subkilowatt Levels

R. J. Conti, Carbon Dioxide Laser Welding, *Weld. J.*, p. 800 (October 1969).

W. W. Duley, "CO_2 Lasers, Effects and Applications," Academic Press, New York 1976, Chapter 6.

J. H. Mason and J. H. Wasko, Spot and Continuous Welding with Solid State Lasers, Soc. of Manufacturing Eng. Tech. Paper MR74-955 (1974).

J. F. Ready, "Effects of High Power Laser Radiation," Academic Press, New York, 1971, Chapter 8.

J. G. Siekman and R. E. Morijn, The Mechanism of Welding with a Sealed-Off Continuous CO_2-Gas Laser, *Philips Res. Rep.* **23**, 367 (1968).

I. P. Shkarofsky, Review on Industrial Applications of High Power Laser Beams III, *RCA Rev.* **36**, 336 (1975).

Staff Report, State of the Art in Continuous Nd:YAG Laser Welding, *Metal Progr.*, p. 61 (November 1970).

W. C. Thurber, Laser Welding: Theory, Status and Prospects, *Record Symp. Electron. ᵗ ᵢn, Laser Beam Technol.*, *11th* San Franciso Press, San Francisco, California, 1971.

J. M. Webster, Welding at High-Speed with the CO_2 Laser, *Metal Progr.*, p. 59 (November 1970).

B. Welding with Multikilowatt Continuous Lasers

E. L. Baardsen, D. J. Schmatz, and R. E. Bisaro, High Speed Welding of Sheet Steel with a Carbon Dioxide Laser, *Weld. J.* **54**, 227 (1973).

W. W. Duley, "CO_2 Lasers, Effects and Applications," Academic Press, New York, 1976, Chapter 6.

F. P. Gagliano, Application of Lasers in Battery Welding, *in Proc. Ind. Appl. High Power Laser Technol.* Vol. 86 (J. F. Ready, ed.), Soc. of Photo–Opt. Instrum. Eng., Palos Verdes Estates, California, 1976.

E. V. Locke, E. D. Hoag, and R. A. Hella, Deep Penetration Welding with High Power CO_2 Lasers, *IEEE J. Quantum Electron.* **QE-8**, 132 (1972).

E. V. Locke and R. A. Hella, Metal Processing with a High Power CO_2 Laser, *IEEE J. Quantum Electron.* **QE-10**, 179 (1974).

G. E. Overstreet, High-Power CO_2 Lasers in Industry, *Ind. Res.*, p. 40 (May 1974).

D. T. Swift-Hook and A. E. F. Gick, Penetration Welding with Lasers, Welding Research Supplement, *Weld. J.*, p. 492-S (November 1973).

C. Spot Welding

J. E. Anderson and J. E. Jackson, Theory and Application of Pulsed Laser Welding, *Weld. J.* **44**, 1018 (1965).

M. I. Cohen and J. P. Epperson, Application of Lasers to Microelectronic Fabrication, *in* "Electron Beam and Laser Beam Technology, Advances in Electronics and Electron Physics" (L. Marton and A. B. El-Karh, eds.), Academic Press, New York, 1968.

M. I. Cohen, Material Processing, *in* "Laser Handbook" (F. T. Arecchi and E. O. Schulz-Dubois, eds.), North Holland Publ., Amsterdam, 1972.

L. P. Earvolino and J. R. Kennedy, Laser Welding of Aerospace Structural Alloys, *Weld. J.*, p. 127-S (March 1966).

R. H. Fairbanks and C. M. Adams, Laser Beam Fusion Welding, *Weld. J.* **43**, p. 97-S (March 1964).

F. P. Gagliano, R. M. Lumley, and L. S. Watkins, Lasers in Industry, *Proc. IEEE* **57**, 114 (1969).

F. P. Gagliano and V. J. Zaleckas, Laser Processing, *in* "Lasers in Industry" (S. S. Charschan, ed.), Van Nostrand–Reinhold, Princeton, New Jersey, 1972.

K. J. Miller and J. D. Nunnikhoven, Production Laser Welding for Specialized Applications, *Weld. J.* **44**, 480 (1965).

A. J. Moorhead, Laser Welding and Drilling Applications, *Weld. J.*, p. 97 (February 1971).

J. F. Ready, "Effects of High Power Laser Radiation," Academic Press, New York, 1971, Chapter 8.

A. O. Schmidt, I. Ham, and T. Hoshi, An Evaluation of Laser Performance in Microwelding, *Weld. J.*, p. 481-S (November 1975).

D. Heat Treatment

E. M. Breinan, B. H. Kear and C. M. Banas, Processing Materials with Lasers, *Phys. Today*, p. 44 (November 1976).

A. L. Bryant and F. A. Koltuniak, Heat Treating with a High Energy Laser, Soc. of Manufacturing Eng. Tech. Paper MR74-964 (1974).

S. L. Engel, Heat Treating with Lasers, *Am. Mach.*, p. 107 (May 1976).

J. W. Hill, M. J. Lee, and I. J. Spaulding, Surface Treatments by Laser, *Opt. Laser Technol.*, p. 276 (December 1974).

E. V. Locke and R. A. Hella, Metal Processing with a High Power CO_2 Laser, *IEEE J. Quantum Electron.* **QE-10**, 179 (1974).

Photon Sources, Inc., Application Guidelines, Laser Surface Hardening (September 1975).

M. Yessik and D. J. Schmatz, Laser Processing in the Automotive Industry, Soc. of Manufacturing Eng. Tech. Paper MR74-962 (1974).

M. Yessik and R. P. Scherer, Practical Guidelines for Laser Surface Hardening, Soc. of Manufacturing Eng. Tech. Paper MR75-570 (1975).

E. Example of Laser Welding Capability

J. A. Barrett, C. R. Hargreaves, E. Bernal G., and J. F. Ready, The Laser Welding of Miniature Explosive Detonators, *Proc. Symp. Explosives Pyrotechnics, 7th*, Franklin Inst. Res. Lab., Philadelphia, Pennsylvania, (September 8–9, 1971).

CHAPTER 16

APPLICATIONS FOR MATERIAL REMOVAL: DRILLING, CUTTING, ELECTRONIC COMPONENT FABRICATION, MARKING

In the last chapter we described applications involving laser-induced heating and melting. For those applications one generally tries to avoid vaporization and material removal. In this chapter we emphasize applications for material removal. Generally this means operation at higher levels of laser power density than for welding. It therefore often implies finer focusing of the beam than would be needed for welding or heat treating.

A. Laser-Induced Material Removal

We have already discussed some of the considerations involving laser interactions and material removal in Chapter 13. In this section we shall emphasize some aspects of material removal as they relate to applications.

There are several different types of lasers that are of interest for material removal. The list of lasers is slightly different from those that are used for welding. This is because one may employ higher levels of power. Thus, the CO_2 TEA laser and the repetitively Q-switched Nd:YAG laser can be useful for material removal. As Chapter 13 describes, the useful range of pulse duration is shorter than that required for welding, but the power density must be somewhat higher. Continuous lasers are less often used for hole drilling or material removal than for welding and heat treating. This is because thermal conduction can spread the heat out too much in the work-piece, unless pulses are used.

The effects of material parameters are also important. The properties that are most important are thermal diffusivity, latent heat of vaporization, and reflectivity. The amount of material that can be removed is limited by the latent heat of vaporization, assuming that the material is all vaporized.

Let us perform a sample calculation. The maximum depth D of material that can be vaporized is given by

$$D = E_0/A\rho \left[c(T_B - T_0) + L \right] \qquad (16.1)$$

where c is the specific heat, T_B the boiling temperature, T_0 the ambient temperature, L the latent heat of vaporization per unit mass, ρ the density, E_0 the laser energy, and A the area to which the beam is delivered. This equation arises from considerations of conservation of energy. It assumes that all the energy in the laser pulse is employed in raising material to the vaporization temperature and in providing the latent heat of vaporization. As such, it represents a maximum amount of material removal. If we take a typical metal, for example, aluminum, the values are

$$\rho = 2.7 \text{ gm/cm}^3, \qquad c = 0.97 \text{ J/gm-deg},$$
$$T_B - T_0 = 2447°C, \qquad L = 10900 \text{ J/gm}.$$

If we consider a ruby laser with an output of 10 J, focused to a spot with an area of 10^{-3} cm^2, use of the above equation leads to a maximum hole depth of 0.28 cm. In practice, the hole depths are influenced by several other factors. Loss of laser energy by reflection from the surface and by thermal conduction into the interior tends to decrease the hole size. Flushing of molten material, which is not completely evaporated, tends to increase the hole depth. The considerations outlined here indicate the general limitations, however.

Data on the amount of material removed by laser pulses are shown in Fig. 16-1 [1]. These curves show an interplay between latent heat and thermal diffusivity. The curves for aluminum and iron cross. At relatively low laser power densities, loss of heat into the interior of the workpiece is an important consideration. Because the thermal diffusivity of aluminum is higher than that for iron, this loss is more severe for aluminum. The depths of holes in the aluminum are less than those in iron. As the laser power density is increased, the material rises to its boiling temperature faster. There is less time for heat to be conducted away, and the loss due to thermal conduction is less serious. In this regime, the dominant factor is the latent heat of vaporization. Since iron has higher latent heat of vaporization than aluminum, more aluminum is vaporized than iron.

The dominant process in material removal by lasers can often be simple vaporization, with absorption of the laser energy at a continually retreating surface. The vaporized material can simply diffuse away into the surrounding atmosphere, without further interaction with the beam. The upper limit for material removed is given by Eq. (16.1). Thermal conduction can remove heat and thus reduce the amount of material removed. The data in

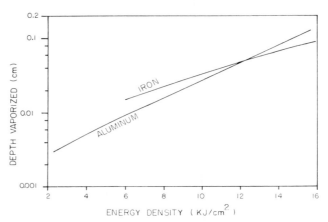

Fig. 16-1 Experimental values of depth vaporized by a 700 μsec pulse from a Nd:glass laser, as a function of energy density on the target. (These data were adopted from measurements of mass removal by V. B. Braginskii, I. I. Minakova, and V. N. Rudenko, *Sov. Phys. Tech. Phys.* **12**, 753 (1967).)

Fig. 16-1 were obtained under conditions where thermal conduction was significant.

Other mechanisms can also be operative, as we have indicated in Chapter 13. Mechanisms which can increase the amount of material removal are flushing of liquid material [2] and particulate emission because of subsurface explosions [3]. Absorption of laser radiation in the blowoff material can shield the surface and reduce the amount of material removal. Thus, if one is investigating an application for laser material removal, an experimental study of the process must be conducted to learn the optimum regime of laser parameters.

The considerations on reflectivity of the surface are similar to those for welding. Laser energy can be lost by reflection from the surface. In the infrared, where surface reflectivity is high, it becomes more difficult to get effective vaporization. CO_2 lasers can accomplish vaporization with reasonable efficiency because the surface reflectivity does drop after the surface begins to be disrupted, and the energy can be coupled into the workpiece reasonably well. This phenomenon has already been described in Chapter 13.

In Fig. 16-2, we see the percentage of light that is absorbed by several materials, as a function of wavelength. This figure emphasizes nonmetallic materials, because laser drilling is often applied to such materials (e.g., ceramics), which are hard to drill by conventional techniques. This figure is plotted in terms of the amount of light absorbed, which is the property of most importance. Losses can occur by reflection or, for the nonmetallic materials, by transmission. Thus, we specify a thickness for the nonmetallic

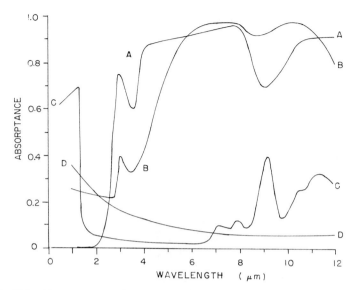

Fig. 16-2 Absorptance as a function of wavelength for several materials. (a) Pyrex glass, 0.125 in. thick. (b) Alumina (AD-5), massive. (c) Silicon, *p*-type, 1.68 mm thick. (d) Iron, massive.

materials. These data depend on many factors, including purity and surface preparation. For example, the small peak near 9.5 μm on the curve for silicon originates from an oxygen impurity in the silicon. Thus, the data in Fig. 16-2 should be taken as being representative of typical conditions, and not as being exact for all circumstances.

The data in Fig. 16-2 indicate how to choose a laser to drill a particular material. For example, glass and plexiglas are quite transparent in the visible and near infrared so that a Nd:YAG laser will not deposit much energy. However, they are highly absorbing at the CO_2 laser wavelength. Such behavior is also typical of the infrared absorption of organic materials, most of which have high absorption near the CO_2 laser wavelength of 10.6 μm. Ceramics are also generally highly absorbing near 10 μm, as the figure illustrates. Silicon has higher absorption in the near-infrared, and a Nd:YAG laser will deliver more energy to silicon than a CO_2 laser. The curve for iron is typical of metals, which have high reflectivity in the far-infrared.

B. Hole Drilling

Laser hole drilling in metals may be useful in a variety of areas, such as production of tiny orifices for nozzles and controlled leaks, apertures for electron beam instruments, and pinholes for optical work. Production of fine orifices with dimensions smaller than 0.001 in. in thin metallic plates is easy. Precise, close patterns can be produced with no heating of adjacent areas and no contamination. Very small specialized holes have been made on a custom basis in samples which conventional methods could not handle well. The heat-affected zone around the hole can be small, typically 0.001–0.002 in. As in welding, material properties influence the amount of metal removed. Pulse lengths in the range of several hundred microseconds (or even less) are generally used. Rarely would a laser pulse length longer than about 1 msec be used for this application.

Table 16-1 shows depths of laser-produced holes for two cases—a short high-power pulse and a longer, lower power pulse. These results were obtained using the same Nd:glass laser [4]. In one case, the Nd:glass laser operated without a Q-switch to produce 600 μsec pulses. In the other case, the laser was Q-switched to produce 44-nsec pulses. The results indicate that the energy supplied by the short pulses is less effective in material removal, as we have already discussed.

TABLE 16-1
Depths of Laser-Produced Holes

Material	Depth for 10^9 W/cm², 44 nsec pulse (cm)	Depth for 5000 J/cm² 600 μsec pulse (cm)
Stainless steel	0.00011	0.061
Brass	0.00025	0.078
Aluminum	0.00036	0.078
Copper	0.00022	0.090
Nickel	0.00012	0.058

The development of the cross section of a hole vaporized in aluminum sheet by a ruby laser pulse focused by a 30 mm focal length lens is shown in Fig. 16-3 [5]. As the energy increases, both hole depth and diameter increase. The figure shows the typical shape of a hole produced by a laser. The hole has a definite taper. The entrance has largest diameter, and the diameter decreases deeper in the hole.

If repeated pulses are delivered to the same area on the target, deeper holes may be drilled. Figure 16-4 shows the relation between the depth of the hole and the number of pulses for a sapphire target [6]. Hole drilling in materials such as sapphire is of interest because such materials are very hard and are difficult to drill by other techniques. The holes represented in Fig. 16-4 were drilled by a ruby laser beam focused by a lens with a 30 cm focal

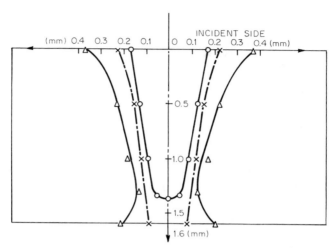

Fig. 16-3 Change in configuration of holes in a 1.6 mm thick aluminum plate as laser output increases. The normal pulse ruby laser beam was focused with a 30 mm focal length lens. Points are as follows: 0, 0.36 J; x, 1.31 J; Δ, 4.25 J. (From T. Kato and T. Yamaguchi, *NEC Res. Dev.* **12**, p. 57 (October 1968).)

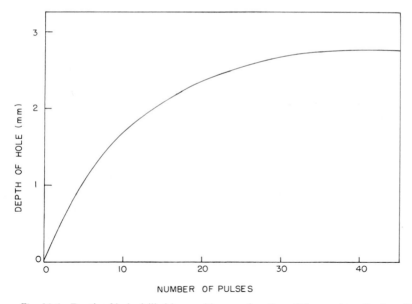

Fig. 16-4 Depth of hole drilled in sapphire as a function of the number of pulses. The data are relevant to 2 J ruby laser pulses focused with a 3 cm focal length lens. (Data from S. Shimakawa, Japan Society for Laser Material Processing.)

length. The laser pulse delivered 2.0 J of energy and lasted 1.5 msec. Fig. 16-4 shows how the hole depth increased with the number of pulses up to about 30. After 30 pulses, the depth of the hole did not increase further.

Hole drilling in metals may be done with pulsed ruby or Nd:glass lasers or with pulsed CO_2 lasers. An example of a cross section of a typical hole in a massive sample appears in Fig. 16-5 [7]. This hole was produced in brass by a single 5-J pulse from a Nd:glass laser focused with a simple lens. Defects illustrated by this hole include rough sides, lack of roundness, taper of the hole, and offset of the center line of the hole from the normal to the surface. These defects are difficult to eliminate and may disqualify the laser from applications where perfection of the hole is important.

Efforts to reduce the taper in laser-drilled holes have centered on good control of the spatial and temporal profile of the beam. The laser should operate in a good TEM_{00} mode, with no "hot spots." In this way, some improvements in shape of laser-drilled holes have been made, but it has not been possible to eliminate the taper completely.

Depths of a few millimeters may be pierced through metallic plates, either by a single pulse of high energy or by a series of pulses of lower energy delivered to the same spot. The maximum depth for which laser piercing appears to be practicable is around 0.5 in. In order to reduce roughness of the sides of the hole pierced through a metal plate, one may blow gas through the hole as it is being formed.

Practical hole drilling with lasers suffers from a number of limitations. The limited depth of penetration due to the limited amount of laser energy is

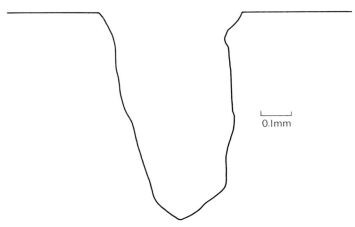

0.1mm

Fig. 16-5   Cross section of hole produced in massive brass by a 5 J pulse from a Nd:glass laser focused on the target surface by a 31 mm focal length lens. (From J. F. Ready, "Effects of High Power Laser Radiation." Academic Press, New York, 1971.)

one drawback. It might be thought that by using a continuous CO_2 laser and allowing it to strike the same area for a long time, this could be over-come. However, the heat then becomes spread over a much larger area and much of the advantage of using lasers is lost. A second limitation occurs in the recondensation of material around the entrance to the hole, which provides a crater-like lip for the hole. This arises because material vaporized and ejected from the hole can easily recondense on the first cool surface that it strikes. The lips around such holes can be removed easily, but this means that there is one more step in the hole drilling process. A third limitation is the general roughness of the walls of the hole. This is usually undesirable. In hard, brittle materials, such as diamond, a final finishing step can be used to smooth the holes that are originally pierced by a laser.

There are a number of advantages associated with lasers which render them useful for hole drilling operations. These advantages include the follow-ing: There is no contact of any material with the workpiece, and therefore no contamination. One can drill hard, brittle materials which are difficult to drill in any other way, e.g., ceramics and gem stones. The heat-affected zone can be very small. Finally, one can produce very small holes. Holes of diameter around 0.002 cm can easily be drilled in thin materials.

Laser machining is readily adaptable to nonmetallic materials. Typically, more material may be evaporated from nonmetallic materials than from metals by a laser with given properties, at least partly because of the difference in thermal conductivities. Organic materials are especially easily vaporized. The CO_2 laser can be employed to advantage because many nonmetallic materials have very high absorption at the wavelength of 10.6 μm. A CO_2 laser operating repetitively pulsed at an average power level around 100 W can effectively drill many nonmetallic materials. Ruby lasers and repetitively pulsed or continuous Nd:YAG lasers have also been used.

A ruby laser can drill narrow deep holes through relatively thick brittle materials, e.g., ceramics or silicon. Holes in alumina with aspect ratios (the ratio of the depth to diameter) exceeding 25 have been produced. This is a greater aspect ratio than is available with conventional techniques for small holes in ceramics. Ceramics as thick as 0.125 in. have been pierced by re-peated ruby laser pulses.

Holes may be produced close together in a chip of brittle material or near the edge of a chip. Lasers can be used for production of holes for attachment of leads to an integrated circuit substrate. Drilling of such materials by conventional techniques is troublesome. Use of a laser elimi-nates problems such as drill bit breakage and wear.

There are several practical applications of lasers in hole drilling that have reached commercial status. One is the use of lasers for drilling holes in gem stones. The original holes in diamonds used to produce diamond dies for extrusion of wire are routinely pierced by lasers. This is a profitable application; the hole drilling was previously very tedious and costly, because diamond is so hard. The initial holes can be pierced by one or a few pulses from a ruby laser. In order to make the sides of the holes smooth, final finishing is still required, but the laser-produced holes are well suited for a final lapping process. This is an application where the laser offers significant advantages over competing techniques. Drilling of holes in ruby watch jewels has become a standard technique.

The ability of lasers to drill holes through ceramics is finding use in the electronics industry. Pulsed CO_2 lasers are commonly used to drill holes in alumina substrates for circuit boards. Figure 16-2 shows that ceramics absorb well at the CO_2 laser wavelength. Laser drilling in hard, high-temperature fired alumina ceramic is attractive because drilling ceramics by conventional means is not simple; it usually requires diamond-tipped, hardened-steel drill bits. Small holes, less than 0.25 mm in diameter, are difficult for current tool technology. Breakage of the drill often results when the thickness of the ceramic is greater than the diameter of the hole to be drilled. Laser drilling easily produces small holes in this brittle material without danger of fracturing.

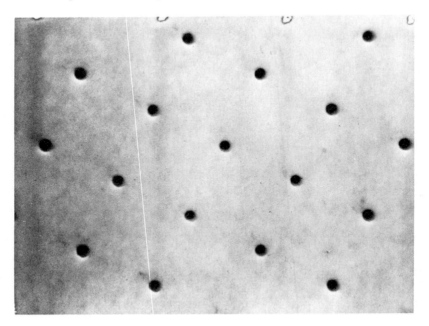

Fig. 16-6 Array of holes drilled in 0.7 mm thick alumina ceramic by a CO_2 laser. The diameter of the holes is 0.3 mm.

When a pulsed CO_2 laser is used with alumina of thickness around 0.025 in., typical hole diameters may be around 0.005–0.010 in., with tolerances of 0.0005 in. For very small holes, the ruby laser or Nd:YAG will be better because of the shorter wavelength. Frequency-doubled Nd:YAG lasers operating at 0.53 μm have been used for fine drilling. Holes as small as 0.0002 in. have been pierced in thin material. Very small holes are produced only with some sacrifice of uniformity. For piercing, the laser pulse length will often be tailored to fall in the 300–700 μsec range.

Because of the brittleness of ceramics, conventional hole drilling is usually performed on the material in the "green" state before it is fired. When it is fired, the dimensions may change. Laser drilling is performed after firing, and thus eliminates this problem.

Automated laser systems to drill complicated patterns of holes in material about 0.6 mm thick are becoming common. The system includes fixturing to hold the ceramic, a stepping motor to move the fixture to predetermined positions at which the holes are to be drilled, and a numerical control (in effect, a small computer) to turn the laser on and off and to drive the motor so that holes are drilled in the correct positions. The operator merely inserts a ceramic plate and starts the operation. The entire process is then automated to yield as a final result the desired pattern of holes in the ceramic. Figure 16-6 shows an array of such laser-drilled holes.

Hole drilling of metals by lasers has not become as common as hole-drilling in ceramics and similar materials. This is because of the limitations on laser hole drilling (taper, roughness, etc.) which we described previously, and because conventional methods for drilling holes in metals are often satisfactory. For materials such as ceramics, which are hard, brittle, easy to fracture and which wear out drill bits quickly, conventional methods often are not satisfactory. Laser drilling of such materials often provides many advantages.

C. Cutting

Lasers may be used to cut or shape a wide variety of nonmetallic materials. By cutting we mean vaporization of material along a line, so as to separate the pieces. Shaping involving scribing or fracturing will be discussed later. Cutting of metals at practical speeds often involves use of a gas jet to burn through. Such oxygen-assisted cutting greatly enhances the laser's capabilities.

Most cutting operations have involved CO_2 lasers, because of their capability for producing high values of continuous power. Cutting of nonmetallic materials proceeds easily in many cases. Because of the higher reflectivity and higher thermal conductivity, cutting of metals generally requires more power.

1. *Cutting of Nonmetallic Materials*

For direct cutting by vaporization, the most favorable combination appears to be the continuous CO_2 laser operating on organic materials. Organic materials practically all have high absorption at 10.6 μm, but many are somewhat transparent at the ruby and Nd:YAG laser wavelengths.

A continuous CO_2 laser emitting 100 W is adequate for many cutting applications. Cutting has been demonstrated on a variety of materials such as paper, rubber, plastics, ceramics, glass, cloth, and wood. As examples, slitting of $\frac{1}{16}$ in. thick acrylic sheet has been demonstrated at a rate of 84 ft/min with a CO_2 laser which leaves the edges clean and slightly fire-polished, and cutting of light-weight paper at a rate in excess of 5 m/sec has been shown to yield cut edges of quality comparable to a mechanical cut.

In order to produce a clean, unscorched edge in laser cutting of flammable materials, it may be necessary to blow a jet of inert gas around the cut.

Advantages of CO_2 laser cutting include lack of tool wear, reduced loss of materials around the cut, and sometimes improved edge quality. Laser cutting of cloth, carpeting, and paper can be economically competitive with mechanical cutting in some cases.

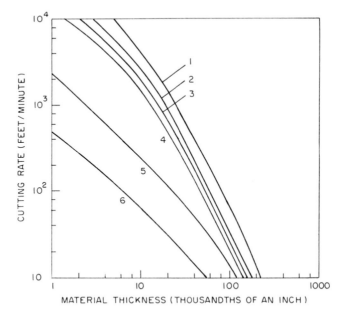

Fig. 16-7 Cutting rate for various plastics cut by a 375 W continuous CO_2 laser, as a function of thickness: curve 1, cellulose; curve 2, acrylic epoxy; curve 3, ABS plastic; curve 4, vinyl; curve 5, polycarbonate; curve 6, poylethylene. (Data from Photon Sources, Inc.)

TABLE 16-2
Cutting of Nonmetallic Materials with CO_2 Lasers

Material	Thickness (in.)	Laser power (W)	Cutting rate (in./min)	Gas assist
Soda lime glass[a]	0.08	350	30	Air
Quartz[b]	0.125	500	29	Not stated
Glass[c]	0.125	5000	180	Yes
Ceramic tile[d]	0.25	850	19	Yes
Plywood[a]	0.19	350	209	Air
Plywood[d]	0.25	850	213	Yes
Hardboard[a]	0.15	300	36	Air
Hardboard[d]	0.19	850	180	Yes
Synthetic rubber[d]	0.1	600	189	None
Plywood[c]	1	8000	60	None
Glass[c]	0.375	20,000	60	None
Boron epoxy composite[c]	0.32	15,000	65	None
Fiberglass epoxy composite[c]	0.5	20,000	180	None

[a] J. E. Harry and F. W. Lunau, *IEEE Trans. on Ind. Appl.* **IA-8**, 418 (1972).
[b] B. Feinberg, *Mfr. Engineering and Development* (December, 1974).
[c] G. K. Chui, *Ceramic Bull.* **54**, 515 (1975).
[d] Data from Ferranti, Ltd.
[e] E. V. Locke, E. D. Hoag, and R. A. Hella, *IEEE J. Quantum Electron.* **QE-8**, 132 (1972).

Figure 16-7 shows data on cutting of various plastics with a 375-W CO_2 laser. The cutting rates can be high enough to have practical application. Table 16-2 presents some other examples of cutting rates measured for a variety of nonmetallic materials. The table is relevant for cutting with CO_2 lasers. The numbers presented in Table 16-2 are a selection of experimental results, obtained under a variety of conditions, and do not necessarily represent optimum values [8, 9, 10, 11].

Let us consider one example of cutting which has reached practical production status [12]. This is a laser cloth cutter which cuts material in a single ply. The laser cloth cutting system consists of two 350 W continuous CO_2 lasers, an optical system to move and focus the beam, two x–y positioning systems, a minicomputer with magnetic tape unit, an aluminum honeycomb cutting surface with vacuum hold down and exhaust and an automatic cloth spreader. The laser cloth cutting system is completely automated under the control of the computer. The material to be cut is spread on the cutting surface. The material is moved to the proper location under the laser head and cutting is performed by the focused laser beam. The laser is stationary and the beam is moved along the perimeter of the patterns by a series of lenses and mirrors. The beam is moved around the periphery of a piece to be cut out of the cloth, under the control of the computer. The diameter of the focal area at the cloth is approximately 0.01 in. After the goods are cut, a conveyer advances and moves the cut parts onto a pickup device. At the same time, uncut material is advanced into position for cutting. Cutting velocities to 40 in./sec are possible, but the average speed is generally less. If one assumes approximately 2000 in. cutting is necessary to cut out the pieces for a man's suit, the laser system with two cutting heads can cut material for 40 to 50 suits per hour.

The material is cut in a single thickness at a time. The advantage of the laser system resides in its speed. The speed makes it possible to cut one thickness of cloth at a time. This allows greater accuracy of cutting than would be possible if the material were cut in multiple plys. The accuracy of the cutting eliminates the necessity of allowances at the end of the cloth. This allows better utilization of the entire width of the cloth and results in significant saving of material. Another advantage of the speed is the capability for quick response to fashion change. The manufacturer can make short production runs with no increase in cutting costs. This can significantly reduce inventories.

The edge quality achieved by laser cutting is good. Material with a high synthetic fiber content gives an excellent seal on the edge of the cut. The edge has less tendency to unravel than edges cut by conventional techniques. The cost of cutting has been estimated as $1.09 per suit. This cost compares favorably with costs for conventional cutting techniques. It gives the manufacturer the flexibility of fast response to fashion change and the reduction of inventory. The laser cloth cutting process was investigated in comparison to a large number of conventional cutting methods, including shears, saber blades, circular knives, turrent chop blades, plasma cutting, and ultrasonics. Laser cutting applies advanced automated technology in an industry which has typically been highly labor-intensive.

Laser cutting has also been applied in production to making ruled dies. Slots may be cut in wood to accommodate the steel rules used in cutting or creasing cardboard cartons. A typical die may be of $\frac{3}{4}$ in. thick maple. Laser cutting of a complicated pattern to hold the steel rules may be completely automated. The cuts are straighter and more uniform than can be produced by a workman with a jig saw. This technique provided a new automated tool for the production of cartons which are used widely in the food industry. Figure 16-8 shows a commercial automated die maker based on a CO_2 laser.

Another successful application of laser cutting has been sawing of alumina substrates of complex shape. (Simple separation of alumina into rectangular substrates with straight line cuts can perhaps be performed more simply by scribing, to be described in Section E.) Many integrated circuit applications require alumina to be cut in a complicated shape to

Fig. 16-8   Photograph of diemaker based on CO_2 laser. (Photograph courtesy of Coherent Radiation, Inc.)

serve as a substrate. Conventional methods were expensive and slow. However, sawing the substrates with a CO_2 laser has allowed cutting of odd forms from standard rectangular blanks, at substantial cost savings.

One system [13] uses a 250 W continuous CO_2 laser to cut 0.6 mm thick alumina in complicated contours, at linear velocities of 1.2 cm/sec. A pantograph system is used to guide the beam around the desired contour. A typical pattern may be cut from the parent blank in less than 30 sec, with good detail and excellent edge quality for the cut. The cost is reduced as compared to conventional cutting. Moreover, the inventory of shapes needed can be much reduced; one needs only the rectangular blanks.

2. Cutting of Metals

In the early history of lasers, cutting of metals developed slowly. The high thermal conductivity and high reflectivity of metallic surfaces demanded high levels of average power in order to cut at a reasonable rate. Developments in high-power lasers have remedied this situation. Also, requirements on the laser power have been reduced by the development of the technique of oxygen-assisted cutting. Because of the need for high average power, CO_2 lasers are most often used in cutting applications.

Laser cutting of metals or of other materials can be enhanced by the use of an oxygen jet. Most of the energy to produce the cut is supplied by an exothermic chemical reaction between the metal and the oxygen. The laser heats the material to its ignition point. The actual cutting occurs by the reaction of the metal with oxygen. Significant thickness of metal can be cut rapidly with a few hundred watts of laser power. The oxygen jet can increase the cutting rate by approximately 40% over the rate when an inert gas is present.

In a typical arrangement, the nozzle for the oxygen jet is coaxial with the laser beam. The CO_2 laser beam is focused by a suitable infrared-transmitting lens (e.g., germanium or alkali halide) through the nozzle and onto the workpiece. Typically, a tapered nozzle of 0.05–0.10 in. diam is used with oxygen supplied at pressures from 15–30 psi.

Figure 16-9 presents data on the cutting rate as a function of laser power for various thicknesses of low carbon steel [14]. The cutting rates are high enough to be of practical industrial significance. In oxygen-assisted cutting, there is a lower limit of cutting rate below which the material burns excessively. Figure 16-10 shows data on the upper and lower limits for cutting of low carbon steel with a 1 kW CO_2 laser. The lower limit represents the minimum rate at which burning of the material does not occur.

Table 16-3 illustrates cutting rates achievable with oxygen-assisted cutting by CO_2 lasers operating at levels of some hundreds of watts [8, 15, 16, 17]. Cutting rates would be much lower without the oxygen assist. The data in Table 16-3 are intended to illustrate some representative values for

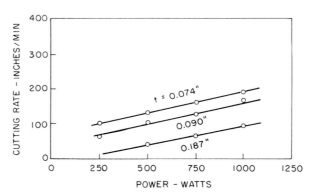

Fig. 16-9   Cutting rate for low carbon steel (thickness *t*) with oxygen assist. (From S. L. Engel, Society of Manufacturing Engineers Technical Paper MR74-960, 1974.)

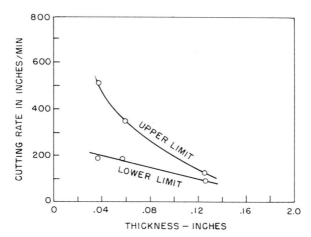

Fig. 16-10   Maximum and minimum cutting rates for low carbon steel at 1 kW laser power. (From S. L. Engel, Society of Manufacturing Engineers Technical Paper MR74-960, 1974.)

experimental cutting rates and are not necessarily optimized values. Moreover, quoted cutting rates may differ somewhat when values from different sources are compared, even for nominally similar conditions. Uncontrolled variables and possible differences in procedures for power measurement may account for the differences. The numbers in Table 16-3 define the order of magnitude capability for oxygen-assisted laser cutting. The values are high enough to be of commercial interest.

Oxygen-assisted cutting is most suitable for reactive metals, such as titanium. All work reported to date has used high-power continuous or phase-continuous CO_2 lasers. The operation yields cuts with narrow heat-affected zones and with small cut widths, as compared to conventional cutting.

When one has multikilowatt CO_2 lasers available, the oxygen assist may no longer be necessary. One may cut with the brute force of the laser beam alone. In fact, a nonreactive shield gas (e.g., helium) may be used to avoid burning. The thickness that can be cut and the cutting speeds are impressive. A tabulation of published values on some parameters for typical cases of multikilowatt CO_2 laser cutting appears in Table 16-4 [12, 18]. These numbers do not necessarily represent optimum values, but are indicative of results that have been obtained in the laboratory. Figure 16-11 shows an example of cutting a pattern of square holes in 0.02 in. thick stainless steel with a CO_2 laser.

Oxygen-assisted laser cutting appears to offer considerable savings in cutting certain types of metal, because of reduced material loss and reduced

TABLE 16-3
Oxgen-Assisted Cutting of Metals with CO_2 Lasers

Material	Thickness (in.)	Laser power (W)	Cutting rate (in./min)
Titanium (pure)[a]	0.02	135	600
Titanium (pure)[b]	0.67	240	240
Titanium alloy (6A14V)[a]	0.05	210	300
Titanium alloy (6A14V)[a]	0.088	210	150
Titanium alloy (6A14V)[a]	0.25	250	110
Titanium alloy (6A14V)[a]	0.39	260	100
Titanium alloy[c]	0.2	850	130
Carbon steel[a]	0.125	190	22
Stainless steel (302)[d]	0.012	200	90
Stainless steel (410)[c]	0.063	250	50
Stainless steel (410)[c]	0.110	250	10
Stainless steel[b]	0.012	350	170
Rene 41[c]	0.020	250	80
Rene 41[c]	0.050	250	20
Zircalloy[a]	0.018	230	600

[a] Booklet, "CO_2 Applications," Coherent Radiation, Inc., Palo, Alto, California 1969.

[b] J. E. Harry and F. W. Lunau, *IEEE Trans. on Ind. Appl.* **IA-8**, 418 (1972).

[c] Data from Ferranti, Ltd.

[d] W. W. Duley and J. N. Gonsalves, *Opt. and Laser Technol.*, p. 78 (April 1974).

[e] J. R. Williamson, *in* "Industrial Applications of High Power Laser Technology" (J. F. Ready, ed.), Vol. 86. Proc. Soc. Photo-Opt. Instrum. Engineers, Palos Verdes Estates, California, 1976.

TABLE 16-4
Metal Cutting with Multikilowatt CO_2 Lasers

Material	Thickness (in.)	Power (kW)	Speed (in./min)
Aluminum (6061)[a]	0.5	10	40
Aluminum[b]	0.04	3.0	250
Aluminum[b]	0.125	4.0	100
Aluminum[b]	0.25	3.8	40
Aluminum[b]	0.50	5.7	30
Inconel[a]	0.5	11	50
Stainless steel (300)[b]	0.125	3.0	100
Steel (low carbon)[b]	0.125	3.0	100
Steel (low carbon)[b]	0.66	3.0	45
Titanium[b]	0.25	3.0	140
Titanium[b]	1.25	3.0	50
Titanium[b]	2.0	3.0	20

[a] E. V. Locke and R. A. Hella, *IEEE J. Quantum Electron.* **QE-10**, 179 (1974).

[b] D. W. Wick, Society of Manufacturing Engineers Technical Paper MR75-491, 1975.

Fig. 16-11 Laser cutting of stainless steel. Squares are being cut in 0.020 in. thick steel with a high-power CO_2 laser. (Photograph courtesy of S. L. Engel and GTE Sylvania).

machining required to clean the cut. Some comparative estimates of costs appear in Table 16-5, for cutting techniques for titanium sheet. The initial capital cost is high for a complete automated laser system. The total cost per foot includes operating costs and secondary cutting costs. Laser cutting offers substantial savings in cost per foot, particularly for thicker material.

Laser cutting has reached full production status in the aerospace industry, for cutting of metals like titanium. One 250-W CO_2 laser system cuts 100,000 ft/yr of material [17]. Sophisticated multiaxis systems are employed for cutting contoured surfaces. Cutting of aluminum has apparently not reached production status in the aerospace industry, but is undergoing intensive scrutiny [19].

TABLE 16-5
Cutting Cost Comparison for Titanium[a]

Method	Capital investment ($)	Total cost ($/ft) for material thickness		
		1/8 in.	1/4 in.	1/2 in.
Band saw	2000	1.52	2.18	4.15
Laser system (250 W)	96,000	0.71	1.19	12.8

[a] From J. R. Williamson, see [17].

D. Electronic Component Fabrication

Lasers are used in the electronic industry in many ways. We have already discussed laser welding and its application in fabrication of electronic components. We shall now describe how laser vaporization can be employed in either fabricating circuit elements directly or in trimming components to desired tolerances. (Scribing of circuit substrates, perhaps one of the most significant laser applications in the electronics industry, will be described in Section E of this chapter.)

Lasers can form circuit elements directly by vaporizing portions of thin film elements on transparent substrates to form the circuit geometry. Resistors and capacitors may be formed directly in thin film structures using laser vaporization. The technique has been demonstrated using a repetitively Q-switched Nd : YAG laser with a rotating mirror Q-switch [20]. The laser emitted a continuous train of pulses with peak power of 1 kW, a pulse duration about 200 nsec, and a pulse repetition rate around 400 Hz. A microscope objective focused the beam to a spot about 8 μm in diameter. When the beam was swept across the workpiece, lines were vaporized in thin metallic films on quartz and sapphire substrates. Lines about 1 μm wide in gold films and 0.4 μm wide in nichrome films were produced, at speeds of the order of 0.1 in./sec. There was relatively little damage to the substrate. The lines were free of metallic debris.

The same laser machined a pattern of fine lines to make gap capacitors in tantalum–chrome–gold thin films. Capacitors could be produced with capacitance around 20 pF, reproducible to good tolerances. The capacitance could be controlled by varying the gap width. The laser could be employed to produce a complete thin film circuit with resistive and capacitive components. Lasers could also be used to produce photomasks for photoetching of microcircuits or for fabricating masks suitable for depositing circuit patterns on a chip.

A complete system for circuit element fabrication, based on an X–Y pattern generator, has been demonstrated [21]. The system employs galvanometer mirrors to scan the beam from a Q-switched Nd:YAG laser. The galvanometer mirrors are controlled by a digital computer. The system has been used to generate conductor paths, masks, and reticle patterns by direct evaporation of metallic films.

Lasers can also trim electronic components by selective material removal. Fast and accurate trimming of resistors is being done to close tolerances with lasers. Trimming of resistors means adjusting the value of a film resistor by making a cut partially through it. The initial resistor is fabricated so that the value of resistance is lower than desired; a cut is then made part way across the resistor. The value of the resistance is monitored, and when the desired value is reached, the laser beam is shut off.

A repetitively pulsed laser with a short pulse duration can trim film resistors by removing material so as to change the dimensions of the resistor or by vaporizing lines part way across the resistor. With real-time measurement of resistance and feedback control of the laser, this technique provides a rapid method for trimming resistors to tolerances better than 0.1%. Although there may be some amount of post-trimming drift of the resistance, the amount of the drift can be reproducible. The resistor is trimmed to a value from which it will drift to the desired value. Trimming to within 0.1% for the final value thus appears practicable.

The laser may be particularly useful in trimming thick film resistors. As compared to abrasive trimming, the laser offers the advantage of cleanliness. Laser trimming can be held to accurate tolerances by real-time monitoring of the resistance. The trimming can be stopped quickly when the desired resistance is reached. The edges of the trim are sealed as the cut is made.

The best laser for resistor trimming may be a repetitively Q-switched Nd:YAG laser; the repetitively Q-switched CO_2 laser is also a reasonable choice. Q-switched operation is desirable to remove the material quickly, without thermal penetration into the substrate. The limited vaporization capability of the Q-switched laser is not a drawback when one is working with films. The Nd:YAG laser offers the advantages of smaller cut width, greater absorption in metallic films, and smaller absorption in the substrates, as compared to the CO_2 laser. Thus, most trimming of thin and thick film resistors has been done with repetitively Q-switched Nd:YAG lasers. Figure 16-12 shows trimming cuts made by a Nd:YAG laser in thick film resistors.

Various geometries for cutting have been studied; one popular geometry is the L-shaped cut. One cuts part way across the resistor, then changes the direction of the cut by 90°. This leads to finer control. If one simply continued to cut straight across, there would be poor control over the tolerance,

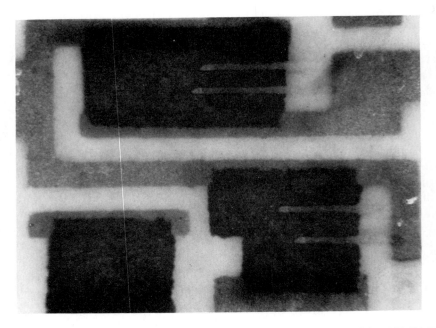

Fig. 16-12 Trimming cuts made in thick film resistors. The cuts were made by a Nd:YAG *Q*-switched laser with dynamic feedback control while the circuit was in operation. The actual area covered by the photograph is 2.3 × 3.1 mm.

especially when the cut reached almost all the way across the resistor. The L shape allows a more gradual approach to the final desired value.

Thermal shock and damage along the cut have been problems which cause drift of the resistor after trimming. Holding laser power down to the minimum needed for a clean cut will much reduce problems of damage and post-trim drift.

Laser resistor trimmers are widely used in industrial production of many types of circuits. The benefits of laser trimming with an automated system including real-time monitoring of resistance and feedback control of the laser have led to its adoption in many industrial facilities. Laser trimming has led to the capability for improved accuracy in resistor values and to higher density in the artwork, and hence to improved circuit design and fabrication.

Another development application of laser technology in electronic fabrication has been the capability to form low resistance ohmic contacts between conductors on circuit chips, after fabrication. One can, for example, make connections between the aluminum and silicon layers at selected positions in MOS structures. A dye laser with nanosecond pulse duration has been used [22]. The laser pulses can open holes in the MOS structure and redistribute conducting material along the edges of the hole so as to form an ohmic contact between the layers without damage to adjacent structures. This can allow random selection of connections on completed circuits. This allows modification or personalization of circuits for specific applications. Process windows of laser parameters can be defined, within which such connections can be made reliably and repeatedly. Although still in a development status, laser personalization offers a powerful approach for selective interconnection. Because the personalization is carried out after normal fabrication is completed, it can be used to reduce inventories and to provide speedy response to desired changes.

It is also appropriate to mention the use of lasers to expose photoresist and to generate patterns. Although this is not an application involving material removal, the introduction of lasers into the pattern generation area has been significant in the electronics industry. It has allowed production of complex and precise circuit patterns. One system, the primary pattern generator [23], may be considered as an archetype. Built around an argon laser, it incorporates mechanical scanning and acoustooptic modulation to form a desired pattern on the photographic plate. The entire system is highly automated under computer scheduling. The primary pattern generator has been in operation for a number of years and represents a considerable advance in artwork generation for mask making.

As an example [24] of a production process employing vaporization by lasers, monolithic crystal filters are trimmed by a Nd:YAG laser trimmer. The laser evaporates gold from coupling stripes between the electrodes, so as to adjust the 8 MHz filter to within 3 Hz.

To summarize the impact of lasers, Table 16-6 lists applications that have reached volume production status in the electronics industry.

TABLE 16-6
Large Scale Laser Applications in the Electronics Industry

Application	Materials	Laser used
Trimming thin film resistors	Ta, Cr, Ni–Cr	Nd:YAG
Trimming thick film resistors	Cermets	Nd:YAG, CO_2
Scribing wafers	Alumina	CO_2
Scribing wafers	Silicon	Nd:YAG
Hole drilling	Alumina	CO_2

E. Additional Processes Involving Material Removal

1. *Marking*

There are many requirements for identification of manufactured parts. Product marking can be carried out for purposes of identification, for imprinting product information, or for theft prevention. Conventional techniques include printing, stamping, mechanical engraving, manual scribing, etching, and sandblasting. In some cases, needs for durable permanent marks stretch the capabilities of the available techniques.

Lasers offer a novel and simple approach to marking of products. Lasers capable of marking manufactured products are usually more expensive than conventional devices like steel stamping machines. One would consider layer marking only when there is some need not well fulfilled by conventional systems. This might occur when the product is small or fragile or when very high speed is needed.

Two different laser-based approaches to product marking are being promoted. In the first method, one scans the focused beam over the surface of the part and modulates it to vaporize a small amount of material at selected positions, thus forming a mark. The mark often consists of alphanumeric characters. The second method, called image micromachining, forms the desired pattern all at once, by delivering a high-power beam already focused in the desired pattern to the surface. We shall describe each of these methods in turn.

The scan technique has been employed for imprinting identification marks on products such as silicon crystals and expensive typewriters. The high-power beam is focused on the surface, scanned in the desired pattern, and pulsed on when it is desired to make a mark. The identifying characters can be formed either in a dot matrix or by bar generation. The Nd:YAG laser has often been employed in this application, because the materials to be marked absorb well at 1.06 μm and because it offers short pulses of high peak power, capable of vaporizing a mark on a surface.

Laser printing of identifying numbers on silicon wafers is attractive because of the brittleness of the material and because subsequent manufacturing operations could destroy printed marks. Laser marking of expensive typewriters offers a distinctive method of identification which cannot be duplicated well by any other method. It also allows marking the typewriters late in the manufacturing process, rather than early.

Image micromachining can produce a complicated pattern all in one laser pulse. A schematic diagram is shown in Fig. 16-13. The optical arrangement is similar to that of a slide projector. A mask which has the desired pattern is inserted in the beam. The lens images the mask pattern on the

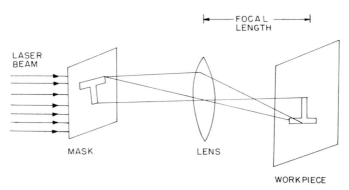

Fig. 16-13 Schematic diagram of image micromachining.

workpiece. If the laser power is high enough, surface material is vaporized, and a permanent image of the mask is imprinted on the surface.

Emphasis has been placed on imprinting fairly large area patterns in a single pulse. Hence, the peak power must be high, and the materials most easily marked are organic materials. Marking of paper and wood products has been studied. The laser most often employed is the CO_2 TEA laser, because the CO_2 laser wavelength is absorbed well by organic materials.

A reasonable size TEA laser can mark an area around 1 cm^2 on typical packaging materials. This allows coding for dating or batch identification or imprinting of specialized logos.

2. Scribing

Scribing is an important method of cutting and shaping without vaporizing all the way through the workpiece. It can be applied to brittle materials such as ceramics, silicon, and glass. Scribing may be performed either by cutting a continuous groove into the surface or by drilling a series of closely spaced holes. The material will then snap easily along the path of the scribed line.

The laser usually used for scribing ceramics is the CO_2 laser. Both continuous and repetitively pulsed lasers have been used. Ceramics usually have high absorption at 10.6 μm. However, if silicon is to be scribed, the higher absorption near 1 μm makes the Nd:YAG laser a better choice.

Direct sawing of complex shapes in alumina with CO_2 lasers has been described in Section C. As compared to sawing, laser scribing requires less energy per unit length, because one does not cut all the way through the material; thus scribing can be carried out for straight-line separations at higher speed than direct cutting and sawing. Also as compared to direct cutting, scribing produces edges with less damage due to heating and with less thermal stress. Because of these considerations, scribing is preferred in operations where only straight cuts are needed, such as dicing rectangular blanks into smaller rectangular pieces. Direct cutting is applicable when sharp corners or complex shapes must be generated. As specific examples, silicon transistor wafers have been scribed at 0.6 in./min by a Nd:YAG laser, and 0.025 in. thick alumina substrates were scribed at 60 in./min by a 100 W continuous CO_2 laser.

Large increases in integrated circuit production are being realized through the application of laser scribing of silicon wafers. The results compare favorably with the diamond saw. The laser beam scribes a straight uniform line with accurately controlled depth. Individual integrated circuits are then produced by snapping each wafer into hundreds of chips along the scribed line. According to one estimate [25], over 2 billion holes are being drilled daily in ceramic scribing operations. Laser scribing of such materials offers the advantage of clean breaks without contamination and without tool wear. The scribe may produce pieces of variable size, thus reducing expenses for substrates of unusual size.

One estimate [26] of the relative costs of laser scribing and diamond scribing indicated a considerable potential saving with laser scribing. The estimate included only operating costs, without allowance for equipment cost. It did include savings due to increased yield. At a volume of 120,000 wafers per year, the costs associated with laser scribing were estimated as less than 20% of those of diamond scribing.

3. Fracturing

One may fracture brittle materials by sudden stress caused by rapid heating. If the fracture can be controlled, this technique can also be used for separating brittle materials, like ceramics or glass. When laser energy is absorbed at the surface of a brittle material, it heats the material rapidly. This produces mechanical stresses which lead to localized fracture. If the material is moved under the laser beam, the fracture will follow along the beam path. The material separates without damage to the surface and without removal of any material.

The technique of controlled fracturing could be used in areas such as separating circuit boards where the materials can be easily damaged by heat or contamination. Experimental demonstrations of controlled fracturing used a CO_2 laser with power less than 50 W [27]. Alumina with a thickness of 0.027 in. was separated at rates over 60 in./min. Separations are made along any desired path, not necessarily only straight lines. No material is lost and the surface remains undamaged. The technique is applicable in separating small, brittle chips for microcircuits.

In comparison to scribing, fracturing has a problem of possible wandering of the fracture line, particularly if one tries to turn a sharp corner. Because of this, scribing has had more general usefulness for controlled separation of brittle materials. Fracturing could be used when the feature of zero material removal is desirable.

Some investigators have described problems in duplicating these results. In one study (cf. [10]), good control of the fracture could be obtained only at low speeds. When speed was increased, the fracture tended to propagate without control.

In the late 1960s, there was enthusiasm for controlled fracturing as a method for separating brittle materials. The enthusiasm has waned somewhat. Controlled fracturing has not been adopted in large scale operations.

4. Balancing

Dynamic balancing can be accomplished by the laser while the workpiece remains in motion. Most conventional balancing operations require determination of the heavy area while the workpice is in motion. One then stops the system, mechanically removes material from the heavy area, and redetermines the heavy portion. This process often is expensive and involves trial and error.

The laser pulse is rapid enough that material can be removed from the heavy spot while the workpiece is rotating in the test fixture. This can provide great simplification and possibly lower cost in dynamic balancing. The laser firing can easily be synchronized with the detection of the heavy portion of the rotating workpiece, so that the laser pulse strikes the heavy region. The unbalance is detected by conventional methods; the unbalance signals are compatible with the laser control circuitry, so that each shot is fired at the unbalanced region.

A fairly ordinary ruby laser is capable of removing 1 mg of metal per pulse, an amount adequate in many balancing operations. The pulse length must be short. At high rotation rates, a 1 msec duration pulse will be deposited around a large angular sector. Thus, pulse lengths of a few tens of microseconds are desirable. With such pulse lengths, rotors can be balanced at rotation rates up to 24,000 rpm. Because of the rotation, the energy is deposited over some length. The amount vaporized will be reduced below the amount removed from a stationary target. Even in such a case, a laser emitting around 10 J per pulse is adequate for dynamic balancing, provided the pulse repetition rate is not too low.

5. Wire Stripping

Industry employs many wire-stripping operations, most of which are satisfactorily accomplished by conventional techniques. In some cases, conventional methods are not adequate. Laser wire strippers can remove certain tough types of insulation, such as are used in the aerospace industry, and eliminate problems of nicking the wire. For example, Kapton insulation, which is used in aerospace applications becuase of its light weight, is not stripped well by either thermal or mechanical strippers.

Laser wire strippers are based on CO_2 lasers, because of the high absorption of organic materials at 10.6 μm. Models have been developed in which the wire is inserted into the stripping head. The laser beam is rotated around the wire to produce a circumferential cut. Then the insulation is slit longitudinally for easy removal. The insulation vaporized is removed cleanly; the high metallic reflectivity at 10.6 μm means that the metallic conductor is not affected.

A continuous CO_2 laser, with output around 50 W, is adequate for wires up to 0.65 in. in diameter. Laser units offer clean repeatable wire stripping for the aerospace industry, in situations where mechanical strippers often nick or cut wires.

F.   An Example—Mask Saving

Laser vaporization is being used to remove small amounts of chromium metal on glass master patterns used in making photographic masks for integrated circuit chips. This use, on a routine production basis, is being

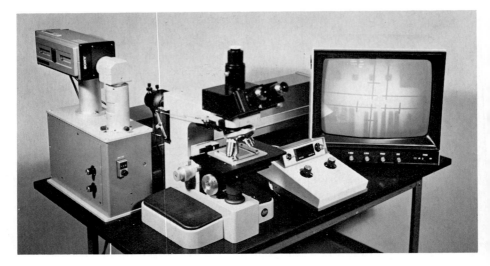

Fig. 16-14   Photograph of laser-based mask cleaner. (Photograph courtesy of Electromask, Inc.)

used to reduce the number of defects per mask and to increase the chip yield.

The laser used is a pulsed xenon ion laser operating at a wavelength of 0.535 μm. This appears to be an optimum choice. The absorptivity of the chrome is high at this wavelength. Longer wavelength lasers are likely to damage the glass. Because of the short wavelength, the beam may be focused to a finer spot than infrared lasers.

The masks are subjected to visual inspection under a microscope, with a display on a TV monitor. The entire surface of the mask is scanned in a pattern. When a chrome spot is located, it is lined up in the center of the display with grid marks to indicate the aiming point. The diameter of the focused beam may be adjusted down to 1 μm to accommodate the size of the particular chrome spot. The laser then is pulsed, delivering a pulse of 0.5 μsec duration and 500 W peak power. As many shots as necessary are delivered to remove the spot, but one pulse is often enough for small spots. A photograph of a mask-saving system is shown in Fig. 16-14.

REFERENCES

[1] V. B. Braginskii, I. I. Minakova, and V. V. Rudenko, *Sov. Phys. Tech. Phys.* **12**, 753 (1967).

[2] M. K. Chun and K. Rose, *J. Appl. Phys.* **41**, 614 (1970).

[3] F. P. Gagliano and U. C. Paek, *Appl. Optics* **13**, 274 (1974).

[4] J. F. Ready, *J. Appl. Phys.* **36**, 462 (1965).

[5] T. Kato and T. Yamaguchi, *NEC Res. Dev.* **12**, 57 (October 1968).

[6] Data from S. Shimakawa, Japan Society for Laser Material Processing.

[7] J. F. Ready, "Effects of High Power Laser Radiation." Academic Press, New York (1971).

[8] J. E. Harry and F. W. Lunau, *IEEE Trans. Ind. Appl.* **IA-8**, 418 (1972).

[9] B. Feinberg, *Mfr. Engineering and Development* (December 1974).

[10] G. K. Chui, *Ceramic Bull.* **54**, 515 (1975).

[11] E. V. Locke, E. D. Hoag, and R. A. Hella, *IEEE J. Quantum Electron.* **QE-8**, 132 (1972).

[12] D. W. Wick, Applications for Industrial Lasercutter Systems, Society of Manufacturing Engineers Technical Paper MR75-491, 1975 .

[13] J. Longfellow, *Solid State Technol.*, p. 45 (August 1973).

[14] S. L. Engel, Society of Manufacturing Engineers Technical Paper MR 74-960, 1974.

[15] Booklet, "CO$_2$ Applications." Coherent Radiation, Inc., Palo Alto, California, 1969.

[16] W. W. Duley and J. N. Gonsalves, *Opt. and Laser Technol.*, p. 78 (April 1974).

[17] J. R. Williamson, *in* "Industrial Applications of High Power Laser Technology" (J. F. Ready, ed.), Vol. 86. Proc. Soc. Photo-Opt. Instrum. Engineers, Palos Verdes Estates, California, 1976.

[18] E. V. Locke and R. A. Hella, *IEEE J. Quantum Electron.* **QE-10**, 179 (1974).

[19] Aerospace Industries Association of America, Report on Project 74.12 (December 1975).

[20] M. I. Cohen, B. A. Unger, and J. F. Milkosky, *Bell System Tech. J.* **47**, 385 (1968).

[21] J. Raamot and V. J. Zaleckas, *IEEE J. Quantum Electron.* **QE-9**, 648 (1973).

[22]   P. W. Cook, S. E. Schuster, and R. J. von Gutfeld, *Appl. Phys. Lett.* **26**, 124 (1975).
[23]   K. M. Poole, *Bell Syst. Tech. J.* **49**, 2031 (1970), and succeeding articles.
[24]   *Laser Focus*, p. 21 (February 1972).
[25]   R. Barber, Society of Manufacturing Engineers Technical Paper MR 74-951, 1974.
[26]   *MicroWaves*, p. 71 (August 1970).
[27]   R. M. Lumley, *Ceramic Bull.* **48**, 850 (1969).

SELECTED ADDITIONAL REFERENCES

A. Laser-Induced Material Removal

S. S. Charschan (ed.), "Lasers in Industry," Van Nostrand–Reinhold, Princeton, New Jersey, 1972, Chapter 3.
M. I. Cohen and J. P. Epperson, Application of Lasers to Microelectronic Fabrication, *in* "Electron Beam and Laser Beam Technology" (L. Marton and A. B. El-Kareh, eds.), Academic Press, New York, 1968.
M. I. Cohen, Material Processing, *in* "Laser Handbook" (F. T. Arecchi and E. O. Schulz-DuBois, eds.), North Holland Publ., Amsterdam, 1972.
F. W. Dabby and U. C. Paek, High-Intensity Laser-Induced Vaporization and Explosion of Solid Material, *IEEE J. Quantum Electron.* **QE-8**, 106 (1972).
W. W. Duley, "CO_2 Lasers, Effects and Applications," Academic Press, New York, 1976, Chapter 4.
F. P. Gagliano, R. M. Lumley, and L. S. Watkins, Lasers in Industry, *Proc. IEEE* **57**, 114 (1969).
F. P. Gagliano and U. C. Paek, Observation of Laser-Induced Explosion of Solid Materials and Correlation with Theory, *Appl. Opt.* **13**, 274 (1974).
U. C. Paek and F. P. Gagliano, Thermal Analysis of Laser Drilling Processes, *IEEE J. Quantum Electron.* **QE-8**, 112 (1972).
J. F. Ready, Effects due to Absorption of Laser Radiation, *J. Appl. Phys.* **36**, 1522 (1965).
J. F. Ready, "Effects of High Power Laser Radiation," Academic Press, New York, 1971, Chapter 3.
I. P. Shkarofsky, Review on Industrial Applications of High-Power Laser Beams III, *RCA Rev.* **36**, 336 (1975).
W. Ulmer, On the Behavior of Different Materials during Treatment with CO_2 Lasers, *Electro-Opt. Syst. Design*, p. 33 (July 1974).
R. E. Wagner, Laser Drilling Mechanics, *J. Appl. Phys.* **45**, 4631 (1974).
L. A. Weaver, Machining and Welding Applications, *in* "Laser Applications," Vol. 1 (M. Ross, ed.), Academic Press, New York, 1971.

B. Hole Drilling

S. S. Charschan (ed.), "Lasers in Industy," Van Nostrand–Reinhold, Princeton, New Jersey, 1972, Chapter 4.
M. I. Cohen and J. P. Epperson, Application of Lasers to Microelectronic Fabrication, *in* "Electron Beam and Laser Beam Technology" (L. Marton and A. B. El-Kareh, eds.), Academic Press, New York, 1968.
M. I. Cohen, Material Processing, *in* "Laser Handbook" (F. T. Arecchi and E. O. Schulz-Dubois, eds.), North Holland Publ., Amsterdam, 1972.
W. W. Duley and W. A. Young, Kinetic Effects in Drilling with the CO_2 Laser, *J. Appl. Phys.* **44**, 4236 (1973).

W. W. Duley and W. A. Young, A Study of Stress and Defect Structures Adjacent to Laser-Drilled Holes in Fused Quartz, *J. Phys. D: Appl. Phys.* **7**, 937 (1974).

W. W. Duley, "CO$_2$ Lasers, Effects and Applications," Academic Press, New York, 1976, Chapter 5.

F. P. Gagliano, R. M. Lumley, and L. S. Watkins, Lasers in Industry, *Proc. IEEE* **57**, 114 (1969).

J. N. Gonsalves and W. W. Duley, Interaction of CO$_2$ Laser Radiation with Solids. I. Drilling of Thin Metallic Sheets, *Can. J. Phys.* **49**, 1708 (1971).

J. N. Gonsalves and W. W. Duley, Interaction of CO$_2$ Laser Radiation with Solids. II. Drilling of Fused Quartz, *Can. J. Phys.* **50**, 216 (1972).

G. Herziger, R. Stemme, and H. Weber, Modulation Technique to Control Laser Material Processing, *IEEE J. Quantum Electron.* **QE-10**, 175 (1974).

J. Longfellow, Production of an Extensible Matrix by Laser Drilling, *Rev. Sci. Instrum.* **41**, 1485 (1970).

Y. Nåkada and M. A. Giles, X-ray and Scanning Electron Microscope Studies of Laser Drilled Holes in Alumina Substrate, *J. Am. Ceram. Soc.* **54**, 354 (1971).

J. F. Norton and J. G. McMullen, Laser-Formed Apertures for Electron Beam Instruments, *J. Appl. Phys.* **34**, 3640 (1963).

I. P. Shkarofsky, Review on Industrial Applications of High-Power Laser Beams III, *RCA Rev.* **36**, 336 (1975).

C. Cutting

D. A. Belforte, CO$_2$ Laser Cuts Metals—And Costs, *Electro-Opt. Syst. Design*, p. 18 (May 1975).

D. A. Belforte, Precision Metal Cutting with a Laser, Proceedings of the Technical Program, *Electro-Opt. Syst. Design Conf., 1975, Anaheim* (November 11–13, 1975).

G. K. Chui, Laser Cutting of Hot Glass, *Ceram. Bull.* **54**, 514 (1975).

W. W. Duley and J. N. Gonsalves, CO$_2$ Laser Cutting of Thin Metal Sheets with Gas Jet Assist, *Opt. Laser Technol.*, p. 78 (April 1974).

W. W. Duley, "CO$_2$ Lasers, Effects and Applications," Academic Press, New York, 1976, Chapter 6.

B. Feinberg, Advent of the Job-Shop Laser, *Manufacturing Engineering and Management* (December 1974).

J. E. Harry and F. W. Lunau, Electrothermal Cutting Processes Using a CO$_2$ Laser, *IEEE Trans. Ind. Appl.* **IA-8**, 418 (1972).

E. V. Locke and R. A. Hella, Metal Processing with a High-Power CO$_2$ Laser, *IEEE J. Quantum Electron.* **QE-10**, 179 (1974).

J. Longfellow, Sawing Alumina Substrates with a CO$_2$ Laser, *Solid State Technol.*, p. 45 (August 1973).

J. F. Ready, "Effects of High Power Laser Radiation," Academic Press, New York, 1971, Chapter 8.

I. P. Shkarofsky, Review on Industrial Applications of High-Power Laser Beams III, *RCA Rev.* **36**, 336 (1975).

L. A. Weaver, Machining and Welding Applications, *in* "Laser Applications," Vol. 1 (M. Ross, ed.), Academic Press, New York, 1971.

D. Electronic Component Fabrication

J. Arruda and W. Prifti, Microelectronic Package Sealing with a Laser, *Electro-opt. Syst. Design* (May 1973).

A. L. Berg and D. E. Lood, Effects of Laser Adjustment on Cr-SiO, *Solid State Electron*. **11**, 773 (1968).

L. Braun and D. R. Breuer, Laser Adjustable Resistors for Precision Monolithic Circuits, *Solid State Technol*., p. 56 (May 1969).

K. R. Bube, Laser-Induced Microcrack in Thick-Film Resistors—A Problem and Solutions, *Ceram. Bull*. **54**, 528 (1975).

R. L. Cadenhead, Production Trade-offs in Laser Trimming Operations, *Solid State Technol*., p. 39 (January 1976).

S. S. Charschan (ed.), "Lasers in Industry," Van Nostrand–Reinhold, Princeton, New Jersey, 1972, Chapter 4.

M. I. Cohen and J. P. Epperson, Application of Lasers to Microelectronic Fabrication, *in* "Electron Beam and Laser Beam Technology" (L. Marton and A. B. El-Kareh, eds.), Academic Press, New York, 1968.

M. I. Cohen, B. A. Unger, and J. F. Milkosky, Laser Machining of Thin Films and Integrated Circuits, *Bell Syst. Tech. J*. **47**, 385 (1968).

P. W. Cook, S. E. Schuster, and R. J. von Gutfeld, Connections and Disconnections on Integrated Circuits Using Nanosecond Laser Pulses, *Appl. Phys. Lett*. **26**, 125 (1975).

P. D. Dapkus, W. W. Weick, and R. W. Dixon, Laser-Machined GaP Monolithic Displays, *Appl. Phys. Lett*. **24**, 292 (1974).

F. P. Gagliano, R. M. Lumley, and L. S. Watkins, Lasers in Industry, *Proc. IEEE* **57**, 114 (1969).

G. A. Hardway, Applications of Laser Systems to Microelectronics and Silicon Wafer Slicing, *Solid State Technol*., p. 63 (April 1970).

R. C. Headley, Laser Trimming Is an Art That Must be Learned, *Electronics*, p. 121 (June 1973).

J. L. Hokanson and B. A. Unger, Laser Machining Thin Film Electrode Arrays on Quartz Crystal Substrates, *J. Appl. Phys*. **40**, 3157 (1969).

D. Maydan, Micromachining and Image Recording on Thin Films by Laser Beams, *Bell Syst. Tech. J*. **50**, 1761 (1971).

T. D. Mino, D. J. Oberholzer, and C. A. Strittmatter, *Solid State Technol*., p. 37 (August 1974).

R. Sard and D. Maydan, A Structural Investigation of the Laser Machining of Thin Bismuth Films, *J. Appl. Phys*. **42**, 5084 (1971).

J. Schmidling, Laser Trimmer Branches Out, *Electro-Opt. Syst. Design*, p. 54 (July 1975).

S. E. Schuster and P. W. Cook, Laser Personalization of Integrated Circuits, in *Proc. Ind. Appl. High Power Laser Technol*. Vol. 86 (J. F. Ready, ed.), Soc. of Photo-Opt. Instrum. Eng., Palos Verdes Estates, California (1976).

E. J. Swenson, Laser Trimming of Hybrid Microelectronics, Soc. of Manufacturing Eng. Tech. Paper MR74-963 (1974).

G. C. Waite, Laser Application to Resistor Trimming, *in Proc. Ind. Appl. High Power Laser Technol*. Vol. 86 (J. F. Ready, ed.), Soc. of Photo-Optical Instrumentation Eng., Palos Verdes Estates, California (1976).

B. A. Unger, Laser Applications in the Fabrication of Thin Film Circuits, Soc. of Manufacturing Eng. Tech. Paper MR74-961 (1974).

R. L. Waters and M. J. Weiner, Resistor Trimming and Micromachining with a YAG Laser, *Solid State Technol*., p. 43 (April 1970).

J. M. Webster, The Microelectronics Horizon for Lasers, *Electronic Packaging and Production*, p. 23 (March 1971).

W. W. Weick, Laser Generation of Conductor Patterns, *IEEE J. Quantum Electron*. **QE-8**, 126 (1972).

D. R. Whitehouse and R. W. Ilgenfritz, Laser Beam Technology for the Microelectronics Industry, *Solid State Technol*., p. 32 (July 1972).

E. Miscellaneous Processes Involving Material Removal

A. J. Beaulieu, Rapid Balancing of Gyroscopes with TEA CO_2 Laser, *Electro-Opt. Syst. Design*, p. 36 (May 1973).

S. S. Charschan (ed.), "Lasers in Industry," Van Nostrand–Reinhold, Princeton, New Jersey, (1972), Chapter 4.

M. I. Cohen, Material Processing, *in* "Laser Handbook" (F. T. Arecchi and E. O. Schulz-Dubois, eds.), North Holland Publ., Amsterdam, 1972).

F. P. Gagliano, R. M. Lumley, and L. S. Watkins, Lasers in Industry, *Proc. IEEE* **57**, 114 (1969).

G. Holzinger, K. Kosanke, and W. Menz, Printing of Part Numbers Using a High Power Laser Beam, *Opt. Laser Technol.* **5**, 256 (1973).

W. F. Iceland, Design and Development of Equipment for Laser Wire Stripping, *Proc. Ind. Appl. High Power Laser Technol.* Vol. 86 (J. F. Ready, ed.), Soc. of Photo-Opt. Instrumentation Eng., Palos Verdes Estates, California (1976).

Laser Microbalances Inertial Gyros, *Electro-Technology*, p. 39 (August 1969).

Laserscribing, Quantronix Corp., Smithtown, New York (1973).

R. M. Lumley, Controlled Separation of Brittle Materials Using a Laser, *Ceram. Bull.* **48**, 850 (1969).

J. A. Nilson, Image Micro-machining with TEA CO_2 Lasers, Soc. of Manufacturing Eng. Tech. Paper MR75-584 (1975).

J. F. Ready, "Effects of High Power Laser Radiation," Academic Press, New York, 1971, Chapter 8.

L. A. Weaver, Machining and Welding Applications, *in* "Laser Applications," Vol. 1 (M. Ross, ed.), Academic Press, New York, 1971.

F. An Example—Mask Saving

F. Lee, Correction of Mask Defects, *Laser Focus*, p. 87 (May 1975).

CHAPTER 17

PRINCIPLES OF HOLOGRAPHY

Holography is a method of lensless three-dimensional photography that was first invented in 1948. However, it was only with the advent of lasers in the early 1960s that holography really developed; the bright, coherent, monochromatic laser was needed before really good holograms could be made.

A. Formation of Holograms

In ordinary photography, one records only the amplitude of the light wave; in holography one preserves both the amplitude and phase of the light. This means that the hologram will produce a true three-dimensional representation of the light wave, just as it would have come from the original object.

Holography involves a two-step process:

(1) Recording of a complex interference pattern which is produced by the superposition of two light waves. The interference pattern is recorded on photographic film.

(2) Reconstruction of an image from the interference pattern. This is done by illuminating the recorded pattern with a light wave, identical to one of the light waves used in the original recording process. The result of the reconstruction is an image which is a duplicate of the original object. It includes the full depth and perspective of the three-dimensional object.

The recording process begins by combining a coherent monochromatic light wave with another beam which results from reflecting part of the light off some object. The arrangement will often be similar to what is shown in

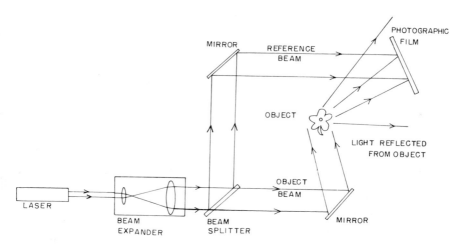

Fig. 17-1 Typical geometry for recording of a hologram.

Fig. 17-1. The monochromatic coherent beam from the laser is expanded by a collimator. It is then divided into two parts by the beam splitter. One of the beams is sent directly to the photographic film by reflection from a mirror. This beam is referred to as the reference beam.

The other part of the beam strikes the object and is reflected from it. This reflected light from the object carries information concerning the object. The reflected light waves contain a pattern which is characteristic of the object, and which will be different for different objects. If one were to view this light with his eye, he would see the object. Some of the light that is reflected from the object will reach the photographic film. The light that travels the path from the object to the film is called the object beam. The light waves from both the object beam and the reference beam are superimposed at the position of the film. There they form a complex interference pattern, which consists of a multitude of fringes produced by constructive and destructive interference. The shape and intensity of these fringes depends on the phase and amplitude of the two different waves.

The photographic film is exposed to the light and then developed using conventional developing techniques. The film preserves the interference pattern produced by the interference of the two beams. This is the basic process of recording. The resulting piece of developed film is the hologram. It appears like a fogged negative; visual inspection reveals nothing that looks like the object.

There are many possible arrangements of object, film, beam splitter, mirrors, and parallel or diverging or converging beams. Different names are applied to different arrangements. However, the most common arrangement is similar to that shown in Fig. 17-1. This is called off-axis holography, because the reference and object beams arrive at the film traveling at an angle to each other.

The arrangement for reconstructing the hologram is shown in Fig. 17-2. The hologram (the developed film) is reilluminated by the reference beam alone. One may remove the object, or may block the path of the object beam with a shutter, or may simply remove the hologram completely to a different place and illuminate the hologram with a laser beam traveling in the same direction as the original reference beam. This beam is now called the reconstructing beam.

When the hologram is illuminated by the laser light at the same angle as the original reference beam, two images of the object are produced by light diffracted by the hologram. One may consider the original process of hologram formation as yielding a type of diffraction pattern, a closely spaced array of interference fringes, whose exact shape depends on the nature of the object. When this diffraction pattern is reilluminated, the diffracted light will have characteristics that are dictated by the original object. Later we shall describe how this process occurs in more detail, and show that the diffracted light will in fact reproduce the light waves that came from the object.

Now let us consider the two different images of the object formed during the reconstruction of the hologram. One image is a virtual image. It is viewed by looking through the hologram. The observer will see an undistorted view of the object, just as if it were still present. The image is three-dimensional, i.e., one may look part way around the object. This is called a virtual image because it requires a lens (the lens of the observer's eye) to form it.

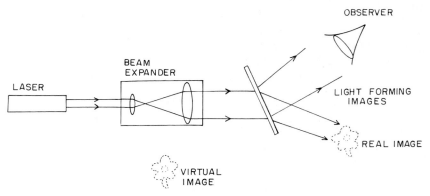

Fig. 17-2 Typical geometry for reconstructing a hologram.

According to elementary optics, one requires a lens to produce a virtual image. The lens of the human eye or the lens of a camera can be employed. Thus, one can look at this distribution of light and one will see an image of the original object.

A second image is formed by light traveling in a different direction. This is the real image; it can be projected directly on a screen and does not need a lens to form it. In fact, the real image must be projected on some surface in order to view it.

There will also be some light transmitted directly through the hologram without being diffracted by the hologram. This undiffracted light will emerge from the hologram traveling in the same direction as the reconstructing beam. This undiffracted light is not of great interest. The two other distributions of light that correspond to the two images will be on opposite sides of the undiffracted beam. The relative orientations are as shown in Fig. 17-2.

Before we describe how the holographic process operates, let us consider briefly the history of holography. Holography was first invented by Professor Dennis Gabor in 1948. Gabor's motivation was to provide a new method of microscopy. The holographic process can involve magnification. The exact value of the magnification will depend on the radii of curvature of the wavefronts used for producing the hologram and for reconstructing the hologram. However, under reasonable conditions, the reconstructed image will be magnified by the ratio of the wavelength of the light used for reconstructing the hologram to the wavelength of the light used for producing the hologram. Gabor's idea was to produce holograms using x rays and to reconstruct them using visible light. This would lead to magnification by a factor of several thousand.

Application of holography to high magnification microscopy was hampered by two factors: (1) the lack of a high quality x-ray source for producing the holograms, and (2) the lack of a bright coherent light source for reconstructing the hologram. After some interest in the late 1940s, holography was relatively dormant until the advent of the laser. The laser provided a bright coherent source that can be employed both for producing and reconstructing holograms. It was early recognized that holography employing lasers could yield high quality three-dimensional images, and beginning in the early 1960s, there was a great revival of interest in holography. This revival of interest was spurred by a further advance which was developed by E. N. Leith and J. Upatnieks at the University of Michigan. In Gabor's original work, the reference beam and the object beam used for making the hologram were both incident in the same direction. Then when the hologram was reconstructed, using a reconstructing beam traveling along the direction of the original reference beam, the three images were all coincident. That is, the real object, the virtual object, and the undiffracted transmitted beam were colinear in space. This led to problems in distinguishing the different images.

The contribution of Leith and Upatnieks was the off-axis geometry, as shown in Fig. 17-1. In this geometry, the reference beam and the object beam are incident at an angle to each other. When the hologram is reconstructed by light traveling in the same direction as the reference beam, the three resulting beams are separated in space. The undiffracted transmitted beam travels in the same direction as the reconstructing beam. The real image and virtual image, however, are produced in directions at an angle to the reference beam, on either side of it. This physical separation makes it much easier to view the images.

Since these developments, there has been a great surge of interest in holography. Many applications are now beginning to develop. Many additional methods and geometrical arrangements for producing holograms have been devised. The original application envisioned by Gabor for microscopy has still not been exploited because there is not yet a good high quality source of x-rays for producing the hologram. Among other reasons, this lack is motivating the search for x-ray lasers.

When the general arrangement of object and reference beams and the recording medium is as shown in Fig. 17-1, the hologram is called a Fresnel hologram. A Fresnel hologram is formed when the photographic plate is placed relatively near the object, more precisely, in the near field of the diffraction pattern of the object. This is a natural way to produce the hologram, because no lenses are required in the process. Variants on the Fresnel process can include using converging or diverging beams and also illumination of the recording medium with the object and reference beams incident from opposite sides. Holograms can be formed that can be viewed either in light reflected from the hologram or in light transmitted through the hologram. It is beyond the scope of this chapter to consider all the various types of holograms.

One geometry that is worth mentioning is the Fourier transform hologram. Such holograms generate plane wave amplitudes at the hologram, which is the Fourier transform of the subject. An arrangement for forming a Fourier transform hologram is shown in Fig. 17-3. There are other methods for producing Fourier transform holograms which do not require the use of a lens. However, the discussion here will show the essential features of the Fourier transform hologram. The Fourier transform hologram is restricted by the fact that the object must lie within a single plane. Thus, the objects for Fourier transform holograms tend to be photographic transparencies. The method is less applicable to subjects which have an extended depth. Fourier transform holograms are of special interest for the case of optical pattern recognition and for optical computer memories. In

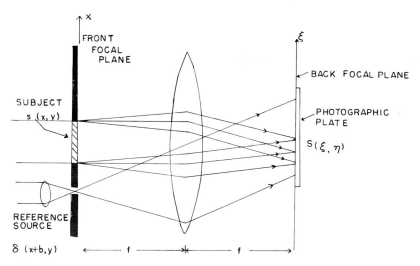

Fig. 17-3 Arrangement for formation of a Fourier transform hologram. The focal length of the lens is f. The spatial coordinates in the front focal plane are x and y; those in the back focal plane are ξ and η. The light distribution of the subject is s; that in the transform plane is S. The reference beam, incident through a pinhole at a distance b from the subject beam, can be represented by a δ-function.

such cases, the presentation of the data in a plane is not particularly a disadvantage.

The Fourier transform hologram takes advantage of the fact that a lens converts a distribution of light in one of its focal planes into the Fourier transform of that distribution in the other focal plane.† Thus, if the distribution of light in the first focal plane of the lens is $f(x, y)$, the distribution of light in the second focal plane will be the Fourier transform of $f(x, y)$. In the reconstructing process, another Fourier transform is performed by a second lens. The net effect of the two Fourier transforms restores the original distribution of light, $f(x, y)$, in the focal plane of the second lens, except that the image is upside down and backwards. The same process converts the conjugate image, which is already reversed, into the same light distribution as the original wavefront. Therefore, the Fourier transform hologram yields two images, reversed in direction from each other, on opposite sides of the undiffracted beam. An advantage of a Fourier transform hologram is that the angular spread of the reconstructing beam can be large [1].

---

† The two-dimensional Fourier transform, $F(p, q)$, of the function $f(x, y)$ is defined by $F(p, q) = \int_{-\infty}^{\infty} \int_{-\infty}^{\infty} f(x, y) \exp i(px + qy) \, dx \, dy$. We may regard x and y as the coordinates in the front focal plane of the lens and p and q as the coordinates in the rear focal plane.

B.   The Holographic Process

We shall now consider how the holographic process operates. We begin by giving a simple heuristic picture of how the hologram is formed and how the reconstruction can be obtained to produce a view of the original object. Consider an object which is a point source. When a coherent plane wave is diffracted by a pinhole, one obtains a spherical wave. When the spherical wave is combined with the plane reference wave, the intensity distribution will consist of annular zones of constructive and destructive interference. This is shown in Fig. 17-4 [2]. This distribution resembles a Fresnel zone plate. As is known from elementary optics, the Fresnel zone plate consists of a series of concentric circles, with radii proportional to the square roots of the even integers. A zone plate can be constructed by drawing concentric circles on paper, with radii proportional to the square roots of the consecutive integers 1, 2, 3, etc. Every alternate annular zone formed by the circles is blackened. Then the figure may be photographically reduced. The resulting transparency is the Fresnel zone plate. The zone plate is known from elementary optics to act as a diffraction grating with focusing properties. It acts both as a positive and a negative lens. If collimated light is incident on the zone plate, some of the diffracted light will converge to a point. This is essentially a real image and represents the positive focusing properties of the zone plate. Some of the diffracted light will appear to diverge from a virtual image behind the zone plate. This represents the negative focusing properties of the zone plate.

In accordance with the positive focusing properties, the zone plate of concentric circles formed as shown in Fig. 17-4 will act as a lens and will

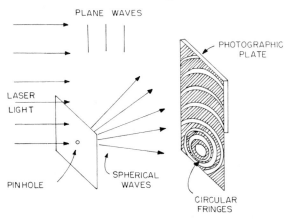

Fig. 17-4   Formation of a Fresnel zone plate. (From W. E. Kock, L. Rosen, and J. Rendeiro, *Proc. IEEE* **54**, 1599 (1966).)

refocus a plane wave to a point, which would correspond to the original, admittedly very simple object. This is then the reconstruction. The Fresnel zone plate produces two beams of light. One is a spherical wave which appears to diverge from the position of the original pinhole. This is the virtual image. If one observes this beam, he will see a bright point of light at the position of the original point. The second beam is a converging bundle of light rays, which corresponds to the real image. If a piece of paper is placed in the beam at the focus, a bright spot of light will be observed on the paper. This is illustrated in Fig. 17-5.

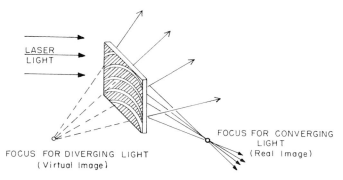

Fig. 17-5 Reconstruction of point images from a Fresnel zone plate. (From W. E. Kock, L. Rosen, and J. Rendeiro, *Proc. IEEE* **54**, 1599 (1966).)

A more complicated object can be considered as a collection of point objects. Each point produces its own zone plate by interfering with the reference beam. All the zone plates add together to form a complicated interference pattern. In reconstruction, the same reference beam is used to recreate all the points comprising the complicated object. Therefore, when one views the reconstruction, one sees exactly the same image including the three-dimensional nature, as one would have seen if he would have looked at the object directly.

The above argument is intended to give an intuitive feeling for how the holographic process operates. It is not meant to be a proof. We shall now proceed with a more accurate discussion, which will require use of more mathematics.

We shall employ Fig. 17-6 for this discussion. Figure 17-6 shows the recording of a hologram. It contains the essential features as shown in Fig. 17-1, namely, a plane reference beam and an object beam intersecting at an angle in the plane of the photographic film. The exact details of where the beams come from (laser, beamsplitter, etc.) have not been included. Also the object is shown as a two-dimensional transparency. At the plane of the

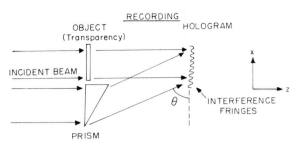

Fig. 17-6   Recording of hologram.

photographic film, it will have some distribution which is characteristic of the object.

We can represent the distribution of the light due to the object as

$$S(x, y) = S_0(x, y)\exp i[\omega t + 2\pi z/\lambda + \phi(x, y)] \qquad (17.1)$$

where x and y are the coordinates in the film plane, z is the direction of propagation of the light, ω the angular frequency of the light, S_0 the amplitude function, and ϕ gives the phase of the light wave.[†] We note that S_0 and ϕ are functions of x and y. Notice also that S will be a complex quantity. S_0 represents the instantaneous distribution of the light amplitude in the film plane.

The instantaneous amplitude of the reference beam, a plane wave, will be

$$R(x, y) = R_0 \exp i[\omega t + (2\pi x/\lambda) \cos \theta + (2\pi z/\lambda) \sin \theta] \qquad (17.2)$$

The amplitude function R_0 is a constant because the reference beam is a plane wave. It has constant amplitude across the wavefront. It is traveling at an angle θ with respect to the x axis. This is shown by the constant phase factor, $(2\pi \cos \theta)/\lambda$. We will define $\alpha = (2\pi \cos \theta)/\lambda$.

In the representations of both the object and reference wave, we have a factor $e^{i\omega t}$. Since the recording process will involve an average over many cycles of the optical field, this term will yield only a constant multiplying factor, which we shall neglect from now on. In addition, if the plane of the film is relatively thin, the value of z will not change much throughout the

---

[†] In comparison, an electric field wave, oscillating at angular frequency ω, can be denoted as $E(x, y, t) = E_0(x, y) \cos[\omega t + \phi(x, y)]$, where E_0 (the amplitude) is the maximum value of the electric field and ϕ (the phase) gives the status of a particular point as it moves through its cyclical variation. We shall employ the complex notation here. Note that $e^{i a} = \cos a + i \sin a$. The amplitude function defined above can be regarded as the amplitude of the wavelike motion of the light wave. Intensity, which gives the power per unit area delivered by the wave, is equal to the square of the absolute value of the amplitude. It is important to understand the distinction. All detectors, including photographic film, respond to intensity, not to amplitude directly. The diffraction phenomena on which holography depends must be described in terms of the wave character of the light, and hence in terms of the amplitude and phase functions.

film. Without loss of generality, we may take $z = 0$. We shall later consider what occurs if the emulsion cannot be regarded as thin.

With these approximations, we have for the representations of the object and reference beams, respectively,

$$S(x, y) = S_0(x, y)e^{i\phi(x, y)} \qquad (17.3)$$

$$R(x, y) = R_0 e^{i\alpha x} \qquad (17.4)$$

The intensity of a light wave is given by the square of the absolute value of the amplitude. Thus, if the object beam alone were incident on the film, the intensity would be

$$|S(x, y)|^2 = [S_0(x, y)]^2 \qquad (17.5)$$

This pattern of intensity would be recorded by a conventional photograph. The photographic film records the intensity, averaged over time. Thus, we expect all the factors involving $e^{i\omega t}$ to drop out. The pattern on the photographic film will preserve the amplitude function S_0, but all information about the phase function ϕ will be lost.

Now consider the result when the film is exposed to the intensity distribution resulting from the sum of the object and reference beams. The intensity distribution $I(x, y)$ recorded by the film will be (for the moment we shall drop the functional notation for R, S, R_0, S_0, and ϕ)

$$\begin{aligned}
I(x, y) &= |R + S|^2 = |R|^2 + |S|^2 + RS^* + R^*S \\
&= R_0^2 + S_0^2 + R_0 S_0(e^{i\phi}e^{-\alpha x} + e^{-i\phi}e^{i\alpha x}) \\
&= R_0^2 + S_0^2 + R_0 S_0[\cos(\alpha x - \phi) + i \sin(\phi - \alpha x) \\
&\quad + \cos(\phi - \alpha x) - i \sin(\phi - \alpha x)] \\
&= R_0^2 + S_0^2 + 2R_0 S_0 \cos(\alpha x - \phi) \qquad (17.6)
\end{aligned}$$

The notations S^* and R^* denote the complex conjugates of S and R, respectively. According to Eq. (17.6), the film records a pattern that has the sum of the squares of the amplitude functions plus an oscillating cosine term. This describes a pattern of fringes. The terms $R_0^2 + S_0^2$ yield a relatively slowly varying background that is modulated by the fringes, which have high spatial frequency. We notice that the recording contains information about both the amplitude function $S_0(x, y)$ and the phase function $\phi(x, y)$.

The next step in the process is development of the photographic film. We shall digress here to review the characteristics of the response of photographic recording media. The characteristics of photographic film are generally described by the so-called H–D curve. This curve is represented in Fig. 17-7. It is a plot of the optical density of the developed film as a

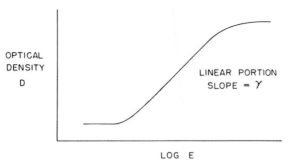

Fig. 17-7   Response of photographic film; optical density D as a function of the logarithm of the exposure E, which is in units of energy per unit area.

function of the logarithm of the exposure (in units of energy per square centimeter). The optical density D is defined by the equation:

$$D = -\log_{10} T \tag{17.7}$$

where T is the transmission of the developed film. Thus, optical density equal to unity means that the transmission is 0.1. Generally, the response of the photographic film depends only on the total energy per unit area that is incident on the film. This phenomenon is called reciprocity. It does not matter whether the energy is delivered slowly at low power or more rapidly at higher power. All that matters is the total energy density delivered. Reciprocity does break down for very short exposures, but it holds over a wide range of exposure times which cover the range of interest for many photographic and holographic purposes.

According to Fig. 17-7, when the recording medium is a conventional photographic film, the result of exposure and development is a change in transmission of the film where the light intensity is high. Silver halide grains in the emulsion are converted to silver atoms during exposure and development. The free silver atoms are most dense where the incident light intensity was high, so that transmission is low. Thus, fringes of light intensity are recorded as absorption fringes for light that is transmitted through the hologram.

The form of the curve in Fig. 17-7 is of interest. At low values of exposure there is relatively little response. This represents the background fog level of unexposed film. Near the right edge of Fig. 17-7, the curve tails over and approaches a constant value. Thus, at high light intensities, the photoresponse saturates. This implies that all the grains of silver in the film have been developed. Generally, for holographic purposes, one tries to work in the region where the H–D curve rises linearly. The slope of the curve in this linear portion is conventionally termed γ. The value of γ is a function of the particular type of film and of the development parameters, including the type of developer used, and the time and temperature of the development. The values of γ under specific conditions are often supplied by the manufacturer of photographic film. For many types of film, γ may be approximately equal to 2. In the linear portion of the H–D curve, the transmission will be approximately proportional to the light intensity incident on the film to the minus γ power. Thus we have the relation that the transmission T is proportional to $E^{-\gamma}$, where E is the exposure (energy per unit area). In a negative ($\gamma > 0$), transmission decreases as exposure increases, as we expect. If the film is printed as a positive, we may consider γ to be a negative quantity.

The transmission described above is the transmission for light intensity. We require the transmission T_a for light amplitude, where we have

$$T_a = T^{1/2} \tag{17.8}$$

Thus T_a is proportional to $E^{-\gamma/2}$. With proper control of the development time, temperature, etc. so as to give $\gamma = -2$ (i.e., printing as a photographic positive), we have a situation where T_a is proportional to E. Thus,

$$T_a = AE \tag{17.9}$$

where A is a constant of proportionality for the given conditions of development. If the development yields a different value of γ, additional terms would have to be added to the mathematical treatment given below. This would yield additional images which could be viewed. Such additional images are in fact often seen in holograms.

This discussion implies development as a photographic positive. If a photographic negative is printed, the regions of high and low transmission would be interchanged. The light and dark fringes would be reversed. However, the holographic process would still operate as we shall describe.

The result of the process (properly controlled) is a piece of developed film that has amplitude transmission given by

$$T_a = A[R_0{}^2 + S_0{}^2 + 2R_0 S_0 \cos(\alpha x - \phi)] \tag{17.10}$$

where we have used the result of Eq. (17.6) in Eq. (17.9).

If we now reilluminate this film with the reference beam alone, the light pattern transmitted by the film will be the reference beam multiplied by the film transmission, i.e.,

$$R(x, y)T_a(x, y) = AR_0 e^{i\alpha x}[R_0{}^2 + S_0{}^2 + 2R_0 S_0 \cos(\alpha x - \phi)] \tag{17.11}$$

where we are still suppressing the functional notation for R_0, S_0, and ϕ. This yields for the light distribution transmitted by the film

$$\begin{aligned} RT_a &= AR|S|^2 + AR|R|^2 + A|R|^2 S + AR^2 S^* \\ &= ARS_0{}^2 + ARR_0{}^2 + AR_0{}^2 S + AR^2 S^* \end{aligned} \tag{17.12}$$

where we have used the first part of Eq. (17.6). Of the various terms in this equation, the first two taken together, $AR(R_0{}^2 + S_0{}^2)$, are equivalent to the reconstructing beam multiplied by the relatively slowly varying factor $A(R_0{}^2 + S_0{}^2)$. This therefore represents undiffracted light traveling in the same direction as the reconstructing beam. Figure 17-8 shows this beam as the beam transmitted straight through the hologram.

The next term,

$$AR_0{}^2S = AR_0{}^2S_0(x, y)e^{i\phi(x, y)} \qquad (17.13)$$

can be recognized as the light distribution from the original object, multiplied by the constant term $AR_0{}^2$. We can recognize that the constant multiplier does not change any of the essential parts of the image, but will only change the brightness of the image. Thus, this term represents a light distribution that is identical to what would have come from the original object. Viewing this light gives a view of the original object. Both the amplitude and phase functions are present, so all the information about the object will be present. This includes the three-dimensional nature of the object. This is a virtual image.

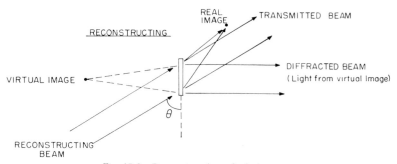

Fig. 17-8   Reconstruction of a hologram.

The light from the virtual image is diverging, apparently from the position where the original object was located during recording. Thus, one must use a lens to collect and refocus this light in order to see the image. The image will appear to be behind the hologram. To view the virtual image, one should relax his eye and look through the hologram. The angular separation between the light in the virtual image and the undiffracted light is shown in Fig. 17-8.

Figure 17-9 shows two photographs of the reconstruction of a virtual image from a hologram. The photographs (two-dimensional, of course) were taken from two different angles of viewing the reconstructed image. The three dimensional nature of the image is apparent in the parallax shown between the two different views.

Fig. 17-9 Photographs of the reconstruction of a hologram. The two photographs, taken at slightly different angles, show the parallax arising from the three-dimensional nature of the hologram. The bright spot near the top is the undiffracted beam.

The emergence of this term, $A R_0^2 S_0(x, y) \exp i\phi(x, y)$, is not merely an artifact of the mathematics. This term represents real light that emerges from the hologram during reconstruction. The presence of this light, with a

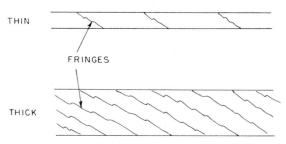

Fig. 17-10   Illustration of thin and thick holographic recording media.

distribution identical to light that came from the original object, is the essential point of the holographic process. Information about the object is stored in the fringe pattern in the film. When the reconstructing beam is diffracted by the fringes, the emerging light is distributed in such a way that the image of the original object is reformed.

The final term in Eq. (17.12),

$$AR^2S^* = AR_0{}^2S_0(x, y) \exp i[2\alpha x - \phi(x, y)] \qquad (17.14)$$

represents light that has the amplitude and phase functions preserved. However, the negative sign in the phase factor means that this light is converging, rather than diverging. This light can form an image on a flat surface placed in the plane in which the light converges. Thus, this represents a real image. The angle θ' at which this light is traveling with respect to the x axis is given by the angular factor $e^{i2\alpha x}$. We will have

$$\cos \theta' = 2\alpha\lambda/2\pi = 2 \cos \theta. \qquad (17.15)$$

Thus, $\cos \theta' > \cos \theta$, where θ is the angle between the x axis and the reconstructing beam. Hence, the real image appears on the opposite side of the undiffracted beam (the opposite side from the virtual image). This is illustrated in Fig. 17-8.

C.   Hologram Types and Their Efficiencies

So far we have discussed the case of a thin recording medium. This will record line traces of the contours of maximum light intensity. However, if the photosensitive medium is relatively thick, the contour surfaces of maximum intensity will be recorded throughout the thickness of the medium. Figure 17-10 indicates the distinction between thin and thick holograms. In the thick hologram, there will be a number of fringes throughout the thickness of the material, as one passes through the material in a direction perpendicular to the plane of the material. The distinction between thick and thin is not an absolute distinction based on the physical thickness of the material. A piece

of film of a given thickness could be either a thin hologram or a thick hologram, depending on the method in which the hologram was recorded. One
can show that if the z-dependent terms in Eqs. (17.1) and (17.2) were retained,
the treatment would be modified so as to yield the following result in place of
Eq. (17.6)

$$|R + S|^2 = R_0{}^2 + S_0{}^2 + 2R_0 S_0$$
$$\times [\cos(\alpha x - \phi) \cos(\beta z - kz) - \sin(\alpha x - \phi) \sin(\beta z - kz)]$$

$$(17.16)$$

where $k = 2\pi/\lambda$ and $\beta = (2\pi \sin \theta)/\lambda$. Thus, in the thick hologram, the contours of maximum intensity will depend on both the x and z position
throughout the film.

The properties of thin and thick holograms are different. In particular,
the efficiency can be much increased in the case of thick holograms. Thick
holograms are sometimes also called volume holograms. The values of
efficiency that are available for different types of holograms will be discussed
later.

There are some methods whereby the recording can be made through
changes in the index of refraction of the emulsion.† Such a hologram is
called a phase hologram. The recording of the fringes can be regarded as
producing regions in which the optical path length through the medium
has changed, because of the change in index of refraction. For such holograms, there is no absorption of the incident reconstructing light, so that
there is a possibility of higher efficiency. There are several types of material
which can be employed to produce phase holograms. These include bleached
photographic film, in which an amplitude hologram is originally produced,
and which then is bleached through chemical processes. In the bleaching, the
silver is removed and the medium appears transparent. However, the index
of refraction has been changed in regions in which the silver grains were
developed. This leads to the production of the phase hologram.

There are, in addition, media which can produce phase holograms
directly without the necessity of the bleaching step. One such medium is
dichromated gelatin, in which a gelatin emulsion film is impregnated with a
chemical such as ammonium dichromate. Interaction of light with the
ammonium dichromate tends to tan the gelatin and changes the index of
refraction. Thus, dichromated gelatin films are directly used to produce
phase holograms without having to employ an intermediate bleaching step.

---

† In the previous discussion, we assumed that the transmission of the film is varied by
the incident light. Thus, such holograms operate by absorption of the reconstructing beam
(or at least of some of it). These holograms are called absorption holograms, or equivalently,
amplitude holograms.

However, dichromated gelatin films are less sensitive to light than conventional photographic emulsion. The spectral sensitivity does not extend into the orange or red, so that helium–neon lasers cannot be used to make holograms on dichromated gelatin films. The energy per unit area required to expose the dichromated gelatin film is much higher than for conventional photographic film.

There is a further distinction in types of holograms. This is the distinction between transmission and reflection holograms. So far, we have discussed only transmission holograms. The fringes of high light intensity are formed perpendicular to the plane of the emulsion. The reference and object beams are incident from the same side of the film and the object is viewed by transmission of the reconstructing beam through the developed film. However, for a reflection hologram, the object beam and reference beam are incident from opposite sides of the original film. The interference surfaces become nearly parallel to the holographic recording medium. A number of fringes will be formed through the thickness of the material and roughly parallel to the plane of the material. Thus, recording of the interference pattern results in a stack of closely spaced planes, which can act as a reflective filter. This can either be an amplitude filter, for example, reflecting planes of developed silver grains, or a phase filter, that is, planes in which the index of refraction varies. We see that of necessity a reflection hologram must be a volume hologram. During reconstruction, the illuminating wave is incident on the same side as the direction of observation, and one views the image as

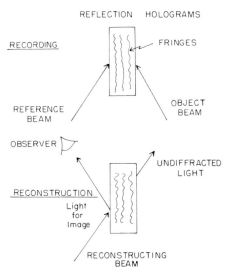

Fig. 17-11   Recording and reconstruction for reflection holograms.

if it were reflected from the hologram. Hence, such holograms are called reflection holograms. Some interesting properties of reflection holograms have been observed. They are selective enough so that white light may be used for reconstruction of the object. The formation and reconstruction of a reflection hologram are shown in Fig. 17-11.

Let us review the relation of these different types of holograms. This is shown in Table 17-1. First, we may have either thick or thin holograms, depending on the relation of the fringe spacing to the thickness of the recording medium. Thick (or volume) holograms can be formed so as to be either reflection or transmission holograms. Thin holograms are necessarily transmission type. Finally, each type may be produced as either a phase hologram or an absorption hologram.

TABLE 17-1
Classifications of Holograms

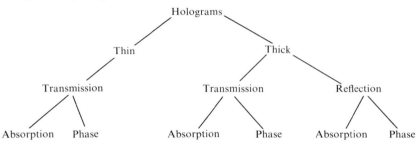

Now let us consider the question of efficiency. Efficiency is defined as the amount of light diffracted into the image, divided by the amount of light in the reconstructing beam. A full discussion of the efficiency of various types of holograms is beyond the scope of this work. However, maximum theoretical values are tabulated in Table 17-2. This table gives the maximum possible efficiency for each of the different types of holograms which we defined in Table 17-1. We see that the maximum theoretical efficiency for

TABLE 17-2
Maximum Efficiency^a of Various Hologram Types

Absorption holograms			Phase holograms		
Thin transmission	Thick transmission	Thick reflection	Thin transmission	Thick transmission	Thick reflection
6.25%	3.7%	7.2%	33.9%	100%	100%

^{a} Maximum theoretical values

phase holograms is relatively high, and always considerably exceeds that of the corresponding type of amplitude holograms. The reason for this is fairly obvious: amplitude holograms rely on absorption of light, and therefore are wasteful of the energy which is used to illuminate them for reconstruction, whereas phase holograms are almost entirely transparent and use the reconstructing light much more efficiently. The theoretical efficiency has been approached in each of the cases defined above. In particular, for thick transmission phase holograms in dichromated gelatin, maximum efficiencies of 90% have been obtained experimentally.

In general, one desires high efficiency. The reconstructed image is brighter and the requirements on the laser power for reconstruction are reduced, if efficiency is high. For a thin amplitude hologram, at most about 6% of the reconstructing light appears in the image, whereas for a thick phase hologram, most of the reconstructing light can appear in phase holograms. Thus, even though they are more difficult to produce, thick phase holograms are important.

D. Practical Aspects of Holography

Now let us consider some of the practical aspects of holography. The light from the laser must be expanded and collimated to cover a suitably large area, so as to cover the entire scene of which the hologram is to be made. This is the reason for the collimator in Fig. 17-1. The light must be spatially coherent over the entire scene. Otherwise, the contrast in the fringes will be degraded. Hence, one usually employs lasers operating in the TEM_{00} mode to ensure good spatial coherence.

In order to produce successful holograms, the monochromaticity of the laser should be good. This is equivalent to saying that the temporal coherence should be high. The coherence length is inversely proportional to the spread of the frequency spectrum of the laser. The frequency spread is broadened by the presence of more than one longitudinal mode in the laser. If the laser emits a single longitudinal mode, the spectral width at half intensity is of the order of 10^5 Hz for a helium–neon laser. The coherence length would be of the order of 10^5 cm, which would be much longer than the optical path lengths in most holographic applications. In this case, one could make good holograms even if the reference and object beams traveled path lengths which were quite different.

However, if more than one longitudinal mode is present in the laser output, the frequency spread is increased. As two different longitudinal modes with slightly different wavelengths traverse a holographic setup with unequal path lengths for the object and reference beams, the two modes will be somewhat out of phase when they arrive at the recording medium. This will reduce the visibility of the fringes in the hologram. The criterion that must be satisfied is that

$$D << \Delta l = c/\Delta f \qquad (17.17)$$

where D is the difference in path length between object and reference beams, Δl the coherence length, and Δf the frequency spread of the laser. If the laser is L cm long, and if N longitudinal modes are in operation, one has

$$\Delta f = (N - 1)c/2L \qquad (17.18)$$

Hence, one must have

$$D << 2L/(N - 1) \qquad (17.19)$$

If a multimode laser is employed, the path length that the object beam and the reference beam travel must be carefully matched. Otherwise, the interference patterns will be washed out because of the dephasing between different ends of the spectrum. If a single mode laser is used, however, this restriction on optical path length differences is removed. Instead, one simply uses the width Δf of a single longitudinal mode in Eq. (17.17).

The maximum value for D may also be interpreted as the possible depth of field that may be preserved in the hologram. In order to produce a hologram of a scene that has depth greater than a few centimeters, one must employ a laser operating in a single longitudinal mode.

If there are N separate longitudinal modes in the laser, the fringe visibility V is given by the equation

$$V = |\sin(N\pi D/2L)/N \sin(\pi D/2L)| \qquad (17.20)$$

where L is the length of the gas laser and D the difference in the path length traveled by the object beam and reference beam in the formation of the hologram. This function is plotted in Fig. 17-12. We see that the fringe visibility decreases if the path difference D becomes greater than a few centimeters, if several longitudinal modes are present. This has the effect of reducing the available depth of field. If a laser operating in several modes is used, the path length between the two beams must be adjusted to within a few centimeters in order to preserve good fringe quality.

An extremely important factor is the elimination of relative motion between the components in the holographic apparatus. If the path length changes by one-half of the wavelength of the light, a fringe will change from a light fringe to a dark fringe. Thus, any positional change which changes the difference in optical path length between the object and reference beams will greatly change the pattern. A change of half a wavelength during exposure will completely wipe out the pattern. A change greater than approximately

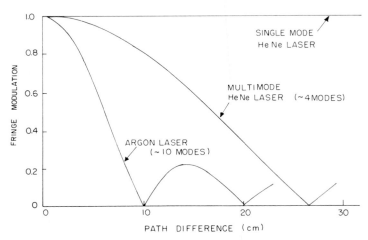

Fig. 17-12   Fringe modulation in a hologram as a function of path difference between object and reference beams for lasers with different numbers of longitudinal modes.

one-tenth of a wavelength will tend to degrade the contrast severely. Thus, one requires rigid mounts, stable positioning, and elimination of vibration. Many holographic tables are constructed of granite slabs, which can be isolated from vibrations by variety of means. One simple means is to support a granite table on a mounting of inflated innertubes. The rigidity of the granite prevents relative motion between the components mounted on it. If the entire granite slab moves as a whole, this will not affect the relative position of the components and will not change the interference pattern. A more expensive mounting involves hydraulic tables, which are available commercially.

Holograms of moving objects can be made by using very short pulses of light, from a Q-switched laser. For example, holograms of bullets in flight have been made by using ruby laser pulses with durations around 10 nsec. During the period of exposure, the bullet will not move a significant fraction of a wavelength.

Air currents and temperature changes can vary the optical path length by changing the index of refraction of the air. This can degrade the quality of a hologram. Often, the holographic equipment will be shielded from air currents.

Some materials that have been employed for holographic recording are listed in Table 17-3. These include several types of photographic film (used both as amplitude and phase media) and several more exotic types of recording media. These more exotic media are included to show the range of materials and effects that have been employed to record holograms. It is by

TABLE 17-3
Recording Materials for Holography

Material	Energy density (erg/cm²)	Maximum efficiency (%)	Resolution (lines/mm)
Kodak 649F film	20,000–80,000	4	2000
Kodak SO-243 aerial film	5–10	6	500
AGFA 8E75 film	110–210	4	3000
AGFA 10E75 film	30–60	3	2800
AGFA 14C75 film	5–10	0.4	1500
Magnetic film (MnBi)	300,000	0.01	500
Metallic film (Bi)	500,000	6	1000
Bleached 649F film (phase)	20,000–80,000	33	2000
Dichromated gelatin (phase)	20,000–50,000	90	5000
Photopolymers (phase)	~100,000	45	>3300
Thermoplastics (phase)	200	10	>1100

no means an exhaustive list. Many other types of materials have been used, at least experimentally. Let us emphasize, however, that photographic film is by far the most common type of holographic recording medium. The other types of recording materials are used experimentally, or else when some special feature (e.g., erasability) is desired.

Table 17-3 lists the energy per unit area to form a holographic exposure. These numbers should be interpreted as being approximate, because they vary with the wavelength of the laser used.

To reduce exposure time, one desires a recording medium with reasonably high sensitivity. The more sensitive the film is, the shorter the exposure that is needed and the less effect vibrations will have. The fringes will be very close together, so the film must be capable of reproducing fine detail. Films with resolution of the order of several thousand lines per millimeter are commonly employed. This high resolution is usually not available with films of extremely high sensitivity, so there must be some compromise between resolution and exposure time. Thus, the fine-grained films listed in Table 17-3 are commonly used, even though there are other coarser-grained films with higher sensitivity.

The types of photographic films listed in Table 17-3 are films with small grain size, in order to provide maximum resolution. The fact that the silver occurs as separate grains with a finite size ultimately limits the resolution, because a recorded fringe must be at least one grain wide. Some other media (e.g., dichromated gelatin) have no grains and thus do not have this limitation on their resolution.

Of the types of film listed in Table 17-3, type 649F has been the standard type used in much of the early history of holography, at least partly because of its high resolution. A comparison with type SO-243 shows how one may sacrifice resolution in order to gain sensitivity. In more recent times, other types of film are becoming available, offering favorable tradeoffs in the sensitivity–resolution relationship.

In order to be in a range of film response which gives maximum change of optical density in the fringes, one must provide a total exposure that lies within the region of linear response of the film. This means that the relatively constant reference beam should provide an exposure that lies near the center of the linear portion of the H–D curve (see Fig. 17-7). This in turn generally requires that the reference beam (at the position of the recording medium) be approximately five to ten times more intense than the object beam. This ratio is often used so as to give maximum fringe modulation.

The top seven entries in Table 17-3 are amplitude recording media, whereas the lower four entries are phase media. The increase in efficiency which arises from use of phase media is obvious.

One process for producing a bleached phase hologram is given in Table 17-4. This process was developed by the Eastman Kodak Company for

TABLE 17-4
Bleaching Process for Phase Holograms

Step	Material[a]	Time	Temperature (°F)
Develop	Solution A plus solution B (mix equal parts immediately before use)	5–8 min	75
Stop	Solution C	15 sec	70–80
Rinse	Running water	1 min	70–80
Bleach	Solution D	2 min	70–80
Wash	Running water	5 min	70–80
Dry	—	As needed	70–80

[a] Solution A : 750 ml water, 8 gm sodium sulfite, 40 gm pyrocate-chol, 100 gm sodium sulfate, plus water to make 1 liter.

Solution B : 750 ml water, 20 gm sodium hydroxide, 100 gm sodium sulfate, plus water to make 1 liter.

Solution C : 1 liter water plus 48 ml acetic acid (28%).

Solution D : 1 liter water, 9.5 gm potassium dichromate, 12 ml concentrated sulfuric acid.

producing high-quality phase holograms from photographic film. After exposure, one processes the film according to the steps in Table 17-4, instead of using conventional photographic development.

Since photographic fims are easily available and their technology is well developed, they are the media most often used for holograms. Sometimes other considerations might dictate the use of some of the other materials in Table 17-3. For example, thermoplastics and MnBi magnetic films offer erasability; dichromated gelatin offers high resolution and the highest efficiency.

These various nonconventional materials employ various physical or chemical effects to record the holographic fringe pattern. The thin films of metallic bismuth are simply vaporized in areas where the light intensity is high. The magnetic MnBi films are heated and demagnetized in areas where the light is most intense; the reconstructed image is then formed through magnetooptical effects. Intense light causes the surface of thermoplastics to deform in the presence of an externally applied electric field. Intense ultraviolet light causes the photopolymers to polymerize. These last two effects simply cause a variation in optical path length where the light was most intense.

Finally, dichromated gelatin is worthy of somewhat more discussion. Table 17-5 gives a procedure for preparing a hologram from a thin coating

TABLE 17-5

Production of Hologram from Thin Gelatin Films on Glass

Step	Material	Conditions
Sensitize plates	4 % aqueous solution of ammonium dichromate plus 1 part in 20,000 Synthrapol N wetting agent	Dip vertically into solution for 2 min.
Dry	—	Hang vertically as necessary.
Store	—	In dark, temperature $<20°$C. Use within 12 hr after sensitizing.
Exposure	—	Argon laser, exposure to 20,000–50,000 erg/cm^2
Reduce chromium	0.5 % aqueous solution of ammonium dichromate followed by Kodak rapid fixer	5 min each solution
Develop	Water with 1 part in 20,000 Synthrapol N	Dip vertically for 1 min. 25–40°C.
Dehydrate	Isopropyl alcohol	70°C for 30 sec
Dry	Dry air	Blow stream of air over hologram as necessary.

(~ 15 μm) of gelatin on a glass backing.† The procedures given in Table 17-5 can be varied to give holograms of slightly different characteristics [3, 4]. The process involves a tanning or crosslinking of the gelatin in regions where the light intensity is high. This in turn changes the refractive index in those areas to produce the phase hologram. The chemistry of these processes is understood only poorly. The reducing step may be omitted entirely; however, if it is omitted the finished hologram tends to appear milky and may scatter light excessively. The development step involves removal of the ammonium dichromate so as to eliminate further reactions with light. The isopropyl alcohol dehydration removes the water from the hologram, and is needed to achieve the highest values of efficiency. The exact chemical concentrations, times, and temperatures are not extremely critical in any of these steps.

The end result of recording a hologram in any of these materials is a recording of a series of fringes in the recording medium. The method of forming the fringes varies (photochemical reduction of silver halides, tanning of gelatin, etc.) and the result may be any of a number of types of hologram (see Table 17-1) according to the way the exposure was made. The result is always the storage of a pattern of fringes in the recording medium. The pattern almost certainly has no obvious relationship to the original object. But when the fringe pattern is reilluminated, it diffracts the light so as to form an image of the object. Moreover, the fringe pattern is spread through the entire hologram so that information about the entire image is spread through the entire hologram. Thus, the entire scene may be reconstructed from any small area of the hologram (or at least from any area large enough to contain a reasonable number of fringes). Thus, one has the amusing property (in contrast to ordinary photography) that if the hologram is cut in half, one has two holograms, each containing the entire picture.

REFERENCES

[1]   A. L. Rosen, *Appl. Phys. Lett.* **9**, 337 (1966).
[2]   W. E. Kock, L. Rosen, and J. Rendeiro, *Proc. IEEE* **54**, 1599 (1966).
[3]   R. G. Brandes, E. E. Francois, and T. A. Shankoff, *Appl. Opt.* **8**, 2346 (1969)
[4]   L. H. Lin, *Appl. Opt.* **8**, 963 (1969).

† One source of gelatin on glass plates is to dissolve in a fixing bath the silver halides from the emulsion of an undeveloped photographic plate. Then the gelatin film is washed in water and methyl alcohol.

SELECTED ADDITIONAL REFERENCES

A. Formation of Holograms

R. J. Collier, C. B. Burckhardt, and L. H. Lin, "Optical Holography," Academic Press, New York, 1971, Chapter 1.

M. Françon, "Holography," Academic Press, New York, 1974.

D. Gabor, A New Microscopic Principle, *Nature (London)* **161**, 777 (1948).

D. Gabor, Holography, 1948–1971, *Proc. IEEE* **60**, 655 (1972).

E. N. Leith and J. Upatnieks, Wavefront Reconstruction with Diffused Illumination and Three-Dimensional Objects, *J. Opt. Soc. Am.* **54**, 1295 (1964).

E. N. Leith and J. Upatnieks, Photography by Laser, *Sci. Am.*, p. 24 (June 1965).

E. N. Leith and J. Upatnieks, Progress in Holography, *Phys. Today*, p. 28 (March 1972).

K. S. Pennington, Advances in Holography, *Sci. Am.*, p. 40 (February 1968).

G. W. Stroke, Lensless Photography, *Int. Sci. Technol.*, p. 52 (May 1965).

B. The Holographic Process

J. C. Brown and J. A. Harte, Holography in the Undergraduate Optics Course, *Am. J. Phys.* **37**, 441 (1969).

R. J. Collier, C. B. Burckhardt, and L. H. Lin, "Optical Holography," Academic Press, New York, 1971, Chapter 2.

J. B. DeVelis and G. O. Reynolds, "Theory and Applications of Holography," Addison-Wesley, Reading, Massachusetts, 1967, Chapters 2 and 3.

W. E. Kock, L. Rosen, and J. Rendeiro, *Proc. IEEE* **54**, 1599 (1966).

H. M. Smith, "Principles of Holography," Wiley (Interscience), New York, 1969, Chapters 2 and 3.

C. Hologram Types and Their Efficiencies

R. J. Collier, C. B. Burckhardt, and L. H. Lin, "Optical Holography," Academic Press, New York, 1971, Chapter 9.

A. A. Friesem and J. L. Walker, Thick Absorption Recording Media in Holography, *Appl. Opt.* **9**, 201 (1970).

E. N. Leith *et al.*, Holographic Storage in Three-Dimensional Media, *Appl. Opt.* **5**, 1303 (1966).

E. N. Leith, White-Light Holograms, *Sci. Am.*, p. 80 (October 1976).

E. G. Ramberg, The Hologram—Properties and Application, *RCA Rev.* **27**, 467 (1966).

T. A. Shankoff, Phase Holograms in Dichromated Gelatin, *Appl. Opt.* **7**, 2101 (1968).

H. M. Smith, "Principles of Holography," Wiley (Interscience), New York, 1969, Chapters 3 and 4.

D. Practical Aspects of Holography

J. J. Amodei and R. S. Mezrich, Holograms in Thin Bismuth Films, *Appl. Phys. Lett.* **15**, 45 (1969).

R. E. Brooks *et al.*, Holographic Photography of High Speed Phenomena with Conventional and Q-switched Ruby Lasers, *Appl. Phys. Lett.* **7**, 92 (1965).

R. J. Collier, C. B. Burckhardt, and L. H. Lin, "Optical Holography," Academic Press, New York, 1971, Chapters 7 and 10.

M. E. Cox and R. G. Buckles, Influence of Selected Processing Variables on Holographic Film Parameters: Kodak SO-243, *Appl. Opt.* **10**, 916 (1971).

C. K. Felber and F. G. Massialas, Design Features of Holographic Apparatus, *Mater. Res. Std.*, p. 19 (September 1971).

J. A. Jenney, Holographic Recording with Photopolymers, *J. Opt. Soc. Am.* **60**, 1155 (1970).

C. H. Knowles, Do It Yourself Laser Holography, *Popular Electron.*, p. 27 (January 1970).

Kodak Plates for Science and Industry, Eastman Kodak Co., Rochester, New York (1967).

R. L. Lamberts and C. N. Kurtz, Reversal Bleaching for Low Flare Light in Holograms, *Appl. Opt.* **10**, 1342 (1971).

T. C. Lee, Holographic Recording on Thermoplastic Films, *Appl. Opt.* **13**, 888 (1974).

L. T. Long and J. A. Parks, Inexpensive Holography, *Am. J. Phys.* **35**, 773 (1967).

R. S. Mezrich, Curie-Point Writing of Magnetic Holograms on MnBi, *Appl. Phys. Lett.* **14**, 132 (1969).

K. S. Pennington and J. S. Harper, Techniques for Producing Low-Noise Improved Efficiency Holograms, *Appl. Opt.* **9**, 1643 (1970).

G. L. Rogers, When to Use Holography . . . and When Not To, *Optical Spectra*, p. 20 (November 1970).

B. Ruff, Pulsed Laser Holography, *Opt. Spectra*, p. 48 (January 1967).

H. M. Smith, "Principles of Holography," Wiley (Interscience), New York, 1969, Chapter 6.

C. L. Stong, How to Make Holograms and Experiment with Them or with Ready-Made Holograms, *Sci. Am.*, p. 122 (February 1967).

J. C. Urbach and R. W. Meier, Thermoplastic Xerographic Holography, *Appl. Opt.* **5**, 666 (1966).

R. H. Webb, Holography for the Sophomore Laboratory *Am. J. Phys.* **36**, 62 (1968).

M. Young and F. H. Kittredge, Amplitude and Phase Holograms Exposed on Agfa-Gevaert 10E75 Plates, *Appl. Opt.* **8**, 2353 (1969).

CHAPTER 18

APPLICATIONS OF HOLOGRAPHY

In this chapter, we will review some of the practical applications that have been suggested for holography. By far the most important application for industrial purposes is holographic interferometry, which can provide a valuable engineering tool for strain and vibration analysis and for defect detection. Thus, most of this chapter will be devoted to holographic interferometry.

It is also instructive to review, at least superficially, the wide range of applications that have been suggested for holography. Some of these (e.g., holographic movies) are still in an early stage of development. Other uses (e.g., use as an art form) are outside the scope of this book. Still, the variety of applications in this introductory survey can be stimulating to the imagination.

After completing the discussion of holographic interferometry, we shall describe briefly some other applications which can have an impact in industry.

A. Holographic Interferometry

Holographic interferometry is the area in which the most significant industrial applications of holography have occurred. Conventional interferometry has commonly been used to determine surface contours for surfaces with relatively simple shapes. One example is the inspection of mirror surfaces and optical flats in the optical industry. Such interferometric measurements had previously been restricted to reflecting surfaces with simple shapes. This restriction is removed by the advent of holographic interferometry. It makes interferometric methods applicable for the testing of surfaces with relatively complicated shapes and for surfaces which are not specularly reflecting.

A hologram can be considered as a device which stores a wavefront representing an image of some object. The stored wavefront is released by the reconstruction of the hologram. In holographic interferometry the wavefront is released and is used to interfere with some other wavefront, so as to form bright and dark fringes in regions of constructive and destructive interference. The two wavefronts can represent an object at different instants of time. For example, one may make a hologram of an object and then change the object slightly. If the reconstructed image, released from the hologram during reconstruction, is allowed to interfere with light from the altered object, the interference pattern can give sensitive information about how much the object has changed. Since one fringe will be formed when the object has its dimensions changed by one wavelength of light, the sensitivity of this measuring technique is high, comparable to the wavelength of the light being used.

There are a number of different ways in which the basic technique of holographic interferometry can be employed, depending on how the different wavefronts representing the object are obtained. We shall describe three distinct types of holographic interferometry. These three types are

(1)   real-time holographic interferometry,
(2)   double-exposure holographic interferometry, and
(3)   time-average holographic interferometry.

1.   *Real-Time Holographic Interferometry*

This method basically involves making a hologram, altering the object and then allowing the light from the reconstructed image stored in the hologram to interfere with light from the altered object. Because alterations in the object can be continuously varied, the changes in the object can be viewed continuously by motion of the fringe system. Thus, this method is sometimes called live fringe holographic interferometry.

The steps in the procedure are as follows. First, a hologram is made of the desired object. The hologram is developed using conventional photographic techniques and then is replaced in the position it occupied when the hologram was made. The image is reconstructed using the original reference beam. Thus, one obtains an image of the original object. This image will be superimposed on the real physical object which has remained in position. Thus, one has two views of the object, one from the object itself and one from its reconstructed holographic image.

The two wavefronts representing these two images will interfere. If the path difference in light coming from the same point in the two different images varies by one wavelength of light, one interference fringe will appear.

If the object has not changed in any way since the hologram was made, the two wavefronts will exactly coincide and there will be no interference fringes. This situation is difficult to achieve in practice. It is difficult to avoid distortion of the hologram. During development, photographic film will change dimensions somewhat. Also, it is difficult to replace the hologram exactly in the position that it occupied during the original exposure. Thus, there will often be some background fringes, even if the object has not changed. With careful technique the number of such fringes can be small. Development of the film in situ can eliminate the repositioning problem.

If the object has changed since the hologram was made, there will be interference fringes visible across the superimposed images, with each fringe representing a cumulative deformation of the object by one wavelength of light.

Let us consider a specific example. If the object is placed under stress and made to deform, the optical distance from the observer's eye to any point of the displaced surface will change. Light rays coming to the observer's eye from the two corresponding points on the image and on the deformed object will thus have a relative phase shift, which produces the observed fringe pattern. The fringe pattern defines the amount of deformation of the surface. If there are any constant features in the object, these will not give rise to fringes. Thus, any constant background will be subtracted out. Holographic interferometry will make small changes in the object stand out.

The motion of the fringe system can be seen immediately as the object is deformed. The fringe pattern will change and spread across the surface of the combined images according to the deformation of the object in real time. Hence, the name real time holographic interferometry is used to describe this process.

A schematic diagram of real-time holographic interferometry is shown in Fig. 18-1. In (a), a hologram is made using holographic techniques as described earlier. In (b), the hologram has been developed and replaced in its original position. The object, shown here as a can, has its end deformed by increasing the air pressure so that the end of the can bulges out. The result will be a circular pattern of fringes as shown in (c). The maximum displacement of the end of the can will be at the center, so that the maximum cumulative number of fringes across the fringe pattern will be largest in the center. If one were to observe this pattern in real-time, one would observe new fringes forming at the center of the can and gradually opening out as the air pressure is increased.

The spacing and motion of the fringes can be measured and related to the amount of deformation of the surface. Thus, real-time holographic interferometry can provide a sensitive tool for measuring strain of objects as

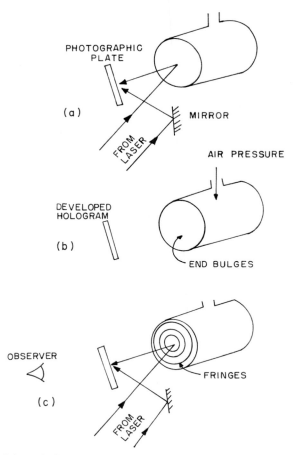

Fig. 18-1   Schematic diagram of holographic interferometry. In (a), a hologram is made of the end surface of a can. In (b), the hologram has been developed and replaced in position, while increased air pressure in the can causes the end to bulge. In (c), interference between the two scenes (the physical object and the holographically reconstructed image) leads to a pattern of circular fringes.

they deform in real time. For example, if there is a weak area in the surface, it will deform more. The fringes will crowd closer together around the weak area. Therefore, this technique could be used for detecting defects or weak areas in a structure.

Real-time holographic interferometry does have several problems in its practical exploitation. One problem involves the replacement of the hologram in exactly the position it occupied during the exposure. It must be repositioned accurately within a fraction of the wavelength of light. In order to achieve this, the hologram mount may be equipped with micrometer drives for precise alignment. The alignment is accomplished by viewing the fringe pattern and varying the position of the hologram until the number of fringes visible across the image is at a minimum. This can be a time-consuming process. To alleviate this problem, the hologram can be developed in place. Development systems have been produced such that the chemicals are added and the photographic film is processed in situ, without any need for moving it.

A second problem occurs because of distortion of the photographic film. Some dimensional change in the emulsion inevitably occurs during the development process. Careful control of the development can minimize distortion but in practice there will always be some distortion remaining. This will give a background shift of a small number of fringes across the field of view.

Another problem is interpretation of the fringe pattern so as to obtain quantitative measurements of the change in the object. If one desires information such as the presence or location of defects, the crowding together of fringes near the defect will easily give the desired result, without the need for much processing of the data. But when complete quantitative descriptions of the changes in the object are desired, data reduction can be tedious. In order to discuss this, we must first consider the concept of fringe localization. Interference fringes in holographic interferometry are localized on some surface. The surface of localization is a surface on which the fringes can be observed with highest contrast, by an observer using an apertured optical system. The location of this surface will depend in a complicated fashion on all the parameters of the experimental arrangement. However, in general it will not be the surface of the object. This fact causes the difficulty. One cannot focus the optical system so as to view the object surface and the high-contrast fringes simultaneously.

Let us first consider a simple case, where the surface of localization is the object surface. One such case in which the fringes are localized on the surface of the object is a pure rotation of the object about an axis in the surface, with observation of the fringes in a direction perpendicular to the surface. In other cases, such as translations of the surface or combinations of translations and rotations, the fringes will be localized on some other surface in space, which may lie anywhere between the original surface of the object and infinity. It is the failure of the fringes to be localized on the object surface which makes the analysis difficult in the general case.

Let us now analyze the simple case in which one has a pure rotation about an axis in the surface and where one is looking in a direction perpendicular to the surface. This geometry is illustrated in Fig. 18-2. The z axis is the direction of illumination and of observation. The phase shift δ between

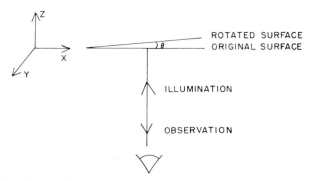

Fig. 18-2 Schematic diagram for interpretation of fringe pattern formed in holographic interferometry when a surface is rotated through an angle θ. Fringe pattern will consist of parallel fringes in the y direction.

the two waves coming from the holographic image and from the original object is given by the equation

$$\delta = 4\pi x\theta/\lambda \qquad (18.1)$$

where x is the distance along the x axis, as illustrated, θ the angle of rotation, and λ the wavelength of the light. The intensity I of the fringes will be given by

$$I = C(1 + \cos \delta) \qquad (18.2)$$

where C is approximately constant over the fringe pattern. Thus, an observer will see straight line fringes varying as a cosine function. If one moves a detector parallel to the surface in the x direction, one will obtain an amplitude variation which has a cosine dependence. The maximum intensity seen by the detector will occur when the phase shift is equal to an integral multiple of 2π. Therefore, for a change of one entire fringe, the phase shift will vary by 2π. According to Eq. (18.1), this means a variation in x of $\lambda/2\theta$. Therefore, if one measures the spatial frequency of the fringes, one easily obtains the angle of rotation of the surface. This example shows how the surface deformation can be obtained in one simple case.

In most cases, the situation is much more complicated. In general, the deformation of the surface at a particular point will be a combination of a translation and a rotation, and the magnitude of the translation and rotation will vary from point to point. In addition, the fringes will not in general be localized on the surface. The position at which the fringes are localized will vary as the direction of observation changes. Some techniques have been developed by which the surface displacement can be derived from measurements of the fringes. Such procedures are relatively complex and we will not derive them here. We will simply summarize one procedure which has been developed.

The relation of surface displacement to fringe observation can be understood with reference to Fig. 18-3, in which an imaging lens is focused on the target surface. The surface on which the fringes are localized is shown. The imaging lens has its axis oriented along the line PA, where P is a point in the surface of localization. The image plane of the lens is apertured so that one observes light originating from only a small area surrounding the point A on the image surface. The smaller the aperture becomes, the greater is the depth in which fringes may be viewed with good contrast. When the aperture is made small enough, the fringes and the object may be viewed simultaneously.

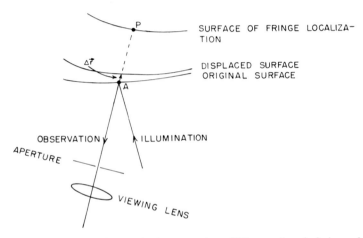

Fig. 18-3 Schematic diagram for interpretation of fringe pattern in holographic interferometry when the fringes are localized at a position not on the object surface. The point A on the surface, which is being viewed, is displaced vectorially by $\Delta\mathbf{r}$. The axis of the viewing system intersects the surface of fringe localization at point P.

The lens focuses through the hologram onto the original and deformed surfaces, which are nearly coincident, and simultaneously allows observation of high-contrast interference fringes. One then determines the displacement of one small area surrounding A on the subject surface. Observations are made on the fringes which appear near A. Suppose that one makes an observation in a direction represented by a unit vector $\mathbf{n}_1$. The phase difference δ_1 between rays from the area A for the two surface positions is given by

$$\delta_1 = (2\pi/\lambda)\,\Delta\mathbf{r}\cdot(\mathbf{n}_0 - \mathbf{n}_1) \qquad (18.3)$$

where $\Delta\mathbf{r}$ is the vector displacement between the two surfaces and $\mathbf{n}_0$ is a unit vector in the direction of illumination. Then the area around A is viewed from a new viewing direction represented by the unit vector $\mathbf{n}_2$. As one changes viewing direction, the number of fringes m passing through A is counted. The direction of illumination is of course unchanged and the phase difference δ_2 is now given by

$$\delta_2 = (2\pi/\lambda)\,\Delta\mathbf{r} \cdot (\mathbf{n}_0 - \mathbf{n}_2) \qquad (18.4)$$

Subtraction yields

$$\delta_1 - \delta_2 = (2\pi/\lambda)\Delta\mathbf{r} \cdot (\mathbf{n}_2 - \mathbf{n}_1) = \pm 2\pi m \qquad (18.5)$$

where $2\pi m$ is the number of radians representing the motion of m fringes through A. The appropriate sign must be determined from other considerations. This equation represents a linear equation in the three unknown components of the displacement $\Delta\mathbf{r}$. If one then carries out two more pairs of measurements, one has two additional equations from which to determine all three components of the displacement $\Delta\mathbf{r}$. This procedure is relatively time consuming and leads to information about the displacement at only one point on the surface. A more general characterization of the surface displacement would involve additional measurements at many other points.

Development of methods for determination of three-dimensional surface motion from the fringes that appear in the reconstruction of a holographic interferogram has been the subject of continuing development for many years. Many methods have been proposed and demonstrated, but as the above example shows, most are fairly cumbersome and involve extensive measurement and calculation. Continuing developments have led to advances and it is anticipated that further work in this area will eventually lead to simplified techniques.

In one particular case, a simplified method which does not require extensive calculation is available [1]. This is the case when the distorted surface is locally flat, is being observed at normal incidence by an optical system with a narrow slit aperture, and the components of object motion lie in the surface of the object. If the long direction of the narrow aperture is the x direction, one obtains

$$\partial L_x/\partial x = \lambda/2d_x\,\partial D_x/\partial x \qquad \partial L_x/\partial y = \lambda/2d_x\,\partial D_x/\partial y \qquad (18.6)$$

where L_x is the component of object motion in the x direction, D_x the distance of the localization plane from the surface, λ the wavelength, and d_x the fringe spacing in the x direction. If one rotates the aperture $90°$ so that measurements are made along the y direction, one has

$$\partial L_y/\partial x = \lambda/2d_y\,\partial D_y/\partial x \qquad \partial L_y/\partial y = \lambda/2d_y\,\partial D_y/\partial y \qquad (18.7)$$

where L_y is the component of object motion in the y direction, D_y the distance from the object surface to the localization plane in the y direction, and d_y the fringe spacing in the y direction. We note that the partial derivatives of D_x and D_y are the slopes of the localization plane. Thus, $\partial D_x/\partial x$ is the slope of the localization plane in the x direction when the aperture is in the x direction. These equations provide enough parameters to characterize in-surface rotation and strain, in terms of parameters which can be measured, namely the fringe spacings and the slopes of the localization plane for the fringes.

Figure 18-4 shows a photograph of the fringe pattern formed in real time holographic interferometry for an object with a relatively complicated shape [2]. The hologram was made with no stress present. When stress was added, fringes were formed and the motion of the fringes could be observed as they moved across the surface. Each fringe can be regarded as representing

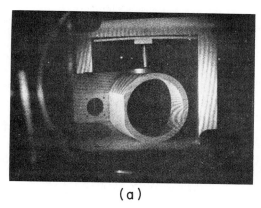

(a)

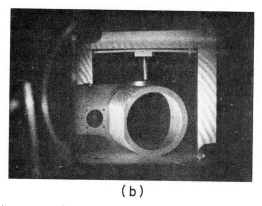

(b)

Fig. 18-4 Fringe pattern formed in real-time holographic interferometry as a complex object is stressed. The stress is larger in (b) than in (a). (From K. A. Haines and B. P. Hildebrand, *Appl. Opt.* **5**, 595 (1966).)

a constant change in the surface so that the holographic interferogram is similar to a contour map. In this example, the deformation of the object could be monitored continuously as the stress was varied. It is clear both that the interferogram shows strain patterns very dramatically and that mathematical reduction to provide quantitative results will be tedious.

2. Double-Exposure Holographic Interferometry

Double-exposure holographic interferometry differs from real-time holographic interferometry in that two exposures of the object are made at different instants of time. For the two exposures the object will be slightly different. Thus, the double-exposure method compares the object in two different conditions. Two separate holographic images are obtained when the hologram is reconstructed, and the interference is between the wavefronts representing the two images. In contrast to real-time holographic interferometry, the object need not be present during the reconstruction. The wavefront which is characteristic of the object in its original condition is stored in the hologram, along with the wavefront representing the altered state of the object.

This method is simpler and easier to carry out than real-time holographic interferometry. The double-exposure method avoids the problem of realignment of the hologram, because both images are stored in the hologram. It may be reconstructed without taking any particular pains for exact repositioning and realignment. Distortion due to emulsion shrinkage is also eliminated, because emulsion shrinkage is identical for both exposures.

In comparison with real-time holographic interferometry, the double-exposure method has the disadvantage that it can compare the original object with only one altered state of the object. Therefore, it is somewhat less versatile than the real-time method and will yield less complete information about the continuous change of the object. In many practical cases, continuous monitoring of the surface deformation will not be necessary. Recording of relative surface displacement at a fixed interval of time can be useful. Because of the lack of the ability to monitor changes continuously, the double-exposure method is sometimes called a frozen fringe method.

The same comments about fringe localization and the same difficulties in interpreting the fringe pattern to obtain quantitative information about the surface are still valid. Analysis of the fringe pattern to obtain relative deformation of a surface is very difficult in the general case.

Figure 18-5 shows how double-exposure holographic interferometry can be applied to defect detection. The example shown here is relevant to the testing of tires. One exposure is made with low air pressure in the tire. The second exposure is made after the air pressure has been increased. Two

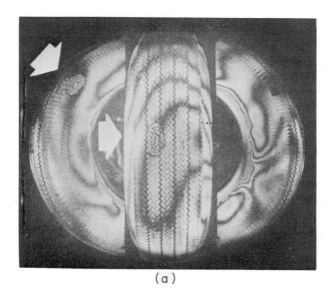

(a)

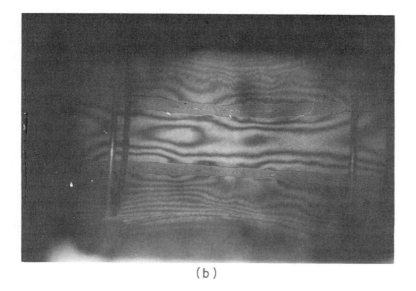

(b)

Fig. 18-5 Double-exposure holographic interferograms of tires, in which air pressure was increased between exposures. In (a), arrows indicate areas of relatively large distortion, where fringes are crowded close together. These areas may represent defects. In (b), the areas where the fringes are close together show belt edge separation in each shoulder of the tire. (Photographs courtesy of GCO, Inc.)

examples of fringe patterns obtained when the double-exposure holograms were reconstructed are shown. A weak area on the tire deforms more when stress is applied. Thus, fringes are crowded closer together in the weak areas. These regions are areas where the tires had low strength and possibly represent defects.

3. Time-Averaged Holographic Interferometry

A general rule applicable to most holographic situations is that the object should remain stationary during the period of exposure of the hologram. This general rule is violated dramatically in the case of time-average holographic interferometry. During the exposure, the object is moving continuously. Time-average holographic interferometry is generally employed to study vibrating surfaces. The hologram which results when the surface is vibrating at high frequency can be considered as the limiting case of a large number of exposures for many different positions of the surface. The mathematics of the situation are complex and we shall not deal with them here. A conceptual idea of the result can be obtained by considering a simplified model. The exposure of a continuously vibrating surface is considered to be similar to a double-exposure holographic interferogram for which the two different exposures represent the positions where the surface spends the most time. These two positions would be the positions of extreme movement of the surface, which are the positions where the speed of vibration is lowest. In this very simplified view, the two wavefronts are considered to originate from the surface at its extreme positions in its vibratory motion. This situation is sketched in Fig. 18-6.

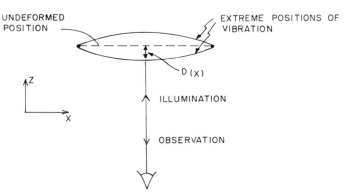

Fig. 18-6 Schematic diagram for interpretation of fringes on vibrating surface in time average holographic interferometry. As one moves across the surface in the x direction, the maximum vibratory displacement of the surface from its undeformed position is $D(x)$.

With this method one can measure the vibrational amplitude of diffusely reflecting surfaces with high precision. The method is simple to employ. It involves making a single hologram while the surface is vibrating. The period of exposure should cover many vibrational periods of the surface. The restriction, usual in holography, of no object motion during exposure has been dramatically removed. The hologram is made just as if the surface had been motionless. After development and reillumination, the resulting fringe pattern is measured to obtain information about the relative vibrational amplitude across the vibrating surface. Measurements of this type can be very useful in determining modes of vibration of complex structures which are difficult to measure by conventional techniques. Time-average holographic interferometry is most useful in its application to vibrational analysis.

The reduction of the observed fringe pattern to obtain quantitative measurements of the amplitude of the surface vibration can be obtained with reference to Fig. 18-6. The light intensity I as a function of position x is given approximately by the equation

$$I = \lambda/4\pi D(x) \cos^2[4\pi D(x)/\lambda - \pi/4] \tag{18.8}$$

where $D(x)$ is the displacement of the surface from its equilibrium or undeformed position, as a function of the transverse coordinate x. In this case, one fringe appears when the quantity $4\pi D/\lambda$ changes by π. We note that the change is not 2π, because the equation involves the square of a cosine function. Therefore, one fringe corresponds to a change of D by one-quarter wavelength of light. Counting fringes across the pattern thus yields the displacement as a function of position. This analysis is valid when

$$4\pi D/\lambda \gg 1 \tag{18.9}$$

Equation (18.8) breaks down when this inequality does not hold. In that case, more complicated mathematical expressions must be employed.

Equation (18.8) allows analysis of fringe patterns to obtain the maximum vibratory amplitude as a function of position across the surface. The amplitude of the fringes, according to Eq. (18.8), decreases as a function of increasing displacement D. This decrease of intensity limits the number of fringes which can be observed. Therefore, there is a maximum displacement of the surface from its equilibrium position for which time average holographic interferometry will be useful.

A specific example of the application of time average holographic interferometry to vibration analysis is shown in Fig. 18-7 [3]. Vibrations of the bottom of a 35 mm film can were produced by a solenoid mounted inside the can and driven by a audio signal generator. In this figure, one type of

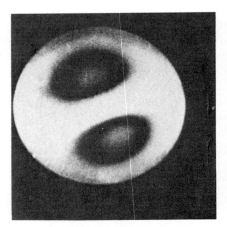

Fig. 18-7   Reconstruction of the pattern of vibration of the bottom of a film can vibrating at an audio frequency. The fringes represent contours of equal vibratory amplitude. The frequency is the same in both photographs, and the amplitude of vibration is larger in the right photograph. (From R. L. Powell and K. A. Stetson, *J. Opt. Soc. Am.* **55**, 1593 (1965).)

vibration is observed when the can is excited at one particular audio frequency. Each closed fringe represents a contour of constant amplitude. Thus, the interferogram could be interpreted like a contour map. The vibration pattern is symmetric about a central line along which the amplitude is zero. As one moves across the surface of the can in a direction perpendicular to this line, the amplitude increases from zero at the edge to a maximum, falls to zero at the center line, rises to another maximum, and falls to zero at the opposite edge. In the right portion of the figure, the voltage applied to the audio frequency generator was increased above the voltage used for the left photograph. The vibration amplitude was greater and more fringes are visible. The exact values of the amplitude could be obtained by application of Eq. (18.8).

The technique of time-average holographic interferometry has proved valuable for analysis of the motion of the surface of musical instruments, for analysis of turbine blade vibrations in the aircraft industry, and for study of vibrations of engines, frames, and brake systems in the automotive industry.

4.   *Summary of Holographic Interferometry*

In the foregoing sections we have discussed some of the leading methods by which holography is applied to sensitive measurements of surface deformation. The specific types of holographic interferometry discussed by no means cover the entire list of possibilities. The discussion does cover some of the more important methods and illustrates the general principles. In order to compare these methods, we summarize the relative advantages and disadvantages and potential areas of application in Table 18-1.

TABLE 18-1
Summary of Techniques for Holographic Interferometry

Technique	Manner of implementation	Advantages	Disadvantages	Typical application
Real-time	Expose hologram, develop, replace, then stress object	Complete information about changes in object	Problems in interpretation, repositioning, and emulsion shrinkage .	Strain analysis Defect detection
Double-exposure	Expose hologram, stress object, reexpose hologram, then develop	Easy to make. No problems of repositioning or emulsion shrinkage	Less complete information than real time. Problems in interpretation	Strain analysis Defect detection Analysis of transient events
Time-average	Expose hologram while object is in motion, then develop	Easy to make. Relatively easy to interpret	Not useful for static surfaces	Vibration analysis

In industrial processes, holographic interferometry is an engineering tool for specialized measurements which are difficult to make by conventional techniques. Holographic interferometry has not reached production use by unskilled operators, but rather is employed by skilled personnel for surface contouring, defect detection, investigations of strain, and vibration analysis.

The application of holographic interferometry does have some limitations. It is an extremely sensitive technique, with potential resolution of the order of a wavelength of light. Many practical situations do not require this accuracy, and in such cases the extreme sensitivity of holographic interferometry may actually be a disadvantage. When the distortion of a piece of material is large, holographic interferometry is probably not the technique of choice. In many cases encountered in practical industrial applications, the motion of the piece covers many thousands of wavelengths. In this case, the fringes would be faint and difficult to count. This sets a practical limitation on the usefulness of holographic interferometry, to those cases where the

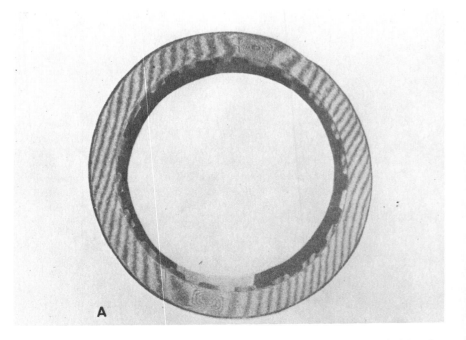

Fig. 18-8 Double-exposure holographic interferogram showing areas of disbonding in an automotive clutch plate. The plate was stressed thermally between exposures. (From R. K. Erf, ed., Holographic Nondestructive Testing. Academic Press, New York, 1974.)

motion is a relatively small number of optical wavelengths. In cases where the motion is larger, moiré techniques are more applicable [4].

Testing of laminates and composite structures can be carried out simply. Thermal stressing is an easy method to apply to metal structures adhesively bonded with resin. If a hologram is made of the structure at ambient temperature and viewed in real time as the structure is heated a few degrees above ambient temperature, the holographic fringes may be viewed in real time. If there are disbonded areas, the fringes will be distorted as they move over these areas. Alternative methods of stressing include the use of pressure and vibration. Such methods have been applied to laminated structures such as automotive clutch plates, simulated uranium fuel elements, graphite epoxy jet engine fan blades, and composite compressor blades. Figure 18-8 shows an example of the detection of disbonding in an automotive clutch plate, adhesively bonded to a steel substrate [5]. Thermal stressing by induction heating was used in this case. Disbonded areas are easily detected.

B.   A Miscellany of Applications

No other application of holography is so well developed for industrial use as holographic interferometry. There are a number of other possibilities, some of which we shall review briefly below. This will by no means represent a complete survey, but will serve to identify a variety of interesting potential uses of holography.

1.   *Microscopy*

The technique of holography was originally suggested by Professor Dennis Gabor for microscopy. The original purpose of the two-step hologram microscope was to use electron waves or x rays in the first stage to form the hologram and optical radiation in the second stage to obtain a high resolution reconstruction. Some progress toward achieving magnification has been achieved using visible light for both steps. Magnification is achieved without lenses by using a diverging instead of a collimated reference beam. The image can be greatly magnified and has depth that cannot be achieved by conventional means. In addition, the magnified image can be three-dimensional. The quality of the image does not compete with well-developed techniques of conventional microscopy for routine uses.

The very high magnification that would result from the full implementation of Gabor's original idea still lies in the future. Coherent x-ray sources are still lacking. Some work has been done toward producing holograms in the vacuum ultraviolet. For example, holograms have been produced at 0.1182 μm, using the ninth harmonic of the Nd:YAG laser line at 1.06 μm

[6]. The recording medium was polymethyl methacrylate, used as type of photoresist in which the ultraviolet light produces a surface relief. Reconstruction of these holograms with a visible laser would yield magnification equal to the ratio $\lambda_v/0.1182$ μm, where λ_v is the wavelength of the visible laser. Magnification up to six would be attainable. Gabor's suggestion envisioned use of much shorter wavelengths for recording the hologram, and hence of much greater magnification.

The practical problems associated with achieving magnification comparable to that of an electron microscope are many; for example, coherent x-ray sources, positioning of the reference source, preparation of the object, and control of hologram aberrations and intensity levels. These are some of the problems yet to be solved before an x-ray holographic microscope with magnification extending possibly to atomic dimensions is experimentally realizable. Development of new and intense x-ray sources could conceivably alter this situation and give to holographic microscopy the same impetus that the development of the laser gave to the field of holography as a whole. This is one reason for interest in development of x-ray lasers.

2. *Study of Events in Depth*

Holographic microscopy has been used for examining three-dimensional records in a way that could not be done before. Because the reconstruction can be studied at leisure, the form and distribution of small objects can be studied fully, even though their positions may have subsequently changed. A hologram made with a pulsed laser freezes the motion of the particles in the sample volume. The film is processed and illuminated so as to produce a reconstructed three-dimensional sample volume with a large depth of field. By focusing on a single plane in the image, details in that plane can be viewed. This technique has had application in the study of aerosols and the measurement of their particle distributions by optical sectioning of the hologram image. Similar techniques have been employed for examination of holographic microscope pictures taken of microscopic flora and fauna. The distribution at an instant of time can be preserved and examined three-dimensionally in a way that was not possible before. Nuclear physics applications in studying bubble chamber tracks also appear promising.

The very first direct use of holography in a practical application was in fact the sizing of small aerosol particles. Holograms of naturally occurring fog particles with a range of sizes between 4 and 200 μm were recorded. When the hologram was reconstructed, the image was magnified and picked up by a TV camera and displayed on a screen. As the sample volume in the reconstructed image was scanned, different particles came into sharp focus at their correct positions within the volume. This allowed the investigators to obtain particle size distributions within the fog by periodically recording holograms of all particles in the sample volume.

Similar work can be useful in the study of the operation of spray devices, nebulizers, and aerosol cans. Current public awareness of possible problems of airborne particulates makes it important for manufacturers to characterize the production of aerosols carefully; holography could make a contribution in this area.

3. Display

Holography offers many possibilities for display. The striking nature of the three-dimensional image produced by holograms makes this obvious. Use of holography for artistic purposes is also important. Holography appears to be developing rapidly as an art form. Holographic jewelry has also made a commercial appearance.

The production of vivid three-dimensional imagery is of considerable significance. When coupled with holographic generation of depth contours on the image, it can provide a valuable tool in such areas as machinery modeling and photographic reconnaissance.

Another less obvious possibility is the application of computer-generated holograms to displays of objects that never existed. The technique has been shown to be feasible. The intensity pattern produced by diffraction from the hypothetical object is calculated by a computer and is plotted by a computer-directed plotter. This plot can then be photographed, reduced and illuminated by a laser beam. The resulting reconstruction shows the image of the object which existed only as a mathematical specification to the computer. At present, such techniques are limited because of the nature of the steps employed in making the plot. If one used erasable recording media which could operate quickly, real-time computer-generated displays could be obtained in three dimensions. This technique offers exciting possibilities for such areas as military displays or air traffic control.

The use of holographic displays for advertising does not seem to have been much exploited.

4. Holographic Movies and Television

Startling applications in three-dimensional television and movies have been suggested. Such uses do not appear practical in the near future, although the long-term possibilities cannot be ignored. True holographic motion pictures (true in the sense that each frame is itself a hologram) have been generated experimentally.

Such applications have involved the use of repetitively Q-switched lasers, operating at a pulse repetition rate compatible with the motion picture framing rate. Such motion pictures have so far not been as pleasing artistically as conventional motion pictures. In addition, there is a problem with viewing of the pictures by a sizable number of people simultaneously.

Three-dimensional holographic television has also been demonstrated experimentally, but considerable advances in the formation of the holograms, their transmission, and the viewing systems are all required before holographic television becomes commercially practical.

5. *Flow Visualization*

Interferometric techniques have been used to examine fluid flow patterns. Variation in refractive index gives rise to optical path differences. Holographic interferometry can be carried out to give a single hologram capable of yielding information comparable to conventional interferometry and Schlieren photography. The double-exposure holograms are recorded with and without flow.

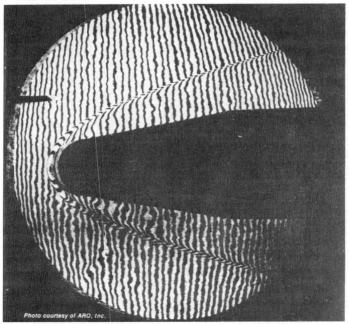

Fig. 18-9 Double-exposure holographic interferogram of a bullet in flight, made with a Q-switched ruby laser. The initial exposure was of the undisturbed air. The discontinuity in the fringe pattern outlines the shock front produced by the bullet. (Photograph courtesy of ARO, Inc.)

Flow visualization using pulsed lasers has also been demonstrated to photograph transient shock waves. In one example, the pulse (from a Q-switched laser) was short enough to freeze the motion of a bullet. This is illustrated in Fig. 18-9. The fringes delineate the shock waves in the air due to the passage of the bullet. The change in refractive index gives rise to optical path length changes. The density of the air and the position of the shock front can be deduced as a function of position from this reconstruction.

6. Acoustic Holography

Holographic recording principles may be extended to ultrasonic analysis of opaque objects or to underwater mapping. In one method, a sonic wave "illuminates" (or, more properly, insonifies) an object in a fluid. There is an object wave and a reference wave, derived from two acoustic signal generators. The two waves interfere at the free surface of the liquid. The hologram is the pattern of ripples generated at the surface. Reconstruction may occur by means of a laser beam which is reflected off this pattern of ripples (see Fig. 18-10). The reconstructed light beam is photographed on a photographic plate. This procedure yields a view of the object immersed in the fluid.

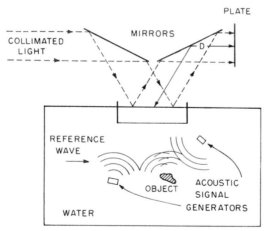

Fig. 18-10 Schematic diagram showing arrangement for acoustic holography of an object immersed in a liquid. The hologram is formed as an amplitude pattern at the liquid surface and reconstructed with visible light.

Pure acoustic holography has also been demonstrated (see Fig. 18-11). An object, shown as a letter, is insonified. A detector scans in a pattern through a plane to detect the diffracted sound waves. Since the sonic detector can respond directly to the amplitude of the sound wave (in contrast to

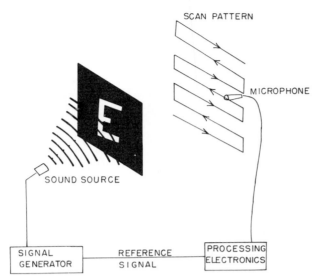

Fig. 18-11   Schematic diagram for purely acoustical holography, in which the reference wave is supplied electrically.

optical detectors which respond to intensity), the reference wave can simply be taken directly from the sonic driver and added electrically to the detector. Thus, a separate reference beam is eliminated. The hologram then consists of an electrical signal which represents the intensity of the acoustic interference pattern as a function of position. The signal could be processed purely electronically.

Acoustic holography has potential applications in detecting and measuring objects immersed in opaque liquids. It is under study for medical applications, for examination of structures internal to the human body, which in some senses is a container of fluid.

7.   *Optical Computing*

Optical computing techniques include performance of Fourier transforms and convolutions by the use of coherent light. A lens produces a Fourier transform of the amplitude distribution of an object located in its front focal plane. The transform appears in the back focal plane of the lens. If an amplitude mask is placed in the back focal plane, then the convolution of the Fourier transform of the mask with the object amplitude distribution appears in the back focal plane of a second lens. This shows that holographic techniques can be used in a natural manner to produce the mathematical operations of Fourier transformation and convolution in two dimensions.

Methods for performing certain other mathematical operations optically have been developed. For example, calculation of correlations and power spectra, optical techniques for addition, subtraction, multiplication, division, differentiation and integration, and the taking of Hilbert transforms have all been described. These methods offer potential promise for optical computations in two dimensions. Data obtained in a two-dimensional form (e.g., photographs) could potentially be processed in a very natural manner. The parallel nature of the processing offers advantages when compared to the sequential nature of electronic processing. This technique is potentially of importance, but many problems have to be solved. Widespread application of this type of computing lies in the future.

8. *Pattern Processing*

Fourier holograms could be used for pattern processing. One makes a hologram of an object, develops it, and replaces it in the Fourier transform plane (the back focal plane) of a lens. Such a hologram is called a complex matched spatial filter. A transparency representing an unknown pattern is placed in the object plane (the front focal plane) of the lens. The complex matched spatial filter will destructively interfere with light from the unknown that does not correspond to the original object. Portions of the light from the object that correspond to the original desired object will be passed by the filter. In the image plane, there will be light at all positions corresponding to places where the desired object appeared in the unknown.

Alternatively, a transparency representing the pattern can be placed in the object plane. In the Fourier transform plane, features with different periodicity will be represented by light in different physical positions. Thus, insertion of filters (opaque stops) can remove this light. When a second Fourier transform is taken (by a second lens), the restored image will lack the features that corresponded to the removed light. Thus, the technique can be used to enhance features or to process data available in the form of photographic transparencies. This technique appears to be very powerful, in principle, for such areas as character recognition, identification of objects in aerial photographs, processing of photographic images, etc. At present there are drawbacks that severely limit its use.

If the light pattern in the input plane of the transparency is a repetitive pattern, the familiar diffraction pattern of an array of dots will be formed. This leads to the possibility of a practical application involving inspection of photomasks, formed by photographic techniques. Very large numbers of patterns can be formed at the same time. Complete inspection is difficult. The mask is an array of the same pattern repeated at the same interval, and therefore it behaves similarly to a diffraction grating. The diffraction pattern is a periodic array of light spots. The filter that would be used would consist of an array of black dots at the position of the light spots. When the image is reformed by the second lens, the result of this operation would remove all information which is repeated at the period corresponding to the basic mask. The result in the image plane contains only things that are not present on all the masks (i.e., it displays only the errors in the mask). All random errors are displayed. This is an application of optical information processing that has reached practical application in industry.

The application of holographic techniques and spatial filtering techniques for pattern processing will be described more fully in Chapter 21.

9. Holographic Computer Memories

A hologram can be formed of a pattern of bits which is to be stored in a computer. The hologram can store this array of bits until they are needed. Then the hologram can be reconstructed and the image projected onto an array of photodetectors to read out the stored information. Thus, the hologram can serve as a computer memory. An entire block of bits is stored or read out at one time (instead of a single bit at a time). Thus, holographic computer memories have the potential for high throughput rates.

The practical problems for holographic computer memory are large, but they are under extensive study at a number of laboratories. The methods and techniques will be described more fully in Chapter 21.

A hologram plate can record more than one image at a time in a thick recording medium, provided that different reference beam directions are used. Each image can be reconstructed separately. The three-dimensional data storage means that the photographic plate can be used in a much more efficient way as a store than if conventional photographs are used. There is considerable interest in recording media which themselves are three-dimensional (e.g., photochromic glass, alkali halides, lithium niobate, and thick emulsions).

When such a medium is used for holography, a three-dimensional fringe pattern is recorded within it. This can be read out only when the reconstructing wave is identical in direction to the recording reference wave. The process involves diffraction from the developed silver grains which lie in parallel planes within the recording medium. The conditions for constructive interference are much more stringent than for holography with thin media. However, a great number of images can be recorded holographically in the same plate and read out separately without interfering with one another. This offers possibilities for use in computer memories with large three-dimensional packing densities.

C. An Example of Holographic Application

We shall describe the use of holographic interferometry to study thermal distortions of transparent materials being developed for use as optical elements for high power lasers. High-power CO_2 lasers require a window material, transparent at 10.6 μm, through which the beam can be transmitted. If any absorption is present in the wsndow, it will heat and deform. The optical thickness is affected by two factors, namely, thermal expansion of the window and the change in refractive index with temperature. The distortion of the optical thickness of the window can seriously degrade the quality of the transmitted high-power beam. Holographic interferometry has proved to be a valuable tool for studying such distortions, with the objective of reducing their effect on the beam. A description of this particular example will demonstrate the capabilities of holographic interferometry in solving difficult problems in materials research.

Holographic interferometry of the optical distortion of transparent materials is more complicated than conventional interferometry, but it offers some advantages since it is a truly differential method. For conventional interferometry, a transparent object whose optical homogeneity is to be evaluated is compared to an essentially perfect homogeneous medium. This means that the interferogram contains information about the surface condition as well as the bulk homogeneity. Thus, measurements of laser-induced changes in optical thickness require surfaces that are flat to a fraction of a wavelength and preferably parallel to a few seconds of arc. In holographic interferometry, one compares the optical path through the sample at the time that the hologram is made to the optical path at a later time, after changes have been made. As an example, one can make a hologram of a sample which is not of uniform thickness and that has scratches on the surface. One compares the original optical thickness point by point with the optical thickness during irradiation of the sample with the beam from a CO_2 laser. In the absence of techniques for producing optical-quality surfaces in these materials, this technique allows one to evaluate changes in samples with imperfect surfaces.

The holographic interferometer used in this work is shown in Fig. 18-12. The light source is a He–Ne laser that operates in three longitudinal modes and has a coherence length of approximately 5 cm. This means that one can have an optical path difference of approximately 5 cm between the two paths of the interferometer without losing information. The beam from the laser is divided into two beams by a dielectric-coated beam splitter and each beam is sent through an expanding telescope and a spatial filter. The telescopes give an expansion of approximately 20 × and the spatial filters are pinholes of 10 μm diameter. The reference beam, which is made to diverge slightly

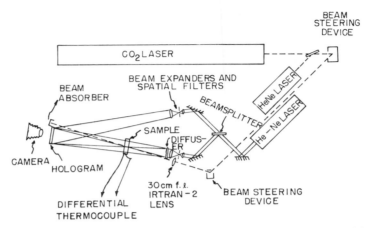

Fig. 18-12   Holographic interferometer for use in measuring changes in optical thickness of transparent materials produced by CO_2 laser irradiation.

after going through the expanding telescope, goes directly to the holographic recording medium. The object beam is allowed to impinge on a diffuser plate after exiting from the expanding telescope. The diffuser is a ground glass plate which has a mean particle size around 5 μm. After being diffused, the light is transmitted through the sample and goes to the holographic recording medium where it interferes with the reference beam to form a hologram. What is actually recorded is a hologram of the diffuser plate as seen through the sample. The diffuser produces, in effect, a very large number of point sources which are used to illuminate the sample, thus allowing a multiplicity of paths between any point in the sample and the recording medium. This allows one to image the sample for later viewing. If a diffuser were not used, the field of view as seen from any finite aperture looking at the holographic reconstruction would be determined by the divergence of the beam exiting from the expanding telescope.

The use of a diffuser does introduce a complication in the photographing of interferograms. This is the problem of speckle or spatial noise in the recording plane. The speckle is a result of the finite size of the particles in the diffuser and it becomes more noticeable as the aperture through which the object is viewed is decreased.

In early work, the recording medium was 4 × 5 in. 649F photographic plates. The plate was placed in a gimbaled holder where the hologram was recorded. It was then removed, developed, and replaced.

When the hologram is illuminated by the reference beam, it reconstructs the wavefront characteristic of the object in the undeformed state. This wavefront then interferes with the wavefront from the sample. In the absence of object deformation, and if the film is replaced in the exact position where the hologram was recorded, one is able to completely extinguish the object in the region of interference. Then, if the object begins to deform, interference fringes appear, corresponding to one-half wavelength of deformation per fringe. However, if the film is not replaced in exactly the same position that it was during recording of the hologram, the plane where the fringe visibility is optimum does not coincide with the plane of the object and can, in fact, be a considerable distance from the object.

In later work, the photographic film has been replaced with an erasable thermoplastic recording medium [7]. The application of thermoplastics as a holographic recording medium has been mentioned in Chapter 17. The thermoplastic is developed in situ, so that the problem of plate repositioning is eliminated. Other advantages of the thermoplastic include reusability, low requirements on recording energy, high diffraction efficiency, and lack of grain noise.

The sample is irradiated by a beam from a 250-W continuous CO_2 laser, which passes through the center of the sample. The second He–Ne laser is boresighted with the CO_2 laser and is used for alignment. The CO_2 laser flux density at the center of the sample can be as high as 13 kW/cm². Absorption of energy from the CO_2 laser by the sample causes heating and optical distortion, which is recorded holographically.

Figure 18-13 shows one example of the sequences of interferograms photographed during the irradiation of windows inside the interferometer at 13 kW/cm² and during cooling after the laser was turned off [8]. This is a polycrystalline sample of pure potassium chloride. It has an absorption coefficient of 5×10^{-3} cm^{-1}, which is high for this type material. It was mechanically polished and was uncoated. The fringes are labeled so the dark fringe nearest the center of the field of view in the undisturbed state is numbered zero. The ones to the left are negative; and the ones to the right are numbered with increasing positive integers. As time progresses, the fringes that were initially in the field of view move out and new ones move in, as indicated by the high-value integers in some of the frames. The laser is turned off between the fourth and fifth frames. The sample then begins to cool, and fringes move back toward the right.

There are two qualitatively different effects that can be seen in the interferograms. They are the uniform motion of the fringe pattern across the sample, indicating a uniform temperature rise, and a localized deviation of the fringes from straight lines in the region near the point where the beam strikes the sample. The uniform fringe motion represents a uniform change in optical thickness of the sample, and it should have no effect on a collimated beam propagating through the material. For a converging or diverging beam, the effect of this change in thickness is to move the location of the focal plane of the optical system without increasing the diffraction-limited spot size. On the other hand, local deviations from straight lines in the form

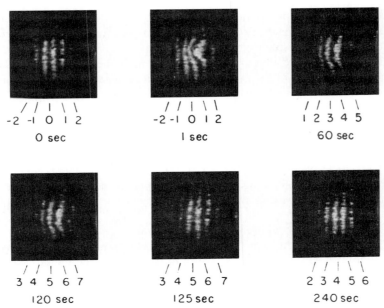

Fig. 18-13 Interferograms showing wavefront distortion introduced by high absorption KCl sample into He–Ne laser beam during irradiation with a continuous flux density of 13 kW/cm² from a CO_2 laser. Laser impinges on sample near the center of the field of view, at location of highest fringe curvature. The numbers under each frame represent the elapsed time, in seconds, since CO_2 laser is turned on. (From E. Bernal G. and T. C. Lee, *in* "Digest of Papers, International Optical Computing Conference." Washington, D.C., April 23–25, 1975.)

of fringe curvature will have a serious effect on the quality of the beam propagating through the window. This example shows how holographic interferometry has been employed in a practical sense for solving a difficult problem in evaluation of window materials needed for high-power lasers.

REFERENCES

[1] K. A. Stetson, *J. Opt. Soc. Am.* **66**, 627 (1976).
[2] K. A. Haines and B. P. Hildebrand, *Appl. Opt.* **5**, 595 (1966).
[3] R. L. Powell and K. A. Stetson, *J. Opt. Soc. Am.* **55**, 1593 (1965).
[4] G. Oster and Y. Nishijima, *Sci. Am.*, p. 54 (May 1963).
[5] R. K. Erf, (ed.), "Holographic Nondestructive Testing." Academic Press, New York, 1974.
[6] G. C. Bjorklund, S. E. Harris, and J. F. Young, *Appl. Phys. Lett.* **25**, 451 (1974).
[7] T. C. Lee, *Appl. Opt.* **13**, 888 (1974).
[8] E. Bernal G. and T. C. Lee, *in* "Digest of Papers, International Optical Computing Conference." Washington, D.C., April 23–25, 1975.

SELECTED ADDITIONAL REFERENCES

A. Holographic Interferometry

J. D. Briers, The Interpretation of Holographic Interferograms, *Opt. Quantum Electron.* **8**, 469 (1976).

J. N. Butters and J. H. Leendertz, Application of Coherent Light Techniques to Engineering Measurement, *Appl. Opt.* **11**, 1436 (1972).

H. J. Caulfield and S. Wu, "The Applications of Holography," Wiley (Interscience), New York, 1970, Chapter XII.

R. J. Collier, C. B. Burckhardt, and L. H. Lin, "Optical Holography," Academic Press, New York, 1971, Chapter 15.

T. D. Dudderar and R. O'Reagan, Laser Holography and Interferometry in Materials Research, *Mater. Res. St.*, p. 8 (September 1971).

A. E. Ennos, Measurement of In-Plane Surface Strain by Hologram Interferometry, *J. Phys. E.* (*J. Sci. Instrum.*) **1**, 731 (1968).

R. K. Erf (ed.), "Holographic Nondestructive Testing," Academic Press, New York, 1974.

P. A. Fryer, Vibration Analysis by Holography, *Rep. Progr. Phys.* **33**, 489 (1970).

K. A. Haines and B. P. Hildebrand, Surface Deformation Measurement Using the Wavefront Reconstruction Method, *Appl. Opt.* **5**, 595 (1966).

R. Levin, Industrial Holographic Applications, *Electro-Opt. Syst. Design*, p. 31 (April 1976).

T. Matsumoto, K. Iwata, and R. Nagata, Measuring Accuracy of Three-Dimensional Displacements in Holographic Interferometry, *Appl. Opt.* **12**, 961 (1973).

R. L. Powell and K. A. Stetson, Interferometric Vibration Analysis of Three-Dimensional Objects by Wavefront Reconstruction, *J. Opt. Soc. Am.* **55**, 612 (1965).

L. W. Riley, Pulsed Holography: Problem Solver in the Lab, *Opt. Spectra*, p. 27 (December 1973).

E. R. Robertson and J. M. Harvey (eds.), "The Engineering Uses of Holography," Cambridge Univ. Press, London and New York, 1970.

R. C. Sampson, Structural Measurements with Holographic Interferometry, *Mater. Res. St.*, p. 26 (September 1971).

H. M. Smith, "Principles of Holography," Wiley (Interscience), New York, 1969, Chapter 8.

K. A. Stetson, Fringe Interpretation for Hologram Interferometry of Rigid-Body Motions and Homogeneous Deformations, *J. Opt. Soc. Am.* **64**, 1 (1974).

K. A. Stetson, Homogeneous Deformations: Determination by Fringe Vectors in Hologram Interferometry, *Appl. Opt.* **14**, 2256 (1975).

C. M. Vest, E. L. McKague, and A. A. Friesem, Holographic Detection of Microcracks, *J. Basic Eng.*, p. 237 (June 1971).

B. A Miscellany of Applications

E. S. Barrekette *et al.*, "Applications of Holography," Plenum Press, New York, 1971.

S. A. Benton, Holographic Displays—A Review, *Opt. Eng.* **14**, 402 (1975).

R. Bexon, M. G. Dalzell, and M. C. Stainer, In-line Holography and the Assessment of Aerosols, *Opt. Laser Technol.* **8**, 161 (1976).

B. B. Brendon, Acoustic Holography, *Opt. Eng.* **14**, 495 (1975).

H. J. Caulfield and S. Wu, "The Applications of Holography," Wiley (Interscience), New York, 1970, Chapters VIII, IX, X, XI, XIII, XIV, and XV.

R. J. Collier, C. B. Burckhardt, and L. H. Lin, "Optical Holography," Academic Press, New York, 1971, Chapters 13, 14 and 16.

D. H. Close, Holographic Optical Elements, *Opt. Eng.* **14**, 408 (1975).

J. B. DeVelis and G. O. Reynolds, "Theory and Applications of Holography," Addison-Wesley, Reading, Massachusetts, 1967, Chapter 8.

A. E. Ennos, Holography and its Applications, *Contemp. Phys.* **8**, 153 (1967).

B. P. Hildebrand and B. B. Brenden, "An Introduction to Acoustical Holography," Plenum Press, New York, 1972.

G. R. Knight, Holographic Memories, *Opt. Eng.* **14**, 453 (1975).

H. M. Smith, "Principles of Holography, "Wiley (Interscience), New York, 1969, Chapter 8.

B. J. Thompson and J. H. Ward, Particle-Sizing, the First Direct Use of Holography, *Sci. Res.*, p. 37 (October 1966).

B. J. Thompson, Applications of Holography, *in* "Laser Applications," Vol. 1 (M. Ross, ed.), Academic Press, New York, 1971.

B. J. Thompson, Holographic Particle Sizing Techniques, *J. Phys. E: (Sci. Instrum.)* **7**, 781 (1974).

C. A Specific Example of Holographic Application

E. Bernal G. *et al.*, Preparation and Characterization of Polycrystalline Halides, *Proc. Conf. High Power IR Laser Window Mater. 2nd*, p. 413, Hyannis, Massachusetts (1972). Published as Air Force Cambridge Res. Lab. Rep. #AFCRL-TR-73-0372 (II).

J. S. Loomis and E. Bernal G., Optical Distortion by Laser Heated Windows, *in* "Laser Induced Damage in Optical Materials" (A. J. Glass and A. H. Guenther, eds.), Nat. Bur. of St. Spec. Publ. 435, U.S. Dept. of Commerce (1976), p. 126.

E. Bernal G. and T. C. Lee, Real-Time Holographic Interferometry of Laser-Induced Thermal Distortion in IR Windows, Digest of Papers, *Int. Opt. Comput. Conf.*, Washington, D.C., April 23–25, 1975, IEEE Catalog No. 75 CH0941-5C (1975).

CHAPTER 19

CHEMICAL APPLICATIONS

The development of intense tunable laser sources has opened up many new areas of research in photochemistry. As a scientific tool, lasers have been used in chemistry for spectroscopy and for flash photolysis, in which the kinetics of chemical reactions are studied. Such studies have yielded much useful information on the progress of chemical reactions. High resolution spectroscopy with a tunable laser source can be a valuable analytical tool.

For direct stimulation of chemical reactions in a desired direction, the laser may be tuned to a wavelength corresponding to a selected molecular absorption. This can stimulate the chemical reaction to proceed in a direction different from the direction it would take without the photoexcitation.

Perhaps one of the most significant possibilities for laser photochemistry involves isotope separation. The slight shifts that occur in the molecular absorption spectrum between molecules containing different isotopes of the desired atom are utilized. The laser is tuned to the resonant absorption for only one of the isotopes. Molecules containing this isotope then undergo a chemical reaction and may be conveniently separated. This process can have great practical significance, particularly for the separation of uranium isotopes.

Most of the applications to be described in this chapter require tunable lasers. Tunable lasers have been described in Chapters 3 and 4. The only tunable laser with easy availability is the dye laser, which may be tuned through the visible and near-ultraviolet portions of the spectrum. There are many other promising techniques for tunable lasers, particularly in the infrared. The status of development of these techniques was also discussed in Chapter 4.

A. Spectroscopy

1. *Absorption Spectroscopy*

A wide variety of different configurations and methods for laser spectroscopy have been developed. The first and most obvious is simply absorption spectroscopy, carried out by tuning the laser. Because of the narrow linewidth and high brightness of the laser, very high resolution spectra can be obtained. The laser beam is transmitted through the sample whose spectrum is desired. The intensity of the transmitted beam is monitored by a photodetector, and the laser is simply tuned through the region of interest. There is no need for gratings, prisms, or any of the other dispersive elements commonly used in spectrometers.

The combination of narrow spectral linewidth, brightness, and tunability available from lasers has led to great improvement in resolution, far beyond what has been available with even the best conventional dispersive spectrometers. The laser is much more than just another light source for spectroscopy. The capabilities of the laser have led to a qualitative revolution in this branch of analytical technique. This technique has already been used in the visible portion of the spectrum, with tunable dye lasers. As the spectral range of tunable lasers increases and easily available tunable lasers cover the infrared spectrum, further applications are to be expected. The expected further advances in lasers as discussed in Chapter 4 should yield tunable lasers in the ultraviolet (in the range from 2000 to 3500 Å) and also more convenient sources of laser radiation in the infrared, tunable from approximately 1 to 30 μm. Availability of such sources would considerably increase the capabilities of lasers for spectroscopy.

An example in which a tunable laser was used in absorption spectroscopy has already been given in Chapter 11. Tuning of a semiconductor laser in the infrared was described in its application to detection of trace elements in automobile exhaust. This can serve as an example of the extremely high resolution which tunable lasers offer for spectroscopy.

A variety of spectroscopic techniques other than absorption spectroscopy are possible with lasers. There are many techniques that have been described in the literature and applied to practical systems. It is not possible to describe all of the different applications in a reasonable space. We shall, instead, describe two of the leading types of laser spectroscopy other than absorption spectroscopy. These two shall serve as examples of the remaining types. The two are Raman spectroscopy and saturation spectroscopy.

2. *Raman Spectroscopy*

The Raman effect involves scattering of light by molecules of gases, liquids, or solids. The Raman effect consists of the appearance of extra spectral lines near the wavelength of the incident line. The Raman lines in the scattered light are weaker than the light at the original, unshifted wavelength. The Raman-shifted lines occur both at longer and at shorter wavelengths than the original light; the lines at shorter wavelengths are usually extremely weak.

The Raman spectrum is characteristic of the scattering material. The lines occur at frequencies $v \pm v_k$, where v is the original frequency, and v_k are the frequencies of quanta of molecular vibration or rotation. Since the values of v_k are characteristic of individual molecules, investigation of the Raman spectrum can provide a sensitive analytical tool.

Raman spectroscopy has long been used (in conjunction with infrared spectroscopy) for qualitative analysis and for identification of characteristic localized units of structure within molecules. The advent of lasers has provided a new source with desirable properties for use in Raman spectroscopy. Raman spectroscopy has also been used for remote detection of pollutants in the atmosphere. This application and the positions of some of Raman-shifted lines of common pollutants have been discussed in Chapter 11.

The availability of lasers with narrow spectral widths and high brightness permits much higher resolution of Raman-shifted spectra than was possible with conventional sources. The variety of laser wavelengths makes it possible to carry out Raman spectroscopy while avoiding interfering absorption bands in the molecule being studied. With a tunable laser, the excitation frequency can be tuned to produce a larger Raman signal. Although a tunable laser is not strictly necessary for Raman spectroscopy, the use of a tunable laser can lead to an enhancement of the Raman scattering. In practice, many of the studies involving Raman spectroscopy with lasers have been carried out with argon lasers.

Laser Raman spectroscopy has been used for identifying drugs, for detecting trace quantities of drugs in blood or urine samples, to detect the metabolic byproducts of drugs, to determine impurity levels in various types of products (e.g., medicines) and to study polymers in solution. The Raman spectroscopic technique is applicable to very small samples, because Raman spectroscopy requires only a sufficient sample size to fill the focused beam of the argon laser, a volume about 10 μm in diameter. Thus, Raman spectroscopy can be useful for small samples of fibers, coatings or finishes. Because Raman spectroscopy is a scattering technique, sample transmission is not required. Therefore, materials such as pills, chemicals, drugs, and coatings can be sampled as received. These capabilities are making the chemical laboratory an important market for laser Raman spectroscopic instrumentation.

3.   *Saturation Spectroscopy*

In ordinary absorption spectroscopy, a limitation is the Doppler effect, the broadening of the spectral lines in gases and vapors because of motion of the atoms or molecules. The Doppler linewidth, which is proportional to the square root of the temperature, can often be larger than the separation between neighboring narrow spectral lines. In particular, it can hide much of the fine structure contained in atomic or molecular spectra. The use of lasers in ordinary absorption spectroscopy does not help this limitation, because even though the laser linewidth is extremely narrow, the observed broadening arises from the motion of the atoms or molecules in the sample. This broadening effect can be eliminated by a technique called saturation spectroscopy. This yields information about the natural width of the narrow spectral lines.

The technique is to split the laser beam into two parts, one strong and one weak. The two beams are introduced to an absorbing gas cell, traveling in opposite directions. The highly monochromatic laser beam interacts only with a small fraction of the Doppler broadened line for any given wavelength to which the laser is tuned (i.e., it interacts only with a small fraction of the atoms or molecules, those which happen to be traveling with such a velocity as to be at the right part of the Doppler broadened curve to absorb the laser light.) The more intense beam can saturate the absorption, that is, it can interact with all the available atoms under the narrow fraction of the Doppler linewidth corresponding to the laser wavelength. Thus, the saturating beam can bleach a path for the weaker probing beam. The weaker interrogating beam is chopped, so that the probe beam light reaching the photodetector will be modulated. This happens only if the two beams are seeing the same molecules. This is the case when the laser is tuned exactly to the center of the resonance line so that the output is absorbed by molecules which have zero component of velocity along the direction of the laser beam. For those molecules, neither beam is Doppler shifted and the two beams are in resonance with the same molecules. In that way, the effects of Doppler broadening can be eliminated.

As an example, Fig. 19-1 shows the hyperfine structure of a line of molecular $^{127}I_2$ [1]. This work was carried out using an argon laser, which could be frequency-tuned over a very small range by a piezoelectric drive on one end mirror. The Doppler width is shown for comparison. In conventional absorption spectroscopy, none of the detail on a scale narrower than the Doppler width would have been observable. The figure shows dramatically the capability of saturation spectroscopy for overcoming the broadening introduced by the Doppler effect.

Saturation spectroscopy techniques take advantage of the distortion of the Doppler line shape which occurs because only molecules within a

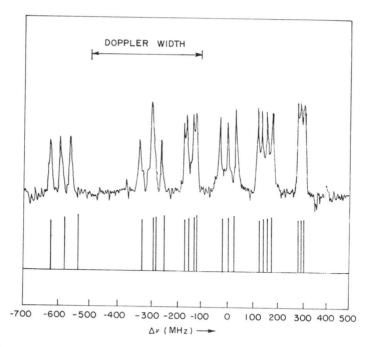

Fig. 19-1 Hyperfine structure of an absorption line of $^{127}I_2$, probed by saturation spectroscopy. The frequency shift from the line center is denoted $\Delta\nu$. (From A. L. Schawlow, *in* "Fundamental and Applied Laser Physics" (M. S. Feld, A. Javan, and N. A. Kurnit, eds.). Wiley, New York, 1973.)

narrow spread of velocity interact with the laser beam. Only those molecules whose molecular transition frequency is Doppler shifted into resonance with the laser will absorb. The linewidth of the narrow resonance is related to the natural linewidth of the level involved. Saturation spectroscopy produces a natural linewidth resonance at the center frequency of the atomic or molecular transition. The two waves of identical frequency but opposite direction interact with molecules with velocity components v_z along the laser beam centered at

$$v_z = \pm c(v_m - v_L)/v_m \qquad (19.1)$$

where v_m is the molecular transition frequency, v_L the laser frequency, and c the velocity of light. The width of these two velocity distributions is given by the natural linewidth of the molecular transition. As the quantity $v_m - v_L$ passes through zero, the waves in both directions interact with the same set of molecules having zero velocity in the z direction. The situation is shown in Fig. 19-2, where the depletion of the Doppler absorption profile is shown.

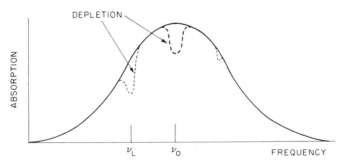

Fig. 19-2 Schematic diagram of depletion of the Doppler absorption profile of a line, centered at frequency v_0, by a laser of frequency v_L.

When $v_m - v_L \neq 0$, the probe beam and saturation beam interact with molecules on opposite sides of the line center. The depletion of absorbing molecules by the saturating beam does not allow transmission of the probing beam. When the laser is tuned to the line center, the depletion shown by the dashed line allows the probe beam to increase in transmission.

B. Photochemical Applications

1. *Selective Excitation of Reactions*

Lasers provide the possibility of selective excitation of chosen excited states of atoms and molecules, or selective breaking of specified chemical bonds. These processes can be very attractive for stimulation of desired chemical reactions. It is possible to tune a laser to a molecular resonance so that only certain chosen chemical bonds are excited. This leads to the possibility of directing chemical reactions in a desired direction. The current status is one of experiment and investigation. This promising field is still in its infancy. However, successful developments can produce revolutionary changes in the entire chemical industry.

Early experiments in laser photochemistry were carried out using the fixed wavelength lasers which were available in the early and middle 1960s. Such experiments studied reactions in which there is a fortuitous coincidence between the laser frequency and a resonant molecular absorption frequency. The development of tunable lasers now permits the excitation and study of almost any desired molecular state. It is apparent that laser photochemistry will rely heavily on the use of tunable lasers.

The application of lasers to produce vibrationally excited molecules offers great possibilities for selective stimulation of chemical reactions. Chemical reactions will occur only when a given activation energy is supplied. Endothermic reactions do not occur without activation. This is because the reactants must be brought very close together in order to react. At short distances, forces of repulsion between the molecules appear, and energy is needed to overcome them. Thermal energy is often employed. The rate of chemical reaction increases with heating. However, the internal energy associated with molecular vibration can also be useful. According to modern ideas, the vibrational molecular energy is even more conducive to promoting chemical reaction than thermal translational energy. Thus, a new approach to chemical reactions is to influence the individual bonds in a molecule by exciting the vibrational states. This appears preferable to exciting the molecule as a whole by providing thermal energy, most of which will occur as thermal energy of translation.

The selective excitation of certain states is produced by tunable lasers, which are tuned to a wavelength that is resonant with the molecular absorption band. This procedure can be very selective. It can be much more conserving of the energy as compared to thermal initiation. The direct vibrational excitation can be more efficient in producing a reaction than the same amount of energy supplied as heat to the reactants. The method of laser initiation can substantially influence some important chemical production processes. It opens the possibility of the direct synthesis of new chemical compounds.

The study of the interactions of lasers in producing chemical reactions is still in a very early stage and only a relatively small number of such reactions have been studied. As yet none of practical commercial importance have been identified. We will present three examples that show the possibilities of the laser in stimulating selected reactions. A considerable amount of work is underway at many laboratories to identify suitable reactions for laser stimulation.

As one example [2], experiments were carried out on mixtures of the gases boron trichloride (BCl_3) and acetylene. There is an overlap between the emission of a CO_2 laser and the absorption of BCl_3. The absorption in the region around 10.6 μm excites an asymmetric stretching vibration of the BCl_3 molecule. The absorption of laser radiation and the subsequent excitation of this vibrational energy level causes an interaction between BCl_3 and acetylene. Acetylene is transparent to the CO_2 laser light and does not usually react with boron trichloride. However, under irradiation by pulsed CO_2 lasers, the following reaction was observed:

$$BCl_3 + C_2H_2 \longrightarrow HCl + HC \vdots CBCl_2$$

This demonstrates that in this laser-induced photochemical reaction, the dominant role belongs to excited vibrational states of molecules that are produced by absorption of the laser light.

In another example [3], CO_2 laser radiation at a wavelength of 10.6 μm interacted with the system N_2F_4–NO. Under ordinary circumstances, the N_2F_4 does not react with NO. However, under irradiation with the CO_2 laser, the vibrational states of the N_2F_4 molecule are excited. Chemical reactions occurred to produce NF_3, NFO, N_2, and F_2. The percentage of each product depended on the experimental conditions, including the pressure and the laser intensity.

The third example [4] involved the set of reactions

$$B(CH_3)_3 + HBr \longrightarrow B(CH_3)_2Br + CH_4$$
$$B(CH_3)_2Br + HBr \longrightarrow BCH_3Br_2 + CH_4$$
$$BCH_3Br_2 + HBr \longrightarrow BBr_3 + CH_4$$

Laser-specific product formation was observed, with control possible by tuning a CO_2 laser. A laser frequency of 970.5 cm^{-1} could drive either the first or the third of the above reactions. A laser frequency of 1039.4 cm^{-1} could drive either the second or the third reaction. A nonthermal nature for the processes was demonstrated by the line-specific action of the CO_2 laser and also by temperature probes.

These experiments indicate that under the action of laser irradiation, vibrationally excited molecules are formed which initiate chemical reactions. These reactions can include normally exothermic reactions, but they can be directed so as to obtain products which differ from those obtained by thermal activation. Thus, resonant interaction of the laser light with the system allows one to direct a chemical reaction to maximize desired reaction products, to obtain products which cannot be produced by simple heating of the reactants, and also to reduce the amount of energy required to stimulate given chemical reactions.

2. Flash Photolysis

Flash photolysis provides a technique for the study of rapid chemical reactions. Flash photolysis basically involves the break-up of a molecular species by an intense flash of light. Then the absorption lines of the resulting products are monitored in order to determine the concentrations of the different chemical species after the flash. In this way, the course of the chemical reactions may be followed.

There are two variations of the flash photolysis technique. In the first technique, a second flash with a variable time delay is used to provide a spectroscopic continuum against which to record the absorption spectrum of the intermediate reaction products produced by the first flash. A photographic record of the spectrum over a broad spectral range is obtained. This technique is usually used for exploratory work, when one does not know the nature of the chemical species which appear. Once the chemical species are identified and their absorption lines are known, the second technique may be employed. The second flash is replaced by an intense continuous background source. Observations are made at a single wavelength using a monochromator and a photodetector. Alternatively a tunable continuous laser could be used. This second technique provides better detailed kinetic studies because it gives a complete time history of the variation of the absorption due to a particular species.

With the use of flash photolysis, it has been possible to generate photochemically and study spectroscopically a wide variety of free radicals and excited molecules. Typical pulse durations available before lasers were in the range of several microseconds. The time resolution of a flash photolysis system is limited mainly by the duration of the initiating flash. Previous attempts, before the development of the laser, to reduce the flash duration without decrease in the available energy met with only marginal success. The earliest laser experiments used a Q-switched ruby laser with a pulse duration around 20 nsec [5, 6]. This immediately had the effect of extending the flash photolysis technique from a microsecond time scale to a nanosecond time scale. The development of the short-pulse, high-power laser provided a new source with great capability for flash photolysis studies. A great many ingenious configurations have been devised to carry out flash photolysis with lasers. The major applications of these techniques have been excited state spectroscopy and kinetics of large organic molecules in solution. Kinetic monitoring has yielded data such as fluorescence lifetimes, information on oxygen quenching of singlet and triplet states of these molecules, and absolute values for absorption coefficients.

The availability of mode-locked lasers has allowed flash photolysis to be pushed into the picosecond regime, which represents a still further advance of three orders of magnitude in time resolution, as compared to the earliest laser experiments using Q-switched ruby lasers.

3. Picosecond Spectroscopy and Reaction Kinetics

The technique of picosecond spectroscopy has also been used to study ultrafast kinetics of chemical reactions. The availability of tunable optical pulses from mode-locked dye lasers, with durations of 10^{-12} sec or less, has opened up entirely new areas of investigation. Such short pulses of tunable, narrow-band radiation are ideal for the excitation of specific molecular energy levels. The decay of these excited levels can be followed either by their fluorescent emission, or in absorption. In this way, chemical kinetic processes can be monitored in a fashion that was not previously possible.

Processes taking place on this time scale, which can now be investigated, include vibrational and orientational relaxation of molecules in liquids, primary radiationless transitions of electronically excited large molecules, solvation of photo-ejected free electrons, and isomerization of photoexcited states of such complex molecules as visual pigments.

One such study involved observation of the first steps in photosynthesis by observing buildups and decay of intermediate chemical compounds on a picosecond time scale [7]. In another experiment using a mode-locked neodymium laser [8], the ultrafast time-resolved emission spectrum of a dye molecule solution was observed. The time resolution of the system was 2 psec. This method of time-resolved picosecond emission spectroscopy yielded a direct measurement of the relaxation in excited electronic states of molecular systems. This technique thus provides the first direct experimental determinations of time- and frequency-resolved vibrational relaxation in excited states of large molecules.

Let us consider a single example which will illustrate some of the techniques and serve to demonstrate the capabilities of picosecond spectroscopy. This experiment [9] involved the formation and decay of a biophysical compound called prelumirhodopsin, which is considered to be the initial step in photo-induced chemical changes which begin the chain of physiological processes constituting vision. Vision begins when rhodopsin, a photosensitive pigment in the retina, is bleached by light. When rhodopsin present in the eye is exposed to light, prelumirhodopsin is known to be formed faster than can be followed by conventional methods.

The basic method used in the experiment was photoexcitation by a picosecond pulse of light at 0.53 μm. The absorption band of prelumirhodopsin at 0.56 μm was monitored by a train of short pulses at this wavelength. The original laser was a mode-locked Nd:glass laser from which one pulse of 6-psec duration was extracted. This pulse was amplified and frequency-doubled in a potassium dihydrogen phosphate crystal to yield 0.53 μm light. Part of this light was used to excite the sample; another part was split off and Raman-scattered to yield picosecond duration pulses of 0.56 μm light. The 0.56 μm light was reflected off an echelon, a stepped optical element which introduced different optical delays into different spatial parts of the wavefront. Thus, the 0.56 μm light was transformed into a series of short interrogating pulses to monitor the desired molecular absorption band. This light was focused onto the same area of the sample as the 0.53 μm exciting light. After emerging from the cell, the 0.56 μm light was imaged onto a camera. Light from each optical segment of the echelon was imaged onto a different area of the film. Since the light from each element has a unique incremented optical path, the temporal pulse separation (about 20 psec) was recorded as a spatial separation. The results of the experiment showed that the risetime of the prelumirhodopsin is less than 6 psec after the start of the 0.53 μm exciting pulse. The data indicated that production of prelumirhodopsin is the primary photochemical event.

Probing the time-resolved absorption and fluorescence of large molecules by picosecond spectroscopy is in a research stage. It can provide important information on molecular processes and intramolecular energy transfer, which in turn can be important for future industrial chemical processes.

C. Isotope Separation

1. *Principles*

The use of lasers in isotope separation may well be the most important of the photochemical applications of lasers. It has particular significance because of the possibility of the separation of uranium isotopes. Processes for separation of the isotopes for many different elements have been suggested. Such processes could have importance in production of isotropically enriched elements for medical, industrial and scientific applications. However, the biggest economic incentive is undoubtedly that of uranium isotope separation.

Laser isotope separation essentially involves the use of a laser tuned to a resonant absorption of a chemical compound containing one isotope of the desired atom. There are small isotopic shifts in the absorption spectra of many molecules. Because of the smallness of the shifts, very narrowband light sources are required to distinguish between them. It is only the laser which can produce both the required narrow linewidths and the required high power. Absorption of the laser light by molecules containing the one desired isotope leads to a chemical reaction involving only the molecules containing that isotope. The products of the chemical reaction may be separated to yield a product which is enriched in the one desired isotope.

There are several things needed for laser-assisted isotope separation. These include

(1) a shift in the absorption spectrum between different isotopes of the desired element;

(2) a tunable laser which can be so tuned as to excite only one of the isotopes;

(3) a chemical or physical process which acts only on the excited species and separates them from the unexcited molecules;

(4) the absence of a scrambling process such as thermal excitation which can undo the effect of exciting only one of the isotopes.

The species may be either atoms of the desired element or molecules containing the desired isotopic species. For atoms, the lines of interest are in the electronic structure and tend to lie in the visible or ultraviolet. For molecules, the spectra of interest are the vibrational and rotational spectra, and the lines of interest tend to lie in the infrared. Some characteristic isotope shifts of some absorption lines for both atoms and molecules are shown in Table 19-1. The isotope shift must be compared to the linewidth. If the lines of the two isotopes overlap, a selective excitation of only one isotope will be impossible. Because spectral lines in solids and liquids are generally broad, usually only gaseous media are considered for photoseparation of isotopes. The linewidth for atomic lines in gaseous media at low pressure is the Doppler linewidth, proportional to the square root of temperature. The isotope shifts of some of the atomic species in Table 19-1 are larger than the Doppler width, which is shown for a temperature large enough to yield a vapor pressure of 1 Torr for the element. In general, the isotope shift is larger than the Doppler shift only for very light or for very heavy atoms.

TABLE 19-1
Isotope Shifts

Atoms	Wavelength (Å)	Isotope shift (cm^{-1})	Doppler Linewidth (cm^{-1})
$^6\text{Li}-^7\text{Li}$	3232.6	0.35	0.026
$^{10}\text{B}-^{11}\text{B}$	2497.7	0.175	0.44
$^{200}\text{Hg}-^{202}\text{Hg}$	2536.5	0.179	0.04
$^{235}\text{U}-^{238}\text{U}$	4246.3	0.280	0.055

Molecules	Wavelength (μm)	Isotope shift (cm^{-1})	Rotational bandwidth (cm^{-1})
$^{10}\text{BCl}_3-^{11}\text{BCl}_3$	10.15	9	25
$^{14}\text{NH}_3-^{15}\text{NH}_3$	10.53	24	71
$^{48}\text{TiCl}_4-^{50}\text{TiCl}_4$	20.06	7.6	18
$^{235}\text{UF}_6-^{238}\text{UF}_6$	16.05	0.55	16

For molecules, the relevant bandwidth is the width of the rotational substructure. This is also given in Table 19-1. The linewidth is usually larger than the isotope shift. This often makes separation using molecules difficult, although some ways have been suggested for evading the problem. One such way is to use absorption lines due to simultaneous excitation of more than one vibrational quantum. Flow cooling of gaseous species can also be used to narrow the linewidth.

There are several possible forms that the separation process can take. Some of them are

(a) *Two-step photoionization* The atom of the desired isotope is excited by absorption at one wavelength. Then excited atoms are ionized by absorption of light at a second wavelength. Unexcited atoms do not

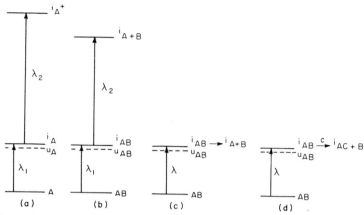

Fig. 19-3 Schematic diagram of possible isotope separation processes. In this figure the left superscript i denotes an isotopically selected species, and the superscript u an unselected species. (a) Two-step photoionization. (b) Two-step photodissociation. (c) Photopredissociation. (d) Chemical reaction of excited species.

absorb the second wavelength. This process is shown schematically in Fig. 19-3a. The ions may then be extracted electrically. This process has the desirable feature of easy extraction, but it is not conservative of light energy, because the second wavelength is usually short and the associated photon energy is high.

(b) *Two-step photodissociation* A molecule containing the desired isotope is excited by absorption of one wavelength, whereas molecules with other isotopes are not. Then the excited molecules are broken up by absorption of a second wavelength. This process is sketched in Fig. 19-3b. The dissociation products must then be extracted by some chemical or physical means. The dissociation products may be free atoms of the selected isotope or possibly fragments of the original molecule containing the selected isotope. This process, like two-step photoionization, is energetically expensive.

(c) *Photopredissociation* Photopredissociation occurs in some molecules which are excited into metastable states. The excited molecule has energy levels corresponding to unbound states, which cross the bound excited states. An excited molecule can make a crossover into the unbound channels and dissociate. The process is shown in Fig. 19-3c. This process is more conserving of energy. It depends on identification of molecules with suitable sets of energy levels. Several such systems have been suggested, such as the levels of the Br_2 molecule. As an example, the separation of bromine isotopes using photopredissociation will be described later.

(d) *Chemical reaction of excited molecules* In this method, the excited molecules undergo a chemical reaction, whereas the unexcited molecules do not. Such selective laser-stimulated chemical reactions have already been discussed in Section 19.B. The process is shown schematically in Fig. 19-3d. The reaction product must then be separated. This process should be conserving of energy, because the second high-energy photon is not needed. However, our present understanding of reactions with excited molecules is limited, so that no generally applicable methods can be described. This separation technique will depend on identification of suitable reactions for specific molecules.

(e) *Deflection of an atomic beam*   This method relies on resonant absorption and reemission of laser light by the desired isotopic species in an atomic beam. The atoms gain momentum in the direction perpendicular to the beam because of the absorption. Two spatially separated atomic beams are obtained, one containing the desired species and the other containing the unselected isotopes. This method has been demonstrated on barium isotopes, as will be described in a later example. This method should in principle be applicable to any heavy element, and should be very conserving of energy. In principle, the reemitted light could be collected by mirrors and reused. The atoms are physically separated in space, so that collection is simple.

(f) *Multiple-step infrared photodissociation*   In an idealized picture, the vibrational energy levels of a polyatomic molecule will be spaced equally in energy. If a laser is tuned to the proper frequency, absorption can take place sequentially one step at a time, driving the molecule up the ladder of equally spaced states until it has absorbed enough energy to dissociate. This is shown schematically in Fig. 19-4. Slight isotopic shifts in the vibrational levels allow isotopic selection. In practice, the states are not equally

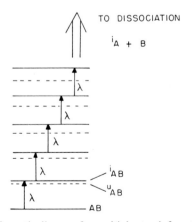

Fig. 19-4   Schematic diagram for multiple-step infrared dissociation.

spaced, so that one would not expect a tuned laser to drive the molecule more than a few steps up the ladder. Yet the method has been observed to work experimentally. Theoretical understanding of this phenomenon is not yet well advanced. This method requires long wavelength tunable lasers, and should be relatively conserving of the laser energy.

There have been many other suggested methods for isotope separation. However, the six methods listed here cover many of the leading contenders.

2. Uranium Isotope Separation

The most important application of laser-assisted isotope separation may be ^{235}U. The application of interest is to provide enriched uranium fuel for light water reactors used for electrical power in the United States and elsewhere. Natural uranium contains 0.71% of ^{235}U, which is the isotope necessary for use in reactors. The remainder is ^{238}U. Commercial light water reactors require about 3% ^{235}U. Thus, natural uranium must be enriched before use as fuel in a light water reactor. The justification for separation of ^{235}U by laser techniques is entirely economic. It is anticipated that in the mid-1980s the United States will run out of capacity to supply the enriched uranium for reactors now operating or under construction. Needs for enriched uranium are satisfied by gaseous diffusion plants built many years ago. The projected needs are equivalent to building a new gaseous diffusion plant every 18 months at a capital investment of between \$2 and \$3 billion apiece. A new technology is required with investment and operational costs much lower than those associated with the present gaseous diffusion technology. There are three alternative enrichment technologies: lasers, ultracentrifuges, and nozzles. Of these, the laser approach appears to offer the lowest cost and highest efficiency.

Some values for cost and performance factors for various methods of uranium enrichment are presented in Table 19-2 [10]. Two possible laser processes are included: an atomic process based on two-photon ionization of uranium atoms and a molecular process based on multiphoton dissociation of UF_6. These processes will be discussed more fully later. The separation factor is the final ratio of the isotopic concentrations ($^{235}U/^{238}U$) divided by the initial ratio. For separation factors only slightly greater than unity, this is approximately equal to the final quantity of ^{235}U divided by the initial quantity. The number of separative steps required to reach the final concentration of 3% ^{235}U needed for light water reactors is also noted for the given separation factor. Both laser methods can exceed this ratio within a single step. The energy requirements are stated in terms of the kilowatt hours required per separative work unit (SWU). One SWU is equivalent to 1 kg of enriched uranium. The capital costs and energy requirements differ

TABLE 19-2
Cost and Performance Factors for Several Types of Uranium Enrichment

	Laser Processes				
	Atomic: 2-photon ionization	Molecular: multiphoton dissociation	Centrifuge	Nozzle	Diffusion
Separation factor	10	33	1.25	1.0118	1.0043
Number of steps needed	1	1	6	120	335
Energy requirement (kWh/SWU)	170	51	210	3500	2100
Capital cost ($/SWU)	195	18	233	240	388
Possible completion date	1986	1986	1982	?	1985

between the two laser approaches. Part of this difference may be the intrinsic difference between an atomic approach and a molecular approach. The latter should require long wavelength lasers and less total energy input per separated ^{235}U atom. The laser approaches appear to offer advantages in almost every category.

It is possible that laser uranium enrichment plants could be operative in the 1980s. Meeting this schedule would require considerable advances in laser technology. Tunable lasers with a narrow linewidth must be developed which operate with high pulse repetition rate, high peak power, and high average power. They must be capable of being tuned to the wavelengths of the desired transitions. Table 19-3 shows some estimates for lasers that would be required in the early 1980s in order to provide prototype demonstrations of laser isotope separation using the molecular approach [11]. Such lasers are well beyond the state of the art in the mid-1970s and would require considerable development.

TABLE 19-3

Requirements for Lasers for Demonstration of Uranium Isotope Separation by Molecular Processes[a]

	Near-term needs	Longer-term needs (1980)
Linewidth	<0.03 cm^{-1}	<0.03 cm^{-1}
Pulse repetition rate	1 pps	200 pps
Possible frequencies		
628 ± 1 cm^{-1} ($\sim 15.9\ \mu$m)	0.1 mJ	1 mJ ⎱ Pulse
823 ± 1 cm^{-1} ($\sim 12.2\ \mu$m)	25 mJ	1 J ⎰ energy
1160 ± 1 cm^{-1} ($\sim 8.6\ \mu$m)	8 mJ	0.5 J ⎱ needed
1294 ± 1 cm^{-1} ($\sim 7.7\ \mu$m)	10 mJ	0.5 J ⎰ at each frequency

[a] Requirements on lasers to be used in production ($\sim$ mid-1980s) would be even greater.

The table gives the pulse energy required for each of four possible frequencies. The energy per pulse is lowest for the line near 16 μm, but it may be easier to fulfill the higher requirements at the other shorter wavelengths. Large research efforts are underway at a number of laboratories to supply tunable lasers that would satisfy the requirements given in the table. We note that lasers needed in the mid-1980s and suitable for production use would have to satisfy even more stringent requirements.

The approach towards laser-assisted uranium isotope separation is following two different technology paths. One involves atomic uranium vapor. A laser whose wavelength can be tuned to an absorption line of ^{235}U in the visible spectrum is directed through the vapor. This causes the ^{235}U atoms to be raised to an excited state. A second laser operating in the blue or ultraviolet then ionizes the excited atoms, which can then be separated by electrical or magnetic means. This process is two-step photoionization, process (a) of subsection 19.C.1. The process does work, as was demonstrated in 1974 when small quantities of enriched uranium were obtained.

A two-step selective photoionization of ^{235}U in uranium vapor has been demonstrated [12]. A collimated beam of uranium atoms from a heated uranium–rhenium alloy was irradiated by the tunable output of a continuous dye laser and simultaneously by ultraviolet light. The absorption spectrum of uranium shows a shift of approximately 7 GHz between many of the absorption lines of ^{235}U and ^{238}U. The Doppler width of these lines at the temperature required for a reasonable vapor pressure of uranium was about 1 GHz. Thus, the lines were well enough separated so that the tunable dye laser, with a linewidth less than 0.05 GHz, could be tuned so as to excite only one of the isotopes. Then this one excited isotope could be ionized by absorption of ultraviolet radiation.

The apparatus is shown in Fig. 19-5. The dye laser was tuned to the wavelength of 5915.4 Å, which corresponds to the proper wavelength for excitation of the ^{235}U. A 2500 W mercury arc lamp was filtered to remove

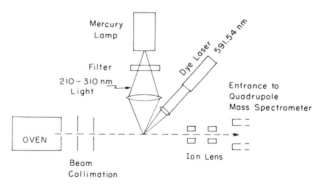

Fig. 19-5   Schematic diagram of apparatus for two-step selective photoionization of ^{235}U.

light of wavelength shorter than 2100 Å. This prevented ionization of the unexcited ^{238}U, but did allow ionization of excited ^{235}U. The photoionized uranium was then collected with an electric field.

Figure 19-6 shows the relevant absorption spectra [13]. (a) shows the absorption spectra of ^{235}U and ^{238}U near 5915 Å. The ^{235}U absorption, with a hyperfine structure consisting of eight peaks, is clearly separated from the ^{238}U absorption, which has no hyperfine structure. The separation

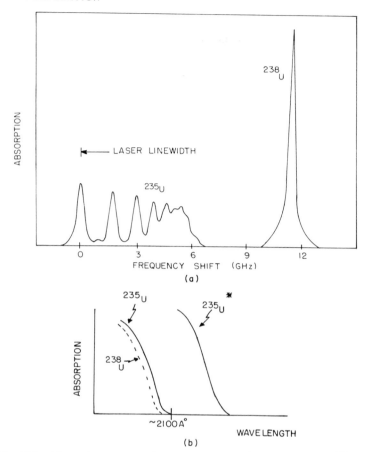

Fig. 19-6 Absorption spectra relevant to two-step photoionization of ^{235}U. (a) Absorption spectra of ^{235}U and ^{238}U near 5915 Å, with the linewidth of a dye laser indicated for comparison. (From B. B. Snavely, *Proc. Conf. Laser 75 Opto-Electron., Munich, June 24–27, 1975* (W. Waidelich, ed.). IPC Business Press, Ltd., Guildford, Surrey (1976).) (b) Absorption spectra of ^{235}U* (excited state absorption), ^{235}U and ^{238}U in the ultraviolet.

is much larger than the laser linewidth. Figure 19-6b shows a portion of the absorption spectra near 2100 Å, for excited ^{235}U* (the asterisk denotes the excited state), for unexcited ^{235}U and for ^{238}U. The ultraviolet light is absorbed by ^{235}U*, but since light of wavelength shorter than 2100 Å has been removed, there is no absorption by ^{235}U or ^{238}U atoms.

A mass spectrometer was used to examine the composition of the material collected by the electric field. With the dye laser tuned to the absorption of ^{235}U, this isotope was much enhanced relative to ^{238}U. When the enrichment started with natural uranium (0.71% ²³⁵U), the collected material was enriched to a concentration of 60%. With the 40 mW continuous dye laser available, the production of ²³⁵U was 10⁶ atoms per second (~10⁻¹⁰ gm/day of uranium enriched to 60% ²³⁵U). This demonstration represented only a first step. In order to reach the status of practical large-scale production, considerable advances in laser power would be required.

Another demonstrated two-step photoionization technique [14] is also based on irradiation of uranium vapor with two short-pulse lasers. The energy level diagram relevant to this work is shown in Fig. 19-7. The excitation to the intermediate ²³⁵U level, separated by 0.32 cm⁻¹ from the ²³⁸U

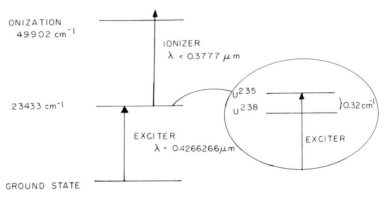

Fig. 19-7 Simplified energy level diagram relevant to two-step photoionization of ²³⁵U.

level, is provided by a dye laser, tuned to overlap the ²³⁵U line and with a linewidth sufficiently small that it does not overlap the ²³⁸U line. The ionization from this intermediate state is provided by a nitrogen laser at a wavelength of 0.3371 μm. The yield of the two isotopes as a function of the dye laser wavelength is shown in Fig. 19-8, from mass spectrometric analysis. This data indicates that at the wavelength of 4266.266 Å the product was enriched to 50% ²³⁵U.

The second laser approach employs uranium hexafluoride (UF₆) vapor rather than atomic uranium. The excitation is a multistep process with an infrared wavelength [process (f) of subsection 19.C.1]. For one possible case, the wavelength for excitation is approximately 16 μm. Other possibilities have been listed in Table 19-3. Absorption of a number of successive infrared photons causes molecules containing the selected isotope to move up through a ladder of approximately equally spaced excited levels, until the molecule eventually dissociates. The dissociation products containing the desired isotope may be scavenged by chemical means. This approach may be more promising because UF₆ gas is easier to handle than corrosive atomic

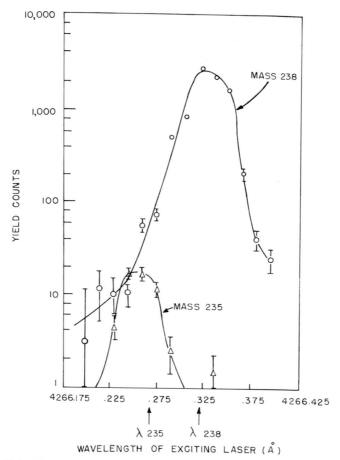

Fig. 19-8 Measurements of two-step laser-induced yield of ^{235}U, as a function of wavelength of dye laser. (From G. S. Janes *et al., IEEE J. Quantum Electron.* **QE-12**, 111 (1976).)

uranium vapor. This process was scheduled for demonstration in 1976. Not all the details of this scheme have been publicly released, but presumably it is similar to work on the molecule SF_6, which is described in the next subsection. Experiments carried out on SF_6 have indicated that rapid deposition of many CO_2 laser photons can yield dissociation, and enrichment in the isotope ^{34}S relative to ^{32}S.

3. Other Isotopes

Separation of other isotopes can also be important in providing inexpensive isotopically selected materials for medicine, research, and industry. Isotopically enriched chemicals are widely used in medicine, both for diagnostic and therapeutic purposes. Radioisotopes are often used as tracers in industrial processes. It has also been suggested that separation of deuterium could reduce the costs associated with heavy water reactors, and that this approach could be a superior use of laser-assisted isotope separation, as compared to separation of uranium. Another suggestion has been separation of ^{50}Ti as a structural material with low thermal neutron cross section, for use in nuclear reactors.

The cost of isotopically pure materials has been high. Laser techniques could reduce costs considerably. In this section we shall consider a few representative examples which will show the range of the techniques. The first example involves the use of multiphoton infrared photodissociation to increase the concentration of ^{34}S in SF_6. This example is similar to the molecular approach for separation of ^{235}U, as was mentioned in subsection 19.C.2.

Mixtures of sulfur hexafluoride and hydrogen were irradiated with focused CO_2 laser radiation, delivering 1–2 J in a 200 nsec duration pulse

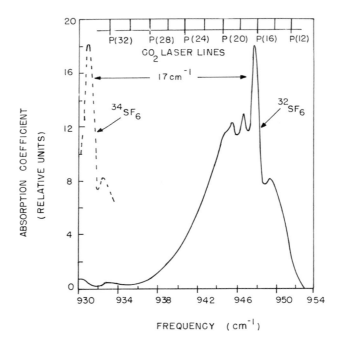

Fig. 19-9 Absorption spectra of $^{34}SF_6$ and $^{32}SF_6$ in the infrared near 10.6 μm. The positions of several CO_2 laser lines are indicated. (From B. B. Snavely, at the 1975 IEEE/OSA Conference on Laser Engineering and Applications, Washington, D.C., May 28–30, 1975.)

[15]. The peak laser power was approximately 6×10^9 W/cm^2. The laser could be line tuned to the P(20) line. This line matches the initial step of the ^{32}SF$_6$ molecule up the vibrational manifold. This increases the transition probability for the selected isotope to walk up the vibrational ladder. The relevant absorption spectra of ^{32}SF$_6$ and ^{34}SF$_6$ are shown in Fig. 19-9 [16]. After irradiation the residual sulfur hexafluoride was analyzed to determine the ^{32}S and ^{34}S content. The initial ratio ^{34}S/^{32}S was 0.044 $\pm$ 0.002. Figure 19-10 [17] shows results for the composition of the starting material and for the residual SF$_6$ after 2000 and after 5000 pulses. These results indicate an increase in the ratio of ^{34}S/^{32}S by a factor of 33. The mechanism seems to be a multiphoton dissociation of ^{32}SF$_6$ with the liberated fluorine being scavenged by the hydrogen. The unselected ^{34}SF$_6$ does not react and progressively becomes a larger component in the residual material. The exact details of the physical phenomena involved, particularly how the molecule gets carried up higher unequally spaced steps in the ladder, have not been fully interpreted.

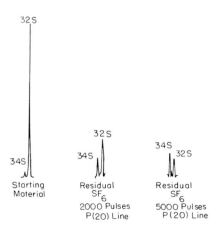

Fig. 19–10   Laser enrichment of ^{34}S by CO$_2$ laser irradiation of SF$_6$. (From C. P. Robinson, *Ann. N.Y. Acad. Sci.* **267**, 81 (1976).)

A second example will demonstrate photopredissociation. The primary step in isotopic separation of bromine isotopes [18] was excitation with a frequency-doubled Nd:YAG laser, which emits in the green near 5580 Å. The laser was tunable over a narrow range (~ 1 cm^{-1}) by rotating an internal etalon in the laser. The laser was tuned to the absorption line of either ^{79}Br$_2$ or ^{81}Br$_2$. The excited bromine molecule could dissociate by crossing over to an unbound state, which crosses the potential curve of the bound excited state. Thus, with the laser tuned to the absorption of $^{81}Br_2$, the reactions

$$^{81}Br_2 + h\nu \longrightarrow {}^{81}Br_2{}^*$$
$$^{81}Br_2{}^* \longrightarrow {}^{81}Br + {}^{81}Br$$

would occur. Chemical scavenging of the ^{81}Br must occur rapidly. A reaction such as

$$^{81}Br + HI \longrightarrow H^{81}Br + I$$

then yields a product which can be separated. Enrichment from a natural abundance of 50% ^{81}Br to an abundance around 80% was demonstrated. The set of reactions should occur so as not to involve the isotope ^{79}Br. However, scrambling reactions such as

$$^{81}Br_2{}^* + {}^{79}Br_2 \longrightarrow {}^{81}Br_2 + {}^{79}Br_2{}^*$$

can occur. If this reaction occurs, the $^{79}Br_2{}^*$ will undergo the remainder of the steps and ^{79}Br will appear in the end product. The presence of scrambling reactions such as this limits the maximum abundance. Thus one desires that the collection process be completed rapidly, before scrambling reactions have much chance to occur.

A third example involves photodeflection of barium atoms [19]. The photodeflection method involves transfer of momentum from the laser beam to an atomic beam. Each time an atom absorbs a laser photon, it acquires momentum $h\nu/c$ in the direction of propagation of the photon, where $h\nu$ is the photon energy and c is again the speed of light. In the experiment the atomic excitation was allowed to decay spontaneously. Since the emission is isotropic on the average, the momentum imparted to the atom by a large number of such emission processes averages to zero. Spatial separation of the desired isotope is achieved when the number N of photons thus scattered is such that $Nh\nu/c$ is greater than or comparable to the (transverse) momentum spread of the atomic beam.

The experimental setup used barium metal, with the natural mixture of isotopes, evaporated in an oven. The vapor was collimated and intersected the laser beam at right angles. It was detected using a mass spectrometer. A 150 mW beam at wavelength 5535.7 Å from a tunable dye laser was used. The laser could be tuned to the correct wavelength for any of the barium isotopes, ^{134}Ba, ^{135}Ba, ^{136}Ba, ^{137}Ba, or ^{138}Ba. One can regard the selected atoms as simply being pushed to the side by the momentum of the light. A separation of 10^{14} atoms per second was demonstrated, enriched by a factor about three over the natural abundance.

These processes are given as examples to show the operation of different aspects of an isotope separation process. There are a wide variety of different photointeractions and chemical interactions that can be employed. Many different isotopes have been considered. The optimum separation process will probably vary from element to element. As yet, these are in experimental stages, and no laser isotope separation process has reached the status of production.

REFERENCES

[1] A. L. Schawlow, *in* "Fundamental and Applied Laser Physics" (M. S. Feld, A. Javan, and N. A. Kurnit, eds.). Wiley, New York, 1973.
[2] N. V. Karlov, *Appl. Opt.* **13**, 301 (1974).
[3] N. G. Basov *et al.*, *in* "Fundamental and Applied Laser Physics" (M. S. Feld, A. Javan, and N. A. Kurnit, eds.). Wiley, New York, 1973.
[4] H. R. Bachmann *et al.*, *Chem. Phys. Lett.* **33**, 261 (1975).
[5] G. Porter and M. R. Topp, *Proc. Roy. Soc.* **A315**, 163 (1970).
[6] J. R. Novak and M. W. Windsor, *Proc. Roy. Soc.* **A308**, 95 (1968).
[7] T. L. Netzel, P. M. Rentzepis, and J. Leigh, *Science* **182**, 238 (1973).
[8] P. M. Rentzepis *et al.*, *Phys. Rev. Lett.* **25**, 1742 (1970).
[9] G. E. Busch, *et al.*, *Proc. Nat. Acad. Sci.* **69**, 2802 (1972).
[10] R. H. Levy, quoted in *Science* **191**, 1162 (1976).
[11] E. A. O'Hair, *Proc. Soc. Photo-Opt. Instrum. Engineers*, 20th Anniversary Technical Symposium, San Diego, August 23–27, 1976.
[12] S. A. Tuccio *et al.*, *IEEE J. Quantum Electron.* **QE-10**, 790 (1974).
[13] B. B. Snavely, *Proc. Conf. Laser 75 Opto-Electron.*, Munich, *June 24–27, 1975*, (W. Waidelich, ed.). IPC Business Press, Ltd., Guildford, Surrey (1976).
[14] G. S. Janes *et al.*, *IEEE J. Quantum Electron.* **QE-12**, 111 (1976).
[15] J. L. Lyman *et al.*, *Appl. Phys. Lett.* **27**, 87 (1975).
[16] B. B. Snavely, at the 1975 IEEE/OSA Conference on Laser Engineering and Applications, Washington, D.C., May 28–30, 1975.
[17] C. P. Robinson, *Ann. NY Acad. Sci.* **267**, 81 (1976).
[18] S. R. Leone and C. B. Moore, *Phys. Rev. Lett.* **33**, 269 (1974).
[19] A. F. Bernhardt *et al.*, *Appl. Phys. Lett.* **25**, 617 (1974).

SELECTED ADDITIONAL REFERENCES

A. Spectroscopy

M. Ahmadjian and C. W. Brown, Petroleum Identification by Laser Raman Spectroscopy, *Anal. Chem.* **48**, 1257 (1976).

J. R. Allkins, Tunable Lasers in Analytical Spectroscopy, *Anal. Chem.* **47**, 752A (1975).

G. E. Busch and P. M. Rentzepis, Picosecond Chemistry, *Science* **194**, 276 (1976).

W. Demtroeder, Laser Spectroscopy, "Topics in Current Chemistry," Vol. 17, Springer, New York, 1976.

M. S. Feld and V. S. Letokhov, Laser Spectroscopy, *Sci. Am.*, p. 69 (December 1973).

T. R. Gilson and P. J. Hendra, "Laser Raman Spectroscopy," Wiley (Interscience), New York, 1970.

E. D. Hinkley, High-Resolution Infrared Spectroscopy with a Tunable Diode Laser, *Appl. Phys. Lett.* **16**, 351 (1970).

S. M. Jarrett, The Use of Dye Lasers in Spectroscopy, *Electro-Opt. Syst. Design*, p. 24 (September 1973).

K. J. Kaufmann *et al.*, Picosecond Kinetics of Events Leading to Reaction Center Bacteriochlorophyll Oxidation, *Science* **188**, 1301 (1975).

P. L. Kelley and E. D. Hinkley, Tunable Semiconductor Lasers and Their Spectroscopic Uses, *in* "Fundamental and Applied Laser Physics" (M. S. Feld, A. Javan, and N. A. Kurnit, eds.), Wiley, New York, 1973.

K. C. Kim *et al.*, Laser Spectroscopy, *Proc. Tech. Program, Electro-opt. Syst. Design Conf.* Anaheim, California (November 11–13, 1975).

K. W. Nill *et al.*, Infrared Spectroscopy of CO Using a Tunable PbSSe Diode Laser, *Appl. Phys. Lett.* **19**, 79 (1971).

H. Preier and W. Riedel, NO Spectroscopy at 100 K with a $PbS_{0.4}Se_{0.6}$ Diode Laser, *Appl. Phys. Lett.* **25**, 55 (1974).

G. J. Rosasco, E. S. Etz, and W. A. Cassatt, The Analysis of Discrete Fine Particles by Raman Spectroscopy, *Appl. Spectrosc.* **29**, 396 (1975).

G. J. Rosasco, E. Roedder, and J. H. Simmons, Laser-Excited Raman Spectroscopy for Nondestructive Partial Analysis of Individual Phases in Fluid Inclusions in Minerals, *Science* **190**, 557 (1975).

J. I. Steinfeld, The Impact of Lasers in Spectroscopy, *Opt. Eng.* **13**, 476 (1974).

C. L. Tang and J. M. Telle, Laser Modulation Spectroscopy of Solids, *J. Appl. Phys.* **45**, 4503 (1974).

B. Photochemical Applications

R. R. Alfano and S. L. Shapiro, Ultrashort Phenomena, *Phys. Today*, p. 30 (July 1975).

K. R. Chien and S. H. Bauer, Laser Augmented Decomposition. 2. D_3BPF_3, *J. Phys. Chem.* **80**, 1405 (1976).

P. Houston, Applications of Lasers to Kinetic Spectroscopy, *Opt. Eng.* **13**, 489 (1974).

D. S. Kliger and A. C. Albrecht, Nanosecond Excited-State Polarized Absorption Spectroscopy of Anthracene in the Visible Region, *J. Chem. Phys.* **50**, 4109 (1969).

V. F. Letokhov, Use of Lasers to Control Selective Chemical Reactions, *Science* **180**, 451 (1973).

L. B. Lockhart, Lasers in Chemistry, *Opt. Laser Technol.*, p. 159 (August 1974).

E. R. Lory, S. H. Bauer, and T. Manuccia, Infrared Laser Augmented Decomposition of Borane-Phosphorous Fluoride, *J. Phys. Chem.* **74**, 545 (1975).

T. J. McIlrath, Absorption from Excited States in Laser-Pumped Calcium, *Appl. Phys. Lett.* **15**, 41 (1969).

C. B. Moore (ed.), "Chemical and Biochemical Applications of Lasers," Academic Press, New York, 1974.

T. L. Netzel, W. S. Struve, and P. M. Rentzepis, Picosecond Spectroscopy, *in* "Annual Review of Physical Chemistry," Vol. 24 (H. Eyring, ed.), Annual Reviews, Palo Alto, California, 1973.

J. R. Novak and M. W. Windsor, Laser Photolysis and Spectroscopy in the Nanosecond Time Range: Excited Singlet State Absorption in Coronene, *J. Chem. Phys.* **47**, 3075 (1967).

S. D. Rockwood and J. W. Hudson, Laser Driven Synthesis of $BHCl_2$ from BCl_3 and H_2, *Chem. Phys. Lett.* **34**, 542 (1975).

A. M. Ronn, Infrared Laser Catalyzed Chemical Reactions, *Spectrosc. Lett.* **8**, 303 (1975).

R. M. Wilson, Prospects in Laser Chemistry, *Ann. N.Y. Acad. Sci.* **168**, 615 (1970).

F. S. Yeung and C. B. Moore, Photochemistry of Single Vibronic Levels of Formaldehyde, *J. Chem. Phys.* **58**, 3988 (1973).

C. Isotope Separation

R. V. Ambartzumian and V. S. Letokhov, Isotopically Selective Photochemistry, *Laser Focus*, p. 48 (July 1975).

D. Arnoldi, K. Kaufmann, and J. Wolfrum, Chemical-Laser-Induced Isotopically Selective Reaction of HCl, *Phys. Rev. Lett.* **34**, 1597 (1975).

E. V. George and W. F. Krupke, Lasers for Isotope Separation Processes and Their Properties, *in Proc. Ind. Appl. High-Power Laser Technol.* Vol. 86 (J. F. Ready, ed.), Soc. of Photo-Opt. Instrumentation Eng., Palos Verdes Estates, California, 1976.

R. W. F. Gross, Laser Isotope Separation, *Opt. Eng.* **13**, 506 (1974).

R. J. Jensen *et al.*, Prospects for Uranium Enrichment, *Laser Focus*, p. 51 (May 1976).

V. S. Letokhov and C. B. Moore, Laser Isotope Separation (Review), *Sov. J. Quantum Electron.* **6**, 129 (1976).

J. L. Lyman *et al.*, Isotopic Enrichment of SF_6 in S^{34} by Multiple Absorption of CO_2 Laser Radiation, *App. Phys. Lett.* **27**, 87 (1975).

J. L. Lyman and S. D. Rockwood, Enrichment of Boron, Carbon and Silicon Isotopes By Multiple Photon Absorption of 10.6 μm Laser Radiation, *J. Appl. Phys.* **47**, 595 (1976).

J. H. McNally, Laser Fusion and Laser Isotope Separation Overview, *Ann. N.Y. Acad. Sci.* **267**, 61 (1976).

E. A. O'Hair and M. S. Piltch, Lasers for Isotope Separation, *in Proc. Ind. Appl. High Power Laser Technology*, Vol. 86 (J. F. Ready, ed.), Soc. Photo-Opt. Instrumentation Eng., Palos Verdes Estates, California (1976).

L. J. Radziemski, S. Gerstenkorn, and P. Luc, Uranium Transitions and Energy Levels Which May Be Useful in Atomic Photoionization Schemes for Separating ^{238}U and ^{235}U, *Opt. Commun.* **15**, 273 (1975).

C. P. Robinson, Laser Isotope Separation, *Ann. N.Y. Acad. Sci.* **267**, 81 (1976).

R. C. Stern and B. B. Snavely, The Laser Isotope Separation Program at Lawrence Livermore Laboratory, *Ann. N.Y. Acad. Sci.* **267**, 71 (1976).

E. S. Yeung and C. B. Moore, Isotopic Separation by Photopredissociation, *Appl. Phys. Lett.* **21**, 109 (1972).

R. N. Zare, Laser Separation of Isotopes, *Sci. Am.*, p. 86 (February 1977).

CHAPTER 20

INTEGRATED OPTICS AND FIBER OPTICS

The term "integrated optics" refers to a variety of techniques that have been employed to produce compact optical circuitry [1]. The techniques are analogous to those employed in integrated electronic circuitry. The components themselves are small and can be integrated on substrate materials which include active components, couplers, filters, and other passive components needed for optical signal transmission and processing. Transmission of the beam over a distance would be carried out in a fiber optic cable. The techniques of production are similar to those in integrated electronic circuitry, with the result that components can be batch-produced economically.

Historically, optical components have involved relatively large devices, with dimensions of the order of centimeters or greater. Typical circuitry might include such components as a laser, modulator, and a detector to form, for example, a communications system. The light beam would propagate freely through the atmosphere and could be periodically recollimated by means of lenses.

The approach of integrated optics and fiber optics relies on waveguide transmission of a light beam. As we shall describe later, a light beam can propagate in a medium which has a higher index of refraction than the surrounding material. The propagation can occur with very low loss and the dimensions of the waveguide can be small. The ability of an optical beam to propagate as a trapped wave in a waveguide is at the heart of integrated optics and fiber optical transmission. Other circuit devices such as lasers, modulators, and detectors can be fabricated in the same material as the waveguide and can couple directly into the waveguide. The dimensions of these components are also small.

The entire approach of integrated optics leads to compact, rugged devices which may be easily and economically fabricated. As compared to propagation of laser beams through the atmosphere, there are advantages such as freedom from variations in the atmosphere and potential savings in size and in cost.

The development of integrated optical circuitry is still in a relatively early stage. In the mid-1970s, integrated optics cannot yet replace the larger individual lasers and circuit elements which have traditionally been used. However, the development of integrated optics and fiber optics is proceeding very rapidly in a number of laboratories, and the potential applications of these techniques are important.

In this chapter we shall briefly survey some of the materials and experiments that are involved in integrated optics and fiber optics. This is not meant to be an exhaustive discussion of all the important developments, but rather a few examples of the approaches that are being considered.

A. Optical Waveguides

Figure 20-1 shows the basic form of an optical waveguide. In its most elementary form, the optical waveguide is simply a small volume of transparent dielectric material with index of refraction higher than the index of the surrounding dielectric material. Thus, in Fig. 20-1, we have $n_2 > n_1$. An optical wave, once trapped inside the medium with higher index of refraction, will propagate by total internal reflection. In the structure shown in the figure, an optical wave coupled into the waveguide at the input

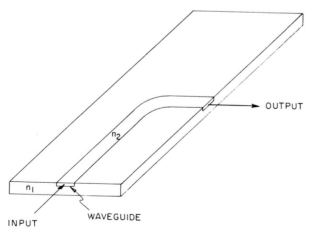

Fig. 20-1 Schematic diagram of optical waveguide in the form of a thin surface film. The index of refraction n_2 in the film is greater than the index of refraction n_1 in the substrate.

position will be transmitted through the waveguide and emerge at the output position. The optical waveguide is analogous to a dielectric rod in air which can support the propagation of microwaves with small attenuation.

One can distinguish between two types of optical waveguide structure.

(1)   The medium in which the light wave propagates is different from the surrounding medium. The two media have rather different indices of refraction. In this case, the waveguide should be rather thin, preferably thinner than one wavelength of light. This case is the one illustrated in Fig. 20-1, in which the waveguide is a thin surface layer, with one side in contact with air. Such waveguides may be formed by sputtering thin films of materials such as zinc oxide or tantalum oxide on glass substrates. Tantalum oxide films in particular have very low optical loss [2].

The structure in Fig. 20-1 will act as a transmission line. The thin film dielectric waveguide is formed by a thin surface layer which has index of refraction higher than the index of refraction of the bulk material. The material in the guiding region may have an index of refraction around 1 % (or less) higher than that of the bulk region. Typical dimensions are of the order of a few micrometers. This leads to high power densities, even for relatively low powers. For example, for a total power of 300 mW and a waveguide with dimensions $5 \times 6 \, \mu m$, one achieves a power density of $10^6 \, W/cm^2$. The structure shown in Fig. 20-1 will guide the light wave around the bend.

(2)   The second general class of optical waveguides involves an embedded guide with a refractive index which is only slightly higher than the index of the surrounding medium. Such a waveguide can be formed by techniques such as diffusion or ion implantation. Because the difference in index of refraction between the waveguide and the surrounding medium is small, the waveguide can have dimensions larger than the optical wavelength. For such optical waveguides, the index of refraction can be only slightly higher ($\sim 0.1 \%$) than that of the surrounding medium. Such a structure is illustrated in Fig. 20-2.

We note that the structures in Figs. 20-1 and 20-2 indicate a sharp discontinuity in index of refraction. This is not necessary. The gradient of the index of refraction can be gradual, and optical waveguide operation will still occur.

Fiber optical cables can be considered as a type of embedded guide. Thin glass fibers surrounded by a cladding of glass with lower index of refraction have been used for many years. A number of fibers can be used to form a flexible transmissive cable. They have been employed in many cases for transmission of light over short distances to inaccessible places.

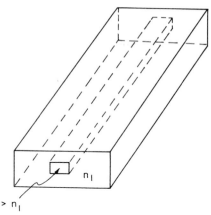

$n_2 > n_1$

Fig. 20-2 Diagram of optical waveguide (index of refraction n_2) embedded in a medium with index of refraction n_1.

They have become particularly familiar in many types of medical instrumentation designed for viewing internal body structures. Advances in glass technology to be described later will make familiar the fiber optic transmission of light over large distances.

The structure of a single fiber is shown schematically in Fig. 20-3. The core, of index n_2, is surrounded by a cladding of index $n_1 < n_2$. In principle, the cladding could be omitted, and one would have a waveguide due to the index difference between the glass and the surrounding air ($n_1 = 1$). In practice, small imperfections on the outer surface of the glass would lead to intolerable light losses, so that the cladding is always used in real devices. The thin glass fibers are flexible and can be formed in very long lengths by drawing glass fibers.

The technology of glass fibers has developed greatly in the early 1970s. Loss is commonly expressed in decibels per kilometer. The use of this unit

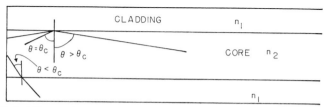

Fig. 20-3 Fiber optic waveguide. The fiber, of circular cross section, has a core with index of refraction n_2 and a cladding of lower index n_1. Rays traveling at angle θ to the normal greater than the critical angle θ_c [$\theta_c = \text{arc }\sin(n_1/n_2)$] are confined by total internal reflection and propagate through the fiber. Rays traveling at angles θ to the normal less than θ_c may escape.

implies that transmission over considerable distances is being considered. Before 1970, the best optical fibers had losses in excess of 1000 dB/km. This figure implies that fiber optic transmission of light over kilometer distances was not feasible. Developments in glass technology in the early 1970s steadily reduced this number, so that in the mid-1970s, glass fibers with losses less than 1 dB/km had been fabricated. For fibers capable of being produced in large quantities, a figure of 5 dB/km is perhaps more realistic. These lowered values of loss imply that fiber optic transmission over relatively large distances, certainly many kilometers, is feasible.

There exists a large body of theoretical work on the propagation of electromagnetic waves in waveguides. We shall not discuss the mathematics in detail here. We simply note that waveguiding can be understood in terms of total internal reflection. When light propagates across an interface between two media with different indices of refraction, n_1 and n_2, the angles θ_1 and θ_2 between the direction of propagation and the normal to the interface are related by Snell's law†

$$n_1 \sin \theta_1 = n_2 \sin \theta_2 \qquad (20.1)$$

If we consider propagation from medium 2 into medium 1, with $n_2 > n_1$, we see that the maximum possible value of $\sin \theta_2$ is given by

$$\sin \theta_2 = n_1/n_2 \qquad (20.2)$$

Thus light rays traveling in medium 2 at an angle θ_2 such that

$$\theta_2 > \arc \sin(n_1/n_2) \qquad (20.3)$$

cannot escape from the medium. Such rays are totally reflected inside medium 2 and cannot escape. This is the basis of optical waveguiding. The angle at which total internal reflection occurs is called the critical angle, and its value is denoted by

$$\theta_c = \arc \sin(n_1/n_2) \qquad (20.4)$$

The situation is sketched in Fig. 20-3. Rays traveling at angles greater than θ_c are confined to the fiber.

Several different structures for fibers are illustrated in Fig. 20-4. Multimode fibers offer ease of coupling of light into the fiber, because of the relatively large core. Multimode fibers have relatively high dispersion because of the variety of optical paths that a light ray may take down the fiber. Thus, a short pulse of light injected into the fiber will undergo a temporal spread as it propagates. This limits the bandwidth of systems based on multimode fibers. Single-mode fibers have small dispersion because

† A discussion of Snell's law can be found in elementary optics texts.

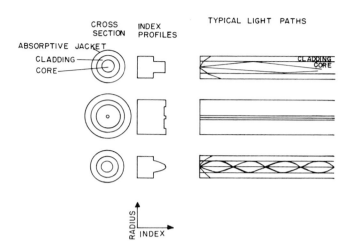

Fig. 20-4 Types of optical fibers, showing cross sections, profiles of index of refraction, and paths of typical light rays. (a) Multimode fiber. (b) Single-mode fiber. (c) Multimode fiber with graded index profile.

light can propagate only at small angles to the axis. Thus there is less spread in the possible path length down the fiber and relatively little distortion of short pulses. Such fibers can be used for systems with large bandwidth. They require greater precision in fabrication and efficient coupling of light into them is more difficult.

Fibers with graded index of refraction offer a compromise. A parabolic radial variation of the index of refraction acts like a continuous focusing element. Rays traveling at an angle to the axis are bent back toward the axis by the gradient in refractive index. A ray traveling at an angle to the axis travels farther than one along the axis, but it spends more time in regions with low index of refraction. Thus, it has higher average velocity than the axial ray. The choice of the parabolic profile tends to keep the different rays in phase and thus keeps dispersion small. Because of the relatively large core, coupling is relatively easy.

There is a minimum tolerable bending wavelength for dielectric waveguide circuitry. The main limitation is radiation loss. A dielectric waveguide will radiate in regions where there is a bending of the field. As an example, for a single-mode waveguide formed with 1% difference of index of refraction, the radiation loss is negligible for bend radii greater than 1 mm. For a bend radius of 0.5 mm, the radiation loss is intolerable [3].

The spectral loss of some modern glass for optical fibers is shown in Fig. 20-5 [4]. The transmission is maximum in the near infrared, so that gallium arsenide lasers are suitable sources for fiber optic systems.

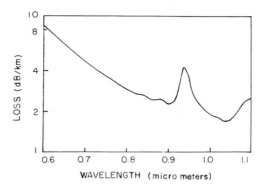

Fig. 20-5 Loss in optical fiber glass as a function of wavelength. (From G. W. Tasker and W. G. French, *Proc. IEEE* **62**, 1282 (1974).)

A few words about fabrication of waveguides are in order. There are numerous techniques for producing thin films and fibers suitable for integrated optics. These include

(1) *Pulling of fibers from heated tips of preforms in the shape of solid rods*   Preforms are often prepared by chemical vapor deposition processes, so that the radial distribution of the chemical composition is controlled. The radial composition in the preform is preserved in the fiber.

(2) *Ion bombardment*   Ion bombardment can produce regions of high index of refraction at the surface of materials such as fused silica.

(3) *Ion exchange*   Diffusion methods have produced films of graded index of refraction on the surface of glass. For example, floating molten glass on molten tin yields the so-called Pilkington float glass, which has a surface layer with higher index of refraction than that of the bulk material.

(4) *Sputtering*   RF sputtering of glass and of semiconductors is a promising method for producing optical waveguide films.

Once surface films have been formed, it is necessary to use masking and etching techniques for use with the films in order to form optical circuits. Conventional photolithographic techniques employ optically exposed photoresist, and have been used to produce integrated optical circuitry in surface films. Photolithography suffers from the fact that it produces an edge which is rougher than desired. Roughness of the edge leads to losses from the waveguide by radiation. Chemical etching has been employed but has not been satisfactory because of clouding of the surfaces.

Use of electron beams to expose photoresist has been used satisfactorily. This technique leads to very high resolution and good edge definition.

The problems of materials and circuit fabrication have not yet been completely solved. Work on materials and techniques is underway at a number of laboratories. We may expect significant advances in this area.

Now let us consider coupling of light into optical waveguides. Suitable methods for bringing light into optical waveguides from the outside are important. Fig. 20-6 shows three methods for coupling optical energy from an external source into a thin film surface optical waveguide. These methods are

(1) prism coupling,
(2) grating coupling,
(3) direct injection.

It is apparent that the same methods can be employed to transform light that is traveling in an optical waveguide into freely propagating light outside the waveguide.

In the prism coupling method (Fig. 20-6a) [5], frustrated total reflection at the base of the prism leads to excitation of a wave in the waveguide by a tunneling process. The gap between the prism and the film must be very narrow, less than the optical wavelength. The prism coupler is simple and versatile, and is capable of transferring more than 70% of an incident light beam into the film.

The grating coupler (Fig. 20-6b) employs frustrated total reflection of the diffracted first-order wave in order to excite the waveguide [6]. The grating coupler is very adaptable in integrated optical circuitry and can be fabricated by techniques that are compatible with the rest of the device. The grating is rugged and stable and is also capable of coupling more than 70% of a

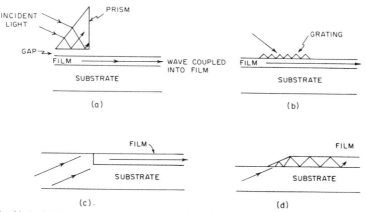

Fig. 20-6   Methods of coupling light into thin film surface waveguides. (a) Prism coupler. (b) Grating coupler. (c) Direct coupling, abruptly terminated waveguide. (d) Direct coupling, tapered waveguide.

light beam into the film. Suitable gratings may be formed by photolithographic techniques, with exposure of photoresist to an interferometric fringe pattern.

The grating coupler and the prism coupler are best applicable when the thin film waveguides are relatively wide. For narrow light guides, these methods are less efficient, because the laser beam has to be focused. In such cases, direct injection (Figs. 20-6c and 20-6d) is more desirable [7]. The beam is inserted into the waveguide through the substrate by directing the beam as shown, at a small angle to the interface between the waveguide and air. The end of the waveguide can terminate inside the substrate (Fig. 20-6c), or there may be a tapered edge (Fig. 20-6d) [8]. The direct injection method is capable of coupling more than 90% of the light into the guide.

Methods for coupling the output of thin film components into fiber optic waveguides are still under development. One method applicable to coupling of the output of light emitting diodes into optical fibers is illustrated in Fig. 20-7 [9]. Coupling has been demonstrated between planar waveguides and optical fibers by means of gratings. The fiber would be in contact with a grating similar to that shown in Fig. 20-6b. The core of the fiber would be in direct contact with the grating. It has been estimated that the efficiency of such coupling could be high, as high as 86% for some cases [10]. Experimental verification of efficient coupling between fibers and planar waveguide structures has lagged behind other areas in development.

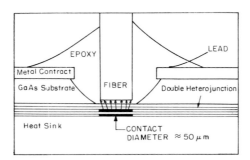

Fig. 20-7 Coupling into fiber optic waveguides. (From C. A. Burrus and B. J. Miller, *Opt. Commun.* **41**, 307 (1971).)

B. Components for Integrated Optics

The entire point of integrated optics is to be able to form all the circuit components, e.g., laser sources, modulators, couplers, etc., in thin film form on the surface of a substrate. The processes must be compatible with large scale economic production and with integration of the various components. So far there have been many demonstrations of individual components.

There have been only a few demonstrations of integration of several components together on the same surface [11].

In what follows, we shall describe several examples of fabrication of circuit components, specifically modulators, thin film lasers, and deflectors. These examples are chosen to show the possible range of fabrication techniques rather than the best possible construction. In this rapidly developing technology, it is impossible to predict the exact form of devices that may reach practical production status, perhaps around 1980.

1. *Modulator*

As an example of how optical circuit elements can be fabricated in integrated optics, we shall discuss experiments on light modulators using $p-n$ junctions in semiconductors. This will also illustrate some of the materials that can be used.

A schematic diagram of a semiconductor junction device demonstrated as a light modulator is shown in Fig. 20-8. The $p-n$ junction in the semiconductor acts as an optical waveguide. The waveguide effect arises from the higher index of refraction in the plane of the junction. Confinement of light waves to the $p-n$ junction region has been known for many years. The operation of such semiconductor junction light guides is very efficient. Waveguides can consist of thin epitaxial layers (approximately 10 μm thick) of semiconductors such as gallium arsenide or gallium phosphide. These are deposited on substrates of the same semiconductor with different resistivity. An alternative method involves a junction in a film of GaAs (about 0.5 μm thick) with epitaxial layers of AlGaAs on both sides of the junction. The

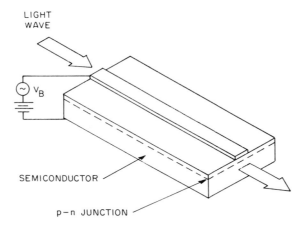

Fig. 20-8 Optical modulator based on semiconductor junction. (From S. E. Miller, *IEEE J. Quantum Electron.* **QE-8**, 199 (1972).)

AlGaAs has lower index of refraction than GaAs and provides good confinement of the light wave near the thin GaAs junction.

The depletion layer in such junctions is birefringent. The birefringence
apparently arises from the linear electrooptic effect within the junction. In
Fig. 20-8, the applied voltage changes the birefringence so that the structure
acts as an optical modulator. Because the light is well confined within the
birefringent material, the operation of the modulator is efficient. For
example, modulation of helium–neon laser light in semiconductor waveguide structures has been demonstrated with modest input power to the
modulator and with wide bandwidth [12, 13]. Modulation of CO_2 laser
radiation has been achieved in GaAs waveguide modulators at 1 GHz
bandwidth [14].

2. Distributed Feedback Lasers

The techniques of integrated optics have been employed in producing
lasers which consist of thin films on the surface of a substrate and in which
the feedback is provided by spatial variation of the refractive index of the
laser medium. Thus, there is no necessity for separate mirrors at the end of
the laser medium. This fact considerably simplifies the construction of the
laser as a thin film. The feedback is produced by Bragg scattering from the
periodic variation of the refractive index of the medium. Such lasers are
termed "distributed feedback lasers" because the feedback is produced
throughout the entire laser medium.

Let us consider the structure of a developmental distributed feedback
laser which employs a thin film of gelatin with a dye incorporated in the
gelatin. The arrangement is shown in Fig. 20-9. A thin gelatin film about
10 mm long and about 0.1 mm wide was deposited on a glass substrate.
Dichromate was added to the gelatin. Two ultraviolet beams from a helium–
cadmium laser were allowed to interfere (Fig. 20-9a). Dichromated gelatin
is a photosensitive medium. After the exposure, the gelatin was developed.
The developed film contained a spatial modulation of the refractive index,
which resulted from the interference pattern produced by the two beams of
light.

Following development of the gelatin, the dye material rhodamine 6G
was added to the gelatin layer. Rhodamine 6G can serve as a material for a
dye laser. A nitrogen laser was then employed to pump the rhodamine 6G.
This is shown in Fig. 20-9b. The fringe pattern (the spatial modulation of
the index of refraction) is indicated. Laser output was obtained at a wavelength around 0.63 μm [15].

Another approach has involved thin film GaAs distributed feedback
lasers [16], which have a corrugated interface between the GaAs active layer

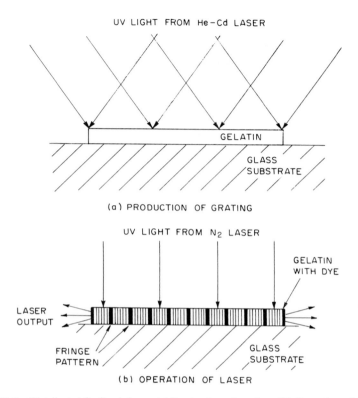

UV LIGHT FROM He-Cd LASER

GELATIN

GLASS
SUBSTRATE

(a) PRODUCTION OF GRATING

UV LIGHT FROM N₂ LASER

GELATIN
WITH DYE

LASER
OUTPUT

FRINGE
PATTERN

GLASS
SUBSTRATE

(b) OPERATION OF LASER

Fig. 20-9 Distributed feedback laser. (a) Production of grating. (b) Operation of laser.

and a $Ga_{1-x}Al_xAs$ layer that acts as the cladding. The corrugation provides a periodic structure which produces the feedback. Such lasers could in principle be formed on a GaAs substrate and integrated with other components on the same substrate. Distributed feedback lasers represent small, rugged structures that are significant advances in the field of integrated optics. Such thin film lasers are easily compatible with other integrated optic structures.

3. *Acoustooptic Deflector*

As a further example of how integrated structures can be formed to perform various functions in optical circuitry, we shall consider an experiment showing acoustooptic light beam deflection [17]. The structure is shown in Fig. 20-10. The waveguide was a thin glass film sputtered on the surface of an α-quartz crystal. A He–Ne laser beam was coupled into the waveguide by a grating coupler.

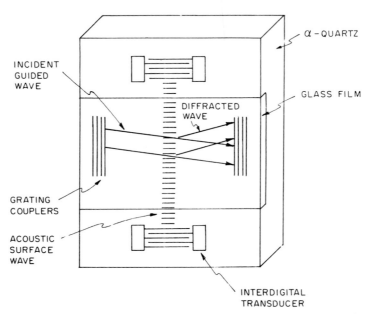

Fig. 20-10   Thin film acoustooptic deflector. (From L. Kuhn *et al., Appl. Phys. Lett.*
17, 265 (1970).)

A surface acoustic wave was produced in the quartz crystal by an inter-
digital transducer. The acoustic wave set up a grating in the glass film. The
optical wave was diffracted by the grating. The compressions and rarefactions associated with the acoustic wave acted as a diffraction grating for
the light. Hence, the light was extracted by the output grating coupler at
different angles, depending on whether the acoustic wave was present or not.
This experiment represented a merging of older techniques employed in
acoustooptic deflection in larger structures with thin film optical waveguide
techniques. It offers a good example of how different structures can be
combined in an integrated optical circuit.

We have presented examples of how several components (laser, modulator, and deflector) can be formed in integrated optical circuitry. This by
no means exhausts the list of components that have been demonstrated.
We shall not attempt to describe all of the different components, nor the
methods of producing them. The field of integrated optics is expanding
rapidly. We shall allow these few components to serve as examples for the
many that have been demonstrated.

An optical thin film waveguide with associated circuit elements is
suitable for control of light beams. There are many practical advantages for
this approach:

(1)   If the thin film is made of electrooptic material, a large electrooptic effect can be obtained with very low applied voltage. Beam modulation and deflection can therefore be achieved with low power consumption.

(2)   Since the beam is confined to a very small region, one can attain very high beam intensity and high electrical field; therefore, nonlinear optical effects can be easily achieved, when nonlinear materials are used as thin films.

(3)   Circuit elements such as resonant cavities, directional couplers, mode converters, filters, modulators, and deflectors can be easily fabricated and can be made integral parts of the system.

(4)   One can employ semiconductor integrated circuit technology to produce thin film waveguide devices. This leads to the so-called technology of integrated optics. Well-established techniques such as photolithography, ion implantation, and sputtering can play important roles in the fabrication of devices. The manufacturing processes can be simple and inexpensive and can yield large quantities of devices. Complex circuit patterns may be formed using photolithographic techniques.

(5)   The laser and associated optical circuitry will be isolated from thermal fluctuations, mechanical vibrations, and airborne acoustic effects, which hinder the operation of separately mounted components for unguided beams.

The current status of integrated optical techniques is mainly developmental. Most practical applications of lasers still involve the use of larger discrete components, with free space propagation of the laser beam. Development of new materials, techniques and components for integrated optics is proceeding rapidly. Communications applications seem particularly favorable and provide much of the motivation for the interest in integrated optics.

Integrated optic transmitters, repeaters, and receivers coupled with low-loss fiber optic cables can offer wide bandwidth systems that will occupy significantly smaller volume than conventional systems based on electrical transmission through wires. Communications companies are considering possible application of integrated optics and fiber optics for communications in crowded areas, where the capacity of current systems is being strained. Communications will be considered in more detail in the next chapter. The prospects for integrated laser circuitry and devices are attractive. Intensive effort to bring these concepts to practical realization is underway.

REFERENCES

[1]   S. E. Miller, *Bell Syst. Tech. J.* **48**, 2059 (1969).
[2]   D. H. Hensler *et al.*, *Appl. Opt.* **10**, 1037 (1971).

[3]   S. E. Miller, *IEEE J. Quantum Electron.* **QE-8**, 199 (1972).
[4]   G. W. Tasker and W. G. French, *Proc. IEEE* **62**, 1282 (1974).
[5]   P. K. Tien, R. Ulrich and R. J. Martin, *Appl. Phys. Lett.* **14**, 291 (1969).
[6]   M. L. Dakss *et al.*, *Appl. Phys. Lett.* **16**, 523 (1970).
[7]   D. Marcuse and E. A. J. Marcatili, *Bell Syst. Tech. J.* **50**, 43 (1971).
[8]   P. K. Tien and R. J. Martin, *Appl. Phys. Lett.* **18**, 398 (1971).
[9]   C. A. Burrus and B. J. Miller, *Opt. Commun.* **41**, 307 (1971).
[10] J. M. Hammer *et al.*, *Appl. Phys. Lett.* **28**, 192 (1976).
[11] *Bell Lab. Rec.*, p. 349 (September 1975).
[12] F. K. Reinhart, D. F. Nelson, and J. McKenna, *Phys. Rev.* **177**, 1208 (1969).
[13] D. Hall, A. Yariv, and E. Garmire, *Opt. Commun.* **1**, 403 (1970).
[14] P. K. Cheo and M. Gilden, *Appl. Phys. Lett.* **28**, 62 (1976).
[15] H. Kogelnik and C. V. Shank, *Appl. Phys. Lett.* **18**, 152 (1971).
[16] W.-T. Tsang and S. Wang, *Appl. Phys. Lett.* **28**, 596 (1976).
[17] L. Kuhn *et al.*, *Appl. Phys. Lett.* **17**, 265 (1970).

SELECTED ADDITIONAL REFERENCES

A. Optical Waveguides

R. L. Aagard, Optical Waveguide Characteristics of Reactive dc-Sputtered Niobium Pentoxide Films, *Appl. Phys. Lett.* **27**, 605 (1975).

M. J. Adams, D. N. Payne, and F. M. E. Sladen, Splicing Tolerances in Graded-Index Fibers, *Appl. Phys. Lett.* **28**, 524 (1976).

D. Chen *et al.*, Multimode Optical Channel Waveguides Induced in Glass by Laser Heating, *Appl. Phys. Lett.* **29**, 657 (1976).

A. G. Chynoweth, The Fiber Lightguide, *Phys. Today*, p. 28 (May 1976).

D. G. Dalgoutte *et al.*, Transition Waveguides for Coupling Fibers to Semiconductor Lasers, *Appl. Phys. Lett.* **27**, 125 (1975).

M. DiDomenico, Wires of Glass, *Ind. Res.*, p. 50 (August 1974).

P. Geittner, D. Kiippers, and H. Lydten, Low-Loss Optical Fibers Prepared by Plasma-Activated Chemical Vapor Deposition (CVD), *Appl. Phys. Lett.* **28**, 645 (1976).

D. Gloge, Optical Fibers for Communication, *Appl. Opt.* **13**, 249 (1974).

R. Gundlach, Fiber-Optic Developments Spark Worldwide Interest, *Electronics*, p. 81 (August 5, 1976).

N. S. Kapany and J. J. Burke, "Optical Waveguides," Academic Press, New York, 1972.

D. Lockie, Fiber Optics—Primed for Take-Off, *Electro-Opt. Syst. Design*, p. 30 (October 1976).

D. Marcuse, "Theory of Dielectric Optical Waveguides," Academic Press, New York, 1974.

R. D. Maurer, Glass Fibers for Optical Communications, *Proc. IEEE* **61**, 452 (1973).

J. L. Merz *et al.*, Taper Couplers for GaAs-Al$_x$Ga$_{1-x}$As Waveguide Layers Produced by Liquid Phase and Molecular Beam Epitaxy, *Appl. Phys. Lett.* **26**, 337 (1975).

S. E. Miller, A Survey of Integrated Optics, *IEEE J. Quantum Electron.* **QE-8**, 199 (1972).

S. E. Miller, E. A. J. Marcatili, and T. Li, Research Toward Optical-Fiber Transmission Systems, *Proc. IEEE* **61**, 1703 (1973).

T. Ozeki and B. S. Kawasaki, Optical Directional Coupler Using Tapered Sections in Multimode Fibers, *Appl. Phys. Lett.* **28**, 528 (1976).

M. D. Rigterink, Better Glass Fibers for Optical Transmission, *Bell Lab. Record*, p. 341 (September 1975).

H. Schonhorn *et al.*, Epoxy-Acrylate-Coated Fused Silica Fibers with Tensile Strengths > 500 ksi (3.5 GN/m^2) in 1-km Gauge Lengths, *Appl. Phys. Lett.* **29**, 712 (1976).

T. Tamir (ed.), "Integrated Optics," Springer-Verlag, Berlin and New York, 1975.

F. L. Thiel and W. B. Bielawski, Optical Waveguides Look Brighter Than Ever, *Electronics*, p. 89 (March 21, 1974).

P. K. Tien, Light Waves in Thin Films and Integrated Optics, *Appl. Opt.* **10**, 2395 (1971).

A. Yariv, Three-dimensional Pictorial Transmission in Optical Fibers, *Appl. Phys. Lett.* **28**, 88 (1976).

B. Components for Integrated Optics

J. C. Campbell, F. A. Blum, and D. W. Shaw, GaAs Electro-Optic Channel Waveguide Modulator, *Appl. Phys. Lett.* **26**, 640 (1975).

W. S. C. Chang, M. W. Muller, and F. J. Rosenbaum, Integrated Optics, *in* "Laser Applications," Vol. 2 (M. Ross, ed.), Academic Press, New York, 1974.

E. M. Conwell, Integrated Optics, *Phys. Today*, p. 48 (May 1976).

R. C. Cunningham, Integrated Optics: 1973–1975, *Electro-Opt. Syst. Design*, p. 26 (June 1975).

R. C. Cunningham, Integrated Optics Update, *Electro-Opt. Syst. Design*, p. 62 (August 1976).

V. Evtuhov and A. Yariv, GaAs and GaAlAs Devices for Integrated Optics, *IEEE Trans. Microwave Theory Tech.* **MTT-23**, 44 (1975).

J. E. Goell and R. D. Standley, Integrated Optical Circuits, *Proc. IEEE* **58**, 1504 (1970).

I. P. Kaminow, H. P. Weber, and E. A. Chandross, A Poly (Methyl Methacrylate) Dye Laser with Internal Diffraction Grating Resonator, *Appl. Phys. Lett.* **18**, 497 (1971).

I. P. Kaminow, Optical Waveguide Modulators, *IEEE Trans. Microwave Theory Tech.* **MTT-23**, 57 (1975).

H. Kogelnik, An Introduction to Integrated Optics, *IEEE Trans. Microwave Theory Tech.* **MTT-23**, 2 (1975).

S. E. Miller, Integrated Optics: An Introduction, *Bell Syst. Tech. J.* **48**, 2059 (1969).

R. E. Nahory and M. A. Pollack, Mixed Crystals Put Lasers and Lightguides on the Same Wavelength, *Bell Lab. Record*, p. 253 (October 1976).

M. Nakamura *et al.*, Optically Pumped GaAs Surface Laser with Corrugation Feedback, *Appl. Phys. Lett.* **22**, 515 (1973).

A. R. Nelson, D. H. McMahon, and R. L. Gravel, Electro-optic Channel Waveguide for Multimode Fibers, *Appl. Phys. Lett.* **28**, 321 (1976).

F. K. Reinhart and R. A. Logan, GaAs-AlGaAs Double Heterostructure Lasers with Taper-Coupled Passive Waveguides, *Appl. Phys. Lett.* **26**, 516 (1975).

R. V. Schmidt and H. Kogelnik, Electro-optically Switched Coupler with Stepped $\Delta\beta$ Reversal Using Ti-Diffused LiNbO$_3$ Waveguides, *Appl. Phys. Lett.* **28**, 503 (1976).

T. Tamir (ed.), "Integrated Optics," Springer-Verlag, Berlin and New York, 1975.

P. K. Tien, Integrated Optics, *Sci. Am.*, p. 28 (April 1974).

P. K. Tien, R. J. Martin, and S. Riva-Sanseverino, Novel Metal-Clad Optical Components and Method of Isolating High-Index Substrates for Forming Integrated Optical Circuits, *Appl. Phys. Lett.* **27**, 251 (1975).

CHAPTER 21

INFORMATION-RELATED APPLICATIONS
OF LASERS

Lasers have a wide variety of potential applications involved with the processing, storage, and transmission of information. Such applications include items like display, computer memories, and communications. In all these areas, lasers may have a large impact in the future. The status of these applications remains developmental, with large-scale usage not likely before the 1980s. Yet the large potential of these applications has attracted continued attention at many laboratories throughout the world.

We shall consider a variety of these potential applications under the generic title of information-related applications. We shall consider specifically the areas of communications, display, laser graphics, optical data storage, optical data processing, video discs and point-of-sale label scanning. These last two items are especially noteworthy because they are possibly the first areas in which lasers will be employed on a large scale in consumer-related products.

In what follows, we shall summarize a few selected examples that will illustrate the broad range of possibilities for lasers in information related fields. We note that the applications described here often make generous use of some of the accessory equipment described earlier, particularly modulators, deflectors, and integrated optical assemblies.

A. Communications

Optical communication has an ancient history, extending back to signal fires and smoke signals, and perhaps reaching a previous peak in the 19th century with the use of the heliograph for military operations in the western United States. Optical communication faded in the late 1800s as electrical communication developed rapidly.

The development of the laser can be recognized as a significant event in optical communication, offering a coherent source with very large bandwidth. Early ideas for laser-based communication systems generally involved transmission through the atmosphere, with the use of separate monolithic components, including lasers, modulators, collimating optics, etc. Because of problems with the turbulent and unclear atmosphere, such systems had relatively little application. In one particular example, a helium–neon laser-based system has been providing a data link between two hospitals in Cleveland for many years. The system provides two-way visual communication between a nurse anesthetist at one hospital and an anesthesiologist 1.2 km away. The laser beam transmits a color television signal, sound, and physiological signals, such as electrocardiograms.

Simple portable communication systems using gallium arsenide lasers were also developed. This application took special advantage of the relative ease of modulation of the gallium arsenide lasers through varying the current. Pulse repetition rates of the order of megahertz are easy to obtain with pulsed power supplies. For example, in one experiment, communication over a 14-km-long path with a pulse rate of 1.25 MHz allowed 24 pulse-code-modulated audio channels. Gallium arsenide laser light will not penetrate fog or cloud, so that such communication systems could be used only in clear conditions.

Space-borne systems, free of the atmosphere, have been employed for data transmission at very high data rates. As one example, laser communication links between synchronous satellites are scheduled for launch in 1979. Systems to be tested include a Nd:YAG laser-based system operating at a data rate of 10^9 bits/sec and a CO_2 laser-based system operating around 3×10^8 bits/sec.

Despite these uses, atmospheric limitations severely restricted the growth of optical communications through the first 15 years of the laser era. It was recognized early that to overcome problems of atmospheric transmission, the laser light would have to be transmitted through an improved medium. Some early systems explored the use of conduits, for example, evacuated pipes, for delivering the light beams from one place to another. If the conduits were filled with gas with a suitable radial temperature gradient sustained by heaters, a gas lens would be formed. This lens could provide a waveguiding action for a light beam traveling down the axis of the conduit, so as to confine the light beam within the conduit. Repeater lenses could improve confinement of the beam.

Such approaches were too cumbersome and expensive to sustain enthusiastic progress toward large scale use of optical communication. The situation has changed with the development of low loss optical fibers, which have been discussed in the last chapter. Fiber optical waveguides have been known for many years, and have been used to transmit light over short distances. Medical instrumentation for viewing portions of the human body internally rely on flexible fiber optic waveguides; such instruments have become well known. However, the losses associated with fiber optic waveguides rendered them impractical for long distance communication. Until the late 1960s, the best optical glasses had attenuation greater than 1000 dB/km. Beginning about 1970, advances in glass technology reduced the losses, until fibers with loss as low as 1 dB/km have been produced. Perhaps a more practical level for large-scale production is around 5 dB/km. It has been estimated that a total loss in excess of 50 dB can be tolerated between the source and receiver. Thus, optical waveguide systems will be able to transmit for several kilometers, before repeater stations are necessary. Splicing techniques are under development, which offer the possibility of splice losses of the order of 0.1 dB. Splicing must become an easy reliable process, which can be done in the field when necessary.

A schematic diagram of an optical communication system is presented in Fig. 21-1, showing the relation between the various components. For each element, a variety of possible selections is possible. The characteristic of the optical data link can vary greatly according to the choices made.

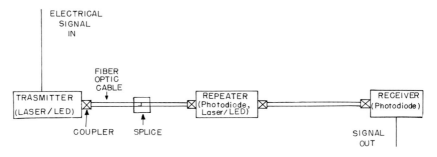

Fig. 21-1 Schematic diagram of optical link based on fiber optic cable.

Some capabilities and design tradeoffs for optically based communications systems are given in Table 21-1, for currently available components [1]. If low bit rates or short distances between repeaters are acceptable, one could employ systems using light-emitting diodes as sources, multimode fibers for the transmission elements, and PIN photodiodes as detectors. As the bit rate or the distance between repeaters goes up, one moves toward laser sources, single-mode fibers, and avalanche photodiodes.

A prototype lightwave communication system is under test in Atlanta, Georgia, in a simulated field environment. The system is based on gallium–aluminum–arsenide lasers coupled into fiber optic lightguides. The lightguides are bundled into a 1/2-in.-diam cable, which can carry the equivalent

TABLE 21-1

Possible Communication Systems Based on Fiber-Optic Transmission

Analog systems			
Bandwidth	4 kHz	1 MHz	
Modulation	PCM	Intensity	
Signal/noise	70 dB	60 dB	
Fiber type	Multimode	Multimode	
Source	Laser	Light-emitting diode	
Detector	Avalanche photodiode	Avalanche photodiode	
Distance between repeaters (for 5 dB/km fiber loss)	20 km	3 km	
Digital systems			
Bit rate	1 Mb/sec	50 Mb/sec	300 Mb/sec
Modulation	PCM	PCM	PCM
Error rate	10^{-9}	10^{-9}	10^{-9}
Fiber type	Multimode	Graded-index multimode	Single-mode
Source	Light-emitting diode	Laser	Laser
Detector	PIN photodiode	Avalanche photodiode	Avalanche photodiode
Distance between repeaters (for 5 dB/km fiber loss)	11 km	12 km	10 km

of 5000 telephone calls. Silicon avalanche photodetectors are used as detectors. Such systems could carry information between telephone switching centers in metropolitan areas. Cable duct space is limited in such areas, and replacement of current cables with fiber optical cables could relieve congestion in such networks.

Development of new materials, techniques, and components for optical communications is proceeding rapidly. Communications applications provide much of the motivation for development of integrated optics. It appears likely that fiber optical communications technology will be in widespread use by telephone companies in the 1980s. Ultimately, such communications systems may be used throughout the entire telecommunications network.

B. Displays

The use of lasers for display of information offers many attractive possibilities. By displays we mean devices which convert electronic signals into visual information. An extremely familiar display is the cathode ray tube, which has had a long and successful history as a display device. New types of displays, which could offer additional features and advantages, could compete in some applications with cathode ray tubes. New types of display have indeed found commercial application in recent years. Light-emitting diodes and liquid crystals are two familiar examples which have reached the marketplace and which have become common.

Other types of display systems are needed which would have desirable features similar to those of cathode ray tubes but which would be smaller, lighter and less expensive. Lasers have potential application in novel types of display systems which can provide such characteristics.

One simple approach using lasers to project displays on a screen involves scanning a visible laser across the screen. The beam is scanned in a raster by a light beam deflector. A modulator varies the intensity of the beam to form the desired picture. This method superficially appears attractive. It could project a picture on a screen which could be viewed simultaneously by a number of people. Optical modulation and deflection techniques have been developed with sufficiently large bandwidth for this technique.

This technique has been used to produce displays large enough to fill a theater screen. As one example, a developmental system filled a 5 ft^2 screen with a 512×512 element display, a resolution comparable to that of commercial television. Seven different colors were obtained from blue and green lines from an argon laser and a red line from a helium–neon laser. Such a display would be useful in situations where many people would be viewing the information at the same time. The system would be similar to a television set with a very large screen, but with no requirement for a vacuum enclosure.

The capabilities for this simple type of projection display have a number of limitations which have kept it from commercial exploitation. The most serious limitation is the brightness of the picture. The laser power that would be required for a system with resolution like commercial television is rather high.

In order to illustrate this, we will calculate the brightness of a screen illuminated by a scanning laser. For a system with resolution comparable to commercial television, one desires about 500×500 resolution elements, or a total of 2.5×10^5 elements, each of which would cover an area of 2×10^{-4} ft^2 on a screen of area 50 ft^2. The beam should return to each element in a time less than 1/30 sec to give the appearance of continuity (i.e., to avoid flickering). This means that the modulator must be able to operate at a rate around $30 \times 2.5 \times 10^5 = 7.5 \times 10^6$ Hz. This rate may be obtained with available light beam modulators.

If the screen is diffusely reflecting, the brightness R of the screen at an angle θ to the normal to the screen is given by Lambert's cosine law,

$$R = f\rho I \cos \theta / \pi = f\rho P \cos \theta / \pi A \tag{21.1}$$

where f is the fractional dwell time of the beam on a resolution element, ρ the reflectivity of the screen, I the power per unit area incident on the screen, P the total power in the beam, and A the area of a resolution element. For the numbers given above, assuming the most favorable case ($\rho = 1$, $\theta = 0$), one obtains

$$R = P/50\pi \text{ W/ft}^2\text{-sr} \qquad (21.2)$$

If we assume an argon laser operating at 5145 Å and convert to photometric units (using the least mechanical equivalent of light as 680 lm/W and the relative photopic luminosity at 5145 Å as 0.61), we obtain

$$R = 0.61 \times 680P/50\pi \text{ lm/ft}^2\text{-sr} = 764P \text{ fL} \qquad (21.3)$$

Since comfortable viewing requires a level around 20 fL, one would need an argon laser emitting around 8 W. Such lasers are available, but would not be compatible with a reasonably small inexpensive system. For less favorable situations (e.g., larger screens, viewing at angles other than the normal, or other wavelengths such as the He–Ne laser wavelength, where the photopic luminosity is lower) the power requirements would be even more stringent.

Thus, we are led to the conclusion that reasonably small and economical present-day lasers cannot produce a brightness on a large screen which is comfortable for viewing under conditions of daylight. In order to produce suitably bright displays, the laser power would have to be considerably higher. This would require advances in laser technology, and the lasers might be too expensive and inefficient to be competitive for the desired market.

A second problem is that with currently available lasers, it is difficult to produce full-color displays with good color resolution throughout the visible spectrum. The best candidates for laser displays appear to be the argon and the krypton lasers, which can provide a number of lines at relatively high power throughout the visible spectrum. Helium–neon and helium–cadmium lasers provide outputs at the red and violet ends of the spectrum, but the available powers are lower. Tunable dye lasers are not rapidly tunable over a wide enough range to be practical for this application. Thus, development of laser displays with good color resolution also awaits further advances in laser technology.

One way of producing laser displays of higher brightness is to use the beam to address a system of light valves. There are a variety of erasable light-beam-addressed light valves which could be an important element in a bright laser-based display system.

One possibility uses a sandwich consisting of a ferroelectric ceramic material covered with a photoconductive layer. One such transparent ferroelectric ceramic material that has been employed in experimental devices is the cermaic lanthanum-modified lead zirconate titanate, often called PLZT.

This material is transparent when it is polished in the form of thin plates. The material also exhibits ferroelectricity. When voltage is applied to it, the sample becomes electrically poled in the direction of the applied voltage. This behavior is similar to ferromagnetic materials, which become magnetically poled when a magnetic field is applied.

The ferroelectric ceramic PLZT also exhibits birefringence, which changes as the direction of electrical polarization changes. Thus, if a PLZT plate is placed between crossed polarizers, it can be switched from transparent to opaque (or vice versa) by a voltage which suitably switches the direction of the ferroelectric polarization.

Figure 21-2 shows a schematic diagram of a switching element which consists of a ferroelectric–photoconductor sandwich. The PLZT ceramic is originally electrically poled with the polarization direction in the plane of the cermaic plate. The voltage drop across the ceramic depends on the intensity of light incident on the photoconductor. The photoconductor could be a material such as cadmium sulfide or the organic photoconductor polyvinyl carbazole. When no light is incident on the device, the photoconductor is nonconducting and most of the voltage appears across the photoconductor. The ceramic remains poled in its original direction. If light strikes the photoconductor in certain areas, the photoconductor will become conducting in those areas. The voltage drop across the photoconductor will drop to a low value. Then the voltage drop appears across the ceramic in those areas where the incident light intensity is high. This will cause the ceramic to switch the direction of its electrical polarization and it will be poled in a direction perpendicular to the plane of the plate. This change in electrical

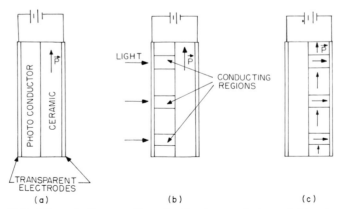

Fig. 21-2 Schematic diagram of ferroelectric–photoconductor switching element. In (a), the polarization vector **P** lies in the plane of the ferroelectric ceramic. In (b), the photoconductor becomes conducting in regions where light is incident. In (c), the voltage has switched the direction of the polarization vector **P** in those regions which the light addressed.

polarization is accompanied by a change in the birefringence. Between the crossed polarizers the material can switch from opaque to transparent.

This light valve effect can be used as the basis for bright displays, with high resolution and low cost. The ceramic PLZT has memory, so that once an area has been switched, it will remain electrically poled in the same direction until it is erased. Erasure can be accomplished by applying uniform voltage across the entire sample, so as to repole it entirely. After repoling, the electrical polarization again will be parallel to the plane of the plate. The entire device will then have switched back to an opaque state.

This type of device has been given the name ferpic, which stands for ferroelectric picture device. Another different type of display using PLZT has also been developed, a device called the cerampic. Unlike the ferpic, the image on the cerampic is created by spatial variations of light scattering rather than by spatial variation of birefringence. When the PLZT switches its direction of ferroelectric polarization, the amount of light transmitted through the material also changes, because of scattering. Therefore, in a sandwich consisting of a photoconductive film deposited on the PLZT substrate, the transmission of the PLZT can be changed by illuminating it simultaneously with application of a voltage across transparent electrodes. The operation of this electrically controlled scattering is shown in Fig. 21-3.

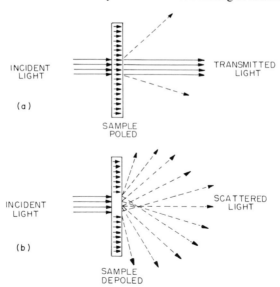

Fig. 21-3 Schematic diagram illustrating electrically controlled scattering. In (a), the sample is uniformly poled perpendicular to the plane of the sample; the material is transmissive. In (b), the polarization is zero in a selected region. The sample has become highly scattering in that region and the transmission is reduced.

In areas where the image is dark, the photoconductive film acts as an insulator and most of the voltage appears across the film. In bright areas, the film acts as a conductor and most of the voltage appears across the ceramic, so that the polarization is switched. However, the effective change in transmission is governed by light scattering, rather than by birefringence. Therefore, the cerampic device has simpler construction, because it does not require polarizers. It can also be switched back to a uniform state by thermal depoling.

The use of a cerampic device in a projection system is shown in Figure 21-4, which shows the basic construction of a light-beam-addressed switch. The laser is scanned in a raster over the device, with intensity modulation creating the desired image. The scene is illuminated by a bright arc lamp, and the image of the light transmitted through the device is projected on a screen. This approach alleviates brightness problems.

Prototype devices have been developed with a resolution of 40 lines per millimeter. It should be possible to form and erase images in the ceramic in raster patterns at rates up to 15,000 lines per second, a rate compatible with a display of televisionlike images. Potential uses of such devices are in generating images, for example, of documents, photographs, or diagrams from signals received by telephone or radio. Such images could be formed in a few seconds. The images would be stored until the device is repoled.

PLZT is one of the leading candidates for application in such light-beam-addressed displays. However, there are a number of additional

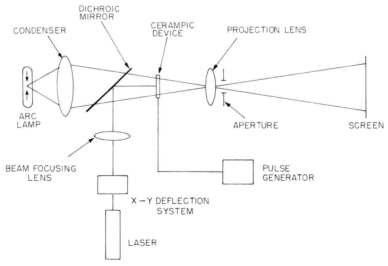

Fig. 21-4 Schematic diagram of a display system based on a cerampic (ceramic picture) device.

materials that are being considered, including materials such as bismuth titanate and lead germanate. In one type of device, the photoconductivity and birefringence are combined in a single monocrystalline material, $Bi_{12}SiO_{20}$, which exhibits photoconductivity and an electrooptic effect.

Such erasable light-beam-addressed light valves can have application in large high-resolution displays suitable for group viewing. They are also suitable for projection. They exhibit inherent memory and can be produced with a low cost per element. They have the disadvantages of being relatively slow and useful only for relatively thick displays.

Another developmental display has used electron beam addressing to switch the birefringence of crystals such as potassium dihydrogen phosphate. Figure 21-5 represents a light valve device using a potassium dihydrogen phosphate plate addressed by a modulated electron beam [2]. The second face of the crystal is coated with a transparent conductive layer which is connected to a grounded electrode. The cathode is operated around −15 kV.

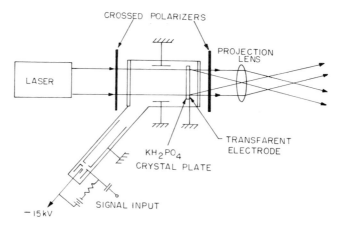

Fig. 21-5 Schematic diagram of electron beam addressed display device. (From G. Marie and J. Donjon, *Proc. IEEE* **61**, 942 (1973).)

The electron beam is scanned in a raster across the crystal, which becomes negatively charged by the electron beam. The amount of negative charge is proportional to the electron beam current, modulated by the input signal. The voltage across the crystal in a particular area is controlled by the electron beam current incident on that area. The voltage switches the birefringence of the potassium dihydrogen phosphate, so that the device operates in a manner similar to that described for the ferpic device. Crossed polarizers are required. Erasure of the image is automatic because the discharge time of the crystal is approximately 0.1 sec at room temperature.

This device essentially is a special purpose cathode ray tube, whose phosphor target is replaced by an electrooptical crystal. The incoming electrical signal modulates the cathode of the electron gun, in synchronization with the deflection of the electron beam. This converts the input image to a charge pattern on the crystal. A collimated laser beam incident on the crystal is modulated point-by-point by the electrooptic effect in the crystal. The electrooptic effect is in turn controlled by the amount of charge deposited on the surface of the crystal by the electron beam. With devices of this type, resolution of 800 lines per inch has been obtained, and operation at 30 frames per second has been obtained in the presentation of video images.

In summary, lasers can be employed in a variety of different devices suitable for display of information. A few of the many possible arrangements have been described here. The best candidates are devices which control the transmission of light and are addressed by either an electron beam or a light beam. The status of such devices is still developmental, but they offer considerable promise for future use.

C. Consumer-Related Products: Video Disks and Point-of-Sale Scanners

Laser specialists have long lamented the fact that most practical laser applications have been in high technology areas, where only small numbers of people are involved. The introduction of lasers into consumer-related products, with the possibility for much larger sales volume and greatly improved recognition has been a long-sought but elusive goal.

Video disks apparently offer one possibility for large scale use of lasers in consumer electronics and for the introduction of lasers into consumers' homes. Video disks would be used for reproducing recorded programs that can be replayed on the consumers' television sets. This would allow the user to watch a particular program at a time that is convenient, in a similar manner to the use of a collection of phonograph records. Several different manifestations of video disks are under development at various laboratories. The information is recorded on the disk and in some systems is read out by a laser beam. Video disks with optical readout appear superior to video disks using mechanical readout because of the noncontact feature. The life of pickups and disks is prolonged and scratching noises are reduced.

One laser-based system is shown schematically in Fig. 21-6 [3]. The disk contains spiral or concentric tracks with indentations of varying length and spacing. The depth is one-quarter of the wavelength of the readout laser, typically a helium–neon laser operating at 0.6328 μm. The laser light is focused onto the disk by a lens. The track width is less than the diffraction limit for the focusing lens. Therefore, the depth of the groove

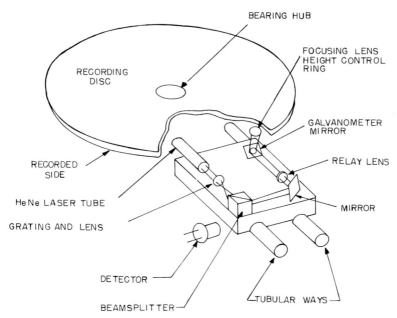

Fig. 21-6 Schematic diagram of a videodisk playback unit. The record is scanned from below by light from a helium–neon laser. (Adapted from K. Compaan and P. Kramer, *Philips Tech. Rev.* **33**, 178 (1973).)

produces destructive interference in the reflected beam, causing a large reduction in the reflected light. The mechanical system must maintain focus and position accurately. Information is recorded on disks having about the same dimensions as phonograph records. Information is recorded in short grooves of variable length and repetition frequency. In the readout, if the laser light falls on the surface of the disk and does not strike one of the grooves, most of the light will be reflected and collected by the objective lens. If the laser beam hits an indentation, the intensity of the reflected light is reduced. The intensity variations are converted into an electrical signal by a photodiode. The width and depth of the bits are arranged to provide information about the television picture, encoded in the bit pattern. Good dimensional control and tracking are required for the laser beam to follow the tracks of bits satisfactorily. Flexure of the disk is compensated by a servo loop which moves the lens with an electromagnet to provide a constant distance between the lens and disk. Such systems have been under development for a number of years. They could provide a significant introduction of laser technology to the consumer market.

A second area where lasers may become commonplace involves point-of-sale scanners, which read the universal product code printed on many items in the supermarket. The scanner will read the code automatically, look up the appropriate price in memory and print a detailed sales slip. The checker can draw the item over the scanner and bag it in one motion, without having to punch keys. The scanner contains four elements: a low-power helium–neon laser, a moving mirror to scan the laser beam in a predetermined pattern, a photodetector to sense variations in the intensity of the reflected light as the beam scans over the product code marking, and electronic logic to interpret the signal from the detector.

The code itself has become a familiar marking on most supermarket items. It is a ten-digit code identifying both the manufacturer and the specific item. It consists of parallel bars of variable width and spacing. When the laser beam scans over this pattern, variations in the intensity of the light reflected from dark and light regions allow identification of the item through the photodetector and the logic. The item may simply be drawn through the scanned beam and immediately bagged in one motion. The system will print a detailed description, including price, for each item.

The beam is projected by the mirror system in a pattern over the product code. A variety of different patterns have been developed, in efforts to make the system insensitive to the orientation of the item as it is pulled through the beam. Commercial systems claim to offer accurate reading of the code, regardless of orientation and up to speeds much higher than are likely to be encountered. The use of a laser is required to provide accurate reading under conditions of ambient light. The use of the low-power helium–neon laser provides eye safety both for checker and consumer.

For the consumer, such scanners offer shorter waits in checkout lines, better accuracy in pricing, and more complete information on sales slips. For the supermarket, scanners offer increased productivity for checkers and possibilities for improved automated inventory control. Early in the development of point-of-sale scanners, it was suggested that supermarkets would no longer mark prices on individual items. Because of consumer resistance and laws in some states requiring prices on individual items, it seems likely that prices will continue to be marked on each item, even when scanner use becomes widespread.

One survey [4] indicated the possibility of cost savings of 2% of gross sales in a large supermarket. Since the survey included cost reductions connected with deletion of prices on individual items, a savings of 1.5% seems more realistic. While this may seem to be a small percentage saving, it could be significant in an industry where profit margins have traditionally been small.

In the mid-1970s, a small number of stores are using laser scanners for checkout. The number should increase rapidly. We should soon see many people interacting routinely with laser-based equipment in the supermarket.

D. Laser Graphics

A wide variety of information-related applications of lasers have developed in the fields of the graphic arts. These applications involve methods to record, reproduce and transmit graphic and textual information. They include systems such as laser-based nonimpact printers, facsimile systems and recording systems. The impact of laser graphics has been most important in the newspaper industry. Applications have included phototypesetting, wirephoto transmission, and laser-based platemaking systems. In what follows, we shall describe a few examples of such systems. Laser graphics can be considered as a form of laser-based recording of information. This recording must be distinguished from the large-scale recording suitable for optical computer memories, which will be discussed in a later section. The recording bandwidth will be lower and the total storage capacity will be smaller in systems developed for laser graphics. In contrast to optical computer memories, which are binary, laser graphic systems will store characters or photographs.

A generalized laser recording system is illustrated in Fig. 21-7. This system scans a laser beam over a recording medium and records a pattern which may represent either printed or pictorial information. As the beam is scanned in a predetermined pattern, the modulator turns the beam on and off to form the desired pattern. Many such systems have been devised, using a wide variety of recording media. Many systems have used dry silver photographic films, which are developed by heat. Such systems obviously place strict requirements on the light beam modulators and deflectors.

The generalized laser graphic system shown in Fig. 21-7 is controlled by an input signal which may derive from a variety of sources, according to

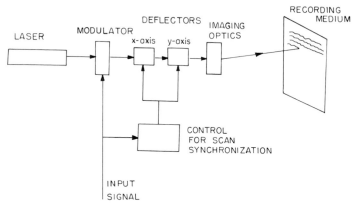

Fig. 21-7   Schematic diagram of system organization for a laser graphic system.

TABLE 21-2
Representative Laser Graphic Systems

Application	Format of recorded data	Recording method	Laser	Recording medium	Speed
Facsimile	Text or Photos on $8\frac{1}{2} \times 11$ in. paper	Xerographic	He–Ne	Plain paper	2 min/page
Facsimile	200 line/in. resolution on $8\frac{1}{2} \times 11$ in. paper	Photographic	He–Ne	Dry silver	10 sec/page
COM (Computer output on microfilm)	36 bit lines in binary	Photographic	CW gas laser	Photographic film	10^7 bits/sec
COM	Characters in 5×7 dot matrix	Photographic	He–Ne	Dry silver	10,000 lines/min
Nonimpact printing	Characters in 11×19 dot matrix	Material transfer of ink or dye	Argon	Plain paper	900 lines/min
Nonimpact printing	Dot matrix characters	Electrophoto-graphic	He–Ne	Plain paper	13,000 lines/min
Phototype-setting	Newspaper page of stroke-generated characters and halftones	Photographic	Argon	Dry silver or typesetting film	1000 lines/min
Platemaking	Text or photo on plate for newspaper printing	Vaporization	He–Ne (scanning); argon (removal of Cu); CO_2 (ablation of plastic)	Layered structure, plastic with Cu film	6 min/plate

the application. In facsimile systems, it will come from a photocell viewing variations in light reflected from the scene to be reproduced. In systems for computer output on microfilm (COM) or for line printers for computer output, it will be generated by the computer.

A variety of representative systems embodying laser graphics are listed in Table 21-2. The list is not meant to be exhaustive but merely to illustrate the forms that such laser-based recording systems can take.

Fascimile systems involve laser scanning of a picture or of textual material which is to be reproduced. A low-power laser beam is scanned in a raster across the surface of the document. There are two lasers in this type of system, the laser for scanning the document and the laser for recording the image. The two laser beams are swept in a synchronous raster.

Light reflected from the surface scanned by the first laser is detected by a photodetector. The output of the photodetector is used as the input signal for the modulator. When scanning laser sweeps over a bright area in the document, the reflected intensity and the detector response will be high. This signal, when fed into the modulator, will allow a high light intensity to reach the recording medium. The bright area will be reproduced. Converse statements apply to dark areas in the original scene.

Facsimile systems can be used to reproduce documents at a distance. The scanning laser, with its deflectors, is at one location. The recording laser, its deflectors and modulator, and the recording medium are at another location. The signal to control the modulator is transmitted between the two locations. For specific applications, this technique forms the basis for a wirephoto system used for newspaper photos. It is being incorporated into telecopier systems for business use.

Laser graphics will find much application in recording computer output, either on microfilm (COM) or for line printers. In such systems, the signal to the modulator is generated directly by the computer. Information can be recorded as binary bit patterns or as alphanumeric characters. The characters can be generated either in a dot matrix or by stroke generation.

Laser phototypesetting represents an advance in phototypesetting, a technology in widespread use in the newspaper industry. Phototypesetting involves photographing characters on film. Phototypesetting machines select, under the control of a keyboard, a negative of a desired character. The image of the character is projected on the photographic film. When the film is developed, it serves as the basis for making the printing surface.

The substitution of a laser as the light source in a phototypesetting operation offers possible cost reductions, and sophisticated laser typesetting machines have been introduced in the newspaper industry. As in nonlaser systems, an operator at a keyboard keys in a character. The electronic logic then controls the laser beam to project the image of the desired character onto the recording medium. Choices of different inputs (e.g., magnetic tape, paper tape, or computer) and different character fonts are possible.

A further extension of laser technology in the newspaper industry allows direct production of the plates for printing, at a considerable reduction in cost and complexity. The plate is made directly from a pasteup of the page, eliminating intermediate steps. Such systems are already employed by a number of newspapers.

One such system employs three lasers, a helium–neon laser for scanning the pasteup, and argon and CO_2 lasers that produce the plate. The recording medium is a layered structure with plastic underlying a thin copper layer. The output from a detector viewing the helium–neon laser light reflected from the pasteup modulates the argon laser beam. The argon laser is focused on the copper film and can vaporize the film. Thus, an image of the pasteup is formed by vaporization of the copper. The CO_2 laser is then scanned over the recording surface. In areas where the copper is intact, the reflectivity is high and there is no effect. In areas where the copper has been removed, plastic is ablated. The result is a plate ready for printing. Such systems can accommodate either text or halftones.

In summary, laser graphics offers a host of possibilities for recording applications. It has had its first major impact in the newspaper industry. Important applications in recording of computer output and in office and business telecopying are rapidly developing. In the future, laser graphics could serve as the basis of information systems in the home. One could envision receiving his daily newspaper, not by hand delivery, but by transmission over a laser graphic system.

E. Information Storage

Techniques using lasers for information storage, particularly in the form of an optical computer memory, have been recognized for years as possible alternatives to conventional magnetic memory technology. Perhaps the most important advantage is the potentially high packing density, which could exceed 10^7 bits/cm^2 on the recording material. The use of an optical beam for recording also allows intertialess addressing. The combination of these factors could lead to computer memories with very large capacity and fast access time.

In the highly developed computer industry, economic considerations and marketing strategy dominate the introduction of new technology. Therefore, the application of optical techniques for computer memories has been developing in a cautious evolutionary fashion. One product that has been developed is a 6.9×10^{11} bit permanent memory. The high packing density is used advantageously, but the other features of optical beam addressable technology have not been employed. In the mid-1970s, no alterable optical memory had been brought to the status of a commercial product. At the same time, magnetic recording technology has continued to advance. Magnetic disk memories appear capable of satisfying the marketplace until around 1980. The status of optical memory technology is that it has been proved technically feasible but not yet economically competitive with existing techniques for large computer memories. Much work still has to be done to reduce the cost. This statement especially applies to the lasers and light beam deflectors that would be used.

The emphasis in this section is on storage of large amounts of information, suitable for large scale computer memories. This application takes advantage of the focusing properties of laser radiation, i.e., the capability to focus the beam to a spot with dimensions comparable to the laser wavelength. This leads to the capability for high packing density, a necessary condition for memories of large capacity and of reasonable physical size. The emphasis in the last section was on recording amounts of information comparable to that contained on a newspaper page, an application where packing density is not a dominant consideration.

Let us consider some possibilities for recording information with lasers. A laser beam of sufficiently high power could be focused to a small spot on an opaque film and a small hole burned in the film. The information represented by the hole in an otherwise opaque film could be read optically. Similarly, materials such as heat sensitive paper, or direct print paper, such as is used in recording oscillographs, can be darkened by exposure to laser radiation. These straightforward techniques have been proposed as the basis of information storage systems using a laser. Their main drawback is their lack of erasability.

High-density recording on photographic film has also been employed. In one system, a helium–neon laser recorded bits on disks coated with emulsion. Total storage in excess of 10^9 bits was obtained, with a packing density of 10^6 bits/cm^2. Photographic film is a well-developed medium offering reliability and good resolution. This system could thus form the basis of a fairly large permanent memory. It has the drawbacks of being nonerasable and of requiring processing in order to develop the film.

The main developmental interest in optical data storage has involved storage materials that can be addressed and erased by light beams. There have been a variety of different systems developed. A computer memory using optical techniques could offer significant improvements in capacity and speed as compared to existing memory devices such as disk files and magnetic tapes. The main advantage offered by optical memories is the large packing density. Because optical techniques can provide higher packing density than the magnetic techniques that are now used, an optical computer memory of very large capacity could be produced with reasonably small physical size.

Figure 21-8 shows the capabilities of some different technologies used for computer memory. The figure plots capacity (i.e., the total number of bits stored) versus access time, that is, the time it takes to retrieve a specified bit upon demand. There is a tradeoff between speed and capacity in available memories. Speed is governed by the method used to reach a given bit. The three most common methods, electronic access, rotation of the medium, and transport of the medium, and their ranges of applicability are indicated in the figure. Solid boxes indicate established technologies. Semiconductor devices used in control memories can be very fast, and in some cases have been used for the "main" internal computer memory. Because of their relatively high cost, they are not suitable for constructing very large memories. Somewhat larger memories are constructed of magnetic ferrite cores, but they are slower. These ferrite cores usually form the "main" memories.

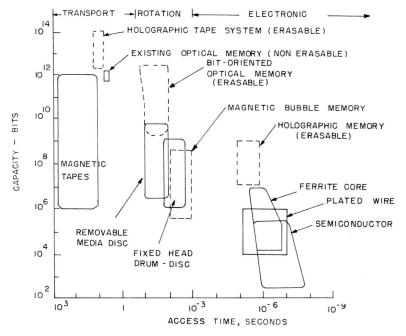

Fig. 21-8 Capacity as a function of access time for various memory technologies. Regions where the accessing is accomplished by electronic means, by rotation of the storage medium, and by transport of the storage medium are indicated. Currently available technologies are indicated by solid boxes and developing technologies by dashed boxes.

External memories outside the computer can be even larger, but require some mechanical motion and have slower access time. Magnetic disk files are commonly used for external memories. Finally, magnetic tapes can store very large amounts of information, but the retrieval time is much slower. Figure 21-8 also shows developing technologies in the dashed boxes. These include the bit-oriented optical memory, the holographic optical memory, holographic tape systems, and the so-called magnetic bubble memory, a rapidly advancing nonoptical technology. These techniques offer the promise of increased size or greater speed.

One may draw a straight line which passes through the current memory technology, a line which defines the capacity–speed tradeoff. One desires memories that move toward the upper right on the diagram, i.e., larger and faster. Optical memories do move from the line of current technology in this direction, as Fig. 21-8 shows.

Another parameter, not shown in Fig. 21-8, is the allowable cost per bit. As the memory capacity increases, the cost per bit must go down. For small semiconductor memories, a cost of several cents per bit may be acceptable. For mass memories with capacity around 10^9 bits, a cost less than 0.01 cents per bit is desired. At 10^{14} total bits, one would want a cost around 10^{-7} cents per bit.

Figure 21-8 indicates a currently available large-capacity optical memory. This is a laser-addressed memory having a capacity around 10^{12} bits. This memory stores information by using a laser beam to burn holes in a storage medium consisting of a metallic film on a plastic base. Readout is based on detecting the presence or absence of holes with a lower-power laser beam. This device is being marketed, but it has the disadvantage that the information stored is not erasable. Research trends are aimed at producing optical memories which can be erased and rewritten.

Optical memories offer potentially high density of bits. The limitation on bit size is set by diffraction, at a value near 0.5 μm. If the spacing between bits is approximately the same as the diameter, one has a limiting bit density of 10^8 bits/cm^2. In practice, it may be necessary to increase the bit spacing and size somewhat, but a density around 2.5×10^7 bits/cm^2 seems attainable. Models of optical memories with bit densities of this order of magnitude have been experimentally demonstrated. In contrast, magnetic disk devices are currently capable of storing information at a density around 10^6 bits/cm^2. Optical technology thus offers the promise of memories which have large capacity, but are of smaller physical size. Optical computer memories will probably compete for the part of the market that is now dominated by magnetic disk files.

The main potential for optical computer memories is for memories of high capacity. Requirements for memories in the region below 10^9 bits total storage capacity are well filled by ferrite cores and by magnetic disks. Rapidly developing bubble memories will also compete in this area. The high packing density available with optical techniques makes optical memories attractive for mass storage applications involving large numbers of bits, in the region of 10^{10}–10^{12} bits total capacity. Costs associated with the laser, modulator, and beam deflection system must be amortized over large numbers of bits.

Two different approaches are envisioned for optical computer memories. One is a bit-by-bit method of storage. This uses the energy of a laser beam focused on a storage medium to record a spot, which represents one bit of data, localized to a specific small area. Recording or reading is done one bit at a time.

The second technique uses holographic methods for data storage. The holographic memory is oriented toward recording and reading a page of data, containing between 10^3 and 10^5 bits at one time. The hologram representing the entire array of bits is stored on an area. As is the case with holograms, the information from any one bit is distributed over the entire area of the hologram.

The optical system and the component requirements are different for bit-oriented recording and holographic recording. A brief description of both systems follows.

1. Bit-Oriented Optical Memories

A block diagram of a bit-oriented memory is shown in Fig. 21-9. The basic system consists of a laser source, a modulator to turn the beam on and

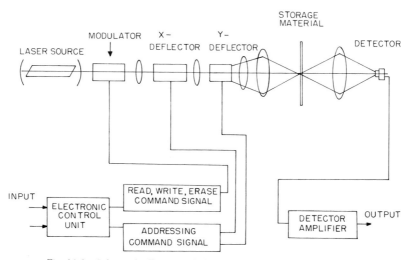

Fig. 21-9   Schematic diagram of bit-oriented optical memory system.

off and to control the intensity of the beam, a deflection system to move the beam to the location of the desired bit, a storage medium to record the bit, and a detection system for readout.

Optical computer memories which use electronic access (deflection of the light beam through electrooptic or acoustooptic effects) could in principle be extremely fast. The random access time to a desired bit of information could be in the microsecond regime. This is in contrast to magnetic disk files, in which rotation of the disk dictates access times of the order of milliseconds.

Exploitation of this potential advantage of optical technology would require development of high-speed, inertialess, light beam deflectors with large numbers of resolvable spots. Present-day light beam deflectors are capable of random access to perhaps 100 spots in each of two perpendicular directions, at megahertz rates. The total number of bits that could be reached in this way would be about 10^4, a value too low to be of interest. Therefore, for first-generation optical memories, mechanical motion will probably be required to address the required number of bits. Memories are envisioned in which the storage medium is in the form of a rotating disk, similar to magnetic disks. The access time will be limited by the rotation speed of the disks. Therefore, for the first generation of optical computer memories, there will probably be no significant advantage in access time, as compared to conventional disk memory systems.

The status of optical computer memory development is limited by the available recording materials. Many materials have been suggested, which employ a variety of physical phenomena for recording. Types of material commonly suggested are

(1) thermomagnetic materials,
(2) ferroelectric materials, and
(3) photochromic materials.

These materials all have binary stable states which can be switched by application of some combination of light and electric or magnetic fields.

Much developmental work for bit-oriented optical computer memories is centered on the use of thermomagnetic materials, and the leading systems at the present time use these materials. We shall only mention ferroelectric and photochromic phenomena briefly. In ferroelectric materials, switching occurs between different stable ferroelectric states. The switching involves large changes in the index of refraction of the material. The change in refractive index is the mechanism by which information is stored. In photochromic materials, a reversible light-induced photochemical process leads to a color change. Photochromic effects can be regarded as a reversible change of a system between two states having different absorption spectra. The change is induced in at least one direction by the action of light.

Thermomagnetic materials are thin films of magnetic materials, which are initially magnetized to saturation in one direction. A beam of laser light is focused on the film and produces localized heating. The heating raises the temperature in a small area of the film above the point where the magnetism is destroyed. When the laser beam is turned off and the material cools, it can become remagnetized if an external magnetic field is present. An external magnetic field is applied simultaneously with the laser heating, and controls the direction of magnetization in the heated area. This magnetic field by itself is incapable of reversing the magnetization when the film is at room temperature. But when the film is heated by the laser, and is momentarily demagnetized, as it cools, it remagnetizes in whichever direction the external field is applied. Remagnetization in one direction can represent storage of a "one" bit in the given area; remagnetization in the opposite direction represents a "zero."

In order to read out the stored information, the laser beam is directed to the same area on the storage medium. The modulator reduces the laser power so as not to erase information. Readout utilizes magnetooptic effects. Magnetooptic effects arise from an interation between polarized light and a magnetized film. Polarized light passing through the magnetized film will have the direction of its polarization rotated. The direction of the rotation depends on the direction in which the film is magnetized. The magneto-optic effect provides a method of sensing the magnetization in the film, and therefore of determining whether the information in a given area corresponds to a zero or to a one.

A variety of thermomagnetic materials have been suggested for this application. Some specific materials which have shown promise include manganese bismuth, europium oxide, and gadolinium iron garnet. A number of laboratories are actively working to develop memory systems that exploit the properties of these materials. Entire systems that demonstrate all the functions of storage and are capable of interfacing with a computer have been constructed. These are prototype models, which have shown the technical feasibility of optical data storage.

Let us consider details of one prototype optical memory system. The properties of characteristics of optical storage on thin films of manganese bismuth have been studied on a rotating disk apparatus. A 50 mW helium–neon laser was used both for storing and reading information. Mirrors directed the laser beam to an electroopitc modulator, then through focusing optics and a galvanometer deflector. The deflector sent the beam to a manganese bismuth film which had been vacuum deposited on a glass disk. The disk was rotated by an air bearing. The deflector was used to select a track of bits around a circle on the film. A copper wire-wound magnetic field coil was located between the final lens used for focusing and the film. This provided an external pulsed magnetic field for writing or erasing. The field was applied coincidentally with the laser pulse. The applied magnetic field extended spatially over an area much larger than a single bit, but only the laser-heated spot was affected by the magnetic field. Readout occurred through the magnetooptic rotation of light reflected from the film. This light traveled back through the focusing optics to a beam splitter which intercepted the light beam and directed it to a detector assembly. The detector assembly contained a polarizing beam splitter to divide the beam between two photodetectors. The amount of light was increased to one detector and decreased to the other, depending on the magnetic orientation of the bit at the focal plane of the light beam. A differential amplifier rejected light intensity fluctuations common to both detectors. The functions of writing, reading, and erasing at a rate around one megahertz were demonstrated. The memory was coupled to a computer and exercised as a memory for the computer, in order to establish technical feasibility for this application.

A photograph of a prototype optical memory system is shown in Fig. 21-10. The 50 mW helium–neon laser is at the rear, with the modulator, tilted at 45°, in front of it. The beam, visible because of scattering, passes

Fig. 21-10 Photograph of prototype optical mass memory system based on storage in manganese bismuth.

through the optical train and is focused on the memory medium, MnBi in this case. During exposure of the photograph, the beam has been scanned over a number of spots by a galvanometer deflector, so that the apparent beam is much wider than the actual beam. Readout is performed on the reflected beam, and readout signals are displayed on the oscilloscope.

2. *Holographic Memory Systems*

The organization of a holographic memory is different from a bit-oriented memory. In a bit-oriented memory, information storage and information readout occur one bit at a time. A holographic memory stores and reads out a large number of bits simultaneously. The basic configuration of a holographic memory is shown schematically in Fig. 21-11. The information is stored as a hologram on some material, the holographic memory medium. The apparatus in Fig. 21-11 contains a holographic recording arrangement, with a reference beam and a signal beam. The object of which a hologram is to be formed is a two-dimensional array of bits. This array is constructed by a device called a page composer. A page composer can be considered as an array of light valves, some of which are open and some closed. The opening and closing may be activated by light, by an electric field or by a combination of both. The open valves will correspond to ones, and the closed valves to zeros. These light valves are arranged in a pattern to represent an array of ones and zeros. It is this array of ones and zeros that is then stored at one time in a hologram located on a particular area of the holographic memory

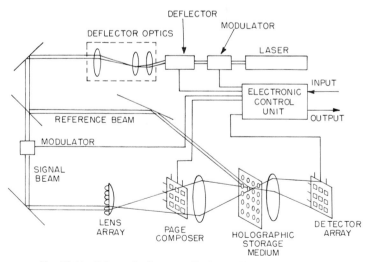

Fig. 21-11   Schematic diagram of holographic memory system.

medium. Because the hologram contains a large number of bits, perhaps around 10,000, one has simultaneous storage of this large number of bits at one time.

The beam passes through the beam splitter and is divided into two parts, the reference beam and the signal beam. The signal beam goes through an optical train and arrives at the page composer where an electronically composed data pattern is set up. This data pattern is imposed on the signal beam. When the signal beam combines with the reference beam to form a hologram on the recording medium, the resulting hologram represents the entire array of bits. The hologram is formed on one particular small area of the storage medium. The area is selected by the light beam deflector. During recording, the modulators allow maximum light intensity in both the signal and reference beams.

In order to store a different hologram in another location on the storage medium, the deflector moves the beam to that location. Movement of the object from one lens to another lens in the lens array changes the position of the hologram on the storage medium. At the same time the reference beam tracks the signal beam, so that both beams reach the same spot in the storage medium.

Readout of data occurs when the hologram is addressed with only the reference beam. The first modulator is partially closed, so that the light intensity reaching the holographic storage medium is reduced. The second modulator is closed.

The deflector directs the beam to the hologram to be read out. An image which represents the array of ones and zeros is produced. This image is focused by the lens next to the recording medium. An image of the data array is projected onto the photodetector array, which has the same relative dimensions as the elements of the page composer. Each bit originally stored in the page composer is incident on one photodetector in the array. The data thus are converted back to an electrical signal. If a particular area in the page composer corresponded to a one. there will be light on the photodetector in that position in the detector array. Thus, the array of bits can be reconstructed and converted to an electrical signal in parallel, with all the bits in the page being recovered at the same time. This feature allows the data readout rate to be high.

Storage of data in holographic form in an optical computer memory offers several advantages compared to a bit-oriented memory. The information about the original array of bits is distributed in a holographic fringe pattern and covers the entire hologram. Therefore, the hologram is not sensitive to small imperfections such as dust particles or scratches. Such imperfections could cause the loss of a bit in a bit-oriented memory, but their only effect on the hologram is to reduce resolution slightly.

A second advantage of holographic storage is that the information is essentially recovered in parallel. A large number of bits are all read out at the same time by the projection of the image of the array of bits directly onto the array of photodetectors. This recovery of a large number of bits at the same time offers possibilities for very high readout rates.

The requirements on light beam deflection are reduced. Each position to which the beam is deflected represents a page of data containing many bits. Thus, for a 10^8 bit memory (10^4 pages of 10^4 bits each) one requires only 10^4 separate locations. This figure lies within the capability of inertialess light beam deflectors. Addressing can be done entirely with nonmechanical light beam deflectors which have random access time less than 10 μsec. Such a holographic optical computer memory could be constructed with no moving parts.

Still another advantage is that the holographic recording and reconstruction is insensitive to the exact positions of the reference or reading beam on the hologram. This is not the case with the bit-oriented memory, for which the beams must be positioned very exactly. This means that the holographic system will be less subject to problems of vibration.

In spite of these advantages, the development of holographic optical computer memories lags behind that of bit-oriented memories. One problem lies in the development of suitable materials for recording and erasing holograms. Large research efforts are being devoted to developing materials which can store holograms and which can be rapidly erased. However, the entire state of development of holographic optical computer memories is behind that of bit-oriented optical computer memories.

The storage medium for holographic recording has very stringent requirements. It should be alterable and very stable. Also, it should have high spatial resolution, low write energy, and above all, it should have a high efficiency for readout. Many different materials have been considered as alterable storage media, but the ones of most interest are thermoplastics, magnetooptic materials, and ferroelectric materials.

Let us consider briefly one alterable holographic recording material, the thermoplastic-photoconductor sandwich. The operation of such a material as a recording medium is shown in Fig. 21-12. The surface must initially receive an electrical charge, as shown in Fig. 21-12a. The charge may be deposited from a corona wire, for example. When light illuminates the system, the photoconductor becomes conducting in the illuminated areas, and negative charge migrates across the photoconductor in those areas. This leads to a charge pattern as shown in Fig. 21-12b.

The material is developed by heat, so that the temperature of the thermoplastic is raised above its softening point. The thermoplastic then deforms in the regions of high charge density as shown in Fig. 21-12c. In this record-

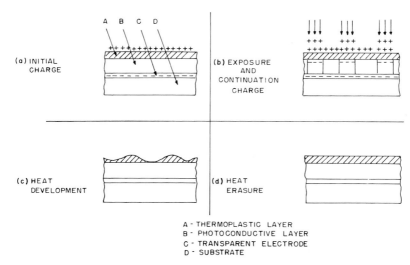

(a) INITIAL CHARGE

(b) EXPOSURE AND CONTINUATION CHARGE

(c) HEAT DEVELOPMENT

(d) HEAT ERASURE

A - THERMOPLASTIC LAYER
B - PHOTOCONDUCTIVE LAYER
C - TRANSPARENT ELECTRODE
D - SUBSTRATE

Fig. 21-12   Recording and erasure in thermoplastic–photoconductor device.

ing, the pattern of light and dark fringes characteristic of the hologram is transformed to a surface relief on the thermoplastic. The thermoplastic thus records a phase transmission hologram, which has high readout efficiency. When it is desired to erase the hologram, the medium is heated above its softening point again, so that the surface returns to its original flat contour, as indicated in Fig. 21-12d. Thus for thermoplastics, both development and erasure are done by heat pulses. During the time of erasure, the page composer can set up the next pattern of information which is to be recorded on the site where the hologram is being erased.

Thermoplastic–photoconductor tapes could be suitable for recording of large amounts of information similar to magnetic tapes but at a higher packing density. A schematic diagram of a proposed holographic tape recording system is shown in Fig. 21-13. Information would be stored in the form of one-dimensional holograms. The page composer would be one-dimensional, i.e., a line composer. The total capacity of such a system could be very large, up to 10^{14} bits. The access time would be limited by mechanical transport of the tape to perhaps 10 sec.

It is too early to state definitely which of the many candidate materials will be most suitable for holographic optical computer memories. Table 21-3 compares characteristics of two representative materials for optical memory: thermomagnetic MnBi films for bit-oriented memories, and thermoplastic–photoconductor materials suitable for holographic materials.

To summarize the status of optical computer memories, only large memories can be considered, because of the requirement of a low cost per

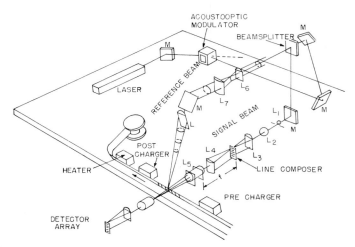

Fig. 21-13   Diagram of one-dimensional holographic recording system based on recording with single track optics on holographic tape.

bit. Continuing improvements in conventional magnetic memory technology have been able to meet the growing need for computer storage at lower development costs. Thus, optical memory technology, which has been proven to be technically feasible, has not yet been demonstrated as economically feasible. Because of the versatile nature of optical recording, many applications are possible. Continued development efforts may be expected. Continuing advances in lasers and in associated components may be expected

TABLE 21-3
Characteristics of Two Representative Materials for Optical Memory

	MnBi Films	Thermoplastic–photoconductor
Format	Bit-oriented	Holographic
Packing density (bit/cm^2)	2.5×10^7	10^7
Writing sensitivity (nJ/μm^2)	1	10^{-4}
Writing speed	250 psec/bit	10 msec/page developing at 60°C (plus 1 sec precharge and postcharge)
Erasure	250 psec/bit at 600 Oe field	100 msec/page at 60°C
Readout	$S/N \approx 7$	10% diffraction efficiency
Reversible cycles tested	$> 10^7$	$\sim 10^3$

to improve the economic situation. Optical technology for data storage may become useful in the 1980s, when present magnetic memory technology apparently will reach its ultimate limitations.

F. Optical Data Processing

Optical methods can be used in a variety of ways in processing of data. Optical data processing can remove certain specified portions of a scene presented in a photographic format, deblur photographs, make specified features in a scene stand out, automatically identify objects of a specified shape (e.g., alphanumeric characters) and identify errors in repetitive photomasks. The application of optical data processing occurs most naturally for data which are easily presented in a photograph format. There is no scarcity of important processing operations which satisfy this requirement.

The most commonly used setup for optical data processing is shown in Fig. 21-14. It consists of a pair of lenses of equal focal length F separated by $2F$. The input plane is the front focal plane of the first lens; the data to be processed can be inserted as a photographic transparency in this plane. The rear focal plane of the first lens (the front focal plane of the second) is called the transform plane. The rear focal plane of the second lens is called the image plane; the processed data is recovered there. One may insert a viewing screen, photographic film, or an array of photodetectors in this plane.

A collimated source of light is required. In practice, the source will almost always be a laser. It is possible to carry out the operations described here with nonlaser sources, but usually the high brightness of the laser makes it the most practical source.

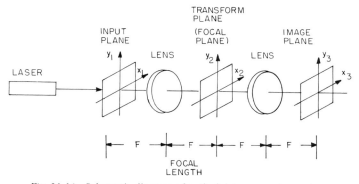

Fig. 21-14   Schematic diagram of optical data processing system.

Applications for optical processing of data involve filtering operations. Such operations are based on capability of lens systems to form the Fourier transform of a distribution of light. Consider the optical system shown in Fig. 21-14. Assume that the input, at a distance F equal to the focal length of the lens, has some distribution of light given by $g(x_1, y_1)$, where x_1 and y_1 are the spatial coordinates in the input plane. The input function g may be produced by a photographic transparency inserted in this plane. It may be shown that the distribution of light $G(p, q)$ in the transform plane, located at distance F behind the lens, is given by the two-dimensional Fourier transform of g, i.e.,

$$G(p, q) = \int\limits_{-\infty}^{\infty}\!\!\int g(x_1, y_1) \exp[i(px_1 + qy_1)]\, dx_1\, dy_1 \qquad (21.4)$$

where p and q are spatial frequency variables related to the spatial coordinates x_2 and y_2 by

$$p = 2\pi x_2/\lambda F \qquad q = 2\pi y_2/\lambda F \qquad (21.5)$$

with λ the wavelength of the light. A multiplicative constant, which does not affect the physical interpretation, is neglected. The spatial frequency variables have dimensions of radians per unit length. The Fourier transform relationship, which we shall not prove, is the basis for the filtering operations which lie at the heart of optical data processing. This relationship is valid under suitable conditions where the bandwidth of the system is sufficient to pass the highest spatial frequencies of the function g. Thus one assumes that the function g is zero outside some suitable spatial limits so that the integration can extend to infinity.

The second lens in the system will take a Fourier transform of the light distribution G in its input plane, thus reforming the original function. The light distribution $g'(x_3, y_3)$ in the image plane will be given by

$$g'(x_3, y_3) = \int\limits_{-\infty}^{\infty}\!\!\int G(p, q) \exp[i(px_3 + qy_3)]\, dp\, dq = g(-x_1, -y_1) \qquad (21.6)$$

This last equation is fundamental to data processing. The operation in Eq. (21.6) is not the inverse Fourier transform, because the sign in the exponential is the same as in Eq. (21.4). Nevertheless, the successive application of two Fourier transforms reforms the original light distribution but with the coordinate system inverted. The inversion of the image is expected from the optical system shown.

The optical-processing operations involve filtering part of the light distribution in the transform plane. The simplest filters are binary amplitude filters, having transmission of either zero or unity, i.e., they are simply stops for part of the light distribution G. Such filters are easy to fabricate. Other filters may operate on both the amplitude and phase of the light distribution G. Such filters may be produced by methods which we will describe below.

The information to be processed is assumed to be in the form of a photographic transparency. It could, for example, represent data from weather satellites, on which screening or pattern recognition operations were to be performed. It could represent aerial photographs for which processing could involve the recognition of features or removal of unwanted features to make later photointerpretation easier. Automated fingerprint identification systems based on optical processing have been devised. Another possible application involves deblurring of blurred photographs.

The properties of the transformation have a number of interesting features. We shall simply describe some of these features here, without giving the mathematical details that are necessary for proof. The light that corresponds to details with fine structure (high spatial frequency) in the original light distribution g appears relatively far from the optical axis in the transform plane. Details corresponding to a slowly varying light distribution, for example, to a relatively constant background, will be focused closer to the axis. This leads to the possibility of removal of unwanted features by physically putting apertures in the focal plane.

If no processing is performed by apertures in the focal plane, the image will be the same as the original transparency. If processing is performed by removing some portion of the light, then the image will have some features removed. The features removed are those that correspond to the light that was blocked.

If, for example, one wanted to remove wrinkles from a photograph of a human face, one would put a low-pass amplitude filter in the transform plane. This filter would pass light near the axis but remove light farther from the axis. The wrinkles correspond to a fine structure which will be focused far from the axis. This light would be removed and would not pass through the aperture. When the image is reformed, the low-frequency portions will be present, but the wrinkles would be absent.

If an opaque stop along the axis is placed in the transform plane, then light corresponding to broad background would be eliminated. This procedure would accentuate details, making fine details stand out in the presence of a slowly varying background.

Let us consider another example. Suppose we have some scene which has an overlapping set of parallel horizontal lines. This is illustrated in Fig. 21-15. The laser light passing through the transparency will have a

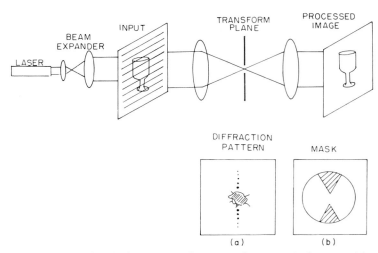

Fig. 21-15 Use of optical data processing system for removal of unwanted horizontal lines in a scene. Inset (a) shows the light distribution in the transform plane. Inset (b) shows the mask required to filter light corresponding to the horizontal lines.

certain distribution g, which contains the desired scene and the undesirable grid of lines. This distribution of light will be Fourier transformed by the lens. The array of horizontal lines will give rise to a familiar form of diffraction pattern. This pattern will be a series of spots in the vertical direction. This is shown in inset (a) in Fig. 21-15. The light corresponding to the remainder of the scene will appear elsewhere in the transform plane. By inserting the proper type of mask in the focal plane, one may remove the light corresponding to the vertical row of spots but allow the light corresponding to the scene to pass. A mask of the proper form is shown in inset (b) in Fig. 21-15. This mask is mounted so that the vertical dots in the diffraction pattern fall on the opaque portion of the mask and are blocked. Other light is allowed to pass. The second lens reforms the image, which is the original scene, with the interference caused by the set of horizontal lines removed. This example is meant to show the utility of optical data processing in dealing with information presented in pictorial format. This application in fact has been used to remove interfering raster scan lines from photographs sent back to earth by spacecraft.

This simple example shows how easily fabricated amplitude filters placed in the transform plane can be used to remove portions of the light corresponding to selected features in the original scene. There are many more sophisticated types of filtering operations that can be performed in the transform plane. These include filters that can operate both on the amplitude and the phase of the light reaching the transform plane. It is beyond the scope of this book to discuss all the variations possible in optical filtering techniques. We should, however, mention the matched complex spatial filter, a filter which provides the possibility for automatic inspection of photographic data and retrieval of specific features in the input scene. Such processing could be used for tasks like searching of aerial photographs for objects of a given shape. It also provides the possibility for optical character recognition. If a matched filter for a given letter, for example, *a*, is inserted in the transform plane and the input consists of alphanumeric characters, the output will be a number of spots of light, one at each position where the letter *a* occurred in the input. This will provide an automatic identification of all the positions on the input page where the letter *a* was present. The light spots may be detected by a photodetector array. Such a system offers the possibility for automatic processing of printed material.

One method of formation of a matched complex spatial filter is shown in Fig. 21-16. The method essentially produces a hologram of the object or character that one desires to recognize. The input is shown as a transparency of the letter *e*. The hologram is formed on the photographic plate, which is then developed and which yields a matched complex spatial filter for the

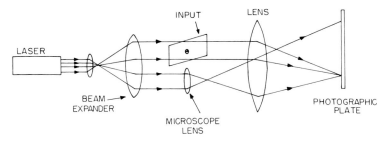

Fig. 21-16   Formation of matched complex spatial filter for the letter *e*.

letter *e*. When this filter is placed in the transform plane of the optical processor and a page of printed data is used as the input, spots of light corresponding to the location of each occurrence of the letter *e* will appear in the output. This is shown in Figure 21-17a [5], which shows the test used as the input. Figure 21-17b shows the detection of the *e*'s by the appearance of sharp peaks of light at positions corresponding to the position of the *e*'s in the original material.

In this system, the image is not faithfully retained, that is, the output is not in the form of the letter *e*. This is because the phase of the light is not faithfully preserved in the transform plane. However, this may still be a useful operation because it is often desirable to detect the presence and position of a signal without having to preserve its fidelity. The advantage of matched filtering comes from the fact that the result gives a sharp peak of light, even if

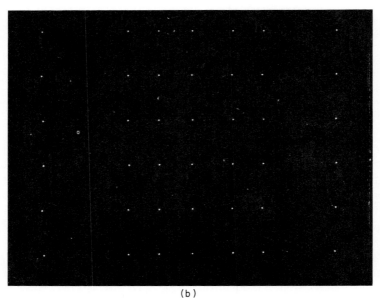

(a)

(b)

Fig. 21-17   Identification of the occurrences of the letter *e*. The input is shown in (a). Detection of the letter *e* is indicated in (b) by sharp autocorrelation peaks. (From S. Lowenthal and Y. Belvaux, *Opt. Acta* **14**, 245 (1967).)

the original signal has no peaks. The use of photodetectors with threshold logic can detect the bright spots of light corresponding to the occurrence of the e's in the lower portion of Fig. 21-17. The broader, more diffuse spots of light representing cross correlations with other members of the character set can be rejected by threshold logic.

Although the fundamentals of optical information processing were developed in the prelaser era and a number of the techniques have been developed with incoherent sources, the creation of matched spatial filters requires a high degree of temporal coherence, and their development therefore rests on the availability of coherent laser sources.

There was great enthusiasm for optical processing as applied to tasks like optical character recognition in the late 1960s. However, the method has several drawbacks that so far have kept it from widespread use. In the mid-1970s the enthusiasm for optical processors as automatic character recognizers has waned considerably. Such processing is very sensitive to the exact size and orientation of the objects. The input must be in the form of a transparency, rather than a printed page. The system requires a large number of matched filters to identify all the characters that may be present in the character set. It is also sensitive to small defects, such as broken letters, and to changes in print fonts. A character in one font may not be well detected by a filter constructed from the same letter from a different font. There are cross correlations between similar letters (for example, O and Q) which can produce signals of similar amplitudes difficult to distinguish with threshold logic.

These drawbacks have kept the technology of optical computing in a relatively early stage of development. There have developed some limited applications of practical significance for industry. One application is the automatic inspection for defects in patterns. This has been applied for rapid inspection of large numbers of similar objects, like repeated patterns on a photolithographic mask. Such masks may contain many hundreds of identical patterns, so that complete inspection by conventional techniques is uneconomical. The mask is an array of the same pattern repeated at the same interval. Thus, it acts like a diffraction grating. In the transform plane, one has an array of bright spots spaced at the fundamental periodic frequency of the pattern. The amplitude filter in the transform plane consists of an array of opaque black dots. The dots are at positions corresponding to the spots characteristic of the periodic structure. The dots will block this light and will remove all information which is repeated at the period corresponding to the fundamental period, that is, the pattern in the mask. The method will not detect mistakes which are repeated in all frames of the mask, but it has been used for rapid detection of random errors which have occurred in the maskmaking.

Another application of practical importance is use for deblurring of blurred photographs. In many cases, a blurred photograph can be represented by a convolution of a function which represents the sharp image with a point spread function which represents the blurring. If g_1 is the sharp image and h is the point spread function characteristic of the blurring process, the light distribution g in the blurred photograph will be given by

$$g(x, y) = \iint g_1(x', y')h(x - x', y - y') \, dx' \, dy' \qquad (21.7)$$

This mathematical formulation is termed a convolution. The deblurring process takes advantage of the theorem that the Fourier transform of a convolution is equal to the products of the Fourier transforms of the functions contained within the integral. Thus, the light distribution in the transform plane can be represented by the product

$$G(p, q) = G'(p, q)H(p, q) \qquad (21.8)$$

where G' and H are the Fourier transforms of g_1 and h, respectively. Thus, if one inserts in the transform plane a filter having transmission characteristics proportional to $1/H$, the result in the image plane will be the function g_1, which represents the sharp image with the effects of blurring removed. Methods for producing holographic Fourier transform division filters representing the function $1/H$ have been devised and have been employed successfully for deblurring of photographs.

Some other suggestions of industrial importance for optical computing systems have involved the inspection of defects in woven goods, in order to establish a grading for pieces of cloth. The diffraction pattern for a perfect piece of cloth exhibits a distinct pattern. Deviations from this pattern cause distinct deformations of the pattern which may be recognized by optical processing techniques. Thus, tedious and error-prone human inspection could be replaced by automated inspection to determine the quality of woven products.

Thus, optical computing systems have already shown promise for a relatively limited number of applications. Optical data processing must be considered as still being in a fairly early stage of development. It has been used for certain specific applications in which the data can be presented in an optical format, for example, as photographic transparencies. So far, practical uses for optical data processing have been fairly small and restricted, although the basic principles have been established for a number of years. Whether optical data processing reaches the areas of large potential payoff, for example automated alphanumeric-character recognition, will depend on further developments in technology.

REFERENCES

[1] T. Li, *Bell Lab. Rec.*, p. 333 (September 1975).
[2] G. Marie and J. Donjon, *Proc. IEEE* **61**, 942 (1973).
[3] K. Compaan and P. Kramer, *Philips Tech. Rev.* **33**, 178 (1973).
[4] *Laser Focus*, p. 48 (May 1976).
[5] S. Lowenthal and Y. Belvaux, *Opt. Acta* **14**, 245 (1967).

SELECTED ADDITIONAL REFERENCES

A. Communications

M. K. Barnoski (ed.), "Fundamentals of Optical Fiber Communications," Academic Press, New York, 1976.

J. D. Barry, Laser Communications in Space: Nd:YAG Technology Status, *Ann. N.Y. Acad. Sci.* **267**, 342 (1976).

S. J. Buchsbaum, Light Wave Communications, An Overview, *Phys. Today*, p. 23 (May 1976).

P. K. Cheo, CO_2 Laser Communications, *Ann. N.Y. Acad. Sci.* **267**, 329 (1976).

J. S. Cook, Communication by Optical Fiber, *Sci. Am.*, p. 28 (November 1973).

B. C. DeLoach, On the Way: Lasers for Telecommunications, Bell Lab Record, p. 203 (April 1975).

D. C. Forster, F. E. Goodwin, and W. B. Bridges, Wide-Band Laser Communication in Space, *IEEE J. Quantum Electron.* **QE-8**, 263 (1972).

R. M. Gagliardi and S. Karp, "Optical Communications," Wiley, New York, 1976.

J. E. Geusic, W. B. Bridges, and J. I. Pankove, Coherent Optical Sources for Communications, *Proc. IEEE* **58**, 1419 (1970).

J. E. Goell, An Optical Repeater with High Impedance Input Amplifier, *Bell Syst. Tech. J.* **53**, 629 (1974).

L. Goldman, The Role of the Laser in Communications Systems for Improved Medical Care, *Ann. N.Y. Acad. Sci.* **267**, 359 (1976).

F. E. Goodwin, A Review of Operational Laser Communication Systems, *Proc. IEEE* **58**, 1746 (1970).

W. S. Holden, An Optical-Frequency Pulse-Position-Modulation Experiment, *Bell Syst. Tech. J.* **54**, 285 (1975).

T. Li, Optical Transmission Research Moves Ahead, *Bell Lab. Record*, p. 333 (September 1975).

J. H. McElroy et al., CO_2 Laser Communication Systems for Near-Earth Space Applications, *Proc. IEEE* **65**, 221 (1977).

C. McIntyre et al., Optical Components and Technology in Laser Space Communications Systems, *Proc. IEEE* **58**, 1491 (1970).

S. Metz, The Avalanche Phototransistor—A Novel Type of Radiation Detector and its Potential Use in Optical Communication Systems, *IEEE Trans. Electron Devices* **ED-22**, 617 (1975).

S. E. Miller, T. Li, and E. A. J. Marcatili, Research Toward Optical Fiber Transmission Systems, *Proc. IEEE* **61**, 1703 (1973).

T. A. Nussmeier, F. E. Goodwin, and J. E. Zavin, A 10.6 μm Terrestrial Communication Link, *IEEE J. Quantum Electron.* **QE-10**, 230 (1974).

E. T. Price, R. L. Cohoon, and R. Pitts, The Case for Laser Communications, *Laser Focus*, p. 49 (June 1974).

M. Ross, Laser Communications, *in* "Laser Applications," Vol. 1 (M. Ross, ed.), Academic Press, New York, 1971.

M. Ross, Optical Communications in Space, *Opt. Eng.* **13**, 374 (1974).

Y. Takasaki and M. Tanaka, Line Coding Plan for Fiber Optic Communication Systems, *Proc. IEEE* **63**, 1081 (1975).

P. K. Tien, J. P. Gordon, and J. R. Whinnery, Focusing of a Light Beam of Gaussian Field Distribution in Continuous and Periodic Lens-Like Media, *Proc. IEEE* **53**, 129 (1965).

G. White, Optical Modulation at High Information Rates, *Bell Syst. Tech. J.* **50**, 2607 (1971).

G. White, A One-Gigabit-per-Second Optical PCM Communication System, *Proc. IEEE* **58**, 1779 (1970).

B. Displays

C. E. Baker, Laser Display Technology, *IEEE Spectrum*, p. 39 (December 1968).

L. Beiser, The Future of Lasers in Information Display, *Electro-Opt. Syst. Design*, p. 68 (January 1970).

E. J. Calucci, Solid State Light Valve Study, Information Display, p. 18 (March/April 1965).

E. I. Gordon and L. K. Anderson, New Display Technologies—An Editorial Viewpoint, *Proc. IEEE* **61**, 807 (1973).

D. W. Kennedy *et al.*, Laser Display Systems, *Instrum. and Control Syst.*, p. 59 (December 1968).

A. Korpel *et al.*, A Television Display Using Acoustic Deflection and Modulation of Coherent Light, *Appl. Opt.* **5**, 1667 (1966).

C. E. Land and P. D. Thacher, Ferroelectric Ceramic Electrooptic Materials and Devices, *Proc. IEEE* **57**, 751 (1969).

J. R. Maldonado and D. E. Fraser, PLZT Ceramic Display Devices for Slow-Scan Graphic Projection Displays, *Proc. IEEE* **61**, 975 (1973).

J. R. Maldonado and A. H. Meitzler, Strain-Biased Ferroelectric-Photoconductor Image Storage and Display Device, *Proc. IEEE* **59**, 368 (1971).

G. Marie, Light Valves using DKDP Operated near its Curie Point: Titus and Phototitus, *Ferroelectrics* **10**, 9 (1976).

A. H. Meitzler, J. R. Maldonado, and D. B. Fraser, Image Storage and Display Devices Using Fine-Grain Ferroelectric Ceramics, *Bell Syst. Tech. J.* **49**, 953 (1970).

D. S. Oliver *et al.*, Image Storage and Optical Readout in a ZnS Device, *Appl. Phys. Lett.* **17**, 416 (1960).

J. R. Packard *et al.*, Electronically Scannable Semiconductor Laser Device, *Opt. Eng.* **14**, 248 (1975).

W. D. Smith and C. E. Land, Scattering-Mode Ferroelectric-Photoconductor Image Storage and Display Devices, *Appl. Phys. Lett.* **20**, 169 (1972).

C. Consumer Products—Video Disks and Point-of-Sale Scanners

K. Compaan and P. Kramer, The Philips VLP System, *Philips Tech. Rev.* **33**, 178 (1973).

E. DeRousse, What to Expect from Laser Scanners, *Optical Spectra*, p. 35 (September 1975).

Videodisk: A Status Report, *Laser Focus*, p. 36 (August 1974).

A. Hildebrand, Reading the Supermarket Code, *Laser Focus*, p. 10 (September 1974).

G. W. Hrbek, An Experimental Optical Videodisk Playback System, *J. SMPTE* **83**, 580 (1974).

J. A. Jerome and E. M. Kaczorowski, Home Video-disk System Creates a New Image on Photographic Film, *Electronics*, p. 114 (April 4, 1974).

L. Mickelson, J. S. Winslow, and K. D. Broadbent, Use of the Laser in a Home Video-disc System, *Ann. N.Y. Acad. Sci.* **267**, 477 (1976).

Y. Tsunoda *et al.*, Holographic Video Disk: An Alternative Approach to Optical Video Disks, *Appl. Opt.* **15**, 1398 (1976).

W. Van den Buseche, A. H. Hoogendijk, and J. H. Wessels, Signal Processing in the Philips VLP System, *Philips Tech. Rev.* **33**, 181 (1973).

D. L. Wright and D. Crane, Laser Becomes a Component for Mass Market Applications, *Electronics*, p. 91 (June 13, 1974).

D. Laser Graphics

C. H. Becker, Laser Facsimile Printing, *Opt. Laser Technol.* **9**, 13 (1977).

H. I. Becker, Visible Lasers in Offset and Letter Plate Making, *Proc. Tech. Program, Electro-Opt. Syst. Design Conf. Anaheim* (November 11–13, 1975).

R. S. Braudy, Laser Writing, *Proc. IEEE* **57**, 1771 (1969).

W. A. Crofut, Laser Recording, *Proc. Tech. Program, Electro-Opt. Syst. Design Conf., Anaheim* (November 11–13, 1975).

D. E. Howarth and S. C. Cummings, Design of a Laser-Based Xerographic Facsimile System, *Proc. Tech. Program, Electro-Opt. Syst. Design Conf., Anaheim* (November 11–13, 1975).

S. C. Johnson, Progress in Laser Graphics, *Ind. Res.*, p. 90 (October 1975).

M. L. Levene, R. D. Scott, and B. W. Siryj, Material Transfer Recording, *Appl. Opt.* **9**, 2260 (1970).

D. Maydan, Micromachining and Image Recording on Thin Films by Laser Beams, *Bell Syst. Tech. J.* **50**, 1761 (1971).

C. J. Palermo, Design of the Laser Facsimile Systems, *Proc. Tech. Program, Electro-Opt. Syst. Design Conf., Anaheim* (November 11–13, 1975).

P. I. Sampath and J. A. Norcross, Non-Silver Photography in Laser Recording, *Electro-Opt. Syst. Design*, p. 30 (October 1973).

R. D. Scott and D. D. Curry, Design Considerations for a Dye Transfer Laser Line Printer, *Proc. Tech. Program, Electro-Opt. Syst. Design Conf., Anaheim* (November 11–13, 1975).

R. W. Wohlford, Nonimpact Printer Uses Laser and Electrophotographic Technologies, *Proc. Tech. Program, Electro-Opt. Syst. Design Conf., Anaheim* (November 11–13, 1975).

E. Optical Computer Memories

R. L. Aagard *et al.*, Experimental Evaluation of an MnBi Optical Memory System, *IEEE Trans. Magn.* **MAG-7**, 380 (1971).

R. L. Aagard, T. C. Lee, and D. Chen, Advanced Optical Storage Techniques for Computers, *Appl. Opt.* **11**, 2133 (1972).

G. A. Bailey, Optical Mass Memories, *Ann. N.Y. Acad. Sci.* **267**, 411 (1976).

A. Bardos, Wideband Holographic Recorder, *Appl. Opt.* **13**, 832 (1974).

C. D. Butter and T. C. Lee, Thermoplastic Holographic Recording of Binary Patterns in PLZT Line Composer, *IEEE Trans. Comput.* **C-24**, 402 (1975).

D. Chen, Magnetic Materials for Optical Recording, *Appl. Opt.* **13**, 767 (1974).

D. Chen, J. F. Ready, and E. Bernal G., MnBi Thin Films: Physical Properties and Memory Applications, *J. Appl. Phys.* **39**, 3916 (1968).

D. Chen and J. D. Zook, An Overview of Optical Data Storage Technology, *Proc. IEEE* **63**, 1207 (1975).

J. T. Chang, J. F. Dillon, and U. F. Gianola, Magnetooptical Variable Memory Based on the Properties of a Transparent Ferrimagnetic Garnet at its Compensation Temperature, *J. Appl. Phys.* **36**, 1110 (1965).

R. J. Collier, C. B. Burckhardt, and L. H. Lin, "Optical Holography," Academic Press, New York, 1971, Chapter 16.

R. W. Damon, D. H. McMahon, and J. B. Thaxter, Materials for Optical Memories, *Electro-Opt. Syst. Design*, p. 68 (August 1970).

M. D. Drake, PLZT Matrix-Type Block Data Composers, *Appl. Opt.* **13**, 45 (1973).

A. A. Eschenfelder, Promise of Magneto-Optic Storage Systems Compared to Conventional Magnetic Technology, *J. Appl. Phys.* **41**, 1372 (1970).

R. S. Eward, Optical Memories Are Alive and Well and Growing, *Opt. Spectra*, p. 25 (August 1975).

G. Y. Fan and J. H. Greiner, Low Temperature Beam-Addressable Memory, *J. Appl. Phys.* **39**, 1216 (1968).

T. K. Gaylord, Optical Memories: Filling the Storage Gap, *Opt. Spectra*, p. 29 (June 1974).

N. Goldberg, A High Density Magnetooptic Memory, *IEEE Trans. Magn.* **MAG-3**, 605 (1967).

E. E. Gray, Laser Mass Memory System, *IEEE Trans. Magnet.* **MAG-8**, 416 (1972).

R. P. Guzik, Optical Data Recording, Frontiers and the Manifest Destiny, *Electro-Opt. Syst. Design*, p. 22 (June 1974).

B. Hill, Some Aspects of a Large Capacity Holographic Memory, *Appl. Opt.* **11**, 182 (1972).

H. Kiemle, Holographic Memories in Gigabyte Region, *Appl. Opt.* **13**, 803 (1974).

T. C. Lee, Reflective Holograms on Thermoplastic-Photoconductor Media, *Appl. Phys. Lett.* **29**, 190 (1976).

W. T. Maloney, Optical Information Channels: Page-Accessed and Bit-Accessed, *Appl. Opt.* **14**, 821 (1975).

D. Maydan, Infrared Laser Addressing of Media for Recording and Displaying of High-Resolution Graphic Information, *Proc. IEEE* **61**, 1007 (1973).

J. R. Packard *et al.*, Electronically Scannable Semiconductor Laser Device, *Opt. Eng.* **14**, 248 (1975).

J. A. Rajchman, An Optical Read–Write Mass Memory, *Appl. Opt.* **9**, 2269 (1970).

H. N. Roberts, J. W. Watkins, R. H. Johnson, High Speed Holographic Digital Recorder, *Appl. Opt.* **13**, 841 (1974).

E. T. Stepke, Optical Mass Memories: The Impossible Dream? *Electro-Opt. Syst. Design*, p. 12 (October 1972).

W. C. Stewart *et al.*, An Experimental Read–Write Holographic Memory, *RCA Rev.* **34**, 3 (1973).

W. H. Strehlow, J. R. Packard, and R. L. Dennison, A Holographic Data Search and Retrieval System, *Proc. Tech. Program, Electro-Opt. Syst. Design Conf., West, San Francisco* (November 5–7, 1974).

K. K. Sutherlin, J. P. Lauer, and R. W. Olenick, Holoscan: A Commercial Holographic ROM, *Appl. Opt.* **13**, 1345 (1974).

K. Tsukamoto *et al.*, Holographic Information Retrieval System, *Appl. Opt.* **13**, 869 (1974).

O. N. Tufte and D. Chen, Optical Techniques for Data Storage, *IEEE Spectrum*, p. 26 (February 1973).

O. N. Tufte and D. Chen, Optical Memories: Controlling the Beam, *IEEE Spectrum*, p. 48 (March 1973).

H. Weider and H. Werlich, Characteristics of GaAs Laser Arrays Designed for Beam Address-able Memories, *IBM J. Res. Develop.* **15**, 272 (1971).

F.   Optical Data Processing

D. A. Ansley, Photo Enhancement by Spatial Filtering, *Electro-Opt. Syst. Design*, p. 26 (July–August 1969).

D. Casasent, The Optical-Digital Computer, *Laser Focus*, p. 30 (September 1971).

D. Casasent, A Hybrid Image Processor, *Opt. Eng.* **13**, 228 (1974).

S. S. Charschan (ed), "Lasers in Industry," Van Nostrand-Reinhold, Princeton, New Jersey, 1972, Chapter 7.

R. J. Collier, C. B. Burckhardt, and L. H. Lin, "Optical Holography," Academic Press, New York, 1971, Chapter 14.

J. B. DeVelis and G. O. Reynolds, "Theory and Applications of Holography," Addison-Wesley, Reading, Massachusetts, 1967, Chapter 8.

M. Eleccion, Automatic Fingerprint Identification, *IEEE Spectrum*, p. 36 (September 1973).

E. B. Felstead, A Simplified Coherent Optical Correlator, *Appl. Opt.* 7, 105 (1968).

A. Flamholz and H. Froot, Mask Tolerance Measurement and Defect Detection with Spatial Filtering, *Electro-Opt. Syst. Design*, p. 26 (May 1973).

S. Iwasa and J. Feinleib, The PROM Device in Optical Processing Systems, *Opt. Eng.* 13, 235 (1974).

N. Jensen, Practical Jobs for Optical Computers, *Machine Design* (February 22, 1973).

H. L. Kasdan and D. C. Mead, Out of the Laboratory and into the Factory—Optical Computing Comes of Age, *Proc. Tech. Program, Electro-Opt. Syst. Design Conf.*, Anaheim (November 11-13, 1975).

M. King *et al.*, Real-Time Electrooptical Signal Processors with Coherent Detection, *Appl. Opt.* 6, 1367 (1967).

S. H. Lee, Mathematical Operations by Optical Processing, *Opt. Eng.* 13, 196 (1974).

S. H. Lee, Optical Computing with Laser Light, *Ann. N.Y. Acad. Sci* 267, 430 (1976).

R. V. Poll and K. S. Pennington, Optical Information Processing, *in* "Laser Handbook," Vol. 2 (F. T. Arecchi and E. O. Schulz-Dubois, eds.), North-Holland Publ., Amsterdam, 1972.

W. J. Poppelbaum *et al.*, On-Line Fourier Transform of Video Images, *Proc. IEEE* 56, 1744 (1968).

K. Preston, "Coherent Optical Computers," McGraw-Hill, New York, 1972.

A. R. Shulman, "Optical Data Processing," Wiley, New York, 1970.

H. M. Smith, "Principles of Holography," Wiley (Interscience), New York, 1969, Chapter 8.

H. Stark, An Optical-Digital Image Processing System, *Opt. Eng.* 13, 243 (1974).

G. W. Stroke, "An Introduction to Coherent Optics and Holography," 2nd ed., Academic Press, New York, 1969.

G. W. Stroke, Optical Computing, *IEEE Spectrum*, p. 24 (December 1972).

G. W. Stroke, A Brief Review of Some Applications of Coherent Optical Processing to Image Improvement, *Proc. IEEE* 63, 829 (1975).

G. W. Stroke and M. Halioua, A New Holographic Image Deblurring Method, *Phys. Lett.* 33A, 3 (1970).

G. W. Stroke and M. Halioua, A New Method for Rapid Realization of the High-Resolution Extended Range Holographic Image-Deblurring Filter, *Phys. Lett.* 39A, 269 (1972).

J. T. Tippett *et al.* (eds.), "Optical and Electro-Optical Information Processing," MIT Press, Cambridge, Massachusetts, 1965.

A. Vander Lugt, Signal Detection by Complex Spatial Filtering, *IEEE Trans. Informat. Theory*, p. 139 (April 1964).

A. Vander Lugt, A Review of Optical Data Processing Techniques, *Opt. Acta* 15, 1 (1968).

A. Vander Lugt, Coherent Optical Processing, *Proc. IEEE* 62, 1300 (1974).

J. C. Vienot *et al.*, Three Methods of Information Assessment for Optical Processing, *Appl. Opt.* 12, 950 (1973).

L. S. Watkins, Inspection of Integrated Circuit Photomasks with Intensity Spatial Filters, *Proc. IEEE* 57, 1634 (1969).

P. M. Will and K. S. Pennington, Filtering of Defects in Integrated Circuits with Orientation Independence, *Appl. Opt.* 10, 2097 (1971).

A LOOK AT THE FUTURE

The development of lasers since the beginning of their history in the early 1960s has been phenomenally rapid. The operating characteristics of lasers, including such items as the available range of wavelengths, the peak powers, and the continuous powers available, have improved almost on a month-to-month basis. New applications have been suggested, tested, and brought to practical utilization. For the future we may expect to see expanded applications of lasers in consumer products. The video disk and the supermarket checkout systems discussed in Chapter 21 are probably the first areas where interaction will occur on a broad basis between laser technology and the consumer. But we may expect such applications to increase dramatically.

Lasers themselves will continue to proliferate and develop. Areas in which further advances are to be expected include tunable lasers. Lasers with high output powers and tunable over the range from the ultraviolet to well into the infrared, at least to 20 μm, may be expected. Continued advances in high average power and peak power systems are to be expected. The development of x-ray lasers is still speculative, but large research efforts at many laboratories may cause this to become a reality. Perhaps the most spectacular advances in power capabilities may occur in the ultraviolet, where excimer lasers now under development offer promise of high peak power. Chemical lasers offer the promise of higher average powers in the infrared.

The impact of lasers on communications may be enormous. Fiber-optic systems carrying messages on wide bandwidth systems offer the capability of easing the congestion in communication links in urban areas.

The continued growth and expansion of laser material processing appears assured, with lasers beginning to displace conventional techniques in some areas where they can be cost competitive. Other areas of potentially

572

large payoffs are perhaps more questionable. Large amounts of research are being devoted toward such items as optical information processing and storage systems, laser-assisted isotope separation, and laser-assisted thermonuclear fusion. Laser-assisted isotope separation has the possibility of coming into commercial use in the late 1980s, either for separation of ^{235}U for use in light water reactors or for separation of deuterium for heavy water reactors.

The potential application of lasers in thermonuclear fusion is perhaps one of the most exciting possibilities. It has strong competition from other techniques, most notably magnetic confinement techniques, which have advanced considerably using the Tokamak concept in recent years. However, laser-assisted thermonuclear fusion does provide an alternative possibility whose potential has not been completely evaluated. At its outer limit, it would offer the possibility of abundant energy for mankind for all the foreseeable future.

We will conclude with a brief discussion of some of the developments aimed toward laser-assisted thermonuclear fusion. The basic concept involves focusing very high power laser radiation on a target. Current thinking envisions that the target may be a glass pellet, approximately 50 μm in diameter, containing a mixture of deuterium and tritium gases. A number of beams would be directed onto the pellet from many directions. Absorption of the laser radiation at the outer edge of the material heats it and causes an implosion, in which a high-pressure wave is driven radially into the material from the periphery. When the implosion reaches the center, the material is raised to a very high temperature and pressure, at which thermonuclear reactions between the colliding atoms occur. Demonstration experiments so far have involved delivery of 10^{12} W in subnanosecond pulses to such targets. Neutrons from thermonuclear reactions in the material have been observed, at levels in excess of 10^9 neutrons per pulse.

This level is much lower than that required for scientific breakeven, which is defined as the level at which the thermonuclear energy release equals the laser energy input. Of course, for a practical system, the net thermonuclear energy release must be much greater than the laser energy input, and the laser efficiency must be reasonably high. Calculations indicate that perhaps 10^{14} W may be necessary in a subnanosecond duration pulse in order to achieve scientific breakeven. Figure E-1 shows a projected schedule for various milestones. Significant thermonuclear burn means reaching approximately 10% of breakeven. The figure also indicates some laser systems and their capabilities. The systems indicated in this figure are all Nd: glass.

The CYCLOPS laser is a single-beam system with a master-oscillator power-amplifier configuration. The ARGUS system, shown in Fig. E-2,

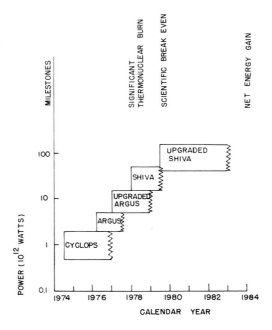

Fig. E-1 Development of power levels from Nd:glass laser systems designed for laser-assisted thermonuclear fusion. Some anticipated milestones are also indicated. (From data presented by Lawrence Livermore Laboratory.)

Fig. E-2 Photograph of two-beam ARGUS Nd:glass laser system. The size can be judged from the chairs visible in the photograph. (Photograph courtesy of C. M. Stickley.)

is a two-beam system. The complexity and size of the lasers being employed in this work is apparent. An idea of the size of the system can be obtained by observing the chairs. The pulse is produced by a master oscillator and then amplified. The oscillator and preamplifier stages are near the center of the photograph. A beam splitter splits the beam and directs each half through a chain of amplifiers. The two chains are at the sides of the photograph.

SHIVA is a 20-beam system in which each beam is amplified in a chain similar to one of the ARGUS beams. The upgraded SHIVA (or SHIVA NOVA) is expected to be able to achieve scientific breakeven. Net energy gain, in which the thermonuclear energy would exceed the total input energy to the laser, will probably not occur before the mid-1980s. Presumably, it would be in the 1990s before prototype operation of a power plant could be demonstrated, and well into the 21st century before commercial power plants could be produced.

The schematic diagram in Fig. E-1 is based on work at Lawrence Livermore Laboratory in which Nd:glass lasers are employed. Work is also proceeding along similar paths at other institutions. For example, parallel work at Los Alamos Scientific Laboratory employs CO_2 lasers.

These two types of laser, Nd:glass and CO_2, are being used because of their relatively high status of development. They are the only lasers that can emit the required levels of high peak power in very short pulses. It is apparent that Nd:glass will not be the final laser, because it does not have the capability of reaching required pulse repetition rates for commercial power generation. The CO_2 laser wavelength may be less favorable than a shorter wavelength. Some considerations indicate that a relatively short wavelength in the ultraviolet or blue portion of the visible spectrum may be desirable. (Recent experimental work indicates that for comparable laser power density, the plasmas produced by CO_2 lasers and Nd:glass lasers have comparable properties [1]. This apparently contradicts the theoretical predictions that short wavelengths will be preferable for laser-assisted thermonuclear fusion.) Thus, part of the thermonuclear research program involves looking for a "Brand X" laser which will have the capability of high peak power, short pulses, and high pulse repetition rate.

Even though the potential payoff of this work is still not certain, it seems definite that the impetus given to laser development in the search for the "Brand X" laser will provide new and exciting developments in the laser field for many years to come.

REFERENCES

[1] D. V. Giovanielli, *Bull. Am. Phys. Soc.* **21**, 1047 (1976).

SELECTED ADDITIONAL REFERENCES

H. G. Ahlstrom and J. F. Holzrichter, More Evidence That Fusion Works, *Laser Focus*, p. 39 (September 1975).

L. A. Booth *et al.*, Prospects of Generating Power with Laser-Driven Fusion, *Proc. IEEE* **64**, 1460 (1976).

K. A. Brueckner and S. Jorna, Laser-Driven Fusion, *Rev. Mod. Phys.* **46**, 325 (1974).

K. A. Brueckner and J. E. Howard, Ellipsoidal Illumination System Optimization for Laser Fusion Experiments, *Appl. Opt.* **14**, 1274 (1975).

P. M. Campbell, G. Charatis, and G. R. Montry, Laser-Driven Compression of Glass Microspheres, *Phys. Rev. Lett.* **34**, 74 (1975).

J. S. Clark, H. N. Fisher, and R. J. Mason, Laser-Driven Implosions of Spherical DT Targets to Thermonuclear Burn Conditions, *Phys. Rev. Lett.* **30**, 89 (1973).

J. L. Emmett, J. Nuckolls, and L. Wood, Fusion Power by Laser Implosion, *Sci. Am.*, p. 24 (June 1974).

W. F. Krupke and E. V. George, Lasers for Thermonuclear Fusion and Their Properties, *in Proc. Ind. Appl. High Power Laser Technol.* Vol. 86 (J. F. Ready, ed.), Soc. of Photo-Opt. Instrumentation Eng., Palos Verdes Estates, California (1976).

J. Nuckolls *et al.*, Laser Compression of Matter to Super-High Densities: Thermonuclear (CTR) Applications, *Nature (London)* **239**, 139 (1972).

J. Nuckolls, J. Emmett, and L. Wood, Laser-Induced Thermonuclear Fusion, *Phys. Today*, p. 46 (August 1973).

S. F. Jacobs *et al.* (eds.), "Laser Induced Fusion and X-ray Laser Studies," Addison-Wesley, Reading, Massachusetts, 1976.

C. E. Thomas, Laser Fusion Target Illumination System, *Appl. Opt.* **14**, 1267 (1975).

J. Wilson and D. O. Ham, Specifications for a Fusion Laser, *Laser Focus*, p. 38 (November 1976).

INDEX

Numbers in italic indicate the primary entry.

A
B
C 8
D 9
E 0
F 1
G 2
H 3
I 4
J 5